J. Mayer, A. Ciechanover,
M. Rechsteiner (Eds.)
Protein Degradation

Further Titles of Interest

J. Buchner, T. Kiefhaber (Eds.)

Protein Folding Handbook

2004
ISBN 3-527-30784-2

J.-C. Sanchez, G. L. Corthals, D. F. Hochstrasser (Eds.)

Biomedical Application of Proteomics

2004
ISBN 3-527-30807-5

K. H. Nierhaus, D. N. Wilson (Eds.)

Protein Synthesis and Ribosome Structure

2004
ISBN 3-527-30638-2

G. Cesareni, M. Gimona, M. Sudol, M. Yaffe (Eds.)

Modular Protein Domains

2004
ISBN 3-527-30813-X

G. Krauss

Biochemistry of Signal Transduction and Regulation

2003
ISBN 3-527-30591-2

John Mayer, Aaron Ciechanover, Martin Rechsteiner (Eds.)

Protein Degradation

Ubiquitin and the Chemistry of Life

Volume 1

WILEY-VCH

WILEY-VCH Verlag GmbH & Co. KGaA

Editors

Prof. Dr. R. John Mayer
University of Nottingham
School of Biomedical Sciences
Queen's Medical Centre
Nottingham, NG7 2UH
UK

Prof. Dr. Aaron Ciechanover
Technion – Israel Institute
of Technology
Department of Biochemistry
Efron Street, Bat Galim
Haifa 31096
Israel

Prof. Dr. Martin Rechsteiner
University of Utah Medical School
Department of biochemistry
50 N. Medical drive
Salt Lake City, UT 84132
USA

Library of Congress Card No.: Applied for
British Library Cataloging-in-Publication Data:
A catalogue record for this book is available from the British Library

Bibliographic information published by Die Deutsche Bibliothek
Die Deutsche Bibliothek lists this publication in the Deutsche Nationalbibliografie; detailed bibliographic data is available in the Internet at ⟨http://dnb.ddb.de⟩.

© 2005 WILEY-VCH Verlag GmbH & Co. KGaA, Weinheim

Printed in the Federal Republic of Germany
Printed on acid-free paper

Typesetting Asco Typesetters, Hong Kong
Printing betz-druck gmbh, Darmstadt
Bookbinding Litges & Dopf Buchbinderei GmbH, Heppenheim

ISBN-10 3-527-30837-7
ISBN-13 978-3-527-30837-8

Contents

Protein Degradation. Edited by J. Mayer, A. Ciechanover, M. Rechsteiner
Copyright © 2005 WILEY-VCH Verlag GmbH & Co. KGaA, Weinheim
ISBN: 3-527-30837-7

Preface

There is an incredible amount of current global research activity devoted to understanding the chemistry of life. The genomic revolution means that we now have the basic genetic information in order to understand in full the molecular basis of the life process. However, we are still in the early stages of trying to understand the specific mechanisms and pathways that regulate cellular activities. Occasionally discoveries are made that radically change the way in which we view cellular activities. One of the best examples would be the finding that reversible phosphorylation of proteins is a key regulatory mechanism with a plethora of downstream consequences. Now the seminal discovery of another post-translational modification, protein ubiquitylation, is leading to a radical revision of our understanding of cell physiology. It is becoming ever more clear that protein ubiquitylation is as important as protein phosphorylation in regulating cellular activities. One consequence of protein ubiquitylation is protein degradation by the 26S proteasome. However, we are just beginning to understand the full physiological consequences of covalent modification of proteins, not only by ubiquitin, but also by ubiquitin-related proteins.

Because the Ubiquitin Proteasome System (UPS) is a relatively young field of study, there is ample room to speculate on possible future developments. Today a handful of diseases, particularly neurodegenerative ones, are known to be caused by malfunction of the UPS. With perhaps as many as 1000 human genes encoding components of ubiquitin and ubiquitin-related modification pathways, it is almost certain that many more diseases will be found to arise from genetic errors in the UPS or by pathogen subversion of the system. This opens several avenues for the development of new therapies. Already the proteasome inhibitor Velcade is producing clinical success in the fight against multiple myeloma. Other therapies based on the inhibition or activation of specific ubiquitin ligases, the substrate recognition components of the UPS, are likely to be forthcoming. At the fundamental research level there are a number of possible discoveries especially given the surprising range of biochemical reactions involving ubiquitin and its cousins. Who would have guessed that the small highly conserved protein would be involved in endocytosis or that its relative Atg8 would form covalent bonds to a phospholipid during autophagy? We suspect that few students of ubiquitin will be surprised if it or a

Protein Degradation. Edited by J. Mayer, A. Ciechanover, M. Rechsteiner
Copyright © 2005 WILEY-VCH Verlag GmbH & Co. KGaA, Weinheim
ISBN: 3-527-30837-7

ubiquitin-like protein is one day found to be covalently attached to a nucleic acid for some biological purpose.

We are regularly informed by the ubiquitin community that the initiation of this series of books on the UPS is extremely timely. Even though the field is young, it has now reached the point at which the biomedical scientific community at large needs reference works in which contributing authors indicate the fundamental roles of the ubiquitin proteasome system in all cellular processes. We have attempted to draw together contributions from experts in the field to illustrate the comprehensive manner in which the ubiquitin proteasome system regulates cell physiology. There is no doubt then when the full implications of protein modification by ubiquitin and ubiquitin-like molecules are fully understood we will have gained fundamental new insights into the life process. We will also have come to understand those pathological processes resulting from UPS malfunction. The medical implications should have considerable impact on the pharmaceutical industry and should open new avenues for therapeutic intervention in human and animal diseases. The extensive physiological ramifications of the ubiquitin proteasome system warrant a series of books of which this is the first one.

Aaron Ciechanover
Marty Rechsteiner
John Mayer

List of Contributors

Dawadschargal Bech-Otschir
MRC Human Genetics Unit
Western General Hospital
Edinburgh
EH4 2XU
UK

Matthias Bochtler
ul. Ks. J. Trojdena 4
02-109 Warszawa
Poland

Hans Brandstetter
Department of Protein Crystallography
Max-Planck-Institute for Biochemistry
Am Klopferspitz 18 a
82152 Martinsried
Germany

Aaron Ciechanover
Vascular and Tumor Biology Research Center
The Ruth and Bruce Rappaport Faculty of
Medicine and Research Institute
Technion-Israel Institute of Technology
Efron St., POB 9649
Haifa, 31096
Israel

Tim Clausen
Institute for Molecular Pathology
Dr. Bohrgasse 7
1030 Wien
Austria

George DeMartino
Department of Physiology
The University of Texas
Southwestern Medical Center at Dallas
5323 Harry Hines Boulevard
Dallas, TX 75390-9040
USA

Wolfgang Dubiel
Department of Surgery
Division of Molecular Biology
Charité – Universitätsmedizin Berlin
Monbijoustrasse 2
10117 Berlin
Germany

Michael J. Eddins
Department of Biophysics and
Biophysical Chemistry
Johns Hopkins School of Medicine
725 N. Wolfe Street, WBSB 605
Baltimore, MD 21205
USA

Michael Groll
Adolf-Butenandt-Institute
LMU München
Butenandtstr. 5
81377 München
Germany

Avram Hershko
Vascular and Tumor Biology Research Center
The Ruth and Bruce Rappaport Faculty of
Medicine and Research Institute
Technion-Israel Institute of Technology
Efron St., POB 9649
Haifa, 31096
Israel

Leigh A. Higa
Department of Genetics
Yale University School of Medicine
333 Cedar Street
New Haven, CT 06520
USA

Kay Hofmann
Bioinformatics Group
Memorec Stoffel GmbH
Soeckheimer Weg 1
50829 Köln
Germany

Robert Huber
Department of Protein Crystallography
Max-Planck-Institute for Biochemistry
Am Klopferspitz 18 a
82152 Martinsried
Germany

Barbara Kapelari
Department of Structural Biology
Max-Planck-Institute for Biochemistry
82152 Martinsried
Germany

Kevin L. Lorick
NCI at Frederick
Building 560, Room 22-103
1050 Boyles Street
Frederick, MD 21702
USA

Nikola P. Pavletich
Howard Hughes Medical Institute
Memorial Sloan-Kettering Cancer Center
Cellular Biochemistry and Biophysics Program
1275 York Avenue
New York, NY 10021
USA

Cecile Pickart
Department of Biochemistry and
Molecular Biology
Johns Hopkins Bloomberg School
of Public Health
615 North Wolfe Street
Baltimore, MD 21205
USA

Martin Rechsteiner
Department of Biochemistry
University of Utah School of Medicine
50 N. Medical Dr
Salt Lake City, UT 84132
USA

Nathaniel S. Russell
Graduate Program in Biochemistry, Cell and
Developmental Biology
Graduate Division of Biological and
Biomedical Sciences
Emory University
Atlanta, GA 30322
USA

Hermann Schindelin
Department of Biochemistry and Cell Biology
SUNY Stony Brook, NY 11794-5215
USA

Yien-Che Tsai
NCI at Frederick
Building 560, Room 22-103
1050 Boyles Street
Frederick, MD 21702
USA

Allan M. Weissman
NCI at Frederick
Building 560, Room 22-103
1050 Boyles Street
Frederick, MD 21702
USA

Keith Wilkinson
Emory University
Department of Biochemistry
Rollins Research Center
Atlanta, GA 30322
USA

Cezary Wojcik
Department of Physiology
University of Texas
Southwestern Medical Center
Dallas, Texas 75390-9040
USA

Yili Yang
NCI at Frederick
Building 560, Room 22-103
1050 Boyles Street
Frederick, MD 21702
USA

Hui Zhang
Department of Genetics
Yale University School of Medicine
333 Cedar Street
New Haven, CT 06520
USA

Ning Zheng
Department of Pharmacology
University of Washington
1959 NE Pacific Street
Box 357280
Seattle, WA 98195-7280
USA

1
Brief History of Protein Degradation and the Ubiquitin System

Avram Hershko

1.1
Introductory Remarks

The reader of this book may be impressed (and possibly overwhelmed) by the enormous recent progress in this field. The ubiquitin system is now known to be involved in basic biological processes, such as the control of cell division, signal transduction, regulation of transcription, DNA repair, quality control in the endoplasmic reticulum, stress response, induction of immune response and inflammation, apoptosis, embryonic development, and circadian clocks, to mention but a few. It has been implicated in diseases such as many types of cancer, neurodegenerative diseases (such as certain types of Parkinson's, Alzheimer's and Huntington's diseases), retroviral infections, certain types of hypertension, mental retardation, and cachexia associated with cancer, renal failure, or sepsis. New functions of ubiquitin and of ubiquitin-like proteins are being reported almost every month, and the number of publications in this field is increasing at an exponential (and bewildering!) rate. It may be therefore instructive to consider briefly the humble beginnings of this field, how significant progress was achieved, and also how at times progress was impeded by wrong dogmas. Important lessons can be learned from both achievements and failures in science.

1.2
Protein Degradation – Does It Exist?

In the first three decades of the twentieth century, a generally accepted theory of protein metabolism was that proposed by Folin [1]. Based on studies on the chemical composition of urine in humans fed protein-rich or protein-free diets, Folin proposed that there are two separate pathways of protein catabolism, which he called "endogenous" and "exogenous" types of protein catabolism. According to this concept, "exogenous" protein catabolism originates from dietary proteins, accounts for the major part of urea excreted under normal conditions, and shows wide variations according to dietary protein intake. By contrast, "endogenous" protein catabolism was thought to originate from tissue proteins, to be mainly represented by excreted creatinine, not to be affected by the amount of dietary protein intake, and to account for a minor part of nitrogenous compounds excreted in the urine. Because of the minor proportion of "endogenous" protein catabolism, it was thought that cellular proteins are predominantly stable, and only a small fraction resulting from "wear and tear" of tissue proteins is subject to catabolism [1].

In spite of its obviously wrong assumptions (such as that creatinine is the end product of protein catabolism), Folin's theory was widely accepted and cited in textbooks of biochemistry until the late thirties. At that time, a breakthrough in the field was achieved by the pioneering studies of Schoenheimer and co-workers, who introduced the extensive use of isotopically labeled compounds in biological studies. In a typical experiment [2], ^{15}N-labeled L-leucine was administered to well-fed rats, and the distribution of the isotope in excreta and in body tissues was examined. According to the concept of Folin, most exogenously administered leucine should have appeared in urinary waste products. This was not the case: less than one-third of the isotope was excreted in the urine, and most of it was found to be incorporated into tissue proteins [2]. Since the weight of the animals did not change during the experiment, it could be assumed that the mass and composition of body proteins also did not change. It was concluded, therefore, that newly incorporated amino acids must have replaced those in tissue proteins in a process of protein turnover. From these studies a new concept has emerged according to which cellular proteins, and some other body constituents, are in a dynamic state of constant and extensive renewal [3].

Schoenheimer's concept of the dynamic state of body proteins did not remain unchallenged. In 1955, Monod and co-workers studied the origin of amino acids utilized for the synthesis of newly induced β-galactosidase in growing *E. coli* [4]. Bacteria were first labeled with ^{35}SO$_4{}^{2-}$ and then were transferred to unlabeled medium containing the inducer methyl-β-D-thiogalactoside. Newly synthesized β-galactosidase was isolated and was found not to contain significant amounts of radioactivity. This result suggested that in growing *E. coli*, the degradation of most cellular proteins is negligibly slow, otherwise newly synthesized β-galactosidase would have contained ^{35}S-labeled amino acids originating from the degradation of pre-existing proteins. Instead of restricting these conclusions to the case of growing *E. coli*, the authors went on to generalize and proposed that cellular proteins are also stable in mammalian tissues. They furthermore suggested that in Schoenheimer's experiments, incorporation of amino acids into tissue proteins might be due to the replacement of cells lost by cell lysis or the replacement of secreted proteins [4]. This was, in effect, a return to Folin's dogma of static cellular proteins.

So great was the authority of Monod at that time that the dynamic state concept of Schoenheimer fell into disfavor, as judged by contemporary review articles [5]. Gradually, however, experimental evidence accumulated which refuted Monod's hypothesis. Using mammalian cells in culture, Eagle and co-workers carefully examined the problem of cellular protein turnover *vs.* cell turnover or protein secretion [6]. In a variety of growing or resting cells in culture, cellular proteins were replaced at a high rate of approx. 1% h^{-1}. This was due to true protein turnover and not to the replacement of secreted proteins or lysed cells, as indicated by the lack of significant amounts of labeled proteins in the culture medium. Further work in several laboratories has shown that protein degradation in animal cells is extensive and is highly selective. Thus, for example, abnormal proteins produced by the incorporation of some amino acid analogues or by certain mutations are selectively recognized and are rapidly degraded in cells [7]. However, it is not correct to state (as is written in some current articles) that, until recently, protein degradation was thought to be mainly a "garbage disposal" system to get rid of abnormal proteins. In the late sixties, it was already evident that normal proteins are also degraded in a highly selective mode. The half-lifetimes of different proteins range from several minutes to many days, and rapidly degraded normal proteins usually have important regulatory functions. These properties of intracellular protein degradation and the importance of this process in the control of the levels of specific proteins were summarized by Schimke and Doyle in 1970 [8].

In retrospect, one can only speculate why the concept of intracellular protein degradation was resisted for such a long time. It is possible that one reason was the difficulty in accepting the idea that cells carry out such a wasteful process. A substantial amount of energy is invested in the formation of peptide bonds in the process of protein synthesis, and all this energy is dissipated when the protein is degraded. A possible explanation is that energy expenditure is used to achieve regulation. Our current knowledge of some of the functions of the ubiquitin system is consistent with this notion.

1.3
Discovery of the Role of Ubiquitin in Protein Degradation

Although the basically important cellular functions of selective protein degradation became evident in the late sixties, the molecular mechanisms involved in this process remained unknown. I became interested in the problem of how proteins are degraded in cells when I was a postdoctoral fellow in the laboratory of Gordon Tomkins in 1969–71. Gordon was mainly interested at that time in the mechanisms by which steroid hormones induce the synthesis of specific proteins. His model system for this purpose was the regulation of the enzyme tyrosine aminotransferase (TAT) in cultured hepatoma cells. Like other regulatory proteins, TAT has a rapid degradation rate. I found at that time, quite by accident, that the degradation of TAT is blocked by inhibitors of cellular ATP production, such as fluoride or dinitrophenol [9]. These results confirmed and extended earlier findings of Simpson [10] on the energy-dependence of the liberation of amino acids from proteins in liver slices. Since ATP depletion also prevented the inactivation of the enzymatic activity of TAT, it was concluded that energy is required at an early step in the process of protein degradation [9].

I was very much impressed by the energy-dependence of intracellular protein degradation because it suggested the involvement of a novel mechanism, different from that of known proteolytic enzymes. One attractive possibility that I considered was that proteins may be modified by some energy-dependent reaction prior to their degradation, and that such modification renders them susceptible to the action of some proteolytic enzyme [11]. To examine the existence of such (or any other) mechanism, a cell-free system was required, which faithfully reproduced energy-dependent protein degradation in the test tube, and which could be subjected to biochemical analysis. A cell-free ATP-dependent proteolytic system from reticulocyte lysates was first established by Etlinger and Goldberg [12]. Subsequently, my laboratory subjected this system to biochemical fractionation, with the aim of isolating its components and characterizing their mode of action. In this work, I was greatly helped by Aaron Ciechanover, who was my graduate student at that time. I have also received a lot of support, help, and great advice from Irwin Rose, in whose laboratory at Fox Chase Cancer Center I worked in a sabbatical year in 1978–79 and for many summers afterwards.

In the initial experiments, we resolved reticulocyte lysates on DEAE-cellulose into two crude fractions: Fraction 1, which contained proteins not adsorbed to the resin, and Fraction 2, which contained all proteins adsorbed to the resin and eluted with high salt. The original aim of this fractionation was to get rid of hemoglobin, which was known to be in Fraction 1, while most non-hemoglobin proteins of reticulocytes were known to be in Fraction 2. We found that neither fraction was active by itself, but ATP-dependent protein degradation could be reconstituted by combination of the two fractions [13]. The active component in Fraction 1 was a small, heat-stable protein; we have exploited its stability to heat treatment for its purification to near homogeneity. We termed this protein at that time APF-1, for ATP-dependent Proteolysis Factor 1 [13]. The identity of APF-1 with ubiquitin was established later by Wilkinson et al. [14], subsequent to the discovery in my laboratory of its covalent ligation to protein substrates, as described below.

The next question was what is the role of this small protein in ATP-dependent protein degradation. It looked smaller than most enzymes, so at first I thought that it might be a regulatory subunit of some enzyme (such as a protein kinase or an ATP-dependent protease) present in Fraction 2. To test this notion, we looked for the association of APF-1/ubiquitin with some protein in Fraction 2. For this purpose, purified radiolabeled APF-1/ubiquitin was incubated with Fraction 2 in the presence or absence of ATP, and subjected to gel filtration chromatography. A marked ATP-dependent association of APF-1/ubiquitin with high molecular weight material was observed [15]. It was very surprising to find that binding was covalent, as indicated by the resistance of the high molecular weight derivative to a variety of denaturing agents [15]. Subsequent work showed that proteins to which ubiquitin is bound are substrates of the ATP-dependent proteolytic system [16]. Based on these findings, we proposed in 1980 that proteins are targeted for degradation by covalent ligation to APF-1/ubiquitin and hypothesized that a protease exists that specifically degrades proteins ligated to ubiquitin [16]. Shortly afterwards, the identity of APF-1 with ubiquitin was established by Wilkinson et al. [14]. Ubiquitin was originally isolated by Goldstein and co-workers in a search for hormones from the thymus, but was subsequently found to be present in all tissues and eukaryotic organisms, hence its name [17]. The functions of ubiquitin were not known, though it was discovered by Goldknopf, Busch, and co-workers that ubiquitin is conjugated to histone 2A in an isopeptide linkage [18].

1.4
Identification of Enzymes of the Ubiquitin-mediated Proteolytic System

In subsequent work in my laboratory, we tried to isolate and characterize enzymes of the ubiquitin-mediated proteolytic system from Fraction 2 of reticulocytes, using a similar biochemical fractionation–reconstitution approach. Over a period of about ten years (1980–1990), we have identified eight different components in Fraction 2, all of which were required for ubiquitin-ATP-dependent protein degradation. Three of these are involved in the conjugation of ubiquitin to protein substrates. These are the ubiquitin-activating enzyme E1 [19], ubiquitin-carrier protein E2 [20] and ubiquitin-protein ligase E3 [20]. We found that E1 carries out the ATP-dependent activation of the carboxy-terminal glycine residue of ubiquitin [21] by the formation of ubiquitin adenylate, followed by the transfer of activated ubiquitin to a thiol site of E1 with the formation of a thiolester linkage [19, 20]. Activated ubiquitin is transferred to a thiol site of E2 by transacylation, and is then further transferred to an amino group of the protein substrate in a reaction that requires E3 [20]. All three types of enzyme were purified by affinity chromatography on ubiquitin-Sepharose [20]. The terms E1, E2, and E3 were suggested by Ernie Rose; "E" stood for enzyme, and not eluate, as stated in some articles. We found that the role of E3 is to specifically bind specific protein substrates [22]. Building on this observation, it was proposed that the selectivity of ubiquitin-mediated protein degradation is mainly determined by the substrate specificity of different E3 enzymes [23]. This notion was verified by subsequent work in many laboratories on the selective action of a large number of different E3 enzymes on their specific protein substrates.

Three other components that my laboratory has identified and partially purified from Fraction 2 of reticulocytes, termed CF1–CF3, are involved in the degradation of proteins ligated to ubiquitin [24]. These are apparently subcomplexes of the 26S proteasome, a large ATP-dependent protease complex first described by Rechsteiner and co-workers [25]. CF3 is identical to the 20S proteasome core particle [26], while CF1 and CF2 may be similar to the "base" and "lid" subcomplexes of the 19S regulatory particle of the 26S proteasome, described more recently by the Finley laboratory [27]. In hindsight, the reason for finding subcomplexes, rather than the complete 26S complex in Fraction 2 was technical: we have routinely prepared Fraction 2 from ATP-depleted reticulocytes [20], under which conditions the 26S proteasome dissociates to its subcomplexes. We found that incubation of the three subcomplexes in the presence of ATP promotes their assembly to the 26S proteasome [24, 26]. The role of ATP in the assembly of the 26S proteasome complex remains unknown.

The last two enzymatic activities that we have described in reticulocytes are ubiquitin C-terminal isopeptidases, which act at the final stages of the ubiquitin proteolytic pathway to release free and reusable ubiquitin from intermediary degradation products. One is an enzyme called isopeptidase T, which preferentially cleaves ubiquitin-Lys^{48}-ubiquitin linkages in polyubiquitin chains [28]. Its main function appears to be the disassembly of polyubiquitin chain remnants following proteolysis of the protein substrate moiety of ubiquitin–protein conjugates by the 26S proteasome complex. Another is a ubiquitin-C-terminal hydrolase that is an integral part of 26S proteasome complex [29]. Its role appears to be to release ubiquitin from linkage to the protein substrate at the final stages of the action of the 26S proteasome. Unlike most ubiquitin-C-terminal hydrolases, this isopeptidase is not inhibited by ubiquitin aldehyde, but is inhibited by the heavy metal chelator o-phenanthroline [29]. It appears to be similar to the Rpn11 metalloprotease subunit of the lid subcomplex, which has been recently identified by the Deshaies laboratory and shown to be essential for substrate deubiquitination and degradation [30].

1.5
Discovery of Some Basic Cellular Functions of the Ubiquitin System

The discovery of the basic biochemistry of ubiquitin-mediated protein degradation opened up the way for significant further progress in the elucidation of the roles of this system in a large variety of biological processes. Such further progress required the additional approaches of molecular genetics and cell biology. Thus, the first indication of the role of the ubiquitin system in cell cycle control was the discovery by Varshavsky and co-workers that the ts85 mammalian cell line, which fails to enter mitosis at the restrictive temperature, is defective in the ubiquitin-activating enzyme E1 [31]. The cloning of various genes of the ubiquitin system in yeast by the same laboratory led to insights into the roles of the polyubiquitin gene in stress response [32] and to the identification of the product of the DNA repair gene *RAD6* as an E2 protein [33]. Shortly afterwards, another E2 protein was identified as the product of the *CDC34* gene, known to be involved in the G1 → S transition in the cell cycle [34]. These early studies on the molecular genetics of the ubiquitin system initiated an avalanche of rapid progress in this field by many laboratories.

The entry of molecular genetics into the ubiquitin field did not signal the end of the usefulness of biochemical approaches. A good example of the power of the combination of biochemistry with genetics is the discovery of the Anaphase Promoting Complex/Cyclosome (APC/C), a large multisubunit ubiquitin ligase essential for exit from mitosis by the degradation of mitotic regulators such as cyclin B. In 1983, Hunt and co-workers discovered cyclin B, the first cyclin, as a protein that is destroyed at the end of each cell cycle in early embryos of marine invertebrates [35]. This discovery not only opened up a new era in cell cycle research, but also kindled interest in the problems of what is the machinery that targets cyclin B for degradation, and why does it act only at the end of mitosis? Though researchers of the cell cycle were at that time searching for a putative "cyclin protease", I thought that a specific ubiquitin ligase might exist that acts on cyclin B only at the end of mitosis. In 1991, independent work from the laboratory of Kirschner [36] and from my laboratory [37] showed that cyclin B is degraded by the ubiquitin system. Both laboratories employed biochemical approaches, using cell-free systems from early embryos of frogs [36] and clams [37]. Initial fractionation of the system in my laboratory [38] showed that in addition to E1, two novel components were required to reconstitute cyclin–ubiquitin ligation. These were a specific E2, termed E2-C, and an E3-like activity, which, in clam extracts, was associated with particulate material. In 1995, rapid progress in this system took place due to the convergence of information from biochemical experiments with genetic analysis in yeast. In work done in collaboration with Joan Ruderman, we solubilized the E3-like activity and partially purified and characterized it [39]. It was found to be a large (~1500 kDa) complex, which has cyclin–ubiquitin ligase activity. The activity of this enzyme is regulated in the cell cycle: it is inactive in the interphase and becomes active at the end of mitosis by phosphorylation. We called this complex the cyclosome, to denote its large size and important roles in cell-cycle regulation [39]. A similar complex was isolated from frog extracts at the same time by the Kirschner lab, and was called the Anaphase Promoting Complex [40]. The identification of subunits of the APC/C was made possible by work from the Nasmyth laboratory, who used an elegant genetic screen to identify yeast genes required for the proteolysis of cyclin B [41]. The products of some of these genes, *CDC16*, *CDC23* and *CDC27*, had been previously shown to be required for the onset of anaphase in budding and fission yeasts. Thus, the genetic work also proved the relevance of the biochemical results on APC/C to its role in exit from mitosis in cells. Subsequent work by several groups showed that APC/C is also involved in the degradation of some other important mitotic regulators, such as securin, an inhibitor of anaphase onset (reviewed in Ref. [42]). In addition, the APC/C is the target of the spindle assembly checkpoint system, a surveillance mechanism that allows sister chromatid separation only after all chromatids have been properly attached to the mitotic spindle [43].

1.6
Concluding Remarks

Several lessons can be learned from our story. One is not to accept authority in science. Monod's statement that there is no protein turnover in animal cells should not have been accepted without examination of the assumptions on which the statement was based. A second is that if you believe that you have a biologically important problem to study, you should pursue it, even if very few other researchers are interested in it. At the beginning, very few scientists were interested in the ubiquitin system (and some of the few who knew about it thought it was all wrong), but being obstinate was rewarding in the long run. If everyone only worked on subjects that are in the current mainstream of science, very few new fields would be discovered. The third lesson, which I keep reiterating with the hope of convincing a few young scientists, is the continued importance of biochemistry in biomedical research. The ubiquitin system could not have been discovered without the use of biochemical approaches, and biochemistry continues to be essential, in combination with molecular genetics and cell biology, in unraveling the myriad cellular functions of this system and their underlying molecular mechanisms.

References

1 FOLIN, O. *Am. J. Physiol.* **1905**, *13*, 66–116.

2 SCHOENHEIMER, R., RATNER, S. and RITTENBERG, D. *J. Biol. Chem.* **1939**, *130*, 703–732.

3 SCHOENHEIMER, R. *The Dynamic State of Body Constituents*, Harvard University Press, Cambridge, MA, **1942**.

4 HOGNESS, D. S., COHN, M. and MONOD. J. *Biochim. Biophys. Acta* **1955**, *16*, 99–116.

5 KAMIN, H. and HANDLER, P. *Annu. Rev. Biochem.* **1957**, *26*, 419–490.

6 EAGLE, H., PIEZ, K. A., FLEISCHMAN, R. and OYAMA, V. I. *J. Biol. Chem.* **1959**, *234*, 52–597.

7 RABINOWITZ, M. and FISHER, J. M. *Biochim. Biophys. Acta* **1964**, *91*, 313–322.

8 SCHIMKE, R. T. and DOYLE, D. *Annu. Rev. Biochem.* **1970**, *39*, 929–976.

9 HERSHKO, A. and TOMKINS, G. M. *J. Biol. Chem.* **1971**, *246*, 710–714.

10 SIMPSON, M. V. *J. Biol. Chem.* **1953**, *201*, 143–154.

11 HERSHKO, A. *Regulation of Gene Expression*, HARRIS, M. and THOMPSON, B., eds., U.S. Government Printing Office, Washington, DC, **1974**, pp. 85–94.

12 ETLINGER, J. D. and GOLDBERG, A. L. *Proc. Natl. Acad, Sci. USA* **1977**, *74*, 54–58.

13 CIECHANOVER, A., HOD, Y. and HERSHKO, A. *Biochem. Biophys. Res. Commun.* **1978**, *81*, 1100–1105.

14 WILKINSON, K. D., URBAN, M. K. and HAAS, A. L. *J. Biol. Chem.* **1980**, *255*, 7529–7532.

15 CIECHANOVER, A., HELLER, H., ELIAS, S., HAAS, A. L. and HERSHKO, A. *Proc. Natl. Acad. Sci. USA* **1980**, *77*, 1365–1368.

16 HERSHKO, A., CIECHANOVER, A., HELLER, H., HAAS, A. L. and ROSE, I. A. *Proc. Natl. Acad. Sci. USA* **1980**, *77*, 1783–1786.

17 GOLDSTEIN, G., SCHEID, M., HAMMERLING, U., BOYSE, E. A., SCHLESINGER, D. H. and NIALL, H. D. *Proc. Natl. Acad. Sci. USA* **1975**, *72*, 11–15.

18 GOLDKNOPF, I. L. and BUSCH, H. *Proc. Natl. Acad. Sci. USA* **1977**, *74*, 864–868.

19 CIECHANOVER, A., HELLER, H., KATZ-ETZION, R. and HERSHKO, A. *Proc. Natl. Acad. Sci. USA* **1981**, *78*, 761–765.

20 HERSHKO, A., HELLER, H., ELIAS, S. and CIECHANOVER, A. *J. Biol. Chem.* **1983**, *258*, 8206–8214.

21 HERSHKO, A., CIECHANOVER, A. and ROSE, I. A. *J. Biol. Chem.* **1981**, *256*, 1525–1528.

22 HERSHKO, A., HELLER, H., EYTAN, E. and REISS, Y. *J. Biol. Chem.* **1986**, *261*, 11992–11999.

23 HERSHKO, A. *J. Biol. Chem.* **1988**, *263*, 15237–15240.

24 GANOTH, D., LESHINSKY, E., EYTAN, E. and HERSHKO, A. *J. Biol. Chem.* **1988**, *263*, 12412–12419.

25 HOUGH, R., PRATT, G. and RECHSTEINER, M. *J. Biol. Chem.* **1986**, *261*, 2400–2408.

26 EYTAN, E., GANOTH, D., ARMON, T. and HERSHKO, A. *Proc. Natl. Acad. Sci. USA* **1989**, *86*, 7751–7755.

27 GLICKMAN, M. H., RUBIN, D. M., COUX, O., WEFES, I., PFEIFER, G., CJEKA, Z., BAUMEISTER, W., FRIED, V. A. and FINLEY, D. *Cell* **1998**, *94*, 615–623.

28 HADARI, T., WARMS, J. V. B., ROSE, I. A. and HERSHKO, A. *J. Biol. Chem.* **1992**, *267*, 719–727.

29 EYTAN, E., ARMON, T., HELLER, H., BECK, S. and HERSHKO, A. *J. Biol. Chem.* **1993**, *268*, 4668–4674.

30 VERMA, R., ARAVIND, L., OANIA, R., McDONALD, W. H., YATES, W. R., KOONIN, E. V. and DESHAIES, R. J. *Science* **2002**, *298*, 611–615.

31 FINLEY, D., CIECHANOVER, A. and VARSHAVSKY, A. *Cell* **1984**, *37*, 43–55.

32 FINLEY, D., OZKANYAK, E. and VARSHAVSKY, A. *Cell* **1987**, *48*, 1035–1046.

33 JENTSCH, S., McGRATH, J. P. and VARSHAVSKY, A. *Nature* **1987**, *329*, 131–134.

34 GOEBL, M. G., JOCHEM, J., JENTSCH, S., McGRATH, J. P., VARSHAVSKY, A. and BYERS, B. *Science* **1988**, *241*, 1331–1335.

35 EVANS, T., ROSENTHAL, E. T., YOUNGBLOOM, J., DISTEL, D. and HUNT, T. *Cell* **1983**, *33*, 289–396.

36 GLOTZER, M., MURRAY, A. W. and KIRSCHNER, M. W. *Nature* **1991**, *349*, 132–138.

37 HERSHKO, A., GANOTH, D., PEHRSON, J., PALAZZO, R. E. and COHEN, L. H. *J. Biol. Chem.* **1991**, *266*, 16276–16379.

38 HERSHKO, A., GANOTH, D., SUDAKIN, V., DAHAN, A., COHEN, L. H., LUCA, F. C., RUDERMAN, J. and EYTAN, E. *J. Biol. Chem.* **1994**, *269*, 4940–4946.

39 SUDAKIN, V., GANOTH, D., DAHAN, A., HELLER, H., HERSHKO, J., LUCA, F. C., RUDERMAN, J. V. and HERSHKO, A. *Mol. Biol. Cell* **1995**, *6*, 185–198.

40 KING, R. W., PETERS, J. M., TUGENDREICH, S., ROLFE, M., HIETER, P. and KIRSCHNER, M. W. *Cell* **1995**, *81*, 279–288.

41 IRNIGER, S., PIATTI, S., MICHAELIS, C. and NASMYTH, K. *Cell* **1995**, *81*, 269–277.

42 HARPER, J. W., BURTON, J. C. and SOLOMON, M. J. *Genes Dev.* **2002**, *16*, 2179–2206.

43 AMON, A. *Curr. Op. Cell Biol.* **1999**, *9*, 69–75.

2

N-terminal Ubiquitination: No Longer Such a Rare Modification

Aaron Ciechanover

Abstract

The ubiquitin–proteasome system (UPS) is involved in selective targeting of innumerable cellular proteins via a complex pathway that plays important roles in a broad array of processes. An important step in the proteolytic cascade is specific recognition of the substrate by one of many ubiquitin ligases, E3s, that is followed by generation of the polyubiquitin degradation signal. For most substrates, it is believed, though it has not been shown directly, that the first ubiquitin moiety is conjugated, via its C-terminal Gly^{76} residue, to an ε-NH$_2$ group of an internal lysine residue. Recent findings indicate that for an increasing number of proteins, the first ubiquitin moiety is fused linearly to the α-NH$_2$ group of the N-terminal residue. An important biological question relates to the evolutionary requirement for an alternative mode of ubiquitination.

2.1
Background

Two distinct structural elements play a role in the ubiquitination of a target protein: (i) the E3 recognition site and (ii) the anchoring residue of the polyubiquitin chain. In most cases, it is believed, though it has been shown for only a few proteins, that the first ubiquitin moiety is transferred to an ε-NH$_2$ group of an internal lysine residue in the substrate. The N-terminal domain of the target protein has attracted attention both as an E3 recognition domain and, recently, as a ubiquitination site.

As for specific recognition, in certain rare cases, the stability of a protein is a direct function of its N-terminal residue, which serves as a binding site for the ubiquitin ligase E3α (Ubr1 in yeast; 'N-end-rule'; [1, 2]). Accordingly, two types of N-terminal residues have been defined, "stabilizing" and "destabilizing". For the Mos protein, it was found that its stability is governed primarily by the penultimate proline residue and by a phosphorylation/dephosphorylation cycle

Protein Degradation. Edited by J. Mayer, A. Ciechanover, M. Rechsteiner
Copyright © 2005 WILEY-VCH Verlag GmbH & Co. KGaA, Weinheim
ISBN: 3-527-30837-7

of serine3 [3]. A mechanistic explanation for the role of the Pro and Ser residues is still missing.

As for the lysine residue targeted, there is no consensus as to its specificity. In some cases distinct lysines are required, while in others there is little or no specificity. Thus signal-induced degradation of IκBα involves two particular lysine residues, 21 and 22 [4]. In the case of Gcn4, lysine residues in the vicinity of a specific PEST degradation signal serve as ubiquitin attachment sites [5]. Mapping of ubiquitination sites of the yeast iso-2-cytochrome c has revealed that the polyubiquitin chain is synthesized almost exclusively on a single lysine [6]. In two other examples, that of Mos (see above; [3]) and the model "N-end rule" substrate X-β-gal (where X is a short fused peptide not encoded by the native molecule [7]), one and two lysines, respectively, that reside in proximity to the degradation signal are required for ubiquitination. In striking contrast, ubiquitination of the ζ chain of the T-cell receptor is independent of any particular lysine residue and proceeds as long as one residue is present in the cytosolic tail of the molecule [8]. Similarly, no single specific lysine residue is required for ubiquitination of either c-Jun [9] or cyclin B [10]: any single lysine residue, even artificially inserted, can serve as a ubiquitin acceptor. Important in this context is that only in a handful of cases it has been shown directly, via chromatographic or mass spectrometric analyses, that ubiquitin is indeed anchored to a lysine residue (see for example Refs. [11, 12]). In most cases studied, and there are not too many, the assumption that an internal lysine serves as the polyubiquitin chain anchor is indirect and based on mutational analyses.

One interesting case involves the artificial fusion protein ubiquitin-Pro-X-β-galactosidase. In this chimera, the ubiquitin moiety was fused to the N-terminal Pro residue of the protein. Unlike other ubiquitin-B-X-β-galactosidase species (where B is any of the remaining 19 amino acid residues), here ubiquitin is not removed by isopeptidases and serves as a degradation signal following generation of a polyubiquitin chain that is anchored to Lys48 of the artificially fused ubiquitin moiety [13]. However as noted, in this case the ubiquitin moiety was fused to the N-terminal residue artificially.

The first substrate that was identified in which the N-terminal residue serves as a ubiquitination target was MyoD. The basic helix–loop–helix (bHLH) protein MyoD is a tissue-specific transcriptional activator that acts as a master switch for muscle development. MyoD forms heterodimers with other proteins belonging to the bHLH group, such as the ubiquitously expressed E2A, E12 and E47. These dimers are probably the transcriptionally active forms of the factor. Association of MyoD with HLH proteins of the Id family (inhibitors of differentiation that lack the basic domain) inhibits its DNA-binding and biological activities. MyoD is a short-lived protein with a half-life of ∼45 min [14, 15]. Degradation of MyoD is mediated by the ubiquitin system both *in vitro* and *in vivo*. Furthermore, the process is inhibited by its consensus DNA-binding site. In contrast, addition of Id1 destabilizes the MyoD–E47–DNA complex and renders the protein susceptible to degradation [15].

2.2
Results

To analyze specific ubiquitination sites in MyoD, we used site-directed mutagenesis to substitute systematically all the lysine residues with arginines [16]. The protein contains nine lysine residues, most of them located within the N-terminal domain of the molecule. The nine residues are in positions 58, 99, 102, 104, 112, 124, 133, 146 and 241. The various proteins were generated either by expression in bacteria followed by purification, or by *in vitro* translation in reticulocyte lysate in the presence of [^{35}S]methionine. Conjugation and degradation of the proteins were monitored in a reconstituted cell-free system or in cells. Proteins were detected by either Western blot analysis or PhosphorImaging. Surprisingly, even a MyoD species that lacked all lysine residues was still degraded efficiently in an ATP-dependent manner *in vitro*. To demonstrate involvement of the ubiquitin system in the process, we followed the degradation of wild-type (WT) and lysine-less (LL) MyoD in the absence and presence of ubiquitin. Similar to the degradation of the WT protein, degradation of the LL MyoD was completely dependent upon the addition of exogenous ubiquitin to an extract that does not contain it (Fraction II). Furthermore, addition of methylated ubiquitin, which cannot form polyubiquitin chains and serves as a chain terminator [17], inhibited the degradation of LL MyoD. The inhibition could be alleviated by the addition of excess of free ubiquitin. These results strongly suggested that polyubiquitination of LL MyoD is necessary for degradation of the protein. Furthermore, they implied that the polyubiquitin chain is synthesized on internal lysine residues of ubiquitin. To demonstrate directly polyubiquitinated LL MyoD, we used *in-vitro*-translated ^{35}S-labeled protein in a partially reconstituted system. We demonstrated that LL MyoD generates high molecular mass ubiquitinated adducts. It should be noted, however, that these conjugates are of somewhat lower molecular mass than those of the WT MyoD. This can be attributed to the role that the internal lysine residues also play in the process (see also below).

To investigate the physiological relevance of the observations in the cell-free system, we followed the fate of the different MyoD lysine-mutated proteins *in vivo*, using pulse-chase labeling experiments in COS-7 cells that were transiently transfected with the different MyoD cDNAs. In agreement with our *in vitro* data, the lysine-less MyoD protein is degraded efficiently in cells as well. However, we could observe a progressive increase in the half-life of the proteins of up to ~2-fold with the gradual substitution of the lysine residues. While the half-life of WT MyoD was ~50 min, that of LL MyoD was ~2 h. Interestingly, we found that the stability of MyoD is not affected by the substitution of any specific lysine residue, and it is the total number of these residues that determines the half-life of the protein. To identify the system involved in the destruction of LL MyoD *in vivo*, transfected cells were incubated in the presence of inhibitors of proteasomal and lysosomal degradation. Chloroquine, a general inhibitor of lysosomal proteolysis, and E-64, a cysteine protease inhibitor that affects lysosomal, but also certain cytosolic proteases, had no effect on the stability of the LL MyoD. In striking contrast, the proteasomal inhibitors MG132 and lactacystin blocked degradation of the LL protein significantly. To demonstrate the intermediacy of ubiquitin conjugates in the degradation of LL MyoD, we incubated COS-7 cells, transiently transfected with either WT or LL MyoD cDNAs, with MG132, and followed generation of ubiquitin-MyoD adducts. Immunoprecipitation with anti-MyoD antibody followed by Western blot analysis with anti-ubiquitin antibody revealed accumulation of high molecular mass compounds in cells transfected with either WT or LL MyoD. A similar analysis of mock-transfected cells clearly demonstrated the specificity of both the anti-MyoD and anti-ubiquitin antibodies.

Based on these results, it was clear that polyubiquitination is essential for targeting MyoD for degradation. The lack of internal lysine residues, the only known targets for ubiquitin modification, made it important to identify the functional group that can serve as an attachment site for ubiquitin. Chemically, several groups can generate covalent bonds with ubiquitin. Ser and Thr can participate in ester bond formation, while Cys can generate a thiol ester bond. However, these bonds are unstable and are hydrolyzed in either high pH (Ser and Thr) or high concentration of −SH groups (Cys). The stability of the MyoD-ubiquitin adducts under these conditions made it highly unlikely that any of these modifications is the one we observed.

A likely candidate, however, was the free amino group of the N-terminal residue of the protein, which can generate a stable peptide bond with the C-terminal Gly residue of ubiquitin. Edman degradation of the N-terminal residue of bacterially expressed, *in-vitro*-translated and cellularly expressed MyoDs, has revealed that the ubiquitin attachment site can be the free, unmodified initiator methionine: the proteins were not modified and the N-terminal residue was not acetylated. To demonstrate a role for the free N-terminal amino group in the degradation of MyoD, we chemically modified this group. Initially, we blocked this group in the LL MyoD protein by reductive methylation. While this procedure blocks all amino groups in a protein in a non-discriminatory manner, in this case, it could have been only the α-NH$_2$ group, which is the only free amino group left in the MyoD molecule. The modification stabilized the protein completely. Whereas a free α-NH$_2$ appears to be sufficient for degradation (probably following ubiquitination) of LL MyoD, it is not clear whether it also plays a physiological role in targeting the WT molecule, which has nine available lysine residues. In order to investigate the role and biological relevance of the free α-NH$_2$ group in the targeting of WT MyoD, we selectively blocked it by carbamoylation with potassium cyanate at low pH. This procedure does not modify ε-NH$_2$ groups of internal lysine residues. Automated Edman degradation along with fluorescamine determination of the extent of remaining free NH$_2$ groups confirmed that the modification affected only the N-terminal group. The modified protein was subjected to *in vitro* degradation and conjugation in cell extract. In contrast to LL MyoD, the N-terminally carbamoylated protein could not be ubiquitinated and was stable. Thus, a free and exposed NH$_2$ terminus of MyoD appears to be an essential site for degradation, most probably because it serves as an attachment site for the first ubiquitin moiety. As an additional control, we selectively modified the internal lysine residues of WT MyoD by guanidination with *O*-

methylisourea. The modification, which does not affect the N-terminal group, generates a protein that is essentially the chemically modified counterpart of the LL MyoD that was generated by site-directed mutagenesis. Similar to the LL protein, this MyoD derivative is degraded efficiently in the cell-free system in a ubiquitin- and an ATP-dependent mode.

To analyze the role of the N-terminal residue of MyoD as a ubiquitination site, we fused to WT MyoD, upstream to the N-terminal residue, a 6 × Myc tag, and monitored the stability of the tagged protein. We showed that it is stable both *in vitro* and *in vivo*. It should be noted that the two first N-terminal residues of the Myc tag, methionine and glutamate, are identical to the first two N-terminal residues in MyoD. In addition, the Myc tag also contains a lysine residue. Thus, altogether, six additional lysine residues were added to WT MyoD in addition to its own nine native residues. Nevertheless, the tag stabilizes it, probably by blocking access to a specific N-terminal residue, and, as became clear later (see below), to its neighboring domain.

Taken together, these findings strongly suggested that MyoD is first ubiquitinated at its N-terminal residue, and the polyubiquitin chain is synthesized on this first conjugated ubiquitin moiety. Internal lysine residues also play a role, probably by serving as additional anchoring sites, whose ubiquitination accelerates degradation. Yet they are not essential for proteolysis to occur. In contrast, ubiquitination of the N-terminal residue plays a critical role in governing the protein's stability [16] (Figure 2.1).

Using a similar, though not a complete, set of experiments, 12 additional proteins have been identified recently that appear to undergo N-terminal ubiquitination: (i) the *h*uman *p*apilloma*v*irus-16 (HPV-16) E7 oncoprotein (18), (ii) the *l*atent *m*embrane *p*rotein 1 (LMP1)(19) and (iii) 2A (LMP2A)(20) of the *E*pstein *B*arr virus (EBV), involved in viral activation from latency, (iv) the *c*ell cycle-*d*ependent *k*inase (CDK) inhibitor p21 (21,22), (v) the extracellular signal-regulated kinase 3, ERK3 (22), the *i*nhibitors of *d*ifferentiation (vi) Id2 (23) and (vii) Id1 (24), two pro-proliferative Helix-Loop-Helix proteins, (viii) hydroxymethyglutaryl-Coenzyme A reductase (HMG-CoA reductase), the first and key regulatory enzyme in the cholesterol biosynthetic pathway (25), (ix) p19^{ARF}, the mouse Mdm2 inhibitor and (x) p14^{ARF}, its human homologue (26), (xi) the HPV-58 E7 oncoprotein, and (xii) the cell cycle regulator p16^{INK4a} (27). As for HMG-CoA reductase, several specific internal lysine residues have also been shown to be important for its targeting, and therefore the essentiality of the N-terminal residue in the process has to be further substantiated. The case of p21 requires further investigation, as one study reported that its degradation by the proteasome does not require ubiquitination [28], while an independent study has demonstrated a role for Mdm2 in targeting p21, also without a requirement for ubiquitination [29]. As we noted for MyoD, substitution of the internal lysines inhibited slightly (up to twofold) both conjugation and degradation of HPV-16, LMP1, and Id2, suggesting that these residues, probably also by serving as ubiquitin anchors, can modulate the stability of these proteins. It is possible to suppose that N-terminal ubiquitination and modification of internal lysines is catalyzed by different ligases that may be even located in different subcellular compartments (e.g. the nucleus and cytosol). Because of the role that internal

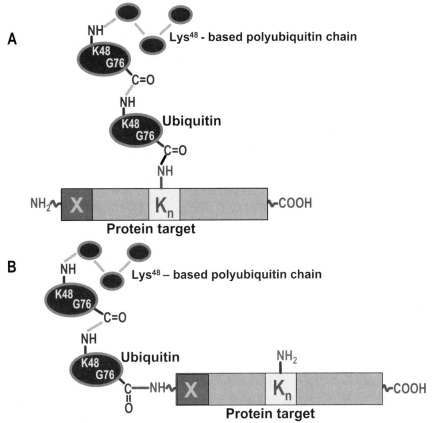

Fig. 2.1. Ubiquitination on an internal lysine and on the N-terminal residue of the target substrate. (A) The first ubiquitin moiety is conjugated, via its C-terminal Gly^{76} residue, to the ε-NH_2 group of an internal lysine residue of the target substrate (K_n). (B) The first ubiquitin moiety is conjugated to a free α-NH_2 group of the N-terminal residue, X. In both cases, successive addition of activated ubiquitin moieties to internal Lys^{48} on the previously conjugated ubiquitin moiety leads to the synthesis of a polyubiquitin chain which serves as the degradation signal for the 26S proteasome.

lysines play in modulating the stability of these proteins, and in order to better understand the physiological significance of this novel mode of modification, it was important to identify proteins whose degradation is completely dependent on N-terminal ubiquitination. An important group of potential substrates for N-terminal ubiquitination is that of *naturally occurring lysine-less proteins* – NOLLPs. Since these proteins cannot use the "canonical" lysine conjugation pathway, in order to be targeted by the ubiquitin system they must use, an alternative site for their tagging. Searching the database, we were able to identify 177 eukaryotic NOLLPs, 14 of which occur in humans. In addition, we have identified 111 viral NOLLPs. We have shown that two of the proteins mentioned above, the human tumor supressor $p16^{INK4a}$ and the viral oncoprotein HPV-58 E7 are degraded via the N-terminal ubiquitination pathway [27]. Interestingly, we demonstrated that $p16^{INK4a}$ is ubiquitinated and degraded only in sparse cells, and is stable in dense cells. Similar findings were reported for the NOLLPs p19^{ARF} and p14^{ARF} (26).

For E7-16 [18], LMP1 [19], and MyoD (unpublished), it has been shown that truncation of a short N-terminal segment of 10–20 residues stabilized the proteins, suggesting that the entire domain beyond the single N-terminal residue plays a role in governing the stability of these proteins. Such a segment can allow the mobility/flexibility necessary for the N-terminal residue to serve as a ubiquitin acceptor. It can also serve as a recognition domain for the cognate E3. There is no homology between the N-terminal domains of these three proteins, suggesting that if the three N-terminal domains serve as recognition motifs, they recognize different components of the ubiquitin system. Interestingly, the LMP2A E3 was identified as a member of the NEDD4 family of HECT domain ligases, AIP4 and/ or WWP2 [20]. A PY motif in LMP2A is recognized by the E3. It resides in the N-terminal domain of the molecule, supporting the hypothesis that in these proteins the E3-binding domain may reside in close proximity to the N-terminal residue ubiquitination site.

Is there any direct evidence for N-terminal ubiquitination? All the different and independent lines of evidence in the various studies strongly suggest that ubiquitination occurs on the N-terminal residue, and any other scenario is highly unlikely. Yet, the only direct evidence must be demonstration of a fusion peptide between the C-terminal domain of ubiquitin and the N-terminal domain of the target substrate. The study on p21 [21] and E7-58 [27] brought us a little closer. Bloom and colleagues [21] transfected cells with N-terminally His-tagged ubiquitin and N-terminally HA-tagged p21 that contained a Factor X proteolytic site immediately after the HA tag and upstream of the p21 reading frame. They then immunoprecipitated and resolved the cell-generated ubiquitin conjugates of p21 and treated the mono-ubiquitin–p21 adduct with Factor-X protease. This treatment released a smaller species of p21 (lacking His-ubiquitin and the HA-tag-Factor X site) and His-ubiquitin-HA-Factor X site, thus demonstrating that the HA-tag-Factor X site, which was previously part of p21, had now become part of the Factor X-cleaved ubiquitin. A similar experimental evidence was brought for the NOLLP HPV E7-58 [27]. Here, Ben-Saadon and colleagues generated two species of the protein containing the eight amino acid sequence of the Tobacco Etch Virus (TEV) protease cleavage site inserted either 21 amino acid residues after the iMet [E7-58-TEV(21)] or immediately after the iMet [E7-58-TEV(1)]. The prediction from this experiment was that if ubiquitin is indeed attached to the N-terminal residue of E7-58, TEV protease-catalyzed cleavage will generate an extended ubiquitin molecule that will also contain the respective N-terminal domain of E7-58 [21 residues or 1 residue, respectively, dependent upon whether the substrate of the reaction is E7-58-TEV(21) or E7-58-TEV(1), and the six amino acids derived from the TEV cleavage site]. Such extended ubiquitin moieties were indeed generated following incubation of the substrate with labeled methylated ubiquitin (which generated mostly the mono-ubiquitin adduct of the E7-58 protein), followed by cleavage of the adduct with TEV. The only conclusion that can be derived from these experiments is that the ubiquitin moiety was fused to any of the amino acid residues of the HA tag-Factor X site at the N-terminal domain of p21, or to any of the first 21 amino acids of E7 or the TEV site (part of it; the protease cleaves after the sixth amino acids out of eight in the complete site). Such internal modification is unlikely, however, as it must require a novel chemistry since none of the residues in the tags, the protease sites or the E7 N-terminal fragment, is lysine. Yet, formally, it is still possible that such a modification occurs. The HA tag contains, for example, three Tyr residues. Thus, the evidence provided by these two experiments clearly limits an unlikely non-peptide bond ubiquitination, such as esterification, to a much smaller zone in the N-terminal domain of the tagged p21 or the TEV-containing E7-58, but does not demonstrate directly that the modification occurs indeed on the N-terminal residue.

As noted, only identification of a fusion peptide between the C-terminal domain of ubiquitin and the N-terminal domain of the target protein will constitute such an evidence. Ben-Saadon and colleagues have recently isolated the long sought after fusion peptide [27]: mass spectrometric analysis of a tryptic digest of the isolated mono-ubiquitin adduct of HPV-58 E7 revealed a peptide of 11 amino acids, GG-MHGNNPTLR which represents the last two C-terminal amino acids of ubiquitin, GlyGly, and the first nine residues of E7, MetHisGlyAsnAsnProThrLeuArg (MHGNNPTLR). It should be noted that WT E7-58 contains an Arg residue in position 2. It was necessary to substitute this Arg with His, since otherwise the digesting enzyme, trypsin, would have generated a tetrapeptide, GG-MR, that would have been difficult, if not impossible, to identify in the MS analysis. MS/MS analysis of the 11-mer, verified its internal sequence. Coulombe and colleagues were also able to isolate and identify the sequence of a fusion peptide between the C-terminal domain of ubiquitin and the N-terminal domain of HA-tagged p21 that also contained, downstream of the tag, a stretch of residues derived from the N-terminal domain of the native substrate, but without the iMet (which was removed during the construction of the tagged protein) [22].

2.3
Discussion

N-terminal ubiquitination is a novel pathway, clearly distinct from the N-end rule pathway [30]. In the latter, the N-terminal residue serves as a recognition and binding motif to the ubiquitin ligase, E3α; however, ubiquitination occurs on an internal lysine(s). In contrast, in the N-terminal ubiquitination pathway, modification occurs on the N-terminal residue, whereas recognition probably involves a downstream motif. It should be mentioned that in yeast, using the model fusion protein ubiquitin-Pro-X-β-galactosidase (where X is a short sequence derived from the λ repressor), a new proteolytic pathway has been described, designated the UFD (**u**biquitin **f**usion **d**egradation) pathway [31]. The stably fused ubiquitin moiety (note that in this exceptional case, with Pro and not any other amino acid residue as the linking residue, the ubiquitin moiety is not cleaved off by ubiquitin C-terminal hydrolases), functions as a degradation signal, where its Lys^{48} serves as an anchor for the synthesized polyubiquitin chain. This pathway involves several enzymes, UFD 1–5, some of which appear to be unique and are not part of the "canonical"

UPS. It is possible that N-terminal ubiquitination is the most upstream event in the UFD pathway – which was discovered using an artificial chimeric ubiquitin-protein model substrate: the N-terminal ubiquitination pathway can function by providing substrates to the UFD pathway.

The physiological significance of N-terminal ubiquitination is still obscure. Naturally occurring lysine-less proteins, NOLLPs, that are degraded by the ubiquitin system must traverse this pathway. Many such proteins, mostly viral, can be found in the database (see above and in Refs. [26,27]). We believe that many additional lysine-containing proteins, will be discovered to be targeted via this novel mode of modification. Of note is that all the proteins that are N-terminally ubiquitinated must contain a free, unmodified N-terminal residue. Such proteins constitute approximately 25% of all cellular proteins, while the remaining 75% are Nα-acetylated. Whether a protein will be acetylated is dependent on the structure of the N-terminal domain of the mature protein. This is determined by the combined activities of **m**ethionine **a**mino**p**eptidases (MAPs) and **N**-terminal **a**cetyl**t**ransferases (NATs), which are dependent on the specific sequence of up to the first four N-terminal residues of the target protein substrates (reviewed in Ref. [32]). Thus, it is possible to predict which proteins will be potential substrates of the N-terminal ubiquitination pathway. Internal C-terminal fragments of Nα-acetylated proteins can also be modified by ubiquitin at their "new" N-terminal residue following limited processing. Many proteins, such as the NF-κB precursors p105 and p100 or caspase substrates are processed initially in a limited manner, generating a C-terminal fragment with a newly exposed N-terminal residue. For all lysine-containing proteins, the intact free N-termini as well as the products of processing, the assumption is that their internal lysines are not easily accessible, for whatever reason, for ubiquitination, and it is only the N-terminal residue that can be modified. Interestingly, most of the substrates identified thus far have a few lysine residues that might not be accessible to the E3s: For example, MyoD has nine (out of 319), E7 two (out of 97), LMP1 has a single lysine residue (out of 440), LMP2A has three (out of 497), Id2 nine (out of 134) and p21 six (out of 164). From the random discovery of thirteen N-terminally ubiquitinated proteins, it appears that their number could well be larger and that many more will be discovered, which will help in the unraveling of the unique characteristics that distinguish this group of substrates.

Acknowledgments

Research in the laboratory of A.C. is supported by grants from Prostate Cancer Foundation Israel – Centers of Excellence Program, the Israel Science Foundation – Centers of Excellence Program, a Professorship funded by the Israel Cancer Research Fund (ICRF) and the Foundation for Promotion of Research in the Technion administered by the Vice President of the Technion for Research. Infrastructural equipment was purchased with the support of the Wolfson Charitable Fund Center of Excellence for studies on *Turnover of Cellular Proteins and its Implications to Human Diseases*.

References

1 BACHMAIR, A., FINLEY, D. and VARSHAVSKY, A. *In vivo* half-life of a protein is a function of its amino-terminal residue. *Science*, **1986**, *234*, 179–186.

2 VARSHAVSKY, A. The N-end rule: functions, mysteries, uses. *Proc. Natl Acad. Sci. USA*, **1996**, *93*, 12142–12149.

3 NISHIZAWA, M., FURUNO, N., OKAZAKI, K., TANAKA, H., OGAWA, Y. and SAGATA, N. Degradation of Mos by the N-terminal proline (Pro2)-dependent ubiquitin pathway on fertilization of *Xenopus* eggs: possible significance of natural selection for Pro2 in Mos. *EMBO J.*, **1993**, *12*, 4021–4027.

4 SCHERER, D. C., BROCKMAN, J. A., CHEN, Z., MANIATIS, T. and BALLARD, D. W. Signal-induced degradation of IκBα requires site-specific ubiquitination. *Proc. Natl Acad. Sci. USA*, **1995**, *92*, 11259–11263.

5 KORNITZER, D., RABOY, B., KULKA, R. G. and FINK, G. R. Regulated degradation of the transcription factor Gcn4. *EMBO J.*, **1994**, *13*, 6021–6030.

6 SOKOLIK, C. W. and COHEN, R. E. The structures of ubiquitin conjugates of yeast Iso-2-cytochrome c. *J. Biol. Chem.*, **1991**, *266*, 9100–9107.

7 CHAU, V., TOBIAS, J. W., BACHMAIR, A., MARRIOTT, D., ECKER, D. J., GONDA, D. K. and VARSHAVSKY, A. A multi ubiquitin chain is confined to specific lysine in a targeted short-lived protein. *Science*, **1989**, *243*, 1576–1583.

8 HOU, D., CENCIARELLI, C., JENSEN, J. P., NGUYGEN, H. B. and WEISSMAN, A. M. Activation-dependent ubiquitination of a T cell antigen receptor subunit on multiple intracellular lysines. *J. Biol. Chem.*, **1994**, *269*, 14244–14247.

9 TREIER, M., STASZEWSKI, L. and BOHMANN, D. Ubiquitin-dependent c-Jun degradation *in vivo* is mediated by the δ-domain. *Cell*, **1994**, *78*, 787–798.

10 KING, R. W., GLOTZER, M. and KIRSCHNER, M. W. Mutagenic analysis of the destruction signal of mitotic cyclins and structural characterization of ubiquitinated intermediates. *Mol. Biol. Cell*, **1996**, *7*, 1343–1357.

11 GOLDKNOPF, I. L. and BUSCH, H. Isopeptide linkage between nonhistone and histone 2A polypeptides of chromosomal conjugate-protein A24. *Proc. Natl. Acad. USA*, **1977**, *74*, 864–868.

12 GRONROOS, E., HELLMAN, U., HELDIN, C. H. and ERICSSON, J. Control of Smad7 stability by competition between acetylation and ubiquitination. *Mol. Cell*, **2002**, *10*, 483–493.

13 JOHNSON, E. S., BARTEL, B., SEUFERT, W. and VARSHAVSKY, A. Ubiquitin as a degradation signal. *EMBO J.* **1992**, *11*, 497–505.

14 THAYER, M. J., TAPSCOTT, S. J., DAVIS, R. L., WRIGHT, W. E., LASSAR, A. B. and WEINTRAUB, H. Positive autoregulation of the myogenic determination gene MyoD1. *Cell*, **1989**, *58*, 241–248.

15 ABU-HATOUM, O., GROSS-MESILATY, S., BREITSCHOPF, K., HOFFMAN, A., GONEN, H., CIECHANOVER, A. and BENGAL, E. Degradation of the myogenic transcription factor MyoD by the ubiquitin pathway *in vivo* and *in vitro*: regulation by specific DNA binding. *Mol. Cell. Biol.*, **1998**, *18*, 5670–5677.

16 BREITSCHOPF, K., BENGAL, E., ZIV, T., ADMON, A. and CIECHANOVER, A. A novel site for ubiquitination: The N-terminal residue and not internal lysines of MyoD is essential for conjugation and degradation of the protein. *EMBO J.*, **1998**, *17*, 5964–5973.

17 HERSHKO, A. and HELLER, H. Occurrence of a polyubiquitin structure in ubiquitin–protein conjugates. *Biochem. Biophys. Res. Commun.* **1985**, *128*, 1079–1086.

18 REINSTEIN, E., SCHEFFNER, M., OREN, M., SCHWARTZ, A. L. and CIECHANOVER, A. Degradation of the E7 human papillomavirus oncoprotein by the ubiquitin-proteasome system:

targeting via ubiquitination of the N-terminal residue. *Oncogene*, **2000**, *19*, 5944–5950.

19 Aviel, S., Winberg, G., Massucci, M. and Ciechanover, A. Degradation of the Epstein-Barr virus latent membrane protein 1 (LMP1) by the ubiquitin-proteasome pathway: targeting via ubiquitination of the N-terminal residue. *J. Biol Chem.*, **2000**, *275*, 23491–23499.

20 Ikeda, M., Ikeda, A. and Longnecker, R. Lysine-independent ubiquitination of the Epstein-Barr virus LMP2A. *Virology*, **2002**, *300*, 153–159.

21 Bloom, J., Amador, V., Bartolini, F., DeMartino, G. and Pagano, M. Proteasome-mediated degradation of p21 via N-terminal ubiquitinylation. *Cell*, **2003**, *115*, 1–20.

22 Coulombe, P., Rodier, G., Bonneil, E., Thibault, P. and Meloche, S. N-Terminal ubiquitination of extracellular signal-regulated kinase 3 and p21 directs their degradation by the proteasome. *Mol. Cell Biol.*, **2004**, *24*, 6140–6150.

23 Fajerman, I., Schawartz, A. L. and Ciechanover, A. Degradation of the Id2 developmental regulator: Targeting via N-Terminal Ubiquitination. *Biochem. Biophys. Res. Commun.*, **2004**, *314*, 505–512.

24 Azar-Trausch, J. S., Lingbeck, J., Ciechanover, A. and Schwartz, A. L. Ubiquitin-proteasome-mediated degradation of Id1 is modulated by MyoD. *J. Biol. Chem.*, **2004**, *279*, 32614–32619.

25 Doolman, R., Leichner, G. S., Avner, R. and Roitelman, J. Ubiquitin is conjugated by membrane ubiquitin ligase to three sites, including N-terminus, in transmembrane region of mammalian 3-hydroxy-3-methylglutaryl coenzyme a reductase: Implications for sterol-regulated enzyme degradation. *J. Biol. Chem.*, **2004**, *279*, 38184–38193.

26 Kuo, M. L., den Besten, W., Bertwistle, D., Roussel, M. F., and Sherr, C. J. N-terminal polyubiquitination and degradation of the Arf tumor suppressor. *Genes and Dev.*, **2004**, *18*, 1862–1874.

27 Ben-Saadon, R., Fajerman, I., Ziv, T., Hellman, U., Schwartz, A. L., and Ciechanover, A. The Tumor Suppressor Protein p16^{INK4a} and the Human Papillomavirus oncoprotein E7-58 are Naturally Occurring Lysine-Less Proteins that are Degraded by the Ubiquitin System: Direct Evidence for Ubiquitination at the N-Terminal Residue. *J. Biol. Chem.*, **2004**, *279*, 41414–41421.

28 Sheaff, R. J., Singer, J. D., Swanger, J., Smitherman, M., Roberts, J. M. and Clurman, B. E. Proteasomal turnover of p21^{Cip1} does not require p21^{Cip1} ubiquitination. *Mol. Cell*, **2000**, *5*, 403–410.

29 Jin, Y., Lee, H., Zeng, S. X., Dai, M. S. and Lu, H. Mdm2 promotes p21$^{waf1/cip1}$ proteasomal turnover independently of ubiquitylation. *EMBO J.*, **2003**, *22*, 6365–6377.

30 Varshavsky, A. The N-end rule: Functions, mysteries, uses. *Proc. Natl. Acad. Sci. USA*, **1996**, *93*, 12142–12149.

31 Johnson, E. S., Ma, P. C., Ota, I. M. and Varshavsky, A. A proteolytic pathway that recognizes ubiquitin as a degradation signal. *J. Biol. Chem.*, **1995**, *270*, 17442–17456.

32 Polevoda, B. and Sherman, F. N-terminal acetyltransferases and sequence requirements for N-terminal acetylation of eukaryotic proteins. *J. Mol. Biol.*, **2003**, *325*, 595–622.

3
Evolutionary Origin of the Activation Step During Ubiquitin-dependent Protein Degradation

Hermann Schindelin

Abbreviations

E1	Activating enzyme of a UbL
E2	Conjugating enzyme of a UbL
E3	Ubiquitin ligase
Moco	Molybdenum cofactor
MPT	Molybdopterin
NEDD8-E1	NEDD8-activating enzyme
Ubiquitin-E1	Ubiquitin-activating enzyme
UbL	Ubiquitin-like protein

Abstract

Ubiquitin and related protein modifiers are activated in an ATP-dependent process, which leads to the initial formation of an acyl adenylate between the C-terminus of the modifier and AMP. The modifier is subsequently transferred onto an active-site cysteine residue in the activating (E1) enzyme via a thioester bond and from there to a conjugating (E2) enzyme. In the case of ubiquitin a large family of ubiqutin ligases (E3 enzymes) primarily ensure specific transfer onto the correct protein substrate. Biosynthesis of the molybdenum cofactor is an evolutionarily conserved pathway present in bacteria, archaea and eukaryotes. The molybdenum cofactor contains a *cis*-dithiolene group and incorporation of the sulfur atoms involves among other proteins, MoaD and MoeB. Structural studies of MoaD revealed that this protein shares the same fold as ubiquitin despite the absence of detectable sequence homology. The crystal structure of the MoaD–MoeB complex in its apo-state defined the structure of MoeB and its corresponding domains in the E1 enzymes. The MoaD–MoeB structures in complex with ATP and after formation of the acyl adenylate identified key residues involved in the catalysis of this enzyme superfamily. The recent crystal structures of the NEDD8 activator confirmed the predictions made on the basis of the MoaD–MoeB complex and describe the more complex architecture of the E1 enzymes. The phylogenetic distribution of the en-

Protein Degradation. Edited by J. Mayer, A. Ciechanover, M. Rechsteiner
Copyright © 2005 WILEY-VCH Verlag GmbH & Co. KGaA, Weinheim
ISBN: 3-527-30837-7

zymes involved in Moco biosynthesis strongly suggest that the two-component systems consisting of a UbL protein and a cognate E1 enzyme, which are present exclusively in eukaryotes, are derived from the simpler and universally distributed MoaD–MoeB pair.

3.1
Introduction

3.1.1
Activation of Ubiquitin and Ubiquitin-like Proteins

The transfer of ubiquitin and related protein modifiers (reviewed in [1]) such as SUMO [2], NEDD8 [3], Apg12 [4], Apg8 [5], ISG15 [6], Urm1 [7] and Hub1p [8] is initiated by an activation step catalyzed by an activating (E1) enzyme, which is specific for the respective modifier [9, 10]. The activation reaction [11, 12] is dependent on ATP, which is hydrolyzed to form an acyl adenylate between the C-terminus of the modifier and AMP. The resulting high-energy mixed anhydride intermediate is nucleophilically attacked by a conserved cysteine residue of the E1 enzyme, leading to the formation of a thioester linkage between this cysteine and the C-terminus of the modifier [11–13]. Subsequently the UbL protein is transferred from the E1 enzyme to a conjugating (E2) enzyme in a *trans*-thioesterification reaction. The thioester linkages preserve the free energy of ATP and facilitate transfer of the modifier onto target proteins in which the modifier is almost always linked with its C-terminus to the side chain of a lysine residue via an isopeptide bond.

In the archetypical transfer of ubiquitin to a target protein via ubiquitin's Lys48 residue and the subsequent elongation of the mono-ubiquitin to an oligo-ubiquitin chain of at least four residues, this modification triggers the proteasome-dependent degradation of the target protein. In addition, ubiquitylation is also involved in DNA repair (summarized in Ref. [14]), receptor endocytosis (reviewed in Ref. [15]), endocytic sorting (reviewed in Ref. [16]), budding of HIV (reviewed in Ref. [17]) and inflammatory responses [18]. While UbLs are generally transferred onto proteins, APG8, which is involved in the process of autophagy, is conjugated to phosphatidylethanolamine [19] through its amino group, thus mimicking the iso-peptide linkage typically observed in other UbL–protein complexes. Ubiquitin and its related modifiers are exclusively found in eukaryotes, but are evolutionarily derived from individual steps in ancient metabolic pathways, which lead to the formation of either the molybdenum cofactor, thiamine or certain types of FeS-clusters.

3.1.2
Molybdenum Cofactor Biosynthesis

The molybdenum cofactor (Moco) is the essential component of a group of redox enzymes [20–22], which are diverse not only in terms of their phylogenetic distribution, but also in their architectures, both at the overall structural level and in their active site geometry, and finally in the wide variety of transformations catalyzed by these enzymes. Some of the better-known Moco-containing enzymes include sulfite oxidase and xanthine dehydrogenase in humans, assimilatory nitrate reductases in plants and dissimilatory nitrate reductases as well as formate dehydrogenases in bacteria. Moco consists of a mononuclear molybdenum coordinated by the dithiolene moiety of a family of tricyclic pyranopterin structures, the simplest of which is commonly referred to as molybdopterin (MPT). Moco biosynthesis is an evolutionarily conserved pathway comprising several interesting reactions. Mutations in the human Moco biosynthetic genes lead to Moco deficiency, a severe disease that leads to premature death in early childhood [23]. The affected patients show severe neurological abnormalities such as attenuated growth of the brain, seizures, and, frequently, dislocated ocular lenses.

Genes involved in Moco biosynthesis have been identified in eubacteria, archaea and eukaryotes. Although many details of Moco biosynthesis are still unclear at present, the pathway can be divided into three universally conserved stages (Figure 3.1A). (1) *Conversion of a guanosine derivative into precursor Z.* This aspect is different from other pterin biosynthetic pathways, since C8 of the purine is inserted between the 2′ and 3′ ribose carbon atoms during formation of precursor Z, rather than being eliminated [24, 25]. (2) *Transformation of precursor Z into MPT.* This process generates the dithiolene group responsible for coordination of the molybdenum atom in the cofactor, and is catalyzed by MPT synthase [26–28]. MPT synthase is composed of two subunits encoded in *E. coli* by the *moaD* and *moaE* genes. In its active form, MoaD contains a thiocarboxylate at its C-terminus, which acts as the sulfur donor for the synthesis of the dithiolene group (Figure 3.1B). MoeB activates MPT synthase by transferring a sulfur atom onto the C-terminus of MoaD generating the thiocarboxylate [29]. MoeB exhibits significant sequence similarity (Figure 3.2A) to two segments of the E1 enzyme for ubiquitin (UBA1); one is located close to the N-terminus, while the other is located near the center of the sequence encoding UBA1. A similar relationship also exists with E1 enzymes, which are heterodimeric such as the NEDD8-E1 (Figure 3.2A). The sequence similarities between MoeB and E1 enzymes, in concert with their functional relationship (Figure 3.2B), have fostered speculation regarding an evolutionary link between the two pathways [30]. (3) *Metal incorporation.* The MogA and MoeA proteins together are responsible for metal incorporation [31–33].

Additional steps occur in the case of eubacteria, namely syntheses of various dinucleotide forms of the cofactor, in which the pyranopterin is linked to a second nucleotide via a pyrophosphate linkage [34].

3.2
The Crystal Structure of MoaD Reveals the Ubiquitin Fold

The high-resolution crystal structure of MPT synthase [35, 36] revealed that the enzyme forms an elongated, heterotetrameric molecule (Figure 3.3A) with overall dimensions of 93 × 28 × 27 Å. The small (MoaD) subunits are positioned on oppo-

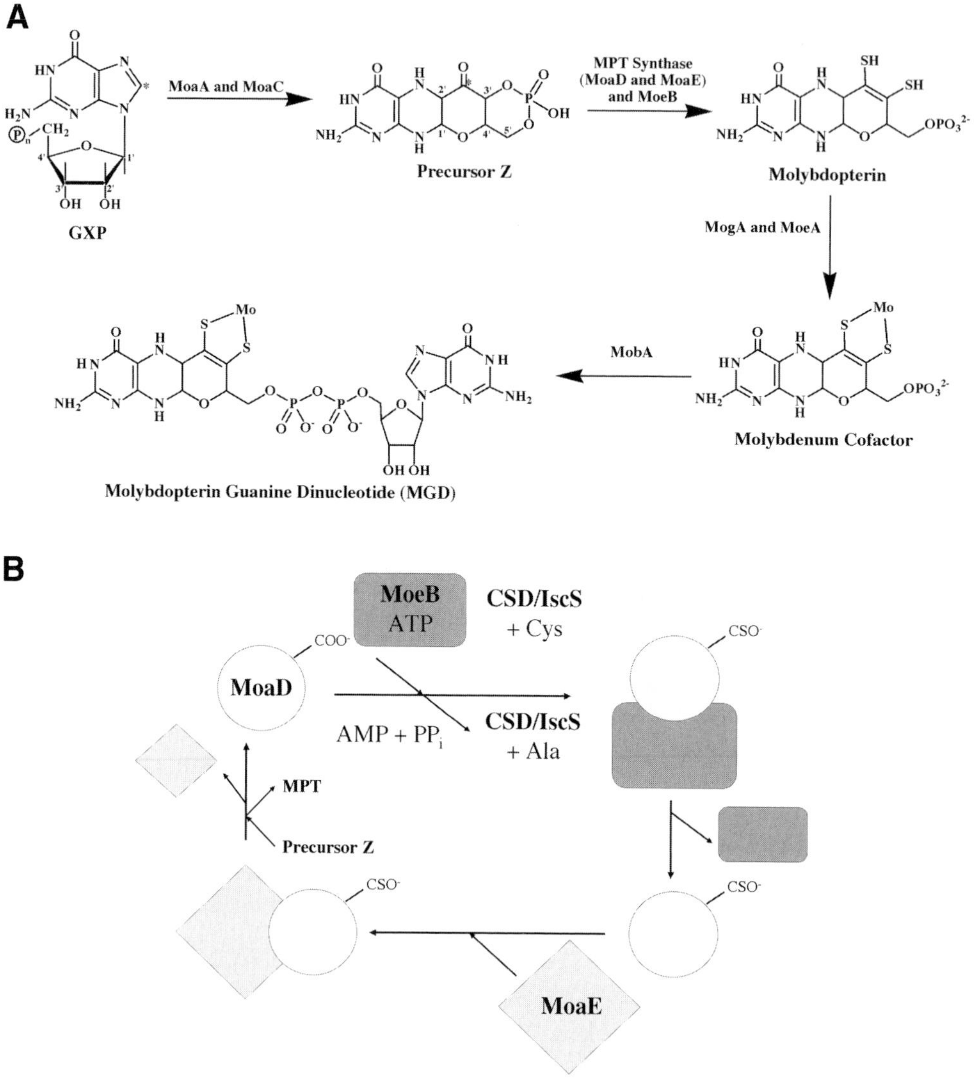
A
MoaA and MoaC
MPT Synthase (MoaD and MoaE) and MoeB
GXP
Precursor Z
Molybdopterin
MogA and MoeA
MobA
Molybdenum Cofactor
Molybdopterin Guanine Dinucleotide (MGD)
B
MoeB
ATP
CSD/IscS
+ Cys
MoaD
AMP + PP
CSD/IscS
+ Ala
MPT
Precursor Z
MoaE

Fig. 3.1. Moco biosynthesis in *E. coli*. (A) The carbon atom at position 8 of the guanosine derivative of unknown composition (GXP), which is the starting structure, is incorporated into precursor Z as indicated by the asterisk. In the mature Moco additional Mo ligands are present besides the dithiolene sulfurs shown here, with the metal being either penta- or hexa-coordinated. (B) Details of step 2, the sulfur incorporation step. MoaD, MoeB and MoaE are represented by a circle, a rectangle and a diamond, respectively.

site ends of the heterotetramer and interact only with the large (MoaE) subunits, which in turn dimerize. The MoaE subunit has an α/β hammerhead fold containing an additional antiparallel 3-stranded β-sheet. The C-terminus of each MoaD subunit is deeply inserted into the active site located in the MoaE monomer. The

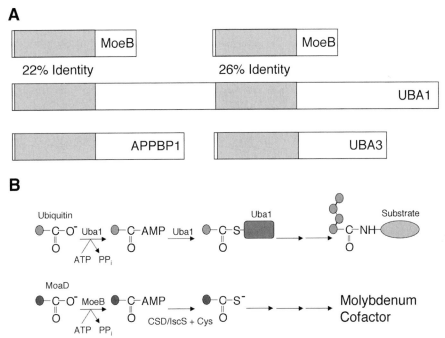

Fig. 3.2. MoeB-E1 sequence relationships. (A) Schematic comparison between MoeB, the ubiquitin-activating enzyme (UBA1) and the heterodimeric (APPBP1–UBA3) NEDD8 activator based on PSI-BLAST [54] sequence analyses. The numbers represent the sequence identities between the first ~170–180 residues of MoeB (shaded) and the corresponding regions in the human ubiquitin-activating enzyme. Residues 3–189 of *E. coli* MoeB can be aligned with residues 47–236 of human ubiquitin E1 (22% identity) and residues 7–175 of *E. coli* MoeB can be aligned with residues 446–620 of human ubiquitin E1 (26% identity). (B) Relationships between the reactions catalyzed by MoeB and the ubiquitin-E1 (UBA1).

interface between the two proteins is quite extensive, burying ~2000 Å² , and is primarily hydrophobic in character. Although primarily restricted to a single MoaE subunit the active site appears to also involve residues from the distal MoaE subunit. Precursor Z binds to this active site and two sulfur atoms are transferred sequentially from the thiocarboxylated MoaD C-terminus to the substrate to form MPT.

The crystal structure of MPT synthase and the simultaneously determined NMR structure of the MoaD-related ThiS protein involved in thiamine biosynthesis [37] unambiguously demonstrated the evolutionary relationship between a subset of enzymes involved in the biosynthesis of S-containing cofactors (e.g. Moco, thiamine and certain FeS-clusters) and the process of ubiquitin activation. MoaD displays significant structural homology to human ubiquitin (Figure 3.3B and C), resulting in a superposition with a root mean square (rms) deviation of 3.6 Å for 68 equivalent Cα atoms out of 76 residues in ubiquitin. The key secondary structure

A

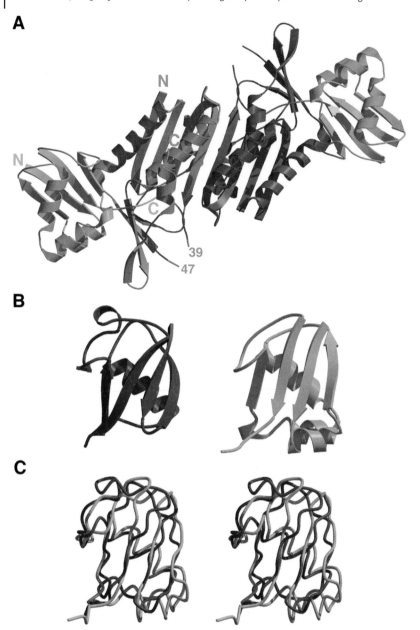

B

C

Fig. 3.3. Structure of MPT synthase. (A) Overall structure of the heterotetramer. MoaD subunits are shown in yellow, MoaE subunits in cyan and magenta. The view is along the two-fold axis of symmetry. N- and C-termini and residues adjacent to a disordered loop are labeled in one of the MoaD–MoaE heterodimers. (B) Side-by-side comparison of MoaD (yellow) and ubiquitin (red) in the same orientation. (C) Stereo diagram of a least-squares superposition of MoaD (yellow) and ubiquitin (red).

elements, a five-stranded mixed β-sheet packed against an α-helix, which runs diagonally across one face of the β-sheet, are present in both structures and also in ThiS. However, there is no statistically meaningful level of sequence conservation between MoaD and ubiquitin (7% identity) with the exception of the C-terminal Gly–Gly dipeptide. This dipeptide is a conserved feature present in almost all UbL proteins, but usually requires proteolytic processing to be liberated. Because of the relationship between Moco biosynthesis and ubiquitin-dependent protein degradation described above, this structural homology was not completely unexpected. Together with the known sequence similarities between MoeB/ThiF and the E1 enzymes, this level of structural homology confirmed that the ubiquitin-like protein modifiers and their activating enzymes are the likely evolutionary offspring of the corresponding proteins involved in Moco and thiamine biosynthesis.

3.3
Structural Studies of the MoeB–MoaD Complex

The crystal structure of the MoeB–MoaD complex (Figure 3.4) was determined by multiple isomorphous replacement in its apo-state at 1.7-Å resolution, with bound ATP at 2.9-Å resolution and after formation of the covalent MoaD-adenylate at 2.1-Å resolution [38]. The latter two structures were obtained by soaking either ATP or Mg-ATP into crystals of the apo-complex.

3.3.1
Structure of MoeB

The structure of MoeB consists of eight β-strands that form a continuous β-sheet surrounded by eight α-helices, which are located on opposite sides of the β-sheet (Figure 3.5A). In the N-terminal half of the sheet all β-strands are parallel and reveal a variation of the Rossman fold [39]: the typical $\beta\alpha\beta\alpha\beta$-topology is interrupted between the second β-strand (β_2) and the fourth α-helix (α_4) by the insertion of two 3_{10} helices. The first of these 3_{10} helices contains five residues that are strictly conserved between MoeB and the E1 enzymes. The loop between β1 and α3 contains a highly conserved glycine-rich motif with the sequence Gly–X–Gly–X–X–Gly, which is reminiscent of the P-loop [40] typically found in the superfamily of ATP and GTP hydrolyzing enzymes. The C-terminal half of MoeB contains an antiparallel β-sheet (β5–β8) and is critical for MoeB dimerization and MoaD binding (see below). Two Cys–X–X–Cys motifs are found in this half of the protein and are responsible for coordinating a zinc atom with tetrahedral geometry through their thiolates. This zinc-binding site is quite distant from the active site, thus suggesting a structural rather than a catalytic role for the metal. Moreover, these zinc-binding motifs are only found in 60–70% of the known MoeB sequences and only in a few E1 sequences, indicating that they are non-essential for the catalytic mechanism.

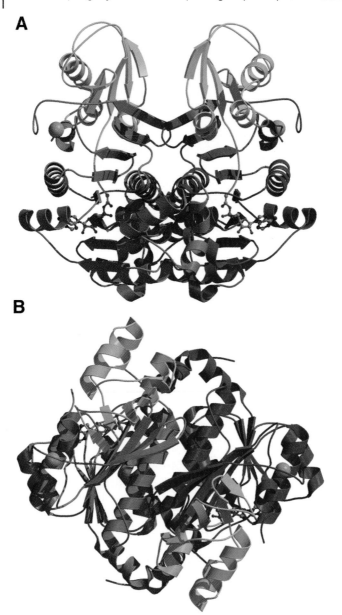

Fig. 3.4. Structure of the MoaD–MoeB complex. (A) Overall structure of the hetero-tetramer. MoaD subunits are shown in yellow, MoeB subunits in red and light blue. The two-fold axis of symmetry is running vertically in the plane of the paper. The AMP is shown in bonds representation together with the cova-lent link to the C-terminal glycine of MoaD. The Zn ion is shown in van der Waals repre-sentation in gray. (B) As in (A) after a 90° rota-tion around the horizontal axis. The view is now along the two-fold axis of symmetry.

A

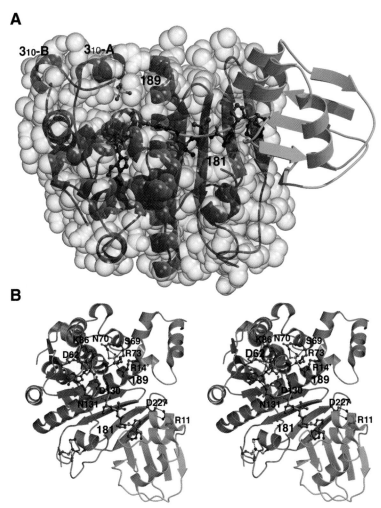

B

Fig. 3.5. MoeB-catalyzed reaction. (A) Structure of the MoeB–MoaD heterodimer with MoaD in yellow and MoeB in red. Atoms in MoeB are shown as gray spheres representing their van der Waals radii and are rendered transparent. Residues 75 to 81 of MoaD, the AMP and sulfate molecule are shown in ball-and-stick representation. (B) Close-up stereo view of the acyl adenylate intermediate formed during MoaD activation in which AMP is covalently linked at Gly 81 to MoaD (yellow). The principal MoeB subunit with which MoaD interacts is shown in red, while the N-terminus of the second MoeB subunit is shown in blue. Selected residues have been labeled and hydrogen bonds are indicated by dashed lines.

3.3.2
The MoeB–MoaD Interface

As in the interface between MoaD and MoaE, the subunits of MPT synthase, the MoeB–MoaD interface is composed to 65% of hydrophobic interactions. In fact, the hydrophobic core between MoeB and MoaD involves the same region of the

MoaD surface buried in the MPT synthase complex, yet it is slightly more extensive in the MoaD–MoeB complex. Three residues found in MoaD, Phe D7, LeuD59 and Phe D75, are common to both interfaces. Residues in MoaD and MoeB are prefixed with D and B, respectively, to allow differentiation between the two chains. Most of the interactions involving MoaD binding are localized to α7, β7 and β8 of MoeB. In addition to the hydrophobic core, a salt bridge and two hydrogen bonds are formed between Asp B227 Oδ1 and Oδ2 and Arg D11 Nη1 and Nη2 at the periphery of the complex. Additional hydrogen bonds involve the C-terminal tail of MoaD (see below).

Perhaps the most striking feature of the MoeB-MoaD interface is the C-terminal extension of residues 76–81 of MoaD into a cleft on the MoeB surface (Figure 3.5A), which is adjacent to the Gly-rich sequence motif. Proper positioning of the C-terminal Gly–Gly dipeptide appears to be accomplished by hydrogen-bonded interactions. Hydrogen bonds are present between the main chain oxygen of Thr D79 and Nε of Arg B135, which is conserved in MoeB and E1 enzymes, and between the oxygen of Glu D80 and the nitrogen of Ala B154. The C-terminus of MoaD extends over β5 of MoeB, which acts as a structural scaffold. Sequence alignments using MoeB and E1 sequences show a preference for small amino acids (Gly, Ala, Ser) at the center of β5, which appear to allow the insertion of the Gly–Gly motif of MoaD and the UbL into the active site of MoeB and E1. The active-site region in MoeB is delineated by residues found in loops connecting β1 and α3 (containing the modified P-loop motif), β2 and 3$_{10}$-A, as well as β4 and α6. The strictly conserved residues in helix 3$_{10}$-A form a region of the active site that binds a sulfate molecule from the mother liquor. In addition, a second sulfate molecule is observed at the active site in close proximity to the incoming Gly–Gly motif of MoaD. Residues 182 to 188 of MoeB are disordered, but owing to the constraints imposed by the adjacent residues (181 and 189) these residues must form a flexible loop crossing over the MoaD C-terminus. One of the residues in this segment, Cys 187, has been proposed to correspond to the active-site Cys in the E1 enzymes and to form a thioester bond with the MoaD C-terminus prior to the formation of the thiocarboxylate [9].

3.3.3
Structure of MoeB–MoaD with Bound ATP

The structure of the MoeB–MoaD–ATP ternary complex reveals that the ATP molecule is bound in a pocket in close proximity to the C-terminus of MoaD. Although the MoaD carboxylate and the α-phosphate are in close spatial proximity, their electrostatic repulsion in the absence of magnesium prevents a nucleophilic attack of MoaD on the α-phosphate. Residues in the modified P-loop motif of MoeB form the floor of the nucleotide-binding pocket. The adenine ring is located in a hydrophobic patch of the pocket, which is created by Phe B63, Leu B109 and Val B134. None of these residues is conserved in MoeB and surprisingly there is only one hydrogen bond to the purine base involving the side chain of Asn 131 and the N7 atom (Figure 3.5B). The ATP appears to be anchored at the active site through the hydroxyl groups of its ribose ring and the triphosphate moiety. O2′ and O3′ form hydrogen bonds with Oδ1 and Oδ2 of Asp B62, which is strictly conserved in MoeB orthologs and E1 enzymes. Additionally, Nζ of the conserved Lys B86 forms hydrogen bonds with Oδ2 of the ribose. The α-phosphate is buried deeply in the pocket and is involved in main-chain contacts with the nitrogen atom of Gly B41 of the P-loop. Additional MoeB-ATP contacts are seen between the side chains of the strictly conserved Arg B73 and one oxygen each of the α- and β-phosphates and between Lys B86 and the β-phosphate. Interactions with the γ-phosphate involve the side chains of Ser B69 and Asn B70. The overall shape of the binding pocket distorts the ATP molecule and induces a kink between the tightly bound α- and β-phosphates. Arg B14′, a residue from the second MoeB monomer, is inserted at the active site of the first monomer and undergoes a significant conformational change compared to the nucleotide-free structure. In the ATP-bound state, the side-chain atom Nε of Arg B14′ is within hydrogen-bonding distance of two oxygens of the γ-phosphate, whereas some atoms of its side chain are displaced by more than 5 Å in the apo-structure. This residue, though conserved in all MoeB sequences, is not present in the MoeB-like central domain of the ubiquitin-E1 and the UBA3 subunit of the NEDD8-E1.

3.3.4
Structure of the MoaD Adenylate

Soaking MoeB–MoaD cocrystals with ATP and Mg^{2+} led to the visualization of a covalent reaction intermediate (Figure 3.5B). The most striking feature of this structure is the presence of an MoaD adenylate at the active site in which Gly D81 is covalently linked to the α-phosphate through a mixed anhydride. Although the pyrophosphate leaving group is not seen in this structure, a bound sulfate molecule from the mother liquor is presumably mimicking one of the phosphates of pyrophosphate. This sulfate is ligated by Ser B69, Asn B70, and Arg B73 and is observed in a position similar to that occupied by the γ-phosphate in the ATP-bound model. The role of the strictly conserved Asp B130 might be to coordinate the divalent Mg^{2+} ion that appears to be necessary for the turnover of ATP, which in turn would be coordinated by oxygen atoms of the α- and β-phosphate. In contrast to glycyl tRNA synthetase, in which the metal remains bound to the α-phosphate after formation of the glycyl adenylate [41], the structure of the MoaD adenylate provides no evidence of a bound Mg^{2+}.

3.3.5
Fate of the Adenylate

After the formation of an acyl adenylate, the similarities between MoeB and E1 appear to come to an end (Figure 3.2B). In the E1 enzymes an active-site cysteine residue attacks the ubiquitin adenylate forming the E1-ubiquitin thioester. *E. coli* MoeB contains nine cysteine residues, four of which are involved in coordinating the zinc atom. Sequence alignments show that among the remaining cysteines only Cys B187 is conserved in all MoeB sequences, and might correspond to the active-site cysteine in the E1 sequence family. It has been postulated [9] that the corresponding residue in MoeB also forms a thioester with the MoaD C-terminus, which is then attacked by sulfide to form the thiocarboxylate. Cys B187 in MoeB is part of the consistently disordered loop region, which is located in close proximity to the active site. There appears to be no obvious reason why the side chain of this residue should not be able to attack the acyl adenylate if a thioester is indeed formed during the reaction. While substitution of this cysteine by alanine has been reported not to impair MoeB activity in an *in vitro* system [42], a more recent study [43] does report a 20% reduction in activity. This latter observation is in agreement with mass spectrometric data on ThiF, the MoeB-related protein involved in thiamine biosynthesis, which demonstrated that the corresponding cysteine forms an acyldisulfide linkage with the C-terminus of ThiS [44], suggesting that this residue does indeed have a somewhat similar activity to the active site cysteine present in the E1 enzymes, despite the fact that no thioester intermediate is formed.

In molybdopterin and thiamine biosynthesis the thiocarboxylate sulfur is derived from cysteine by a cysteine desulfurase. Recent findings have implicated IscS in the biosynthesis of thiamine [45]. IscS mobilizes sulfur from cysteine forming an IscS persulfide that is subsequently shuttled to ThiI, a rhodanese-like enzyme, to form a putative ThiI persulfide. This moiety is responsible for attacking the ThiS adenylate, which is bound at the ThiF active site, leading to the formation of the thiocarboxylate product. IscS has also been implicated in Moco biosynthesis together with the related CSD protein [29], although ThiI is not involved. It is interesting to note that some MoeB orthologs, including human MoeB, contain an additional C-terminal domain, which shares distant sequence relationships with rhodaneses. In fact, a C-terminal truncation of the *A. nidulans* MoeB homolog, CnxF, has been shown to abrogate enzyme function as evidenced by the lack of MPT production [46]. Recently, the rhodanese-like domain of MOCS3, the human MoeB ortholog, was shown to be able to transfer the sulfur required for thiocarboxylate formation of MOCS2A, the small subunit of human MPT synthase [43]. Furthermore, site-directed mutagenesis revealed that the conserved cysteine residue in the rhodanese domain is essential for activity. The fact that residues 182 to 188 of MoeB are disordered could indicate that they are involved in protein–protein interactions with either the rhodanese domain or the cysteine desulfurase. Cys187 of MoeB could either cleave the acyldisulfide between the cysteine of the rhodanese domain and the MoaD C-terminus as suggested by Matthies et al. [43], or could itself form an acyldisulfide bond with the MoaD C-terminus prior to formation of the MoaD thiocarboxylate.

3.4
Structure of the NEDD8 Activator

NEDD8 (Rub1 in yeast) is a UbL, which is attached to cullins following activation by a specific E1 and transfer by an E2 enzyme [47]. Cullins are subunits of the SCF

(*S*kp1-*C*ullin-*F*-box protein) family of ubiquitin ligases (E3) and this modification results in an increase in the ubiquitin ligase activity of these enzymes [48, 49]. The E1 enzyme responsible for activation of NEDD8 is a heterodimer composed of the APPBP1 (534 residues in humans) and UBA3 (442 residues in humans) subunits, which share homology (Figure 3.2A) with the N- and C-terminal regions of ubiquitin's E1 enzyme (UBA1). The reactions catalyzed by the E1 for ubiquitin, SUMO and NEDD8 are more complex than in the case of MoeB as established for ubiquitin's E1 [11–13]. While the initial formation of the acyl adenylate between ubiquitin and AMP is a conserved feature, the ensuing covalent attachment of the UbL to an active-site cysteine is specific to the E1 enzymes. This covalent E1–ubl complex subsequently binds another UbL and ATP leading to the formation of a second acyl adenylate. After transfer of the first UbL onto an E2 enzyme, the second adenylated UbL is transferred to the active-site cysteine thereby re-forming the thioester linkage. The crystal structures of the heterodimeric APPB1–UBA3 NEEDD8 activator in its apo-state [50] and in complex with NEDD8 and ATP [51] were reported in 2003.

3.4.1
Overall Structure of the NEDD8-E1

The crystal structure of the heterodimeric NEDD8-specific E1 in the absence of NEDD8 has been determined at 2.6-Å resolution [50]. The complex consists of three structural entities (Figure 3.6A): (1) An adenylation domain formed by the MoeB-like repeats in both APPBP1 and UBA3. As predicted on the basis of the MoaD–MoeB crystal structure the two MoeB-like repeats are arranged in exactly the same way as they are in the MoeB dimer. However, as will become evident shortly, only one active site is present in this heterodimer. In addition, APPBP1 contains a four-helix bundle domain (residues 407 to 485, shown in dark blue in Figure 3.6A), which replaces the disordered loop of MoeB and occupies the region in which the MoaD subunit is located in the MoaD–MoeB complex. This feature contributes to the fact that the APPBP1 subunit is catalytically incompetent. (2) A catalytic domain responsible for thioester formation, which contains the active-site cysteine (Cys 216). This domain is formed by ∼80 residues from UBA3 (residues 209–287, shown in light green in Figure 3.6A) including the essential cysteine, and an additional ∼225 residues from APPBP1 (residues 169–393, shown in dark green in Figure 3.6A). Both of these domains are α-helical; the segment belonging to UBA3 folds into four α-helices and is inserted at the position of the disordered loop in MoeB (residues 182 to 188), while the additional domain in APPBP1 contains eleven α-helices and is inserted between the β-strand and α-helix corresponding to β6 and α7 of MoeB. (3) Finally, a small, UbL domain (starting at residue 348, shown in orange in Figure 3.6A) is present at the C-terminus of UBA3. These three domains are arranged such that a large groove is formed with the adenylation and catalytic domains on opposite sides and the C-terminal UbL domain on the UBA3-side of the groove. With the exception of a crossover loop, which connects the adenylation domain and the catalytic domain of UBA3, the groove is fully ac-

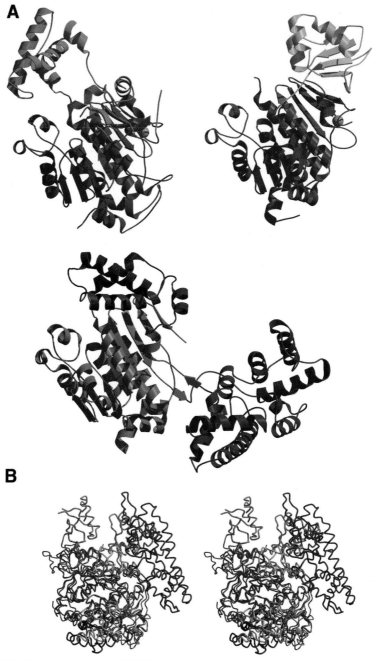

Fig. 3.6. Structure of the NEDD8 activator. (A) Subunit structure of the NEDD8 activator. The UBA3 subunit (upper left) is shown with the MoeB-related adenylation domain in red, the helical insertion containing the active-site cysteine in light green and the C-terminal UbL domain in orange. The APPBP1 subunit (bottom) is shown with the MoeB-related

cessible along its entire length. The groove can be subdivided into two clefts, each of which is of sufficient size to allow binding of either two NEDD8 molecules or one E2 enzyme. Cleft 1 contains the glycine-rich loop involved in ATP binding, while cleft 2 points toward the catalytic cysteine of UBA3. This residue is located at the center of the catalytic domain, in an interface between the segments originating from the UBA3 and APPBP1 subunits, although it is at a substantial distance from the cleft.

Subsequently, the crystal structure of the complex between APPBP1/UBA3 and NEDD8 was determined at 3-Å resolution [51]. NEDD8 is bound in the middle of the groove formed by APPBP1/UBA3 and interacts with the adenylation domain of UBA3 and the large extra-domain of APPBP1 (Figure 3.6A). NEDD8's extended C-terminus is threaded underneath UBA3's crossover loop and extends towards the Gly-rich loop of UBA3. A total of ~3400 Å^2 or 34% of NEDD8's surface area is buried in the complex, and residues involved in these interactions can be mapped to two different areas on the surface: (1) A hydrophobic area including Leu 9, Ile 44, His 68, Val 70, Leu 71 and Leu 73 on NEDD8 interacts with the adenylation domain of UBA3 in a manner analogous to that observed in the MoaD–MoeB interface. This region includes the C-terminus of NEDD8, which contacts the crossover loop and extends into the active site of UBA3. (2) A charged surface patch containing residues Arg 25, Glu 28, Arg 29, Glu 31 and Glu 32, located in the lone α-helix of NEDD8 on the opposite side, contacts the large helical domain of APPBP1. This interaction motif appears to be specific for E1 proteins and is not found in the MoaD–MoeB complex, which explains why significantly less surface area is buried in the MoaD–MoeB complex.

NEDD8 binding is accompanied by conformational changes in NEDD8 and the UBA3 subunit of the activator. The most notable change in NEDD8 involves the C-terminal tail, which rotates around Leu 69 by about 30°. Binding reduces the inherent flexibility, leading to a visualization of the last three residues that contrasts with the structure of free NEDD8 [52]. Conformational changes in the activator are evident in the domain containing Cys 216, which moves away from the adenylation domain, thereby facilitating NEDD8 binding in the widened groove. At the same time the crossover loop moves closer to the floor of the groove (by 2.5 Å) and clamps down on the NEDD8 C-terminus.

The structure also provides insights into how different E1 enzymes discriminate between different UbLs. Ala 72 of NEDD8 appears to be a key determinant of spe-

---

domain in light blue, the helical extension, which forms part of the catalytic sub-domain in dark green and the additional helical sub-domain in dark blue. The MoaD–MoeB complex (upper right) is shown in the same orientation as the UBA3 and APPBP1 subunits with MoeB in red and MoaD in yellow. The Zn atoms in the UBA3 and MoeB subunits are shown as gray spheres. (B) Stereo diagram of a superposition of the APPBP1–UBA3–NEDD8 (color-coded as defined in (A)) and hetero-tetrameric MoaD–MoeB complexes (MoaD subunits are shown in yellow and MoeB subunits in gray). While the MoaD subunit bound to the MoeB subunit corresponding to UBA3 (red) fits reasonably well, the second MoaD subunit overlaps dramatically with the four-helical insertion (dark blue) in APPBP1.

cificity. This residue participates in van der Waals interactions with Leu 206 and Tyr 207 of UBA3's crossover loop. Ala 72 is replaced by arginine in ubiquitin and when modeled in the APPBP1–UBA3–NEDD8 complex this residue would be forced into close contact with Arg 190 of UBA3. The residue corresponding to Arg190 in the E1 for ubiquitin, however, is a glutamine, which on the basis of the APPBP1–UBA3–NEDD8 complex is predicted to interact favorably with Arg 72 of ubiquitin, a prediction that was confirmed by biochemical studies [51].

In addition to the ternary APPBP1–UBA3–NEDD8 complex the quarternary complex with bound ATP has been described at 3.6-Å resolution [51]. Because Cys 216 was changed to alanine and Mg^{2+} was not present in the crystals, ATP hydrolysis could not take place, thus preventing formation of a NEDD8 adenylate and subsequent thioester formation. ATP is bound in the nucleotide-binding pocket of UBA3 in close proximity to the NEDD8 C-terminus. The adenine base interacts with several hydrophobic residues (Met 80, Ile 127, Leu 145 and Ala 150), which are at least type-conserved in other E1 enzymes. In contrast to ATP binding by MoeB an adenine-specific hydrogen bond is formed between the exocyclic amino group and a glutamine residue (Gln 128 in UBA3); however, this residue is not conserved in other UbL activators. The remaining interactions involve residues that are highly conserved between E1 enzymes and MoeB. Of particular interest are Asp 146 (Asp 130 in MoeB), which is proposed to ligate the Mg^{2+} ion and two Arg residues: Arg 90 (Arg 73 in MoeB) and Arg 15 from APPBP1 (Arg 14 in MoeB). The latter residue is the only one in the ATP-binding site of the NEDD8 E1 that originates from APPBP1. In MoeB this residue is contributed from the other monomer of the MoeB dimer where it contacts the ATP in a similar fashion. As already envisioned based on the NEDD8-E1 structure alone, the active site cysteine is at the significant distance of ~ 32 Å from the α-phosphate.

3.4.2
Comparison with the MoaD–MoeB Complex

The adenylation domains of APPBP1 and UBA3 are in fact remarkably conserved with MoeB (Figure 3.6A and B). In the case of APPBP1 the regions corresponding to MoeB involve residues 6–168 at the N-terminus (corresponding to residues 5–167 of MoeB), residues 394–404 (169–179 of MoeB) following the larger first insertion and, at the C-terminus, residues 486–534 (189–238 in MoeB) following the smaller second insertion. Overall, the Cα atoms of 220 out of 240 structurally observed residues of MoeB can be superimposed with an rms deviation of 1.6 Å resulting in 19% overall sequence identity. A similar picture is evident when MoeB and UBA3 are compared: at the N-terminus UBA3 residues 12–208 correspond to MoeB's residues 4–181 (an insertion of 12 residues in UBA3 is primarily responsible for the offset in the numbers at the C-terminal ends of the corresponding stretches), while the structural similarities at the C-terminus, following the insertion of the domain containing the catalytic cysteine, involve residues 288–347 of UBA3, which align with residues 189–248 of MoeB. In this case the rms deviation after superposition of the Cα atoms for 230 out of 240 structurally observed residues are 1.9 Å with 23% overall sequence identity. Another feature that is conserved between MoeB and the UBA3 subunit of the NEDD8 activator is the presence of a bound Zn ion, which is ligated in an analogous manner by the thiolates of four Cys residues.

3.4.3
Conformational Changes during the Formation of the Acyl Adenylate

The available structures of the NEDD8 activator and the MoaD–MoeB complex provide valuable insights into the universally conserved adenylation step catalyzed by members of the E1 enzyme superfamily. The following discussion focuses on the available MoaD–MoeB structures, but also applies to the E1 enzymes. A view into the active site of the apo, ATP-bound, and acyl adenylate forms of the MoeB–MoaD complex reveals subtle conformational changes in the protein (Figure 3.7A). Interestingly, the active sites of the apo and acyl adenylate models have remarkably similar structures. Both contain a sulfate molecule from the mother liquor interacting with Arg B14′, Ser B69, Asn B70 and Arg B73. The positions of the sulfates in each model are nearly identical and correspond to the γ-phosphate in the ATP-bound complex. The only noticeable difference in the two models is seen in the conformation of the MoaD C-terminus where Gly D80 and Gly D81 clearly adopt different conformations as a result of the covalent linkage between the C-terminal glycine and the α-phosphate. The active site of the ATP-bound model shows the most pronounced structural changes. The side chain of Arg B14′, which complements the active site across the MoeB dimer interface, exhibits the largest conformational change. Additionally, the side chains of Ser B69 and Lys B86 adopt different conformations compared to the apo and acyl adenylate models. Finally, the C-terminal Gly–Gly motif of MoaD adopts a different conformation owing to a flip of the peptide bond between Thr D79 and Gly D80.

Multiple sequence alignments using various MoeB and E1 sequences reveal a remarkable degree of conservation for the residues surrounding the active site. In light of the structural data on the MoaD–MoeB and APPBP1–UBA3–NEDD8 complexes, it is possible to assign functional roles to most of the residues in these conserved regions. The loop regions between β_1 and α_3 (secondary structure elements and residue numbers refer to MoeB, but the corresponding regions/residues are also present in the APPBP1–UBA3–NEDD8 complex) consist of a glycine-rich nucleotide-binding motif that facilitates entry of ATP into the respective active sites. The loop region between $\beta2$ and helix 3_{10}-A is critical for binding the ribose of ATP. The highly conserved residues forming helix 3_{10}-A are essential for binding the β- and γ-phosphates of ATP and, more importantly, stabilize the pyrophosphate leaving group upon attack by the MoaD or UbL carboxylates. Residues in the loop between $\beta4$ and $\alpha6$ are responsible for the proper positioning of Asp B130 adjacent to the α-phosphate of ATP, and this residue is predicted to be involved in Mg^{2+}-ligation. Arg B135 found inside helix $\alpha6$ properly orients both the incoming C-terminal extension of MoaD and strand $\beta5$ of MoeB, which serves to support the C-terminal Gly–Gly dipeptide.

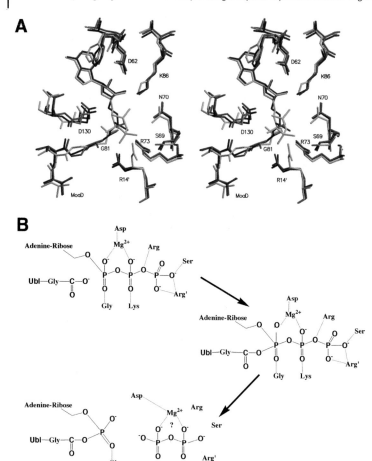

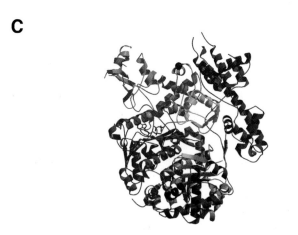

A reaction mechanism for UbL activation can be formulated based on the MoaD–MoeB and the APPBP1–UBA3–NEDD8 structures (Figure 3.7B). After Mg^{2+}-ATP and UbL binding at the active site, the carboxylate oxygen of the terminal glycine of the UbL attacks the α-phosphate of ATP creating a transient pentacovalent intermediate. The divalent metal appears to be required to overcome the electrostatic repulsion between the UbL C-terminus and the α-phosphate. Subsequently, the bond between the α- and β-phosphates is cleaved, which could be facilitated by the strained conformation of the triphosphate observed in the ATP-bound structures of the MoaD–MoeB and APPBP1–UBA3–NEDD8 complexes. Furthermore, the developing negative charge on the β-phosphate appears to be stabilized by two arginine residues, which in MoeB correspond to Arg B73 and Arg B14′ of the second MoeB monomer. In the E1 enzymes the second arginine, corresponding to Arg B14′ of MoeB, is contributed from the second MoeB-like repeat in the case of the single subunit E1s such as the ubiquitin activating enzyme, or the second subunit as is the case for the APPBP1 subunit of the heterodimeric NEDD8-E1. The importance of Arg14, Arg73 and Asp130 of MoeB has been demonstrated by site-directed mutagenesis and a nitrate-reductase overlay assay [38].

Following the formation of the acyl adenylate the reactions will proceed along different paths. In the MoaD–MoeB complex a thiocarboxylate will be formed and the current knowledge regarding this step has been summarized (Section 3.3.5.). In contrast, the E1 enzymes will form the covalent thioester linkage between their active site cysteine and the UbL C-terminus. A remarkable observation of the NEDD8–APPBP1–UBA3–ATP complex in this context has been the large distance ($\sim$32 Å) between this cysteine and the α-phosphate of ATP, which apparently requires substantial conformational changes for the reaction to proceed. As mentioned earlier a second UbL protein will bind to the E1 enzymes and this process could trigger the necessary conformational changes leading to the formation of the thiocarboxylate. Another issue that is poorly understood is binding of the cognate E2 enzyme and the resulting transfer of the UbL to E2. Clearly more structural and mechanistic studies are required to understand the complex mechanism of the E1 enzymes in detail.

Summary

Recent structural studies of the bacterial MoaD–MoeB system have demonstrated that the E1-catalyzed activation of UbL proteins is derived from a more ancient and

Fig. 3.7. Adenylation reaction. (A) Stereo representation of a superposition of the MoaD–MoeB complex in its apo-state (red), in complex with ATP (yellow) and after formation of the acyl adenylate (blue). (B) Proposed reaction scheme for the formation of the acyl adenylate. Arg′ refers to the second arginine originating either from the second subunit in case of MoeB and heterodimeric E1 enzymes, or the N-terminal MoeB-repeat in single subunit E1 enzymes. (C) The structure of the NEDD8 activator in complex with NEDD8 (yellow) and ATP (all-bonds representation) is shown with the same subunit and domain color code defined in Figure 3.6A. The active-site cysteine and the bound Zn are indicated as black and gray spheres, respectively.

widespread step during Moco biosynthesis, namely the temporary incorporation of sulfur as a thiocarboxylate. The MoaD protein is thus the evolutionary ancestor of ubiquitin and other UbLs and due to its near-universal presence in all phylogenetic kingdoms occupies a position ubiquitin was envisioned to assume when it originally received its name and was thought to be present in all kingdoms of like [53]. On the other hand, the MoeB protein, as already deduced by sequence comparisons, is the ancestor of the E1 enzymes. All E1-catalyzed reactions including MoeB involve the formation of an acyl adenylate with the C-terminus of a small protein, but subsequently diverge with the formation of either a thiocarboxylate or a thioester. The factors dictating whether a thiocarboxylate or a thioester is formed are not fully understood at present. After the activation step was adopted to fulfill secondary functions in eukaryotes an additional transfer step, catalyzed by the E2 enzymes, was presumably added and, in the case of ubiquitin, also the complex diversity of the E3 enzymes evolved, which ensures that only selected proteins are degraded at the appropriate time by the proteasome.

Acknowledgments

The author would like to thank Drs Michael W. Lake, Michael J. Rudolph and Song Xiang for their important contributions to this project as well as Drs M. M. Wuebbens and K. V. Rajagopalan for their remarkable collaborative efforts. This work was supported by NIH grant DK 54835.

References

1 SCHWARTZ, D. C. and HOCHSTRASSER, M. A superfamily of protein tags: ubiquitin, SUMO and related modifiers, *Trends Biochem Sci* **2003**, *28*, 321–328.

2 HOCHSTRASSER, M. Ubiquitin-dependent protein degradation, *Annu Rev Genet* **1996**, *30*, 405–439.

3 KUMAR, S., YOSHIDA, Y., and NODA, M. Cloning of a cDNA which encodes a novel ubiquitin-like protein, *Biochem Biophys Res Commun* **1993**, *195*, 393–399.

4 MIZUSHIMA, N., NODA, T., YOSHIMORI, T., TANAKA, Y., ISHII, T., GEORGE, M. D., KLIONSKY, D. J., OHSUMI, M., and OHSUMI, Y. A protein conjugation system essential for autophagy, *Nature* **1998**, *395*, 395–398.

5 KIRISAKO, T., ICHIMURA, Y., OKADA, H., KABEYA, Y., MIZUSHIMA, N., YOSHIMORI, T., OHSUMI, M., TAKAO, T., NODA, T., and OHSUMI, Y. The reversible modification regulates the membrane-binding state of Apg8/Aut7 essential for autophagy and the cytoplasm to vacuole targeting pathway, *J Cell Biol* **2000**, *151*, 263–276.

6 LOEB, K. R. and HAAS, A. L. The interferon-inducible 15-kDa ubiquitin homolog conjugates to intracellular proteins, *J Biol Chem* **1992**, *267*, 7806–7813.

7 FURUKAWA, K., MIZUSHIMA, N., NODA, T., and OHSUMI, Y. A protein conjugation system in yeast with homology to biosynthetic enzyme reaction of prokaryotes, *J Biol Chem* **2000**, *275*, 7462–7465.

8 DITTMAR, G. A., WILKINSON, C. R., JEDRZEJEWSKI, P. T., and FINLEY, D.

Role of a ubiquitin-like modification in polarized morphogenesis, *Science* 2002, *295*, 2442–2446.

9 HOCHSTRASSER, M. Evolution and function of ubiquitin-like protein-conjugation systems, *Nat Cell Biol* 2000, *2*, E153–157.

10 PICKART, C. M. Mechanisms underlying ubiquitination, *Annu Rev Biochem* 2001, *70*, 503–533.

11 HAAS, A. L. and ROSE, I. A. The mechanism of ubiquitin activating enzyme. A kinetic and equilibrium analysis, *J Biol Chem* 1982, *257*, 10329–10337.

12 HAAS, A. L., WARMS, J. V., HERSHKO, A., and ROSE, I. A. Ubiquitin-activating enzyme. Mechanism and role in protein-ubiquitin conjugation, *J Biol Chem* 1982, *257*, 2543–2548.

13 CIECHANOVER, A., ELIAS, S., HELLER, H., and HERSHKO, A. "Covalent affinity" purification of ubiquitin-activating enzyme, *J Biol Chem* 1982, *257*, 2537–2542.

14 PICKART, C. M. DNA repair: right on target with ubiquitin, *Nature* 2002, *419*, 120–121.

15 HICKE, L. and DUNN, R. Regulation of membrane protein transport by ubiquitin and ubiquitin-binding proteins, *Annu Rev Cell Dev Biol* 2003, *19*, 141–172.

16 KATZMANN, D. J., ODORIZZI, G., and EMR, S. D. Receptor downregulation and multivesicular-body sorting, *Nat Rev Mol Cell Biol* 2002, *3*, 893–905.

17 PORNILLOS, O., GARRUS, J. E., and SUNDQUIST, W. I. Mechanisms of enveloped RNA virus budding, *Trends Cell Biol* 2002, *12*, 569–579.

18 DENG, L., WANG, C., SPENCER, E., YANG, L., BRAUN, A., YOU, J., SLAUGHTER, C., PICKART, C., and CHEN, Z. J. Activation of the IkappaB kinase complex by TRAF6 requires a dimeric ubiquitin-conjugating enzyme complex and a unique polyubiquitin chain, *Cell* 2000, *103*, 351–361.

19 ICHIMURA, Y., KIRISAKO, T., TAKAO, T., SATOMI, Y., SHIMONISHI, Y., ISHIHARA, N., MIZUSHIMA, N., TANIDA, I., KOMINAMI, E., OHSUMI, M., NODA, T., and OHSUMI, Y. A ubiquitin-like system mediates protein lipidation, *Nature* 2000, *408*, 488–492.

20 HILLE, R. Molybdenum enzymes, *Subcell Biochem* 2000, *35*, 445–485.

21 KISKER, C., SCHINDELIN, H., BAAS, D., RETEY, J., MECKENSTOCK, R. U., and KRONECK, P. M. A structural comparison of molybdenum cofactor-containing enzymes, *FEMS Microbiol Rev* 1998, *22*, 503–521.

22 SCHINDELIN, H., KISKER, C., and RAJAGOPALAN, K. V. Molybdopterin from molybdenum and tungsten enzymes, *Adv Protein Chem* 2001, *58*, 47–94.

23 REISS, J. and JOHNSON, J. L. Mutations in the molybdenum cofactor biosynthetic genes MOCS1, MOCS2, and GEPH, *Hum Mutat* 2003, *21*, 569–576.

24 WUEBBENS, M. M. and RAJAGOPALAN, K. V. Investigation of the early steps of molybdopterin biosynthesis in *Escherichia coli* through the use of *in vivo* labelling studies, *J. Biol. Chem.* 1995, *270*, 1082–1087.

25 RIEDER, C., EISENREICH, W., O'BRIEN, J., RICHTER, G., GOTZE, E., BOYLE, P., BLANCHARD, S., BACHER, A., and SIMON, H. Rearrangement reactions in the biosynthesis of molybdopterin – an NMR study with multiply 13C/15N labelled precursors, *Eur J Biochem* 1998, *255*, 24–36.

26 WUEBBENS, M. M. and RAJAGOPALAN, K. V. Mechanistic and mutational studies of Escherichia coli molybdopterin synthase clarify the final step of molybdopterin biosynthesis, *J Biol Chem* 2003, *278*, 14523–14532.

27 PITTERLE, D. M., JOHNSON, J. L., and RAJAGOPALAN, K. V. *In vitro* synthesis of molybdopterin from precursor Z using purified converting factor, *J. Biol. Chem.* 1993, *268*, 13506–13509.

28 PITTERLE, D. M. and RAJAGOPALAN, K. V. The biosynthesis of molybdopterin in *Escherichia coli*, *J. Biol. Chem.* 1993, *268*, 13499–13505.

29 LEIMKUHLER, S. and RAJAGOPALAN, K. V. A sulfurtransferase is required in the transfer of cysteine sulfur in the in vitro synthesis of molybdopterin from precursor Z in Escherichia coli, *J Biol Chem* 2001, *276*, 22024–22031.

30 RAJAGOPALAN, K. V. Biosynthesis and processing of the molybdenum cofactors, *Biochem Soc Trans* **1997**, *25*, 757–761.

31 STALLMEYER, B., SCHWARZ, G., SCHULZE, J., NERLICH, A., REISS, J., KIRSCH, J., and MENDEL, R. R. The neurotransmitter receptor-anchoring protein gephyrin reconstitutes molybdenum cofactor biosynthesis in bacteria, plants, and mammalian cells, *Proc Natl Acad Sci USA* **1999**, *96*, 1333–1338.

32 HASONA, A., RAY, R. M., and SHANMUGAM, K. T. Physiological and genetic analyses leading to identification of a biochemical role for the moeA (molybdate metabolism) gene product in Escherichia coli, *J Bacteriol* **1998**, *180*, 1466–1472.

33 JOSHI, M. S., JOHNSON, J. L., and RAJAGOPALAN, K. V. Molybdenum cofactor biosynthesis in *Escherichia coli mog* mutants, *J. Bact.* **1996**, *178*, 4310–4312.

34 MEYER, O., FRUNZKE, K., TACHIL, J., and VOLK, M. *The Bacterial Molybdenum Cofactor.* Edited by STIEFEL, E. I., American Chemical Society, Washington, DC, **1993**.

35 RUDOLPH, M. J., WUEBBENS, M. M., RAJAGOPALAN, K. V., and SCHINDELIN, H. Crystal structure of molybdopterin synthase and its evolutionary relationship to ubiquitin activation, *Nat Struct Biol* **2001**, *8*, 42–46.

36 RUDOLPH, M. J., WUEBBENS, M. M., TURQUE, O., RAJAGOPALAN, K. V., and SCHINDELIN, H. Structural studies of molybdopterin synthase provide insights into its catalytic mechanism, *J Biol Chem* **2003**, *278*, 14514–14522.

37 WANG, C., XI, J., BEGLEY, T. P., and NICHOLSON, L. K. Solution structure of ThiS and implications for the evolutionary roots of ubiquitin, *Nat Struct Biol* **2001**, *8*, 47–51.

38 LAKE, M. W., WUEBBENS, M. M., RAJAGOPALAN, K. V., and SCHINDELIN, H. Mechanism of ubiquitin activation revealed by the structure of a bacterial MoeB-MoaD complex, *Nature* **2001**, *414*, 325–329.

39 ADAMS, M. J., FORD, G. C., KOEKOEK, R., LENTZ, P. J., McPHERSON, A., JR., ROSSMANN, M. G., SMILEY, I. E., SCHEVITZ, R. W., and WONACOTT, A. J. Structure of lactate dehydrogenase at 2–8 A resolution, *Nature* **1970**, *227*, 1098–1103.

40 WALKER, J. E., SARASTE, M., RUNSWICK, M. J., and GAY, N. J. Distantly related sequences in the alpha- and beta-subunits of ATP synthase, myosin, kinases and other ATP-requiring enzymes and a common nucleotide binding fold, *Embo J* **1982**, *1*, 945–951.

41 LOGAN, D. T., MAZAURIC, M. H., KERN, D., and MORAS, D. Crystal structure of glycyl-tRNA synthetase from Thermus thermophilus, *Embo J* **1995**, *14*, 4156–4167.

42 LEIMKUHLER, S. and RAJAGOPALAN, K. V. In vitro incorporation of nascent molybdenum cofactor into human sulfite oxidase, *J Biol Chem* **2001**, *276*, 1837–1844.

43 MATTHIES, A., RAJAGOPALAN, K. V., MENDEL, R. R., and LEIMKUHLER, S. Evidence for the physiological role of a rhodanese-like protein for the biosynthesis of the molybdenum cofactor in humans, *Proc Natl Acad Sci USA* **2004**, *101*, 5946–5951.

44 XI, J., GE, Y., KINSLAND, C., McLAFFERTY, F. W., and BEGLEY, T. P. Biosynthesis of the thiazole moiety of thiamin in Escherichia coli: identification of an acyldisulfide-linked protein–protein conjugate that is functionally analogous to the ubiquitin/E1 complex, *Proc Natl Acad Sci USA* **2001**, *98*, 8513–8518.

45 MUELLER, E. G. and PALENCHAR, P. M. Using genomic information to investigate the function of ThiI, an enzyme shared between thiamin and 4-thiouridine biosynthesis, *Protein Sci* **1999**, *8*, 2424–2427.

46 APPLEYARD, M. V., SLOAN, J., KANA'N, G. J., HECK, I. S., KINGHORN, J. R., and UNKLES, S. E. The Aspergillus nidulans cnxF gene and its involvement in molybdopterin biosynthesis. Molecular characterization and analysis of in vivo generated mutants, *J Biol Chem* **1998**, *273*, 14869–14876.

47 LIAKOPOULOS, D., DOENGES, G., MATUSCHEWSKI, K., and JENTSCH, S. A novel protein modification pathway related to the ubiquitin system, *Embo J* **1998**, *17*, 2208–2214.

48 OSAKA, F., KAWASAKI, H., AIDA, N., SAEKI, M., CHIBA, T., KAWASHIMA, S., TANAKA, K., and KATO, S. A new NEDD8-ligating system for cullin-4A, *Genes Dev* **1998**, *12*, 2263–2268.

49 LAMMER, D., MATHIAS, N., LAPLAZA, J. M., JIANG, W., LIU, Y., CALLIS, J., GOEBL, M., and ESTELLE, M. Modification of yeast Cdc53p by the ubiquitin-related protein rub1p affects function of the SCFCdc4 complex, *Genes Dev* **1998**, *12*, 914–926.

50 WALDEN, H., PODGORSKI, M. S., and SCHULMAN, B. A. Insights into the ubiquitin transfer cascade from the structure of the activating enzyme for NEDD8, *Nature* **2003**, *422*, 330–334.

51 WALDEN, H., PODGORSKI, M. S., HUANG, D. T., MILLER, D. W., HOWARD, R. J., MINOR, D. L., JR., HOLTON, J. M., and SCHULMAN, B. A. The structure of the APPBP1-UBA3-NEDD8-ATP complex reveals the basis for selective ubiquitin-like protein activation by an E1, *Mol Cell* **2003**, *12*, 1427–1437.

52 RAO-NAIK, C., DELACRUZ, W., LAPLAZA, J. M., TAN, S., CALLIS, J., and FISHER, A. J. The rub family of ubiquitin-like proteins. Crystal structure of Arabidopsis rub1 and expression of multiple rubs in Arabidopsis, *J Biol Chem* **1998**, *273*, 34976–34982.

53 GOLDSTEIN, G., SCHEID, M., HAMMERLING, U., SCHLESINGER, D. H., NIALL, H. D., and BOYSE, E. A. Isolation of a polypeptide that has lymphocyte-differentiating properties and is probably represented universally in living cells, *Proc Natl Acad Sci USA* **1975**, *72*, 11–15.

54 ALTSCHUL, S. F., MADDEN, T. L., SCHAFFER, A. A., ZHANG, J., ZHANG, Z., MILLER, W., and LIPMAN, D. J. Gapped BLAST and PSI-BLAST: a new generation of protein database search programs, *Nucleic Acids Res* **1997**, *25*, 3389–3402.

4

RING Fingers and Relatives: Determinators of Protein Fate

Kevin L. Lorick*, Yien-Che Tsai*, Yili Yang, and Allan M. Weissman

4.1
Introduction and Overview

As recently as 1998, the RING finger was a structure without known function, and was often confused with the zinc finger. Three years later this compact structure was rapidly becoming one of the most widely studied protein modules because its presence in proteins has strong predictive value for ubiquitin ligase activity. An exhaustive treatment of RING finger ubiquitin ligases would warrant an entire volume by itself. Therefore, this chapter will provide general information on the RING finger and its structural relatives. This is followed by brief synopses of several well-studied families of RING finger proteins implicated in cell regulation and signaling. These examples are Siahs, IAPs, TRAFs and Cbls. These are intended to illustrate the biological importance and complexities of RING finger proteins. We will then provide more detailed discussions on two RING finger proteins: Parkin, associated with autosomal recessive juvenile Parkinson's disease and Mdm2, the most well-known and extensively studied cellular ubiquitin protein ligase for the tumor suppressor p53.

4.1.1
Historical Perspective

Ubiquitylation of proteins, i.e. their conjugation with ubiquitin, has dramatic effects on their fate and function. Its most well-described role involves proteasomal degradation as a consequence of modification with K48-linked chains of ubiquitin, but it has other roles not linked to proteasomal degradation. These include enhancement of endocytosis and targeting to lysosomes (vacuoles in yeast), DNA repair, transcriptional regulation and kinase activation. Ubiquitylation is a hierarchical, multi-step process generally involving, at a minimum, enzymes known as ubiquitin-activating enzyme (E1), ubiquitin-conjugating enzymes (E2s) and ubiquitin protein ligases (E3s). E1 forms a thiolester linkage with the C-terminus of ubiquitin via an ATP-dependent reaction. Ubiquitin is then transferred to one of

* Equal contributors.

Protein Degradation. Edited by J. Mayer, A. Ciechanover, M. Rechsteiner
Copyright © 2005 WILEY-VCH Verlag GmbH & Co. KGaA, Weinheim
ISBN: 3-527-30837-7

over 30 (in mammals) E2s where a second thiolester linkage is formed. E3s interact with E2 and substrate, which then mediate the transfer of ubiquitin to substrate [1–5].

Because of the large number of known ubiquitylation substrates and since E3s recognize specific proteins or modified forms of proteins, the number of E3 specificities is necessarily enormous. However, until recently few E3 had been molecularly characterized. By 1992, for example, 13 E2s had been described in yeast [6]. Of these, 11 function with ubiquitin and one with each of two different ubiquitin-like (UbL) proteins, Sumo and Nedd8/Rub1 [2]. However, while several yeast and metazoan E3s had been identified biochemically by that time, the primary amino acid sequence was known only for a single E3 [7]. This was the *Saccharomyces cerevisiae* Ubr1p, which had been studied extensively by Varshavsky and co-workers in delineation of the N-end rule [8, 9]. Its mammalian ortholog, E3α, was the prototypical E3 used in the description of the basics of the ubiquitin pathway by Hershko and Ciechanover [1, 10].

A major advance in the identification of E3s came with the discovery by Howley and co-workers of E6-AP (*E6* protein-*a*ssociated *p*rotein). This cellular protein functions as an E3 for p53 in human papilloma virus (HPV)-infected cells. Specifically, oncogenic strains of HPV (HPV-16 or -18) express a protein, E6, that usurps the E3 function of E6-AP [11, 12]. E6-AP was found to resemble other deduced protein sequences in cDNA databases. This led to the identification in 1995 of a family of proteins characterized by a C-terminal 350 amino acid HECT (*h*omologous to *E6*-*A*P *c*arboxyl *t*erminus) domain [13]. Many of these have now been shown to be E3s. HECT E3s generally recognize substrates through their N-terminal halves. HECT domains interact with E2s and form transient thiolester linkages with ubiquitin before transferring ubiquitin to target substrates (see Chapter 5). Based on searches using the BLAST program [14, 15], there are 27 HECT proteins encoded within the human genome, not considering splice variants.

As HECT domain function was being elucidated, studies in the mid-1990s on cell-cycle regulation led to the initial description of SCF E3s. These were first characterized as containing *S*kp1, *C*ullin-1, and an *F*-box-containing protein [16, 17]. Also identified during the mid-1990s was the APC (*a*naphase *p*romoting *c*omplex – also known as the cyclosome), another multi-subunit cullin-containing E3 that mediates ubiquitylation of mitotic cyclins [18, 19]. Mdm2, initially thought to be a HECT domain variant, was shown in 1997 to have E3 activity towards p53 *in vitro* [20]. For the non-HECT E3s no common structural feature had been detected. Thus, by 1997 the only defined molecular signature for E3 activity was the HECT domain.

In 1999, the world of E3s began to change dramatically. Using a yeast two-hybrid approach, a protein of unknown function, AO7, was identified in the Weissman laboratory as a binding partner for a human E2, UbcH5B, and was found to mediate UbcH5B-dependent self-ubiquitylation [21]. These properties depended upon a motif known as the RING finger, which has the general sequence $CX_2CX_{(9-39)}CXHX_{(2-3)}C/HX_2CX_{(4-48)}CX_2C$ (Figure 4.2).

The term RING finger was first coined by Freemont and colleagues referring to a gene product in the MHC class I locus [22]. Their apparent exuberance led to the

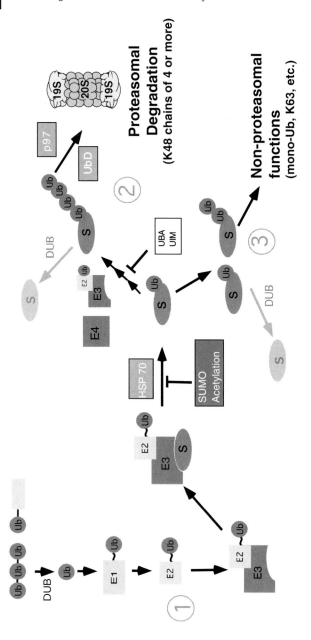

descriptor **R**eally **I**nteresting **N**ew **G**ene 1 (RING1). Little did they know how interesting it would be! Determination of the structure of several RING fingers [23–29] revealed that the RING finger consists of two Zn binding sites with a total of eight cysteine and histidine ligands. They may be ordered either as C3H2C3 (RING-H2) or C3HC4 (RING-HC) with the resultant fold assuming a cross-braced type arrangement (Figures 4.2 and Figure 4.3). AO7 was determined, specifically, to be a RING-H2 protein. Strikingly, each of four other RING-H2 proteins, for which there was no prior evidence for roles in ubiquitylation, were also demonstrated to have RING finger-dependent E3 activity. The RING-HC proteins Brca1 (**Br**east **c**ancer **a**ssociated protein 1) and Siah1 (**s**even **in** **a**bsentia **h**omolog 1) were found to behave in a similar manner [21].

Coincidentally, while initial E3 studies were ongoing, the Ashwell lab (next door to the Weissman lab), had evidence for a potential role for proteasome activity in IAP (**i**nhibitor of **ap**optosis) function. A connection became clear with the determination that most IAPs include RING-HC fingers and have E3 activity [30]. It shortly became apparent through studies by a number of groups that many proteins whose functions had been associated with ubiquitylation, such as Cbls, Mdm2, Ubr1p/E3α and Hrd1p/Der3p are RING finger E3s [31–34]. Similarly, the presence of a small RING finger protein, Apc11, within the multi-subunit APC took on new significance [35].

Remarkably, at the same time, several groups independently determined that there was a previously undiscovered component to the SCF complex – a small non-canonical RING finger protein known variably as Rbx1, Roc1 or Hrt1 [36–

Fig. 4.1. Fundamentals of the ubiquitin system. Adapted from Ref. [5]. Figure 4.1 shows the fundamentals of the ubiquitin system. (1) Ubiquitin is synthesized in linear chains or as the N-terminal fusion with small ribosomal subunits that are cleaved by de-ubiquitylating enzymes to form the active protein. Ubiquitin is then activated in an ATP-dependent manner by E1 where a thiolester linkage is formed. It is then transthiolated to the active-site cysteine of an E2. E2s interact with E3s and with substrates and mediate either the indirect (in the case of HECT E3s) or direct transfer of ubiquitin to substrate. A number of factors can affect this process. We know that interactions with Hsp70 can facilitate ubiquitylation in specific instances and competition for lysines on substrates with the processes of acetylation and sumoylation may be inhibitory in certain instances. (2) For efficient proteasomal targeting to occur chains of ubiquitin linked internally through K48 must be formed. This appears to involve multiple cycles of E3-mediated transfer of ubiquitin or in some cases other factors known as E4s may play a role in facilitating the processivity of polyubiquitin-chain formation. Interactions with proteins containing UbDs (**ub**iquitin **d**omains), including some E3s, may facilitate targeting to the proteasome. For a number of subrates an ATPase known as p97 (also known as VCP or in yeast as Cdc48p) facilitates transport to proteasomes. (3) The factors that influence the balance between K48 chains and mono-ubiquitin or other linkages, such as K63, are poorly understood. However, ubiquitin-binding domains such as the UBA or UIM could influence this balance in cells by blocking K48 on ubiquitin and thus favoring chain termination or other linkages [82] – this is a point that is far from being established with certainty. Anywhere along the pathway deubiquitylating enzymes may reverse the process, including at the level of the proteasome itself.

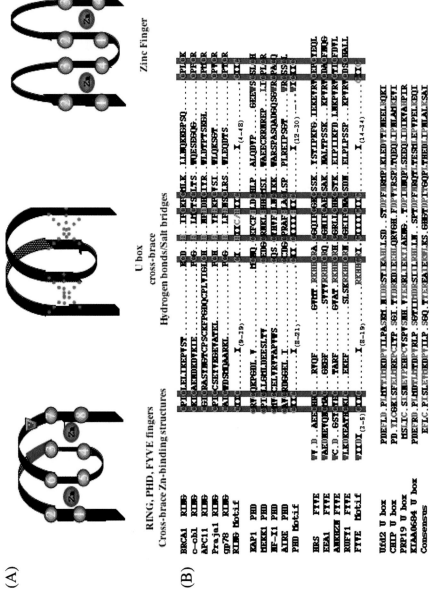

39]. It was also found that Rbx1 was central to the E3 activity of the newly described CBC^{VHL} complex [40, 41]. During this time, others demonstrated binding of E2s to RING finger proteins – although in these cases the connection with ubiquitylation was not made [42]. So, by mid-2000 there was a new family of over 300 additional potential human E3s.

Things did not stop with the identification of the RING finger. The PHD (*p*lant *h*omeo*d*omain) or LAP (*l*eukemia-*a*ssociated *p*rotein) domain represents a variation on the RING finger [43, 44]. A second more distant RING finger relative has been identified as the U-box (see below) [45]. Members of both of these have now been shown to have E3 activity [46–51]. Thus, from 1998 to 2002 we had gone from a family of perhaps 27 potential human E3s, which could be discerned based on their primary amino acid sequence, to over 400. Additionally, substantial heterogeneity in substrate recognition is provided by the cullin-containing E3s.

4.2
RING Fingers as E3s

4.2.1
General Considerations

It is reasonable to divide RING finger E3s into two major types, cullin and non-cullin E3s, the latter being the primary focus of this chapter. However, to put the full range of RING finger E3 specificities into perspective some mention of the cullin E3s is required. cullin-containing E3s are remarkable in having the capacity to bind to and ubiquitylate multiple different targets through use of different substrate-recognition subunits. These E3s have a small RING finger protein (in most cases Rbx1/Roc1/Hrt1) associated with a specific member of the cullin family.

◀

Fig. 4.2. Comparison of RING finger and RING finger-related motifs. (A) Schematic representation of RING, PHD and FYVE fingers (left), U-box (center), and tandem classical zinc fingers for comparison (right). (B) Multiple alignments of representative RING, PHD, and FYVE, fingers. Adapted from Ref. [46]. In Figure 4.2A, numbered residues in RING fingers represent metal coordinating residues. A canonical RING finger has histidine in position 4, cysteine in positions 1–3, 6–8. RING fingers are classified as RING-H2 or RING-HC depending on whether position 5 is occupied by histidine (-H2) or cysteine (-HC). Canonical PHD finger consensus has histidine in position 5. FYVE finger includes (R/K) (R/K)(HHCR) insertion indicated in purple.

PHD finger includes invariant tryptophan two amino acids before the seventh coordinating residue indicated in blue. In the U-box the predicted conformation is conferred by hydrogen bonding and salt bridges, indicated schematically by dotted lines. In Figure 4.2B, predicted metal-coordinating residues are indicated in blue boxes. Consensus RING, PHD and FYVE finger motifs are indicated below each grouping. The position of tryptophan conserved in PHD fingers is indicated in blue. The FYVE finger insert is in purple. The canonical RING, PHD and FYVE finger motif is indicated under each set of alignments. For the U-box, conserved residues are indicated in red.

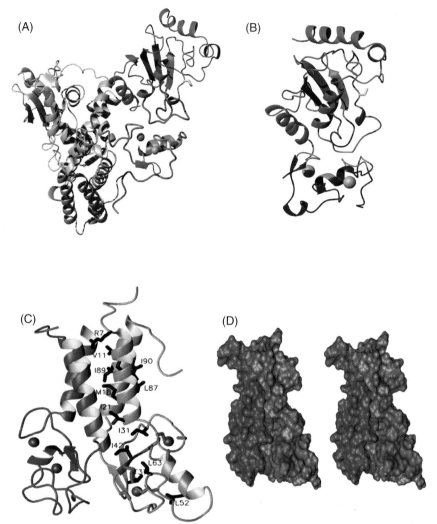

Fig. 4.3. Structures of RING-E2 complexes and RING dimers. (A) c-Cbl and UbcH7 based on crystal structures [23]. (B) Cnot4 and UbcH5B based on NMR solution structures and molecular modeling [28, 70]. (C) Solution structure of Brca1–Bard1 dimer [25, 77]. (D) Space-filling model of the Brca1–Bard1 dimer showing (right) sites of interaction with UbcH5C with which it functionally interacts, and (left) less extensive sites of interaction with UbcH7, for which there is no evidence of functional interaction. (C and D adapted from ref. 25 and ref. 77 with permission). In Figure 4.3A, Cbl is shown in silver, its RING finger region is shown in blue, and E2 in red. Points of interaction of UbcH7 on c-Cbl (RING finger and linker α-helix) are indicated in cyan and Cbl-interacting sites on E2 are shown in green. Note that most interactions involve a linker α-helix of Cbl (shown in orange), loops including the cysteine pairs that coordinate the first (shown in pink) and last (shown in orange) Zn molecule and the central α-helix of the RING finger. In Figure 4.3B, E2, RING, active sites, contacts and Zn residues are colored as in Figure 4.3A. Interactions involve the same general regions of the E2 and RING finger with no evidence of involvement of regions outside

cullin-containing E3s include the SCF E3s, where F-box proteins recognize substrates and Skp1 connects the Cul1 core to F-box proteins [17, 24, 52–55]. For the CBC (elongin *C*, elongin *B*, *C*ul2 or Cul5) E3s the dimer of elongin C and elongin B play an analogous role to Skp1. VHL (*V*on-*H*ippel-*L*indau), the first defined substrate-recognition element for the CBC complex, or other members of the SOCS (*s*uppressor *of* *c*ytokine *s*ignaling) box proteins, serve to recognize the substrate [40, 56, 57]. More recently the Cul-3 BTB/POZ (*B*ric-a-Brac *T*ramtrack and *B*road Complex/*Pox* virus and *Zn* finger) family of E3s has been identified. In these complexes BTB proteins both recognize substrate and link to Cul3, combining the roles played by the Skp1 and F-box proteins of the SCF [58–60]. The most complex cullin-containing E3 is the anaphase promoting complex, APC. Apc2 is a cullin, and Apc11 is a small RING finger protein. The APC complex also includes at least nine other essential subunits, a large number of which include TPR (*t*etratrico*p*eptide *r*epeat) protein–protein interaction domains. Tandem TPR repeats generate a right-handed helical structure with an amphipathic channel that is thought to accommodate an α-helix of a target protein [61]. These subunits may therefore serve to hold the APC complex together or to bind substrates. Other known APC substrate-recognition elements include Cdc20 and Cdh1/Hct1 [62–64]. In general, the cullin-containing E3 is written with the family name followed by the substrate-recognition element in superscript. Thus, the SCF E3 for IκB may be written as SCF$^{\beta TRCP}$, while the E3 that recognizes hypoxia inducible factors, HIF-1α and HIF-1β, is CBC^{VHL}. As shorthand, cullin-containing E3s are often referred to as "multi-subunit E3s" to underscore the need for multiple subunits for discernable activity. This is clearly an over-simplification, as single subunit, non-cullin, E3s can also function in the context of more complex assemblies. For more detail on cullin-containing E3s see Chapters 6 and 7.

An obvious question is whether all consensus RING fingers mediate ubiquitylation. Our view, based on the literature, our experience and discussions with colleagues, is that the large majority do. A few RING proteins persist in showing no E3 activity, as defined by the minimal criteria of E2-dependent self-ubiquitylation *in vitro*, despite experimental evaluation by multiple laboratories. However, almost all of these persistent negatives influence the ubiquitin pathway. The best example of this is Bard1 (*B*rca1-*a*ssociated *R*ING *d*omain 1). Inactive as an E3 by itself, Bard1 dimerizes with Brca1 and greatly enhances the latter's E3 activity [21, 65, 66] (Figure 4.3C and D). Furthermore, failure to express Bard1 markedly destabilizes Brca1 [67]. Until recently, MdmX had not been reported to be an E3. However, there is now evidence that MdmX has at least a small amount of ligase activity

---

the RING finger. Figure 4.3C shows Brca1 in silver, Bard1 in blue. Zn-binding residues are indicated in green. Known mis-sense mutations of Brca1 are labeled. Mutations of residues marked in red have been linked to familial cancers; other mutations (in black)

have been observed in patients with sporadic breast or ovarian cancer. Extensive E2-interacting regions on Brca1 extend beyond the RING finger to include regions in the α-helices N- and C-terminal to the RING that form the four-helix dimerization interface.

[68]. Perhaps, analogous to Bard1, MdmX may modulate the E3 activity of Mdm2 towards p53 through dimerization (see below). Ironically, PML (*pro*myleocytic *leu*kemia), one of the first RING finger proteins to be characterized has, to date, not shown E3 activity (see below).

4.2.2
Structural Analysis and Structure–Function Relationships

How do RING fingers function to mediate ubiquitylation? Unlike HECT domains, which function as catalytic intermediates in the transfer of ubiquitin from E2 to substrate, there is no evidence that RING fingers play such a role. A minimalist view is that RING fingers act as E2 docking sites. As such, they would position the complex of E2 and ubiquitin allowing for nucleophilic attack from a substrate lysine onto the C-terminus of ubiquitin [69]. In this model, the RING finger is not expected to affect the catalytic reaction required for ubiquitin transfer. Consistent with a docking role, the crystal structure of the c-Cbl RING finger with an E2, UbcH7, provides no evidence for a catalytic role in RING finger function (Figure 4.3A). This is similarly the case for the solution structure of the Cnot4 RING finger with UbcH5B. The Cnot4 RING finger is atypical in having only cysteine residues coordinating Zn, but otherwise has a consensus RING finger sequence [70] (Figure 4.3B). For both of these the active site cysteine of the E2 projects away from, and is not in close proximity to, the RING finger. Thus, based on these structures, there is little to suggest direct interactions between the RING finger and the site of ubiquitin attachment on E2.

Despite the structural data, this docking-site model may be an over-simplification of how RING fingers function. There is evidence that E2 binding without an intact RING finger is not sufficient to target substrates for degradation. This is based on the analysis of other E2 binding domains, including those found within the yeast protein Cue1p [34, 71] and in gp78 [72]. The latter, also known as AMFR (the *a*utocrine *m*otility *f*actor *r*eceptor), is a mammalian E3 involved in ERAD (*e*ndoplasmic *r*eticulum *a*ssociated *d*egradation). It includes both a RING finger and a distinct E2 binding site [72], both of which are required for proper E3 activity in the presence of its cognate E2. Also contrary to the minimalist view are instances where E2–RING finger interactions occur but lack functional activity. This is the case for UbcH7 and the Brca1/Bard1 heterodimer (see below) [23, 25]. It is, therefore, reasonable that the RING finger might act as an allosteric co-factor whose binding to E2-Ub alters the E2 so as to facilitate transfer of ubiquitin to substrate.

4.2.2.1 **RING finger–E2 Interactions**
For both Cnot4 and c-Cbl there are several points of contact between the E2 (shown in red) and the RING finger (shown in blue in Figure 4.3A and B). These include residues in the first α-helix of the E2 and two loops that extend down from the core

of the E2 into the RING finger (shown in green). One precedes and one immediately follows the active-site cysteine (yellow side chain). The first E2 α-helix contacts residues close to the first cysteine pair of the RING finger (shown in cyan around the first Zn molecule, which is depicted in pink). The first loop of the E2 interacts with the central α-helix of the RING finger at one or more points (contacts in cyan), while the second loop of the E2 contacts residues near the final cysteine pair (contacts in cyan). However, there is variation between the two E2–RING finger pairs as to the specific residues involved. In addition, c-Cbl also contacts the first α-helix of the E2 through a linker α-helix immediately N-terminal to the RING finger (indicated in orange). On the other hand, for Cnot4 all of the E2 contacts are confined to the RING finger. While Cnot4 exhibits robust activity with UbcH5B *in vitro*, it is notable that there is little evidence for direct functional interactions between c-Cbl and UbcH7.

A third RING finger protein for which E2 interaction data is available is Brca1. Notably, mutations of Zn-coordinating and other residues in the Brca1 RING finger are associated with familial breast and ovarian cancers [73–76]. By itself, Brca1 has weak E3 activity, which is dramatically enhanced by dimerization with Bard1, which also contains a RING finger. While the two RING fingers are juxtaposed in the dimeric structure (Figure 4.3C), the major determinants of dimerization are α-helices N- and C-terminal to each RING finger. These form a four α-helix bundle [25]. NMR studies have examined the binding of the Brca1-Bard1 heterodimer to both UbcH5C, with which this E3 functions, and UbcH7, which has no activity with the heterodimer as assessed by self-ubiquitylation [77].

Although the Brca1-Bard1 heterodimer contains two RING finger subunits, UbcH5C and UbcH7 interact only with the Brca1 subunit. One reason for this may be the lack of a central α-helix in the Bard1 RING finger. These central α-helices may be one essential element required for E2 contact. UbcH5C contacts the Brca1 RING finger through an extended 17 amino acid interface on the RING finger (Figure 4.3C and 3D). Points of contact on the Brca1 RING finger are more extensive than those seen for c-Cbl and Cnot4, although similar regions of the RING finger are involved. Interactions with UbcH5C extend beyond the Brca1 RING finger forming additional interfaces on the two adjacent Brca1 α-helices. In contrast to UbcH5C, UbcH7 forms a single, less extensive interface involving 10 Brca1 RING finger residues. Despite the differences in interactions, both E2s show only weak affinity for the Brca1 RING finger, and NMR titration experiments found no large difference in affinity between the two E2s. These observations may be explained if UbcH5C and UbcH7 bind to Brca1 in similar, but not identical, orientations.

Interestingly, Ile26, which immediately precedes the second cysteine of the Brca1 RING finger, is a contact point for both UbcH5C and UbcH7. While not required for the coordination of Zn, this residue is also essential for maintaining E3 activity. Conversely, mutations in other E2-contact residues in the RING finger, particularly those in the central α-helix of Brca1 can be altered with little functional consequence [77]. The residues corresponding to Ile26 in both Cnot4 (Leu16) and c-Cbl (Ile83) are also included in E2 interaction interfaces. Again, however, there is substantial variation in E2 contact points on the central α-helices of these RING fingers. Consistent with the structural observations, inspection of a number of RING proteins (Figure 4.2) reveals a high degree of conservation of the residue immediately preceding the second Zn-coordinating cysteine. Similar conservation is found surrounding other Zn-coordinating residues. With exceptions [23, 31], this is not the case for amino acids corresponding to the central α-helix of RING fingers.

While the major site of interaction for each E2–E3 pair is within the RING finger, there are additional binding surfaces for E2 on c-Cbl and Brca1 outside of this region. These contacts vary between RING finger proteins. For example, those c-Cbl residues implicated in E2 binding that are N-terminal to its RING finger, do not correspond to those non-RING finger residues implicated in Brca1–UbcH5C interactions. The structural data summarized in Figure 4.3 also provides mechanistic insights into the sequential actions of E1, E2 and E3 enzymes. In addition to interacting with RING fingers (and HECT domains), the first α-helix of E2 is also a major site of interaction with E1 [78]. This supports a mechanism where E2 must receive ubiquitin from E1 prior to association with E3. In considering the structures depicted in Figure 4.3, one should note that all were determined using E2 without bound ubiquitin. The possibility that E2 loaded with ubiquitin will provide different results cannot be overlooked.

As noted above for Brca1, E2–RING finger interactions do not need to be of high affinity to be productive. While UbcH5B binding is easily detectable for the RING finger protein AO7, stable binding of E2s to RING fingers is not generally the norm. In considering this generality one needs to keep in mind that generation of poly-ubiquitin chains on proteins requires repetitive transfers of ubiquitin from E2 to substrate. As such, E2s exist in both ubiquitin-bound and -unbound states. It follows that an ideal RING finger, whether as an E2 binding platform or as a means to decrease the activation energy for ubiquitin transfer, must not bind the E2 so tightly as to limit its dissociation after transfer of ubiquitin to the target protein. Consistent with this notion are observations that mutations outside the RING of AO7 that reduce E2 binding correlate with increased self-ubiquitylation (Lister, K. M., Lorick, K. L., Jensen, J. P. and Weissman, A. M., unpublished observations) and that release of E2-ubiquitin by the SCF complex E3 appears to be an important first step in substrate ubiquitylation [79].

One point that arises from study of these E2–E3 structures, is that there is a degree of specificity in their physical interactions. The biochemical data extends the idea of physical specificity in E2–E3 interactions to functional specificity. An example of this is the Brca1-Bard1 heterodimer, which interacts physically with UbcH7 and UbcH5C, but is only active with UbcH5C. There appear to be two consequences of E2 specificity: variability in the strength of a particular E3 response and variability in the type of ubiquitin modification.

The Snurf (small *nuclear* **R**ING *f*inger) protein has E3 activity using at least six different E2s [80], all of which appear to have different patterns of ubiquitin modification. On the other hand AO7 and Brca1 only appear active with members of the

UbcH5 family [21, 80, 81]. Presumably, the ability to interact productively with a smaller number of E2s will restrict E3 activity to situations where the RING finger protein and the E2 are co-expressed and co-localized.

Because ubiquitylation is not limited to degradation of substrate proteins – a process primarily utilizing K48-linked poly-ubiquitin chains [82] – the use of different E2s may control the type of ubiquitin modification or ubiquitin chain formed. Brca1-Bard1 has been shown to form K6-linked and K29-linked ubiquitin chains when employing UbcH5C as the E2 [83]. When employing the closely related but distinct UbcH5B, the same E3 can form K63-linked chains [66]. The RING finger protein, Rad5 binds to Mms2-Ubc13 to form K63-linked ubiquitin chains in the DNA repair process [84, 85]. This linkage requires the complex of another RING finger protein, Rad18, and an additional E2, Rad6 [85]. While there is no data to suggest the E3 proteins involved in this process form other ubiquitin linkages, Rad18 has been implicated in proteasomal degradation of Ho endonuclease in yeast [86], implying that it may form K48-linked chains.

4.2.3
Other Protein–Protein Interaction Motifs in RING finger Proteins

A remarkable feature of RING fingers is their small (up to 70 amino acids), compact nature. This is in contrast to the considerably larger elongated HECT domain (∼350 amino acids). Perhaps one consequence of this small size is its inclusion into a large number of proteins having a number of different protein-interaction modules. The RING finger thereby provides ubiquitin ligase activity to a wide range of otherwise functionally divergent proteins. Accordingly, the means by which RING finger E3s interact with substrates are highly variable and run the gamut of protein–protein interactions (http://home.cancer.gov/lpds/weissman). As discussed above, small RING finger proteins can interact with a large number of substrates in the context of a cullin-containing complex. For others, the size and complexity of the ligase presents the possibility for many different interactions. Well-characterized examples of this include Cbl proteins [87] (see below) and the Brca1-Bard1 heterodimer [88]. Protein interaction domains found in this dimeric E3 include: the Brca1 RING finger, which, in addition to interacting with E2, has been reported to bind the de-ubiquitylating enzyme Bap1 [89]; BRCT (*Br*ca1 *c*arboxyl-*t*erminal) domains, found in both Brca1 and Bard1, which bind to basal transcription machinery [90]; the large non-conserved central region of Brca1, whcih binds DNA repair enzymes and transcription factors [91–93]; and the Bard1 ankyrin repeats, which may interact with a number of different proteins and may induce apoptosis [94].

RING finger proteins may have other domains associated with signal transduction, such as SH3 and STAT domains and domains that bind and effect hydrolysis of nucleotides in signal transduction or for other purposes. These include ATPases, ATP synthases, serine/threonine kinases, GTP-binding domains, ADP-ribosylation domains and AAA-superfamily ATPases (http://home.cancer.gov/lpds/weissman).

One example that encompasses two of these in one protein is Ard1 (***ADP-r***ibosylation factor ***d***omain 1), a 64-kDa protein with an ADP-ribosylation factor domain linked to an N-terminal GTPase domain [95, 96]. The Ard1 GTPase domain physically binds its ADP-ribosylation factor domain, stimulating hydrolysis of bound GTP. It has been suggested that this protein plays a role in vesicular trafficking, and it is also a member of the TRIM family of RING finger proteins (see below). The RING finger is found in other proteins associated with organelle transport, including those with kinesin motor domains, peroxisome domains (Pex3, Pex10 and Pex12), and clathrin heavy chain repeats (Vps11) (http://home.cancer.gov/lpds/weissman).

The RING finger motif is also found in numerous proteins having domains associated with nuclear functions (e.g. helicases, DNA repair ezymes). It is also found in proteins having domains associated with the establishment and maintenance of intracellular and extracellular matrices (e.g. scaffold/matrix specific factors, Band4.1, ezrin/radixin homologs). A number of other protein-interaction domains in RING finger proteins, such as WD40 repeats, PDZ domains, sterile α motifs or TPR domains, are not easily pigeonholed as being either organelle- or function-specific. Curiously, RING fingers are frequently found in proteins with a variety of other Zn-binding structures (Table 4.1). In many cases these cysteine-rich domains have no effect on E3 activity, but may mediate interactions with proteins and nucleic acids.

There are a number of RING finger E3s that contain transmembrane domains, including some without characterized substrates, such as Kf-1 and Trc8 [21]. One example for which a direct substrate interaction has been determined is Rnf5, which interacts with and mediates ubiquitylation of paxillin [97]. However, for other transmembrane RING finger E3s, the means of interaction with a substrate does not involve easily traceable direct or even indirect protein–protein interactions. The best example of this is in ERAD, where a single E3 has the potential to target multiple structurally unrelated substrates. For one such RING finger E3, the yeast Hrd1p/Der3p, there is evidence for specific indirect interactions through Hrd3p in targeting yeast HMGCoA reductase [98]. For other Der3p substrates, including the test substrate CPY*, there is little evidence for substrate–ligase interactions [99]. This is also the case for the mammalian transmembrane ERAD E3, gp78 and its substrates such as the T cell antigen receptor CD3-δ subunit and Apolipoprotein B [72, 100]. Whether ERAD substrates are recognized by co-localization in membrane sub-domains, through adaptors or chaperones, or a combination of these remains to be determined.

Some RING finger and related E3s possess regions related to ubiquitin, known as ubiquitin domains (UbD). Examples of these are Parkin and Hoil-1, both of which also contain two RING fingers (see below). There is increasing evidence that at least one function of UbDs is to mediate interactions with proteasomes [101–103]. There is also now a growing list of protein domains that bind ubiquitin. Of these the structurally related UBA (***ub***iquitin-***a***ssociated) domain and Cue domains are found in two of the Cbl family members (see below) and in gp78, respectively.

Tab. 4.1. Alternative zinc-binding domains in RING proteins.

Description (CD ID); Consensus	Function	Examples in RING proteins (Genebank Accession)	Zn-binding domain References
ZnF-RanBP (smart00547) CPACTFLNFASRSKCF ACGAP	Zn-coordinating RNA-binding domain, in RAN-binding protein binds RAN-GDP.	MdmX (NP_002384), Mdm2 (NP_002383), UbcM4-IP 3 (NP112506)	357, 358
Rad18-like CCHC Zn finger (smart00734) LVQCPVCFREVPENLI NSHLDSCL	Yeast Rad18p functions with the RING finger protein Rad5p in error-free post-replicative DNA repair.	Yeast Rad18 (NP_009992); human Xpcc (NP_004619)-Xeroderma pigmentosa complementation group C protein	359, 360
B-Box-type Zn finger (smart00336) QAAPKCDSHGDEPAE FFCEECGALLCRDCD EAEHRGHTVVLL	B-boxes have seven potential Zn-coordinating residues but only four bind zinc. Characteristic of TRIM proteins.	Human Murf2 (NP_908975); (Ifp1) Inerferon-responsive finger (NP_569074)	361, 362
TRAF-type Zn finger (pfam02176) HEKTCPFVPVPCPNK CGKKILREDLPDHLSA DCPKRPVPCPFKVYG CKVDMVRENLQ	Found in Trafs. Protein–protein interactions	Human Traf4 (NP_004286); Traf5 (NP_004610); Traf6 (NP_004611)	363, 364
ZZ Zn finger (smart00291) VHHSVSCDTCGKPIV GVRYHCLVCPDYDL CESCFAKGGHH GEHSM	Zn-binding domain, present in CBP/p300 and Dystrophin. In Dystrophin, domain is implicated in Calmodulin binding. Mis-sense mutation of conserved cysteine correlates with Duchenne muscular dystrophy.	Mindbomb (NP_065825)-involved in neural development; SWIM domain containing protein 2 (NP_872327)	365, 366
C1 domain (smart00109); HHHVFRTFTGKPTYC CVCRKSIWGSFKGGL RCSWCKVKCHKKCA PKVPKPC	Protein kinase C conserved region 1 (C1) domains (Cysteine-rich domains); Some bind phorbol esters and diacylglycerol. Some bind RasGTP.	Mll3 (NM_170606)-mixed lineage leukemia protein	367, 368
ZnF_NFX (smart00438); CGIHTCEKLCHEGDC GPVSCRC	Found in the transcriptional repressor NF-X1, a PHD finger/RING finger protein.	NF-X1 (NP_002495)-MHC class 1 X-box binding factor	369

Tab. 4.1. *(continued)*

Description (CD ID); Consensus	Function	Examples in RING proteins (Genebank Accession)	Zn-binding domain References
ZnF_UBR1 (smart00396) CTYKFTGGEVIYRCK TCGLDPTCVLCSDCF RSNCHKGHDYSLKTS RGSGICDCGDKEAWN EDLKCKAH	Domain is involved in recognition of N-end rule substrates.	Yeast Ubr1(NP_011700); Human E3-α I (NP_777576) and II (NP_056070)	370
ZnF_UBP (smart00290); RCSVCGTIENLWLCLI CGQVGCGRYQLSHA LEHFEETGHPLVVKL GTQRV	Found in ubiquitin hydrolases and other proteins. In BRAP2, this domain binds ras-GTP.	Brap2/Imp (NP_006759)- Brca1-associated protein, impedes Raf signaling.	365, 366
FYVE finger (smart00064) CMGCGKEFNLTKRR HHCRNCGRIFCSKCS SKKAPLPKLGNEKPV RVCDDCYENLNG	Implicated in endosomal targeting. Recent data indicates that these domains bind PtdIns(3)P	Sakura/Fring (NP_476519); Riff/Momo (NP_919247) a suspected inibitor of apoptosis.	371–374
In-between-Ring fingers (IBR) domain (smart00647); KWCPAPDCSAAIIVTE EEGCNRVTCPKCGFS FCFRCKVEWHSPVSC	The IBR (C6HC or DRIL) domain is found to occur between pairs of RING fingers (pfam00097). The function of this domain is unknown.	Ariadne1 (NP_005735), parkin	375
PHD Zn-finger (smart00249) YCSVCGKPDDGGELL QCDGCDRWYHQTCL GPPLLlEEPDGKWYCP KCK	Found in nuclear proteins and implicated in chromatin-mediated transcriptional regulation.	Tif1-α (NP_003843)	43
Zn finger C3H1 (smart00356); KYKTELCKFFKRGNC PYGDRCKFAHPL	Implicated in DNA binding. Found in proteins controlling cell cycle or growth. Shown to interact with the 3′ UTR of mRNA. Often found in tandem.	Zfp 183 (NP_849192), makorin1 (NP_038474)	376, 377
C2HC Zn finger (smart00343) KCYNCGKPGHIARDC PS	Found in the Nucleocapsid protein of retrovirus. Also found in eukaryotic RNA- or ssDNA-binding proteins	Human Rb-BP 6 (NP_008841)- retinoblatoma-binding protein 6	378, 379
ZnF_C2H2 (smart00355) YRCPECGKVFKSKSA LQEHMRTH	First identified in the *Xenopus* transcription factor TFIIIA. Found in numerous nucleic acid-binding proteins.	Rag1 (NP_000439), Strin/ Rfp138 (NP_057355)	380

In addition to containing protein–protein interaction motifs, E3–substrate specificity may be affected by post-translational modifications. In particular, phosphorylation can alter E3–substrate interactions. One example is p53 where certain phosphorylations inhibit its direct binding to Mdm2, while others indirectly enhance their association by promoting nuclear localization of p53 [104–106]. Phosphorylation also directly enhances substrate interactions, as exemplified by the Cbls, which include phospho-tyrosine binding domains (see below) [107].

4.2.4
Variations on the RING Finger

In addition to RING-HC and RING-H2, there is one example of a RING-CC, Cnot4. Cnot4, part of a transcriptional repressor complex, has potent E3 activity that is highly specific for UbcH5B (Figure 4.3B). In addition, there are other variations on the RING finger that also demonstrate ubiquitin ligase activity. These include the PHD finger and the U-box.

The PHD finger closely resembles the RING finger in having eight cysteines and histidines that bind Zn in a cross-brace pattern (Figure 4.2). Differences with the RING finger include a variation in spacing between the coordinating residues; a cysteine in the fourth and a histidine in the fifth (C4HC3) coordinating residues; and an invariant tryptophan two amino acids before the seventh Zn-coordinating residue. NF-X1 is an example of a protein with co-linear consensus sequences for PHD and RING fingers. We have shown that the PHD arrangement – not the RING finger – is essential for *in vitro* E3 activity [46]. Another PHD finger E3 is Mekk1 (*m*itogen-activated protein kinase/*E*RK *k*inase *k*inase 1) [48]. Mekk1 functions not only as an activating kinase for Erk (*e*xtracellular signal *r*egulated *k*inase) and JNK (*J*un *N*-terminal *k*inase) through its kinase domain, but also as a negative regulator by targeting Erk1/2 for degradation. Similarly, the AIRE (*a*uto*i*mmune *re*gulator) protein, which contains two PHD fingers, has ubiquitin ligase activity. This E3 activity requires only the first PHD finger sequence, which, of the two, more closely resembles the PHD consensus [49]. The murine γ-herpesvirus-68 K3 was originally described as a PHD/LAP finger protein [51]. It was later suggested that this is a RING finger variant distinct from the PHD [108]. K3 localizes to the endoplasmic reticulum membrane and binds the cytoplasmic tail of nascent MHC class I H-2D(b), targeting it for ubiquitylation and degradation [50, 109, 110]. Similar PHD finger variant proteins from Kaposi Sarcoma Herpes virus, MR1 and MR2, lead to internalization and lysosomal degradation of cell-surface molecules including Class I MHC, B7.2 and ICAM-1 [51].

The relation of the U-box motif to the RING finger is far less obvious than that of the PHD finger. The first U-box protein implicated in ubiquitylation was CHIP (*c*arboxy-terminus *H*sc70-*i*nteracting *p*rotein) [111–113]. Subsequently, a yeast protein, Ufd2p, was shown to enhance the processivity of the ubiquitin chain formed by the HECT E3 Ufd4p. This led Jentsch and co-workers to coin the term E4 in reference to this function of Ufd2p [114]. Aravind and Koonin then identified a motif common to CHIP, Ufd2p and more than 10 other proteins. They predicted that the

U-box would conform to a cross-brace structure similar to the RING finger, although it lacks the canonical cysteine and histidine residues for Zn-binding. This motif is predicted to fold using salt bridges and other hydrophilic interactions in order to achieve the structure provided to the RING finger by its Zn coordinating residues [45]. The E3 activity of this family was subsequently verified in the same way that the activity of RING fingers was established, although there is yet to be direct structural verification of its predicted RING finger-like topology [47, 115].

It may be that not all domains that resemble RING fingers are E3 modules. The FYVE finger binds phosphoinositides to effect protein transport. It bears similarity to the RING finger in the use of eight Zn-coordinating residues and in its cross-brace structure [116–118]. The FYVE finger is, however, quite distinct. It employs only cysteines to coordinate Zn, has a different spacing for its Zn-binding residues and contains a short basic amino acid residue sequence preceding the third Zn-coordinating cysteine (Figure 4.2) [119]. FYVE fingers have not, so far, been shown to have E3 activity.

4.2.5
High-order Structure of RINGs – TRIMs

RING fingers exist in structural contexts crucial to their function. One example is the relatively small but medically relevant family of TRIAD proteins, which will be discussed in a separate section in the context of Parkin. The largest family of higher order RING finger-containing proteins are the tripartite motif (TRIM) or RING/B-box/Coiled Coil (RBCC) proteins. An example, Midline1 (Mid1) is shown in Figure 4.4A. TRIM proteins constitute up to one-fifth of the nearly 300 known RING finger proteins in the human genome. As expected, RING finger domains of TRIM proteins can serve as ubiquitin ligase modules.

The two B-box Zn-finger domains (Table 4.1) bind phosphoproteins and, similar to TFIIA Zn fingers, they may also bind nucleic acids. The binding of TRIM proteins to DNA through B-boxes may aid in the modification of transcription factors and histones by ubiquitin. Consistent with this, several TRIMs are known transcriptional regulators. For example, Trim24/Tif1-α is believed to bind to the AF2 (*a*ctivation *f*unction 2) region of the estrogen, retinoic acid and vitamin D receptors [120]. The *e*strogen-responsive *f*inger *p*rotein (Trim25/Efp) is a transcription factor that is postulated to mediate estrogen action in breast cancer [121, 122]. Another example of a DNA-interacting TRIM protein is Trim32/Hta (*H*IV *T*AT-*a*ssociated), which translocates from cytoplasm to nucleus in HIV-infected cells. In the nucleus, Hta appears to aid in regulating TAT-mediated transcription [123]. On the other hand, Trim22/Staf50 (*S*timulated *t*rans-*a*ctivation *f*actor 50 kD) down-regulates transcription from the HIV-1 LTR promoter region in response to interferon. Staf50 might mediate interferon's antiviral effects [124].

The coiled-coil domain frequently mediates hetero- or homodimerization. About two-thirds of a large number of TRIM proteins tested dimerize [125]. As a number of other RING finger proteins either homo- or hererodimerize, dimerization may facilitate optimal E3 activity of TRIMs. While most proteins that carry the TRIM

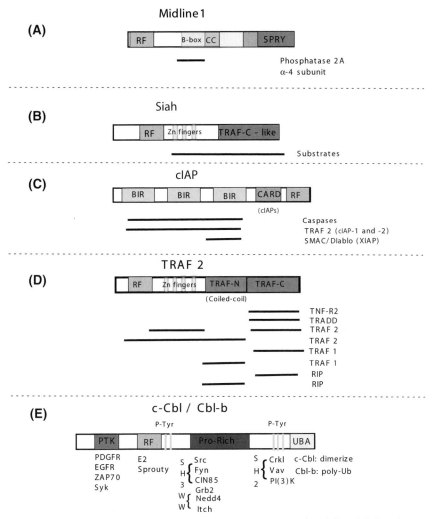

Fig. 4.4. Representative domains and interactions of members of RING finger families. In (A–D) underlines indicate areas of interactions with other proteins. In (E) representative inter- acting proteins are listed directly below the various domains. RING finger indicated by RF. CARD domain not found in XIAP.

designation have RING fingers, there are variants that retain all of the TRIM domains except the RING finger. In some cases these form heterodimers with RING finger-containing TRIMs via their coiled-coil domains. TRIMs without RING fingers may help modulate substrate interactions, or serve as substrates themselves. An excellent example for study is the RING fingerless TRIM29 or ATCD (*a*taxia *t*elangiectsia *c*omplementation group *D* protein). ATCD forms heterodimers with RING-containing TRIMs including Trim1/Mid2, Trim10/hematopoietic RING finger, Trim22/Staf50 and the ret finger protein [125]. The significance of these interactions is not yet known. However, ATCD mutations found in ataxia telangiect-sia patients makes alteration of E3 activity an attractive mechanism for disease pathogenesis.

A number of other TRIM-containing proteins are associated with genetic disorders. For example, mutations of TRIM37 (Mul1) correlate with mulibrey (***muscle-liver-brain-eye***) nanism, an autosomal recessive disorder involving tissues of mesodermal origin [126, 127]. Although Mul1 E3 activity is yet to be established, it is known to affect induction of NF-κB by TRAF2 and TRAF6 [128]. Mid1 (Figure 4.4A) mutations are associated with the X-linked Opitz syndrome, which is characterized by severe midline abnormalities. Mid1 forms both homo- and heterodimers with Mid2 (TRIM1), which also contains a RING finger [129]. Mid1/Mid2 dimers are involved in the formation of microtubule anchors. The ***p***rotein ***p***hosphatase 2A (Pp2A) regulatory subunit, α4, is attached to microtubules by the B-box regions of Mid1/Mid2. Pp2A is also a substrate for this E3 [130]. While not yet implicated in human disease, Murf-1 (***Mu***scle-specific ***R***ING ***f***inger 1) is a TRIM E3 whose overexpression is associated with muscle atrophy in rodents. MURF1$^{-/-}$ mice exhibit protection from muscle atrophy [131].

PML is a TRIM protein that is an exception to the rule – one of the few RING finger proteins for which no E3 activity has yet been detected. There are fourteen TRIM-containing PML splice variants. Phosphorylated PML localizes to PML nuclear bodies, where numerous roles have been ascribed to it. These roles include transcription factor [132], tumor suppressor [133] and regulator of p53 response to oncogenic signals [134]. Though PML is not known to be a sumo ligase, it is sumoylated at multiple sites, including on its RING finger. This modification affects its ability to bind Mdm2 and presumably regulates p53 response [135].

4.3
RING Fingers in Cell Signaling

Families of RING finger proteins play important roles in cell regulation, signal transduction and apoptosis. Some of the more prominent families are the Siahs, IAPs, TRAFs, and Cbls (Figure 4.4). Short summaries of these are presented below.

4.3.1
Siahs

The Siah mammalian RING finger E3s are homologs of *Drosophila* seven-in-absentia (Sina) and are represented in plants by the Sinat family. Before RING fingers were known to be E3s, Sina was implicated in Tramtrack degradation in *Drosophila* [136]. Further, Siahs were found to be involved in the proteasomal degradation of the DCC (***d***eleted in ***c***olorectal ***c***arcinoma) gene product, the Netrin receptor [137]. In humans, there are two family members, Siah1 and Siah2. In mice, in addition to Siah2, there are two highly homologous forms of Siah1 (a and b). In plants, there are five Siah relatives, Sinat1–5. Siah proteins are generally ~280 amino acids and characterized by variable N-terminal extensions followed by a

RING finger (Figure 4.4B). The RING finger is followed by a cysteine-rich Zn-finger-containing region. Siahs are known to dimerize, and there is biochemical and structural evidence that this primarily involves the Zn-finger region and part of the more C-terminal coiled-coil domain [138–140]. A role for dimerization in Sinat5 E3 activity has been established [141]. Consistent with biochemical studies, the crystal structure of the C-terminal region, lacking the RING finger, has revealed that it exists as a dimer [142]. Interestingly, this domain bears substantial similarity to the C-terminal regions of TRAF proteins.

The C-terminal domain of Siah proteins is also referred to as the substrate-binding domain (SBD). Siahs target a wide array of divergent substrates for degradation. Directly recognized substrates include the netrin receptor, c-Myb, Bob/Obf1, Peg3/Pw1, Synphillin-1 and TRAF2 [137, 143–147]. However, Siahs can also exist in the context of an SCF-like complex that includes Skp1, Ebi, Sip (*Siah-interacting protein*) and the adenomatous polyposis coli protein. This complex serves as an alternative to SCF$^{\beta TRCP}$ in targeting β-catenin for ubiquitylation [148]. In addition, a consensus Siah-binding sequence, RPVAxVxPxxR, has been identified with the core sequence PxAxVxP. This sequence is found in Sip as well as a number of Siah substrates, including netrin receptor. Sequences slightly degenerate from this are found in other Siah substrates including *n*uclear receptor *corepressor* (Nco-R), Kid motor protein and Numb. However, other Siah substrates, such as adenomatous polyposis coli protein, Synaptophysin, and group 1 metabotropic glutamate receptors contain no similar sequence, revealing the complexity of Siah interactions [145].

The biological consequences of Siah E3 activity, apparent from the number of its substrates, are significant. Siahs are implicated in mitosis and meiosis [149, 150]. In plants, Sinat5 targets the transcriptional activator Nac1, thereby attenuating auxin-mediated signaling and modulating lateral root development [141]. In mouse, deletion of both Siah1 α and Siah2 results in embryonic lethality [147]. Siahs also modulate tumor necrosis factor receptor (TNFR) function by promoting degradation of TRAF2 [151].

Most recently Siah2 has been shown to target prolyl-hydroxylase family members for degradation. Furthermore, Siah2 is transcriptionally induced in response to hypoxia. Proline hydroxylation has been observed on Hif-1α and Hif-2α (hypoxia inducible factor-1α and -2α) during normoxia. Thus, hypoxia may result in increased Siah2 and decreased prolyl-hydroxylases [152]. This would lead to decreased targeting of Hif-1α and -2α by CBC^{VHL} and, consequently, an increase in levels of VEGF (*V*ascular *e*ndothelial *g*rowth *f*actor) and other Hif-1α and -2α targets.

4.3.2
IAPs

The *sine qua non* of the IAPs (*i*nhibitors *of a*poptosis) is one or more copies of a Zn-finger-containing domain referred to as a BIR (*B*aculovirus *I*AP *R*epeat; Figure 4.4C). This name was derived from their initial identification in baculovirus, where they prevent host-cell apoptosis, allowing viral replication. At least 10 mammalian IAPs have been described. Many have C-terminal RING fingers [153]. A major function of IAPs, including XIAP, cIAP-1 and cIAP-2, all of which are active E3s [30], is to bind to and inhibit the tonic activation of caspases. The most extensively studied and potent mammalian IAP is XIAP, which binds and inhibits processed Caspase 9 as well as activated Caspase 7 and Caspase 3 [154] – the latter being an important effector caspase common to both the intrinsic (mitochondrial) and extrinsic (death receptor initiated) apoptotic pathways. Our understanding of how inhibition of caspases by XIAP is regulated has been greatly assisted by the identification of two XIAP-interacting proteins Smac/Diablo and HrtA2/Omi. These are released from mitochondria in response to permeability changes induced by pro-apoptotic stimuli. Also released on disruption of mitochondrial integrity is cytochrome c, which together with Apaf-1 and Caspase 9 leads to the activation of the effector caspase, Caspase 3. Smac/Diablo and HrtA2/Omi bind XIAP, and possibly other IAPs, resulting in the release of caspases from the IAP [155–157]. This may explain the paradox that, in certain XIAP-expressing cells, death-receptor-mediated apoptosis (i.e. the extrinsic pathway) depends on pro-apoptotic Bcl-2 family members that increase mitochondrial outer membrane permeability. In *Drosophila*, in addition to Smac/Diablo and HrtA2/Omi, induction of apoptosis requires three other proteins: Reaper, Hid and Grim. Upon over-expression, these result in excess cell death, which is suppressed by co-expression of the *Drosophila* IAP, DIAP1 [158–160]. Reaper, Hid and Grim bind DIAP1 and promote its ubiquitylation and degradation in a RING finger-dependent manner. This enables them to promote self-ubiquitylation and degradation of the IAPs. These findings are consistent with the initial observation of IAP E3 activity, where activation of apoptosis via the intrinsic pathway resulted in degradation of IAPs and activation of caspase activity [30]. Both of these functions are RING finger dependent, as expressing RING fingerless IAPs delays apoptosis. Thus, a scheme emerges where IAPs continually inhibit caspases until their dissociation is promoted by proteins such as Smac/Diablo. In this context, self-ubiquitylation is possibly potentiated by Reaper, Hid and Grim and their mammalian orthologs. Ubiquitylation of Caspase 9 by XIAP and of Smac/Diablo by cIAP1 and cIAP2 has also been reported. While such findings are consistent with anti-apoptotic roles of the IAPs, their overall significance is unclear [30, 161–163].

It should be stressed that IAPs are not just caspase inhibitors. cIAP-1 and cIAP-2 are recruited to TNFRs in response to activation by *t*umor *n*ecrosis *f*actor (Tnf) [164]. As discussed below, cIAPs apparently play roles in attenuating TNF signaling by contributing to TRAF2 ubiquitylation. At face value this might be construed as a pro-apoptotic role in that it can contribute to down-regulation of NF-κB activation. However, we are just beginning to scratch the surface as to the roles of the IAPs so judgments should probably be kept in reserve.

4.3.3
TRAFs

TRAFs (*T*NF *r*eceptor *a*ssociated *f*actors) are a family of signaling molecules characterized by a conserved C-terminal TRAF domain (Figure 4.4D) [165], which is divided into TRAF-N and TRAF-C. TRAF-N includes a coiled-coil domain. TRAF-C is structurally similar to the C-terminal region of Siahs. TRAF domains mediate many TRAF interactions, including their association with receptors, oligomerization (e.g. trimerization of TRAF2), and interaction with IAPs. The name TRAF was coined because the two prototypic members of the family, TRAF1 and TRAF2, were first found to associate directly with type II TNFR (TNFR-2) [164]. Interactions with TNFR-1 occur through the adaptor molecule Tradd. Six mammalian TRAFs (TRAF1–6) have been identified. They function as critical signal transducers for the TNFR family and the IL-1/Toll-like receptor family and, therefore, affect a wide range of biological processes, such as embryonic development, innate immunity, inflammation and bone homeostasis [166]. TRAFs2–6 all contain RING fingers at their N-termini, which under many circumstances are required for receptor-mediated signaling.

TRAF6 promotes its own ubiquitylation *in vitro* [167]. This ubiquitylation has been shown to utilize an E2 consisting of Ubc13 (also know as Bendless) dimerized with Mms2/Uev1a (*U*biquitin *E*2 *v*ariant 1A). Mms2, like Tsg101 (below), includes a core 14-kDa UBC domain common to E2s but lacks the canonical active-site cysteine. Association of TRAF6 with this dimer results in K63-linked polyubiquitin chains. It has more recently been shown that ubiquitylation of TRAF6 with K63-linked polyubiquitin chains is required for activation of the kinase Tak1. Tak1, in turn, activates IκB kinase as well as the kinase(s) that activates the JNK and p38 kinase systems [168–170].

The large amount of data available for TRAF2 in comparison to TRAF6 makes its story more complex, but perhaps also more informative. CD40 engagement results in redistribution of TRAF2 to lipid rafts. This correlates with its ubiquitylation and subsequent degradation. Presumably, TRAF2 degradation prevents prolonged JNK activation. Expression of EBV-Lmp1 (*E*pstein–*B*arr *v*irus *l*atent *m*embrane *p*rotein 1) also results in redistribution of TRAF2. However, Lmp1-associated redistribution does not lead to TRAF2 degradation. This may help explain the prolonged activation of downstream signaling pathways by this viral protein [171, 172]. It has not been determined directly whether ubiquitylation of TRAF2 is a function of its intrinsic activity or due to ubiquitylation by other E3s. In fact, cIAP-2 can ubiquitylate TRAF2 following TNFR stimulation and Siah2 can bind to and promote TRAF2 ubiquitylation under stress conditions [151, 173]. Furthermore, there is little evidence for *in vitro* activity of TRAF2 in a purified ubiquitylation system.

Interestingly, recent reports indicate that ubiquitylation of TRAF2 in the presence of Ubc13 and Mms2 is also required for activation of the downstream kinases Gckr (*g*erminal *c*enter *k*inase-*r*elated) and JNK [174, 175]. This suggests that K63-linked polyubiquitylation of TRAFs may be a common mechanism for members of this family to activate kinases. This story becomes more intriguing since the presence of the TRAF2 RING finger domain is necessary for JNK, p38 and NF-κB activation. However, of the three, only JNK activation requires intact E3 activity of TRAF2 and expression of Ubc13. This TRAF2- and Ubc13-dependent activation of JNK correlates with redistribution of ubiquitylated TRAF2 into an insoluble compartment, reminiscent of CD40-mediated signaling [175]. Thus, it seems that JNK activation may involve K63-linked polyubiquitylation and redistribution of TRAF2 to selective membrane microdomains or insoluble compartments. On the other hand, activation of the p38 and NF-κB pathways requires neither K63-linked chains nor TRAF2 E3 activity. However, it does require the physical presence of the RING finger region of TRAF2 [175]. It may be that TRAF2 and TRAF6 mediate assembly of K63-linked polyubiquitin chains that are either directly or indirectly required for activation of some of their targets. However, interactions with IAPs and Siahs may be required to synthesize K48-linked polyubiquitin chains necessary for TRAF2 and TRAF6 proteasomal degradation. Between these two ubiquitin-generating events there may also be a requisite disassembly of the K63-linked chains by de-ubiquitylating enzymes. While we await elucidation of the details, what is emerging is a complex set of interrelationships between the Siahs, IAPs and TRAFs in mediating signaling though TNFRs and related receptors.

4.3.4
Cbls

The Cbls play crucial roles in signaling by determining the fate of tyrosine kinase receptors as well as non-receptor tyrosine kinases. The importance of Cbls first became apparent with the discovery that the *v-Cbl* oncogene (*C*asitas *B*-*l*ineage Lymphoma) is transduced by the Cas NS-1 retrovirus, resulting in lymphomas of B cell lineage and fibroblast transformation. v-Cbl corresponds to the N-terminal region of c-Cbl. v-Cbl includes the phospho-tyrosine binding domain but lacks other domains including the RING finger [176]. The Cbls were first implicated in cell signaling with the finding that the *Caenorhabditis elegans* Cbl protein, Sli1, rescued a loss of function phenotype of the EGFR (*E*pidermal *G*rowth *F*actor *R*eceptor) ortholog Let-23 [177]. In mammals there are three Cbls. c-Cbl is the cellular ortholog of v-Cbl [178], Cbl-b is highly homologous to c-Cbl [179] and the most recently characterized family member is a shorter cousin known as Cbl-3 or Cbl-c [180, 181].

Cbls share a common architecture [107]. All members of the family have an N-terminal *p*hospho-*t*yrosine *b*inding (PTB) site that includes a four α-helix bundle, an EF hand and an atypical SH2 domain. This is followed by a RING finger. C-terminal to the RING finger are proline-rich regions – more extensive in c-Cbl and Cbl-b than in the shorter Cbl-3. The proline-rich region provides interaction sites for SH3 proteins including constitutive interactions with Grb2 and others (Figure 4.4E). Moving further towards the C-terminus, the two longer members of the family include a number of sites for tyrosine phosphorylation that bind heterologous SH2 domains. c-Cbl and Cbl-b include C-terminal UBA domains. As with a subset of other UBA domains including Rad23, the c-Cbl UBA domain mediates homodimerization, apparently functioning akin to a leucine zipper [182]. Recently, the UBA of Cbl-b has been shown to bind polyubiquitin without evidence for a role in dimerization [183].

Because of their many interactions, Cbls were logically thought of as scaffolds for signaling and endocytosis. It is now apparent that Cbls target many tyrosine kinases for ubiquitylation. Extensively characterized substrates include receptor tyrosine kinases such as EGFR and PDGFR (*p*latelet-*d*erived *g*rowth *f*actor *r*eceptor) [31, 184–186], and non-receptor tyrosine kinases, such as Src and Lck [186–189]. Generally, recognition of receptor tyrosine kinases occurs through the PTK-binding domain. Interactions with Src-family tyrosine kinases may occur between the proline-rich regions of Cbls and SH3 domain of the kinases [107, 190]. Cbl may also play ubiquitin-independent roles related to endocytosis. This has been postulated for the SH3-dependent interactions of c-Cbl with Cin85, which may provide a bridge from receptors to endophillins [191, 192].

The first clue to a relationship between Cbl and ubiquitylation came from Stanley and co-workers who demonstrated that c-Cbl was mono-ubiquitylated in response to CSF-1 (*c*olony *s*timulating *f*actor-1) [193]. Yarden and colleagues demonstrated an association between Cbl recruitment to receptors and their ubiquitylation and down-regulation. C-Cbl E3 ligase activity was subsequently demonstrated by several groups [31, 184, 186].

Cbl-mediated ubiquitylation illustrates several important points related to the varied effects of ubiquitylation. The first is that transmembrane receptors targeted for degradation by Cbl proteins are largely targeted to lysosomes [194–196]. Second, as with ubiquitylation of yeast cell surface transporters and receptors by the HECT E3 Rsp5, this targeting does not require formation of K48-linked polyubiquitin chains [197]. Additionally, in response to receptor–ligand binding, multiple components of the receptor signaling complex are targeted for degradation by Cbls. Included among these are the Cbls themselves and the associated proteins, Grb2 and Shc [198]. In contrast to transmembrane receptors, ubiquitylation of non-receptor tyrosine kinases appears to be associated with proteasomal degradation [186, 199, 200; A. Magnifico, S. Lipkowitz, A. M. Weissman, unpublished observations]. Further, there is evidence to suggest that Cbl-mediated ubiquitylation of PI3 kinase does not target it to either lysosomes or proteasomes but leads to its redistribution to CD28 or to T-cell antigen receptors and results in the attenuation of its activity [201]. Thus, Cbls provide an example of the range of different effects that can be mediated by a single ubiquitin ligase. Additionally, all three members of the Cbl family bind members of the WW domain class of HECT E3s including Nedd4 and Itch and are subject to ubiquitylation by these HECT E3s, providing yet another level of regulation [202].

There are still a number of questions regarding Cbl proteins. Where in the process leading from movement off the cell surface to lysosomes does Cbl-mediated ubiquitylation have its effects? Some studies suggest a role distal to internalization [189, 203] while others suggest a role at the cell surface [204]. Another question pertains to why receptor degradation in response to Cbl-mediated ubiquitylation is abrogated by inhibitors of both proteasomes and lysosomes. Is there a short-lived proteasomal protein essential to the process or are some components of the activated receptor complex targeted to proteasomes while the receptor itself is degraded in lysosomes? Is there an interdependent process of degradation of the signaling complex with some substrates "peeled off" and sent to proteasomes while others go on to lysosomes and multi-vesicular bodies? Finally, why is there a requirement for three distinct family members? A partial explanation may be their differential transcription in various tissues. Both c-Cbl and Cbl-b are widely expressed and, from analysis of mouse models, appear to have at least partially redundant developmental functions [205]. However, it is also clear that there are tissue-specific differences. c-Cbl is found to a greater extent in immature thymocytes and Cbl-b is found in more mature T cells [205–207]. In contrast, Cbl-3 exhibits a much narrower range of distribution [180]. There is much more to be learned about this family of complex regulatory proteins.

4.4
Multi RING finger Proteins

4.4.1
Mindbomb and TRIADs

Oligomerization is a common feature of RING finger proteins; examples include Brca1-Bard1 [208], Siahs [142], Mdm2-MdmX [209] and c-Cbl [182]. Additionally, some evidence suggests a propensity for RING fingers to form higher order aggregates *in vitro* [210]. A number of proteins contain multiple RING fingers in the same polypeptide. The most striking example of this is Mindbomb (Mib), one of several E3s implicated in Notch signaling. In *Drosophila* this protein plays a role in ubiquitylation and internalization of Delta, a Notch co-receptor. Mib includes three RING finger motifs, only the most C-terminal of which has been shown to have E3 activity. Mutations in the second RING finger have adverse developmental effects, but the relationship of this to ubiquitylation is as yet unclear [211]. Searches of Genbank additionally reveal at least three families of proteins, including homologs of the Icbp90 transcription factor (Np95), where a RING is found in the same polypeptide chain as a PHD domain. Similarly, the protein KIAA0860 contains both a RING finger and a U-box.

The most well characterized and apparently largest group of multi-RING proteins is the family of 13 human and at least 16 *Arabidopsis thaliana* E3s that generally include two RING fingers or RING finger-like consensus sequences with an intervening cysteine-rich sequence. This intervening sequence is referred to as the IBR (*in between RINGs*). These proteins are variously referred to as TRIAD (*Two RING fingers and Intervening-Associated Domain*), DRIL (*double RING-finger linked*), or RBR (*RING-between RINGs-RING*). The TRIAD motif has a general, although far from absolute, pattern of conserved residues, C3HC4-C6HC-C3HC4.

The most extensively studied TRIAD is Parkin, shown schematically in Figure 4.5A. The N-terminus contains a region homologous to ubiquitin called the *ubiq*-*uitin domain* (UbD), which interacts directly with proteasomes. The C-terminus contains two RING fingers (R1, R2) separated by a cysteine-rich *in-between RING* (IBR) region. This TRIAD motif mediates E3 activity and interacts with molecular chaperones. The last three amino acids of Parkin interact with a PDZ domain and possibly function to anchor Parkin to lipid microdomains.

Other multi-RING proteins with demonstrated E3 activity include HHARI, the

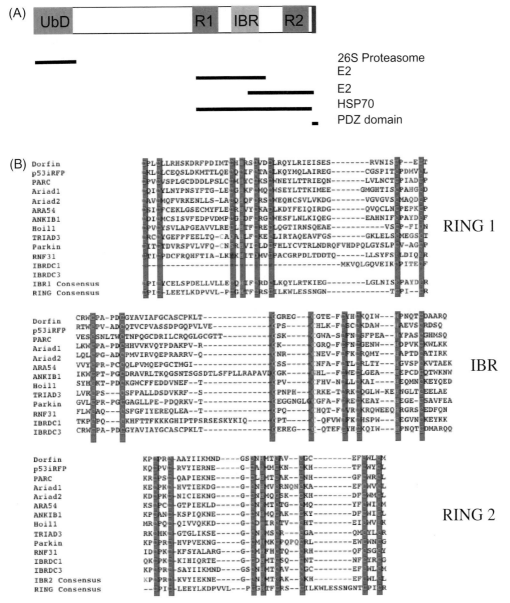

Fig. 4.5. Parkin and RBR proteins (A) Schematic of Parkin. (B) Alignment of RING-IBR-RING domains of RBR proteins.

human homolog of the Drosophila retinoic acid-inducible protein, Ariadne1. HHARI promotes ubiquitylation and degradation of a protein homologous to translation initiation factor 4E [212]. Parc is a TRIAD protein that binds p53 in the cytoplasm and shows E3 activity *in vitro*, although its *in vivo* targets are unknown [213]. Dorfin promotes the ubiquitylation of Synphilin-1, which is also a Parkin substrate [214]. Dorfin also promotes the ubiquitylation and degradation of mutant copper/zinc superoxide dismutase (Sod-1) [215]. In familial ALS and transgenic models, mutant Sod-1 is misfolded and aggregates in inclusions that also contain ubiquitin, proteasome and Hsc70 [216–219]. Hoil-1 is a ubiquitin ligase for iron regulatory binding protein 2 (Irp2), which binds to iron responsive elements in RNA to alter the stability of RNAs encoding ferritin and transferrin receptor. Hoil-1 recognizes and ubiquitylates Irp2 under conditions of high cellular iron [220–221]. Some other members of the TRIAD family are androgen receptor-associated protein 54 (Ara54), which interacts with and functions as a co-activator for androgen receptor and Ariadne2, one of the first family members described.

Analysis of the primary amino acid sequence of the thirteen TRIAD proteins reveals striking differences compared to consensus RINGs in both the first and second RING fingers (Figure 4.5B). Two (Parkin, Triad3) lack the predicted Zn-coordinating cysteine at the third position of the first RING, although other potential coordinating residues can be identified for both. Three lack identifiable first RING fingers (Ibrdc1, Ibrdc3, Rnf31). Seven of the remaining eight (Ara54, Ariadne1, Ariadne2, Parc, p53iRfp, Ankib1, Hoil-1) also have atypical first RINGs in having one or two extra amino acids inserted between the seventh and eighth predicted Zn-coordinating cysteines. In addition, p53iRFP has a cysteine rather than the canonical histidine at position four. Of the thirteen, only Dorfin has a first RING finger motif that fits the general consensus.

The second RING finger is also divergent from RING finger consensus sequences. Ara54, Triad3, Hoil-1 and p53iRfp all lack a histidine in the fourth coordination site with only a lysine, arginine, tryptophan and glutamine respectively available to substitute – none would be predicted to be a good Zn ligand. Ariadne1 and Parkin have two extra amino acids inserted between the fifth and sixth predicted coordinating residues. While the remaining seven proteins may be described as having a consensus RING, the distance between C6 and C7 is abbreviated – only four amino acids as compared to a consensus of ten randomly selected, non-IBR RING finger proteins, where the shortest stretch is eight amino acids. Based on structures of RING fingers this might be expected to impact on coordination of the second Zn. Accordingly, a solution structure of the second RING region of Ariadne1 lacks classic RING finger topology and there is no evidence of a second coordinated Zn. Strikingly it can still mediate ubiquitylation [222]. These findings underscore that when one is dealing with RING finger-like structures, ultimately it is function rather than variation from canonical structures or consensus sequences that counts.

The differences between TRIAD proteins and simpler RING fingers, and among the TRIAD proteins themselves, suggest distinct functions for the two RING finger domains in the context of the complete RBR, and perhaps different functions from non-IBR RINGs. Consistent with this, the proximal HHARI1 RING cannot be substituted for by the c-Cbl RING finger. However, the proximal RING finger of Parkin does not restore function either [223]. Although exceptions exist, RING finger domains isolated from these proteins are generally not sufficient for E2 interactions: the Parkin-UbcH7 interaction requires the entire RBR domain [224], while the interaction of HHARI with UbcH7 or UbcH8 requires the proximal RING finger and part of the IBR [41]. Similarly, the association of the distal RING finger of Parkin with UbcH8 [225] is enhanced by the IBR. E2s other than UbcH7 and UbcH8 also interact with RBRs. Ara54 binds UbcH6, Ube2E2 and Ube2E3 but not UbcH7 [226]. Functional interactions with Parkin have been detected with other E2s, including the mammalian orthologs of the yeast ERAD E2s, Ubc6p and Ubc7p (MmUbc6/Ube2J2 and MmUbc7/Ube2G2) and with UbcH5B [224]. How the RBR functions as a unit for each of the family members remains to be determined. Whether the two RING fingers bind two E2s simultaneously or sequentially to facilitate chain formation, or whether they function more like Brca1-Bard1, where one RING finger enhances the E3 activity of the other without directly binding an E2, also remains an open question.

4.4.2
Parkin and Parkinson's Disease

Parkin has been extensively studied because of its linkage to autosomal recessive juvenile-onset Parkinsonism (ARJP). It may be the most commonly mutated gene in familial Parkinson's disease (PD) [229]. Most disease-associated mutations result in an inactive E3. In addition to being a TRIAD protein, Parkin also includes an ubiquitin domain (UbD), which binds proteasomes [230]. A UbD mutation that may affect proteasome interactions is linked to early-onset PD [231], suggesting that this is crucial to normal function. Notably Hoil-1 also has a UbD. Interestingly, Bag-1 also binds the proteasome via its UbD. Bag-1 recruits a complex consisting of CHIP (a U-box E3) and the chaperone Hsp70. This complex has been proposed to mediate efficient degradation of misfolded proteins [232, 233]. This may be enhanced through non-canonical K33-linked polyubiquitin chains assembled on Bag-1 by CHIP [234].

Loss-of-function mutations in the *Parkin* gene correlate with ARJP with an average age of onset of 26 years. ARJP shares clinical features with autosomal dominant and sporadic PD, which generally present much later in life. Additionally, both diseases respond to L-Dopa treatment. However, while there is loss of dopaminergic neurons in the *substantia nigra* of ARJP patients, there is a general absence of Lewy bodies, the inclusion bodies characteristic of adult PD [235–238]. As a result, Parkin-linked PD is often assumed to lack inclusion bodies. However, there is little autopsy material available from the brains of patients with ARJP and there is at least one case where Lewy bodies have been observed [239]. Thus, it may be too early to draw conclusions.

It has also been proposed that *Parkin* is a tumor suppressor gene [240]. It is included in the fragile site FRA6E [241] and down-regulation of Parkin protein due to exon duplication or deletions has been observed in several cancer cell lines and tumors [241].

(A)

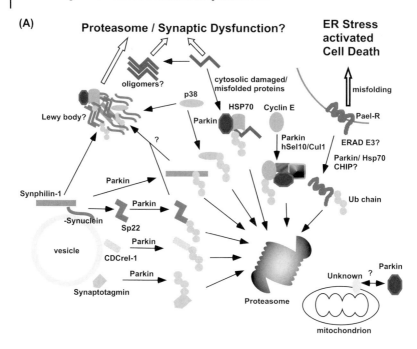

(B)

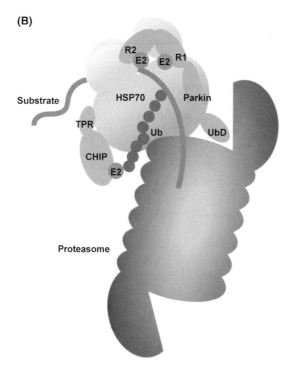

4.4.2.1 Parkin Substrates

There is great interest in identifying Parkin substrates (Figure 4.6A). The first of these, CDCrel-1, was identified as a yeast two-hybrid binding partner [225]. CDCrel-1 is a septin involved in vesicle cycling. CDCrel-1 over-expression inhibits dopamine release in PC12 cells and causes dopamine-dependent degeneration in rats [242]. However, CDCrel-1 null mice show no defect in development or dopamine release [243]. A second vesicle-related substrate is Synaptotagmin (Syt) XI [244], which interacts with Parkin via domains common to many Syt family members, suggesting that Parkin may target multiple Syts [244]. Although Parkin is largely cytosolic, there is evidence for association with synaptic vesicles mediated through Parkin's C-terminal PDZ-domain-binding motif. The PDZ-binding domain of Parkin has also been shown to interact with Cask, a lipid raft-associated protein [245].

α-Synuclein is a cytoplasmic vesicle-associated protein first found in autosomal-dominant PD. It is a prominent component of Lewy bodies and mutations are associated with genetic PD. A rare 22-kDa form of O-glycosylated α-synuclein (αSp22) is a Parkin substrate [246] and has also been shown to accumulate in ARJP. However, this finding remains to be generally established. The modified form represents a relatively small sub-population of α-synuclein and there is little evidence that the more common non-glycosylated form is a substrate. Interestingly, Parkin over-expression protects dopaminergic neurons in *Drosophila* over-expressing α-synuclein [247] and also has protective effects in cell-culture models of α-synuclein over-expression [248]. In these models no αSp22 has been observed. Parkin also promotes ubiquitylation and proteasomal degradation of Synphilin-1 [146], an α-synuclein interacting protein that associates with Parkin through its ankyrin repeats [249]. When over-expressed with α-synuclein, Synphilin-1 promotes the formation of inclusion bodies. Notably, over-expression of Synphilin-1 alone results in aggresome formation in HEK293 cells, suggesting Synphilin-1 itself is an aggregation-prone protein [250]. The significance of Synphilin-1 ubiquitylation in PD remains to be determined. Interestingly, at least two other RING E3s, Siah1 and Dorfin, can mediate Synphilin-1 ubiquitylation.

Vesicular trafficking and inclusion body formation are both dependent on the integrity of microtubules and other cytoskeletal components. Parkin has been shown to target misfolded tubulin for degradation [251] (Figure 4.6B) and to interact with centrosomes upon proteasomal inhibition [252]. Whether this reflects association with specific substrates or co-localization with proteasomes in centrosomes re-

Fig. 4.6. Models of Parkin Function. (A) Parkin substrates. (B) A model for Parkin-dependent degradation of misfolded proteins. Parkin recruits a complex containing molecular chaperones and the unfolded substrates to the proteasome. Degradation may be facilitated by direct coupling of ubiquitylation and degradation of the substrate with chaperone assistance. The complex may also recruit CHIP, a U-box protein, to enhance substrate degradation.

mains an open question, as does the significance of its association with actin [253]. The range of membrane-associated Parkin substrates is not limited to vesicles. Pael-R (*P*arkin-*a*ssociated *E*ndothelin-*l*ike *R*eceptor) [228] is a putative G-protein-coupled receptor that is enriched in dopaminergic neurons. When overexpressed, Pael-R misfolds and forms insoluble aggregates in the endoplasmic reticulum causing endoplasmic reticulum stress-activated apoptosis. Parkin promotes Pael-R ubiquitylation and degradation from the endoplasmic reticulum. Pael-R also accumulates in some ARJP brains. In *Drosophila*, over-expression of human Parkin protects dopaminergic neurons from toxicity induced by Pael-R [247].

Other cytoplasmic Parkin substrates have been identified. Parkin promotes ubiquitylation of the p38 subunit of aminoacyl-tRNA synthetase [254], which can be found in Lewy bodies. When over-expressed in COS-7 cells, p38 results in aggresome-like inclusions that include Hsp70 and Parkin. Parkin also promotes the ubiquitylation of polyglutamine (polyQ)-expanded Ataxin-3 *in vitro* and facilitates degradation of an Ataxin-3-derived fragment containing an expanded polyQ tract [230]. Over-expression of Parkin suppresses cell death and ameliorates proteasome inhibition and Caspase12 activation induced by expression of the polyQ-expanded proteins. Another putatative Parkin substrate is an, as yet, unidentified, protein on mitochondrial membranes that may modulate mitochondrial permeability. Parkin has also been demonstrated to facilitate the targeting of Cyclin-E for ubiquitylation. In this case Parkin exists as a part of a novel SCF-like E3 complex along with the F-box/WD40 protein hSel-10 and Cul-1 [255]. This raises the provocative possibility that Parkin may be associated with other, as yet to be defined, SCF-like E3 complexes.

4.4.2.2 Parkin Animal Models

At present, there are two reports of Parkin$^{-/-}$ mice. One study reports that Parkin$^{-/-}$ mice have motor impairments but no dopaminergic degeneration [256]. No accumulation of CDCrel-1, Synphilin-1 or α-synuclein [256] is observed. Intriguingly, however, they exhibit increased striatal extracellular dopamine and reduced striatal neuronal excitability. This is consistent with a model in which loss of Parkin results in altered dopamine re-uptake or clearance and post-synaptic down-regulation. How this might happen is unclear. However, since Parkin is involved in the degradation of misfolded proteins [230, 257], oxidatively damaged synaptic proteins may accumulate in Parkin null animals. Alternatively, there may be an indirect effect of substrate accumulation with failure of proteasomal function due to aggregation/accumulation of misfolded proteins. In a different study, Parkin mutants show reduced body weight and learning deficits [258]. Diminished motor activation with amphetamine suggests a reduced cytoplasmic pool of dopamine. Increased oxidation of dopamine is also observed, enhancing the potential for free-radical-mediated damage. Interestingly, there is increased glutathione in the striatum, which may partially compensate for increased oxidative stress. No degeneration of dopaminergic neurons is found. However, there are reduced levels of dopamine and vesicular monoamine transporters. These could either be contributory to the abnormalities observed or be an early indicator of degeneration of nigrostriatal terminals. Recently, the quaking (viable) mice, which show demyelination in the central nervous system, have been shown to have spontaneous deletions of *Parkin* and *Parkin co-regulated gene* (*PACRG*) in addition to the *Quaking* gene (qKI) [259, 260]. Again, the quaking (viable) mutants show no dopaminergic degeneration or α-synuclein accumulation.

Loss of a *Parkin* gene ortholog in *Drosophila* results in flies with a reduced life span and male sterility due to a defect in late spermatogenesis [261]. However, there is no overall neuronal degeneration or dopaminergic cell loss, even though dopaminergic neurons in the dorsomedial cluster show cell-body shrinkage and reduced tyrosine hydroxylase staining in proximal dendrites of aged flies [261]. Interestingly, this is the area most affected by α-synuclein transgenics [262]. Parkin null alleles also confer locomotor defects in climbing and flight due to loss of muscle integrity. This is associated with swollen mitochondria with disrupted cristae [261]. In PC12 cells, Parkin over-expression is associated with delayed mitochondrial swelling and cytochrome c release upon exposure to C2-ceramide [263] and there is evidence for Parkin association with the mitochondrial outer membrane where it may protect cells from ceramide toxicity [263]. This protective effect is dependent on Parkin's E3 activity and intact proteasome function. One may expect Parkin null animals to be more susceptible to oxidative stress, which in *Drosophila* may manifest as muscle defects.

Although these animal models show abnormalities, they fail to reproduce the PD phenotype. Still, they suggest a role for Parkin in synaptic function and maintenance of mitochondrial integrity.

4.4.2.3 Possible Pathogenic Mechanisms in ARJP

While possibly providing clues regarding Parkin's role in ARJP, animal models have led to few clear insights. The brain differs from most other organs in its lack of regenerative capacity, so that the damage accumulated from low-level chronic insults may be increased. All forms of PD are primarily manifested in a cell type prone to oxidative insults from decades of dopamine production. It therefore becomes evident why short-lived animals, such as mouse and fly, are of limited utility in testing the role of specific proteins as etiologic factors of this disease. These limits apply even for ARJP, which does not manifest clinically, on average, until the third decade. Additionally, the range of Parkin substrates identified to date, which include proteins associated with PD and other neurodegenerative disorders, are not necessarily derived from unbiased approaches.

The association of Parkin both with dopamine-containing vesicles and with mitochondria is consistent with a role in quality control within axons and dendrites – disposing of proteins that have been subject to oxidative damage. How does Parkin recognize a set of structurally diverse proteins? One mechanism might be through direct recognition of oxidatively-induced modifications or exposed epitopes indicative of misfolded or damaged proteins. Alternatively, Parkin may not directly recognize substrates but rather function together with chaperones, particularly Hsp70, that directly associate with altered proteins (Figure 4.6A and B). Support for indirect recognition includes the finding that although the proximal RING finger of Parkin is implicated in some substrate interactions, there is little to suggest that these are direct [249]. However, this RING finger does directly bind Hsp70 [230, 264]. Additionally, for some substrates, such as polyQ-expanded Ataxin 3, Hsp70 is crucial for ubiquitylation [230]. Other Parkin substrates including Pael-R, Synphilin-1 and p38 are also Hsp70-associated [254, 264; Y. C. Tsai, unpublished result]. There are striking parallels between Parkin and the U-box E3 CHIP. CHIP interacts with the UbD protein Bag-1 to target multiple Hsp70-bound proteins for proteasomal degradation [112, 113, 233]. Also, there is evidence that Parkin functionally interacts with CHIP to enhance Parkin-dependent degradation of Pael-R [264], with CHIP playing an E4-like role. Thus, Parkin may be a fusion of chaperone-binding, ubiquitin-ligase and proteasome-targeting motifs.

It is possible that a subset of TRIAD proteins may play overlapping roles in intracellular protein quality control. In this regard, it is provocative that *Drosophila* Ariadne-1 mutants show motor impairments and defects in muscle and neuronal development [265]. Also in neuronal cultures, over-expression of Hsp70 suppresses aggregation of the Dorfin substrate, mutant Sod-1 [266]. Furthermore Dorfin, like Parkin, has the capacity to target Synphilin-1 for proteasomal degradation. Thus, multiple members of the TRIAD family might function as crucial intermediaries between chaperones, the ubiquitylation machinery and proteasomal degradation in protein quality control. While *Parkin* mutations are associated with ARJP, it remains to be determined whether damaged Parkin contributes to disease progression in other forms of PD. Notably, Parkin and the other TRIAD proteins are cysteine-rich and may be particularly susceptible to oxidation or nitrosylation on these residues. This could result in dysfunction of protein quality control. The atypical nature of TRIAD RING fingers, with evidence for diminished stable Zn coordination, may further increase the propensity of TRIADs to be damaged. One can therefore envision an amplifying effect where failure to maintain vesicular or mitochondrial integrity by mutant or damaged Parkin leads to increased potential for oxidative damage of other cellular proteins.

4.5
Regulation of p53 by Mdm2 and other RING finger Proteins

4.5.1
Mdm2

p53 is maintained at low levels in normal proliferating cells. Increased p53 levels and altered transcription from p53-responsive genes are associated with response to cellular stress and DNA damage [267, 268]. The role of p53 in regulating genes that effect cell-cycle arrest, allowing for DNA repair if possible, and in regulating genes that induce apoptosis in tumor cells, has earned it the apt title of "guardian of the genome" [269]. The significance of losing p53 activity in cancer is underscored by the finding that up to 50% of cancers have inactivating mutations in the *p53* gene [270]. Many other cancers have wild-type p53 yet there is still a failure to activate a p53 response. We now know there are several mechanisms responsible for this. The one that has attracted the most attention is increased levels of the cognate p53 ubiquitin ligase Mdm2 [106]. This E3 binds to the trans-activation domain of p53 through its N-terminus and mediates p53 ubiquitylation through its C-terminus [271].

Mdm2, which was one of the first characterized p53-responsive proteins, was identified in a spontaneously transformed cell line, 3T3DM, as the product of a gene amplified on the mouse double-minute (MDM) chromosome [272]. Its onco-genic property was demonstrated by the finding that over-expression immortalizes rodent primary fibroblasts [271, 273]. The discovery that this 90-kDa protein binds to p53 and inhibits p53 transactivation [274] demonstrated a feedback loop between Mdm2 and p53 [275, 276]. The importance of regulating p53 by Mdm2 is underscored by the observation that early embryonic lethality of Mdm2 deficient mice is rescued by simultaneous deletion of p53 [277, 278]. Moreover, the *Mdm2* gene is amplified in approximately one-third of human sarcomas, and is over-ex-pressed, with or without gene amplification, in a wide range of human tumors [279, 280]. A majority of these cancers retain wild-type p53, which is inactivated by Mdm2, thus allowing for survival of transformed cells.

Since Mdm2 binds the p53 transactivation domain, it directly interferes with the interaction between p53 and basal transcription factors. This leads to a block of transcription. However, additional mechanisms may contribute to inhibition of p53-dependent trans-activation. The N-terminal region of Mdm2 has intrinsic tran-scriptional repression activity [281] and more recent studies found that Mdm2 re-cruits a transcriptional co-repressor, CtBP2 (*C-t*erminal *b*inding *p*rotein 2) [282]. Interestingly, this association is abolished by NADH, which changes the conforma-tion of CtBP2 [282]. Since NADH is increased in hypoxia, it is conceivable that dis-sociation of CtBP2 from Mdm2 may contribute to hypoxia-induced p53 activation.

Besides its role in transcription, Mdm2 also promotes the ubiquitylation and degradation of p53. This function is central to the role of Mdm2 in maintaining the low levels of p53 that allow cell proliferation [283, 284]. The initial study dem-onstrating Mdm2 ubiquitin-ligase activity *in vitro* provided evidence that it could be a HECT domain variant [20]. However, we now know that its E3 activity towards p53 is dependent on its atypical RING finger [32, 285] (Figure 4.7). Moreover, the intrinsic ligase activity of Mdm2 induces its own ubiquitylation and degradation [32]. This self-ubiquitylation of Mdm2 may allow for proper regulation of the p53 response.

Mdm2 is expressed at low levels throughout embryonic development and in most adult tissues [286]. It is transcribed from two promoters. The more distal, located between the non-coding exon I and exon II, is p53 responsive [275, 276]. Over 40 alternatively and aberrantly spliced variants of *Mdm2* mRNAs have been detected [287]. Many encode proteins that are deficient in binding to and therefore regulating p53. Others may function as dominant negative regulators, preventing the inhibitory action of Mdm2 toward p53 [288–290]. Still others promote cell

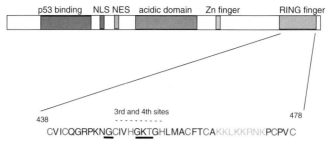

Fig. 4.7. Schematic representation of Mdm2 domains and RING finger. Mdm2 binds p53 as well as p63 and p73 through an N-terminal region of the protein. An acidic central domain, not found in MdmX, is essential for function as is its C-terminal RING finger. Nuclear import and export signals are indicated in Figure 4.7. The RING finger of Mdm2 includes 10 residues that could potentially represent sites of Zn coordination for this atypical RING finger (red), including four that have been variously proposed to represent the third and fourth Zn coordinating residues. The nucleolar localization sequence that is revealed upon binding of Arf or upon adenine nucleotide binding is shown in green and the Walker A site of nucleotide binding is underlined.

growth and tumor formation in a p53-independent manner [291]. The contribution of these to tumor development and prognosis is unknown.

The Mdm2 RING finger is unusual for the spacing of its putative Zn coordinating residues. Four amino acids have been implicated, in one way or another, as being the third and fourth coordinating residues [32, 292, 293]. These include a threonine, in addition to cysteine and histidine residues (Figure 4.7). This raises the possibility that the Mdm2 RING may exist in multiple conformations. Elucidation of the structure of its RING finger will hopefully clarify this. The RING finger of Mdm2 also includes a nucleolar localization signal in the region between the sixth and seventh coordinating residues. This region is apparently revealed when Mdm2 binds to the Arf (alternative reading frame) protein [294]. Interestingly, the Arf–Mdm2 interaction does not directly involve the RING finger (see below) (Figure 4.7). Further, it has recently been determined that this nucleolar localization signal can also be exposed in an Arf-independent manner by binding of adenine nucleotides through a Walker A motif within the RING finger (Figure 4.7) [295]. These findings reinforce the idea that the RING finger may exist in multiple states. The novel nature of the Mdm2 RING finger is underscored by the finding that substitution of a heterologous RING finger results in a chimera that promotes self-ubiquitylation but not ubiquitylation of p53 [32]. The specificity for the Mdm2 RING in p53 ubiquitylation suggests that the RING finger may have to be oriented in a specific manner to present E2-Ub to the substrate. Consistent with this, the ubiquitylation sites on p53 appear limited to a cluster of six lysine residues located near its C-terminus [296]. These also represent potential acetylation sites, which would preclude ubiquitylation, further adding to the myriad ways by which p53 is regulated [297]. The specificity in ubiquitylation sites may provide an explanation for the finding that Mdm2 also binds to p53 family members p63 and p73, but does not promote their ubiquitylation and degradation [296]. These two family members lack the lysine-rich region near the C-terminus [298, 299].

Much attention has been focused on whether Mdm2 can mediate formation of polyubiquitin chains on p53. Studies using purified p53, Mdm2 and ubiquitin with K48 or K63 mutations found that Mdm2 only catalyzes the addition of single ubiquitin to multiple lysine residues of p53 [300]. Another *in vitro* study, however, suggested that Mdm2 induces both mono- and polyubiquitylation of p53, depending on the level of Mdm2 [301]. It is unclear to what extent these *in vitro* observations reflect the *in vivo* setting. Regardless, there is now evidence to suggest that additional factors may work together with Mdm2 to promote polyubiquitylation and proteasomal degradation of p53. p300 has now been reported to have ubiquitin-ligase activity for p53 or to facilitate polyubiquitylation of p53 together with Mdm2 [302]. This suggests that p300 could be an E4 for p53. Since p300 is a transcription co-activator, a possibility is that Mdm2-p300 promotes the selective polyubiquitylation and degradation of p53 that is actively engaged in transcription.

4.5.2
Pirh2

Pirh2 is a RING finger protein that binds p53 and has been shown to mediate its ubiquitylation, independent of Mdm2 [303]. Down-regulation of endogenous Pirh2 increases p53, whereas expression of Pirh2 leads to a decrease in p53. Again, this leads to repression of p53-induced trans-activation and reversal of growth inhibition. Interestingly, the Pirh2 binding site on p53 (residues 82–292) does not overlap the Mdm2 binding site (residues 1–51). Thus, both Mdm2 and Pirh2 might bind a single p53 molecule and cooperate to mediate its ubiquitylation. The existence of an additional E3 for p53 may explain the observation that JNK activation can lead to p53 ubiquitylation and degradation in an Mdm2-independent manner [304]. Like Mdm2, Pirh2 is also transcriptionally up-regulated by p53 [303]. Another RING finger protein that interacts with p53 is a TRIAD protein, Parc, which binds p53 in the cytosol [213]. Whether Parc contributes to p53 ubiquitylation remains to be determined.

4.5.3
MdmX

MdmX (also known as Mdm4) is a close relative of Mdm2. Analogous to Mdm2, MdmX binds p53 through its N-terminus and includes a RING finger at its C-terminus [305]. In addition to direct binding of p53, MdmX heterodimerizes with Mdm2. This interaction appears to be mediated through their RING fingers. Although MdmX does not effectively promote p53 ubiquitylation [209], MdmX$^{(-/-)}$ mice, like those negative for Mdm2, undergo embryonic death that is rescued by loss of p53 [277, 278, 306–308]. Thus, MdmX is also an essential negative regulator of p53 activity. Accordingly, the *MdmX* gene is amplified and over-expressed in certain human malignant gliomas that express wild-type p53 [309]. Therefore, it is possible that MdmX functions as a critical regulator of Mdm2 to regulate the level and activity of p53. There are several lines of evidence supporting this. Down-regulation of MdmX causes a decrease in Mdm2 and an increase in p53. This leads to a subsequent increase in the sensitivity of cells to UV-induced apoptosis [307, 310]. In p53$^{-/-}$/MdmX$^{-/-}$ MEFs, re-expression of MdmX increases the half-life of co-transfected Mdm2 and enhances the degradation of re-expressed p53 [310]. These results suggest that the formation of Mdm2–MdmX heterodimers selectively inhibits Mdm2 self-ubiquitylation and stabilizes Mdm2, leading to increased ubiquitylation of p53. Over-expression of MdmX appears to reverse nuclear export of p53 promoted by Mdm2 [209]. Intriguingly, Mdm2 also promotes the ubiquitylation and degradation of MdmX [311, 312]. It is conceivable that this represents another mechanism for cells to down-regulate Mdm2 activity, ensuring the increase of p53 in response to stress stimuli.

The structural similarities and functional distinctions between Mdm2 and MdmX lend themselves to "mix and match" studies. MdmX alone has very low E3 activity towards either itself or p53. Interestingly, when the RING finger of MdmX is replaced with that of Mdm2, the chimeric molecule can not ubiquitylate p53 despite binding to p53 and mediating its own self-ubiquitylation [313]. However, if the central acidic domain of Mdm2 is also fused to the N-terminal portion of MdmX along with the Mdm2 RING, the resultant chimeric protein now becomes competent for p53 ubiquitylation. The importance of this acidic domain is further indicated by the finding that Mdm2 lacking this domain does not ubiquitylate p53. However, co-expression of the acidic domain *in trans* restores the ability of mutant Mdm2 to ubiquitylate p53 [313]. The underlying mechanism for this complementation is not yet clear. If nothing else it underscores the complexities and nuances of substrate ubiquitylation *in vivo*.

4.5.4
Arf and Other Modulators of Mdm2 Activity

Another member of the Mdm2–MdmX–p53 cast already mentioned above is Arf. This is a small basic protein (pI > 12) encoded by the *Ink4a* locus, which also encodes the cyclin-dependent kinase inhibitor p16Ink4a [314]. Shortly after their identification, both human and mouse Arf (p14Arf and p19Arf respectively) were found to interact with Mdm2 in a region N-terminal to the RING finger, between amino acids 235 and 289 in the central acid domain [315]. This interaction blocked Mdm2-mediated p53 degradation [316–319]. Inhibition of p53 degradation by Arf was the result of its direct inhibition of Mdm2 ubiquitin-ligase activity [320, 321]. However, Arf may also inhibit Mdm2 by promoting Mdm2-mediated ubiquitylation of MdmX [322], thereby lessening the activity of Mdm2 towards p53. Additionally, over-expressed Arf resides in the nucleolus and appears to reveal a nucleolar localization signal contained within the Mdm2 RING finger, although its binding site is N-terminal to the RING finger [294]. This change in Mdm2 localization correlates with inhibition of p53 degradation, presumably by separating Mdm2 from p53 or possibly by preventing nuclear export of the Mdm2–p53 complex [323–325]. However, relocation of Mdm2 to the nucleolus is apparently not essential for the inhibition of Mdm2 by Arf, although it may contribute to the suppression of Mdm2 under certain circumstances [326–328]. Further adding to the complexity of its function is the observation that Arf interacts with the Sumo E2 UbcH9, and mediates sumoylation of Mdm2 [329]. How sumoylation, which is frequently correlated with nuclear transport, might contribute to the observations obtained with Arf is another question to be answered.

Interacting with many other proteins, such as ribosomal components, steroid receptors and tumor suppressor gene products, provides additional ways to modulate the activity of Mdm2. Ribosomal proteins L5, L11 and L23 have all been found to bind Mdm2 and the Mdm2–p53 complex [330–332]. Enforced expression of L11 inhibits Mdm2-induced p53 ubiquitylation and degradation, leading to accumulation and activation of p53. L11 also stabilizes expressed Mdm2, suggesting that it may act by inhibiting the E3 activity of Mdm2 [333]. This prediction remains to be directly demonstrated through *in vitro* experiments. Low concentrations of actinomycin D disrupt ribosomal function and increase the levels of L11 bound to Mdm2. It is therefore conceivable that L11 plays an important role in stabilizing p53 in response to perturbations of ribosome integrity and activity. Tsg101 (*tumor susceptibility gene 101* product) is an E2-like (Uev) protein, which includes a UBC core structure that, like Mms2, lacks an active site cysteine. Tsg101 also binds to and regulates the function of Mdm2 [334]. Its over-expression results in an increase in Mdm2 and a reciprocal decrease of p53 in cells. Tsg101$^{-/-}$ mice accumulate p53 and exhibit early embryonic death [335]. Notably, Tsg101 plays an important role in endosomal trafficking and down-regulation of membrane receptors [336]. The significance of this in relation to Mdm2 regulation is unclear. Whether regulation of Mdm2 contributes to the proposed tumor-suppressor function of Tsg101.

The level and activity of Mdm2 are regulated by a variety of signals. The p53 inducer nitric oxide down-regulates Mdm2 at a post-transcriptional level [337]. Phosphorylaton plays major roles in Mdm2 regulation. In response to DNA damage, Atm (*ataxia-telangiectasia mutated*) phosphorylates Mdm2 on Ser395, impeding Mdm2-mediated nuclear export and degradation of p53 [338]. There is also evidence indicating that Atm activates c-Abl, which phosphorylates Mdm2 at Tyr394 and prevents its interaction with p53 [339]. However, phosphorylation may also enhance the activity of Mdm2. For example, the growth-factor-activated kinase Akt (*AKR* mouse strain *thymoma*) phosphorylates Mdm2 at Ser166 and Ser186, which promotes its nuclear translocation, leading to increased p53 ubiquitylation and degradation [340, 341]. This may contribute to the anti-apoptotic action of growth factors such as IGF1 and EGF. Consistent with this, the tumor suppressor Pten (*p*hosphatase and *ten*sin homolog), a phosphatase that dephosphorylates the Akt activator PIP3, protects p53 from Mdm2 and enhances p53-mediated transcription [342, 343]. Recently, Merlin, the product of *Neurofibromatosis 2* tumor-suppressor gene, was also found to down-regulate Mdm2. This led to inhibition of Mdm2-mediated p53 degradation and an increase of p53 transcriptional activity [344]. Therefore, blocking Mdm2 E3 activity toward p53 appears to be a common mechanism utilized by a number of tumor suppressors.

4.5.5
Other Potential Mdm2 Substrates

In addition to p53 and MdmX, a number of other proteins have been identified as potential Mdm2 substrates. Included among these are β-arrestin, β2-adrenergic receptor, androgen receptor, glucocorticoid receptor, histone acetyl transferase Tip60 and PCAF [322, 345–349]. While the physiological or pathological significance of the ubiquitylation of these proteins by Mdm2 remains to be further explored, these findings are consistent with the notion that Mdm2 has p53-independent functions in cells.

4.5.6
Mdm2 and Therapeutic Intervention in Cancer

Given its importance to cancer, a thorough understanding of how p53 levels and transcriptional activity are regulated is of practical significance. Accordingly, interventions that disrupt the capacity of Mdm2 to modulate both activity and levels of p53 become clinically important. Reagents have now been identified that block the physical interactions of these two proteins [350] and we and others have taken an interest in identifying small molecule inhibitors that might block Mdm2's E3 activity [106, 351, 352]. Whether reagents that inhibit Mdm2's ubiquitin-ligase activity will have therapeutic utility, especially if the self-ubiquitylation activity of Mdm2 is similarly inhibited, remains to be determined. The reason this becomes an issue is that such inhibition may result in the accumulation of p53 bound to Mdm2, which would be incapable of its crucial trans-activation functions. Thus, combinations of blockers, such as the recently identified Nutlins [353], which bind to p53 and block interactions with Mdm2 and *bona fide* specific Mdm2 E3 inhibitors are an attractive combination.

However, the complexity of p53 regulation increases the likelihood that therapies aimed at particular targets in this pathway could give unexpected results. p53 exists as a tetramer, each tetramer can potentially directly bind a combination of four Mdm2 and MdmX molecules and each of these has the potential to bind another Mdm2 or MdmX through RING finger-mediated dimerization. The stoichiometry of these p53–Mdm2–MdmX arrangements could potentially alter the balance between p53 stabilization and degradation. Adding to this complexity are the roles played by p300 and Pirh2 as well as other regulators such as Arf and ribosomal proteins. As if this were not enough, the issue of which E3 adds ubiquitin to p53 and how many ubiquitins each E3 adds awaits elucidation. Also, whether ubiquitin modification of p53 occurs in the nucleus, cytosol or both is unknown.

Underscoring the nuances involved in p53 regulation are studies on the deubiquitylating enzyme HAUSP (*h*erpesvirus-*a*ssociated *u*biquitin-*s*pecific *p*rotease) – newly published as this chapter was being completed. This deubiquitylating enzyme was first found to stabilize p53 presumably by reversing the effects of Mdm2 [354]. HAUSP has effects on p53 when its expression is ablated, either through RNA interference or in HAUSP null mice. However, the predominant effect is the opposite of what was expected. The primary target of HAUSP appears to be Mdm2 and not p53. Thus, loss of HAUSP stabilizes p53 and enhances the presumed self-ubiquitylation and consequent proteasomal degradation of Mdm2 [355, 356]. As with most things related to p53, each additional piece of information pertaining to control of its degradation alerts us to the complexities of regulating this "guardian of the genome".

4.6
Conclusion – Perspective

At the beginning of the 1990s there was only a rudimentary appreciation of the importance of protein degradation as a means to control protein levels in a temporally and spatially defined manner. We now understand that degradation of regulatory proteins plays important roles in almost all cellular processes. Similarly, degradation of misfolded proteins, unassembled proteins or proteins without useful functions is crucial to normal cellular processes. Further, an increasing number of disease states are found to be associated with dysfunction of these degradative processes. While the common final pathway by which most non-cell-surface proteins are degraded is via ubiquitin modification in the 26S proteasome, it is clear that ubiquitin ligases are indispensable regulatory arbiters for both proteasomal and lysosomal targeting. It is through their recognition and targeting of substrates that the destiny of proteins, and by extension the fate of the cell, is decided. Ubiquitn ligases, however, are proving to be even more important than simple arbiters of protein destruction. Non-degradative ubiquitin modification appears to play additional roles in kinase activation, DNA repair and other cellular pathways.

The largest class of E3s, by far, is represented by the RING finger and its structural relatives the PHD finger and the U-box. Among these, further division can be made between the multi-subunit cullin-containing complexes and those non-cullin E3s in which protein–protein interaction domains and the RING, PHD or U-box co-exist in the same polypeptide. Non-cullin E3s, which constitute the large majority of RING finger proteins, have the flexibility to recognize substrates either directly or in the context of protein complexes. In some cases, these RING finger proteins can target specific individual substrates. In other cases, such as CHIP and possibly Parkin, these E3s target multiple proteins through cellular quality control systems. Further, there is now evidence to suggest that E3s not usually thought of as being components of SCF complexes, such as Siah and Parkin, may under certain circumstances function in this manner. The interaction of ubiquitin ligases with specific subsets of E2s may contribute to the processivity and type of ubiquitin modification that the E3 generates. It is also apparent that, in many cases, common features can be found in E3s that contribute to their activities or specificities, such as heterologous Zn-binding domains, coiled-coil domains, UbDs and ubiquitin-binding domains. How the RING finger and each of these contributes to substrate selection, the type and length of ubiquitin chain formed, E2 interactions, and approximating the substrate–E3 complex with the proteasome or other cellular structures are all questions that require analysis. Further insights into the remarkable regulatory pathways mediated by ubiquitylation will emerge as we begin to develop an increased understanding of individual substrate–E3 pairs.

Acknowledgments

We wish to thank Rachael Klevitt, Peter Brzovic, Cyril Dominguez, Rolf Boelens and R. Andrew Byrd for their assistance in analyzing E2–E3 structural data.

References

1 HERSHKO, A. and A. CIECHANOVER, The ubiquitin system. *Annu Rev Biochem*, **1998**, *67*, 425–79.

2 WEISSMAN, A. M., Themes and variations on ubiquitylation. *Nat Rev Mol Cell Biol*, **2001**, *2(3)*, 169–78.

3 PICKART, C. M., Mechanisms underlying ubiquitination. *Annu Rev Biochem*, **2001**, *70*, 503–33.

4 GLICKMAN, M. H. and A. CIECHANOVER, The ubiquitin-proteasome proteolytic pathway: destruction for the sake of construction. *Physiol Rev*, **2002**, *82(2)*, 373–428.

5 FANG, S. and A. M. WEISSMAN, A field guide to ubiquitylation. *Cell Molec Life Sci*, **2004**, 61.

6 JENTSCH, S., The ubiquitin-conjugation system. *Annu Rev Genet*, **1992**, *26*, 179–207.

7 BARTEL, B., I. WUNNING, and A. VARSHAVSKY, The recognition component of the N-end rule pathway. *Embo J*, **1990**, *9(10)*, 3179–89.

8 VARSHAVSKY, A., The N-end rule: functions, mysteries, uses. *Proc Natl Acad Sci USA*, **1996**, *93(22)*, 12142–9.

9 VARSHAVSKY, A., The N-end rule. *Cell*, **1992**, *69(5)*, 725–35.

10 HERSHKO, A. and A. CIECHANOVER, Mechanisms of intracellular protein breakdown. *Annu Rev Biochem*, **1982**, *51*, 335–64.

11 HUIBREGTSE, J. M., M. SCHEFFNER, and P. M. HOWLEY, Cloning and expression of the cDNA for E6-AP, a protein that mediates the interaction of the human papillomavirus E6 oncoprotein with p53. *Mol Cell Biol*, **1993**, *13(2)*, 775–84.

12 SCHEFFNER, M., et al., The HPV-16 E6 and E6-AP complex functions as a ubiquitin-protein ligase in the ubiquitination of p53. *Cell*, **1993**, *75(3)*, 495–505.

13 HUIBREGTSE, J. M., et al., A family of proteins structurally and functionally related to the E6-AP ubiquitin-protein ligase. *Proc Natl Acad Sci USA*, **1995**, *92(7)*, 2563–7.

14 ALTSCHUL, S. F., et al., Basic local alignment search tool. *J Mol Biol*, **1990**, *215(3)*, 403–10.

15 ALTSCHUL, S. F., et al., Gapped BLAST and PSI-BLAST: a new generation of protein database search programs. *Nucleic Acids Res*, **1997**, *25(17)*, 3389–402.

16 SKOWYRA, D., et al., F-box proteins are receptors that recruit phosphorylated substrates to the SCF ubiquitin-ligase complex. *Cell*, **1997**, *91(2)*, 209–19.

17 BAI, C., et al., SKP1 connects cell cycle regulators to the ubiquitin proteolysis machinery through a novel motif, the F-box. *Cell*, **1996**, *86(2)*, 263–74.

18 KING, R. W., et al., A 20S complex containing CDC27 and CDC16 catalyzes the mitosis-specific conjugation of ubiquitin to cyclin B. *Cell*, **1995**, *81(2)*, 279–88.

19 SUDAKIN, V., et al., The cyclosome, a large complex containing cyclin-selective ubiquitin ligase activity, targets cyclins for destruction at the end of mitosis. *Mol Biol Cell*, **1995**, *6(2)*, 185–97.

20 HONDA, R., H. TANAKA, and H. YASUDA, Oncoprotein MDM2 is a ubiquitin ligase E3 for tumor suppressor p53. *FEBS Lett*, **1997**, *420(1)*, 25–7.

21 LORICK, K. L., et al., RING fingers mediate ubiquitin-conjugating enzyme (E2)-dependent ubiquitination. *Proc Natl Acad Sci USA*, **1999**, *96(20)*, 11364–9.

22 LOVERING, R., et al., Identification and preliminary characterization of a protein motif related to the zinc finger. *Proc Natl Acad Sci USA*, **1993**, *90(6)*, 2112–6.

23 ZHENG, N., et al., Structure of a c-Cbl-UbcH7 complex: RING domain function in ubiquitin-protein ligases. *Cell*, **2000**, *102(4)*, 533–9.

24 ZHENG, N., et al., Structure of the Cul1-Rbx1-Skp1-F boxSkp2 SCF ubiquitin ligase complex. *Nature*, **2002**, *416(6882)*, 703–9.

25 BRZOVIC, P. S., et al., Structure of a BRCA1-BARD1 heterodimeric RING-RING complex. *Nat Struct Biol*, **2001**, *8(10)*, 833–7.

26 BORDEN, K. L., et al., The solution structure of the RING finger domain from the acute promyelocytic leukaemia proto-oncoprotein PML. *Embo J*, **1995**, *14(7)*, 1532–41.

27 GERVAIS, V., et al., Solution structure of the N-terminal domain of the human TFIIH MAT1 subunit: new insights into the RING finger family. *J Biol Chem*, **2001**, *276(10)*, 7457–64.

28 HANZAWA, H., et al., The structure of the C4C4 ring finger of human NOT4 reveals features distinct from those of C3HC4 RING fingers. *J Biol Chem*, **2001**, *276(13)*, 10185–90.

29 BARLOW, P. N., et al., Structure of the C3HC4 domain by 1H-nuclear magnetic resonance spectroscopy. A new structural class of zinc-finger. *J Mol Biol*, **1994**, *237(2)*, 201–11.

30 YANG, Y., et al., Ubiquitin protein ligase activity of IAPs and their degradation in proteasomes in response to apoptotic stimuli. *Science*, **2000**, *288(5467)*, 874–7.

31 JOAZEIRO, C. A., et al., The tyrosine kinase negative regulator c-Cbl as a RING-type, E2-dependent ubiquitin-protein ligase. *Science*, **1999**, *286(5438)*, 309–12.

32 FANG, S., et al., Mdm2 is a RING finger-dependent ubiquitin protein ligase for itself and p53. *J Biol Chem*, **2000**, *275(12)*, 8945–51.

33 XIE, Y. and A. VARSHAVSKY, The E2-E3 interaction in the N-end rule pathway: the RING-H2 finger of E3 is required for the synthesis of multiubiquitin chain. *Embo J*, **1999**, *18(23)*, 6832–44.

34 BORDALLO, J. and D. H. WOLF, A RING-H2 finger motif is essential for the function of Der3/Hrd1 in endoplasmic reticulum associated protein degradation in the yeast Saccharomyces cerevisiae. *FEBS Lett*, **1999**, *448(2–3)*, 244–8.

35 LEVERSON, J. D., et al., The APC11 RING-H2 finger mediates E2-dependent ubiquitination. *Mol Biol Cell*, **2000**, *11(7)*, 2315–25.

36 SKOWYRA, D., et al., Reconstitution of G1 cyclin ubiquitination with complexes containing SCFGrr1 and Rbx1. *Science*, **1999**, *284(5414)*, 662–5.

37 OHTA, T., et al., ROC1, a homolog of APC11, represents a family of cullin partners with an associated ubiquitin ligase activity. *Mol Cell*, **1999**, *3(4)*, 535–41.

38 SEOL, J. H., et al., Cdc53/cullin and the essential Hrt1 RING-H2 subunit of SCF define a ubiquitin ligase module that activates the E2 enzyme Cdc34. *Genes Dev*, **1999**, *13(12)*, 1614–26.

39 TAN, P., et al., Recruitment of a ROC1-CUL1 ubiquitin ligase by Skp1 and HOS to catalyze the ubiquitination of I kappa B alpha. *Mol Cell*, **1999**, *3(4)*, 527–33.

40 KAMURA, T., et al., The Elongin BC complex interacts with the conserved SOCS-box motif present in members of the SOCS, ras, WD-40 repeat, and ankyrin repeat families. *Genes Dev*, **1998**, *12(24)*, 3872–81.

41 KAMURA, T., et al., Rbx1, a component of the VHL tumor suppressor complex and SCF ubiquitin ligase. *Science*, **1999**, *284(5414)*, 657–61.

42 MOYNIHAN, T. P., et al., The ubiquitin-conjugating enzymes UbcH7 and UbcH8 interact with RING finger/IBR motif-containing domains of HHARI and H7-AP1. *J Biol Chem*, **1999**, *274(43)*, 30963–8.

43 SCHINDLER, U., H. BECKMANN, and A. R. CASHMORE, HAT3.1, a novel Arabidopsis homeodomain protein containing a conserved cysteine-rich region. *Plant J*, **1993**, *4(1)*, 137–50.

44 SAHA, V., et al., The leukemia-associated-protein (LAP) domain, a cysteine-rich motif, is present in a wide range of proteins, including MLL, AF10, and MLLT6 proteins. *Proc Natl Acad Sci USA*, **1995**, *92(21)*, 9737–41.

45 ARAVIND, L. and E. V. KOONIN, The U box is a modified RING finger – a common domain in ubiquitination. *Curr Biol*, **2000**, *10(4)*, R132–4.

46 FANG, S., et al., RING finger ubiquitin protein ligases: implications for tumorigenesis, metastasis and for molecular targets in cancer. *Semin Cancer Biol*, **2003**, *13(1)*, 5–14.

47 HATAKEYAMA, S., et al., U box proteins as a new family of ubiquitin-protein ligases. *J Biol Chem*, **2001**, *276(35)*, 33111–20.

48 LU, Z., et al., The PHD domain of MEKK1 acts as an E3 ubiquitin ligase and mediates ubiquitination and degradation of ERK1/2. *Mol Cell*, **2002**, *9(5)*, 945–56.

49 UCHIDA, D., et al., AIRE Functions As an E3 Ubiquitin Ligase. *J Exp Med*, **2004**, *199(2)*, 167–72.

50 BONAME, J. M. and P. G. STEVENSON, MHC class I ubiquitination by a viral PHD/LAP finger protein. *Immunity*, **2001**, *15(4)*, 627–36.

51 COSCOY, L., D. J. SANCHEZ, and D. GANEM, A novel class of herpesvirus-encoded membrane-bound E3 ubiquitin ligases regulates endocytosis of proteins involved in immune recognition. *J Cell Biol*, **2001**, *155(7)*, 1265–73.

52 LISZTWAN, J., et al., Association of human CUL-1 and ubiquitin-conjugating enzyme CDC34 with the F-box protein p45(SKP2): evidence for evolutionary conservation in the subunit composition of the CDC34-SCF pathway. *Embo J*, **1998**, *17(2)*, 368–83.

53 PATTON, E. E., A. R. WILLEMS, and M. TYERS, Combinatorial control in ubiquitin-dependent proteolysis: don't Skp the F-box hypothesis. *Trends Genet*, **1998**, *14(6)*, 236–43.

54 SPRUCK, C. H. and H. M. STROHMAIER, Seek and destroy: SCF ubiquitin ligases in mammalian cell cycle control. *Cell Cycle*, **2002**, *1(4)*, 250–4.

55 DESHAIES, R. J., SCF and Cullin/Ring H2-based ubiquitin ligases. *Annu Rev Cell Dev Biol*, **1999**, *15*, 435–67.

56 KAMIZONO, S., et al., The SOCS box of SOCS-1 accelerates ubiquitin-dependent proteolysis of TEL-JAK2. *J Biol Chem*, **2001**, *276(16)*, 12530–8.

57 ZHANG, J. G., et al., The conserved SOCS box motif in suppressors of cytokine signaling binds to elongins B and C and may couple bound proteins to proteasomal degradation. *Proc Natl Acad Sci USA*, **1999**, *96(5)*, 2071–6.

58 GEYER, R., et al., BTB/POZ domain proteins are putative substrate adaptors for cullin 3 ubiquitin ligases. *Mol Cell*, **2003**, *12(3)*, 783–90.

59 VAN DEN HEUVEL, S., Protein degradation: CUL-3 and BTB – partners in proteolysis. *Curr Biol*, **2004**, *14(2)*, R59–61.

60 HARPER, J. W., J. L. BURTON, and M. J. SOLOMON, The anaphase promoting complex: it's not just for mitosis any more. *Genes Dev*, **2002**, *16(17)*, 2179–206.

61 DAS, A. K., P. W. COHEN, and D. BARFORD, The structure of the tetratricopeptide repeats of protein phosphatase 5: implications for TPR-mediated protein-protein interactions. *Embo J*, **1998**, *17(5)*, 1192–9.

62 PRINZ, S., et al., The regulation of Cdc20 proteolysis reveals a role for APC components Cdc23 and Cdc27 during S phase and early mitosis. *Curr Biol*, **1998**, *8(13)*, 750–60.

63 SHIRAYAMA, M., et al., APC(Cdc20) promotes exit from mitosis by destroying the anaphase inhibitor Pds1 and cyclin Clb5. *Nature*, **1999**, *402(6758)*, 203–7.

64 PFLEGER, C. M., E. LEE, and M. W. KIRSCHNER, Substrate recognition by the Cdc20 and Cdh1 components of the anaphase promoting complex. *Genes Dev*, **2001**, *15(18)*, 2396–407.

65 CHEN, A., et al., Autoubiquitination of the BRCA1*BARD1 RING ubiquitin ligase. *J Biol Chem*, **2002**, *277(24)*, 22085–92.

66 XIA, Y., et al., Enhancement of BRCA1 E3 ubiquitin ligase activity through direct interaction with the BARD1 protein. *J Biol Chem*, **2003**, *278(7)*, 5255–63.

67 MEZA, J. E., et al., Mapping the functional domains of BRCA1. Interaction of the ring finger domains of BRCA1 and BARD1. *J Biol Chem*, **1999**, *274(9)*, 5659–65.

68 BADCIONG, J. C. and A. L. HAAS, MdmX is a RING finger ubiquitin ligase capable of synergistically enhancing Mdm2 ubiquitination. *J Biol Chem*, **2002**, *277(51)*, 49668–75.

69 WU, P. Y., et al., A conserved catalytic residue in the ubiquitin-conjugating enzyme family. *Embo J*, **2003**, *22(19)*, 5241–50.

70 DOMINGUEZ, C., et al., Structural model of the UbcH5B/CNOT4 complex revealed by combining NMR, mutagenesis, and docking approaches. *Structure (Camb)*, **2004**, *12(4)*, 633–44.

71 BIEDERER, T., C. VOLKWEIN, and T. SOMMER, Role of Cue1p in ubiquitination and degradation at the ER surface. *Science*, **1997**, *278(5344)*, 1806–9.

72 FANG, S., et al., The tumor autocrine motility factor receptor, gp78, is a ubiquitin protein ligase implicated in degradation from the endoplasmic reticulum. *Proc Natl Acad Sci USA*, **2001**, *98(25)*, 14422–7.

73 BOWCOCK, A. M., Molecular cloning of BRCA1: a gene for early onset familial breast and ovarian cancer. *Breast Cancer Res Treat*, **1993**, *28(2)*, 121–35.

74 STRUEWING, J. P., et al., The risk of cancer associated with specific mutations of BRCA1 and BRCA2 among Ashkenazi Jews. *N Engl J Med*, **1997**, *336(20)*, 1401–8.

75 GAYTHER, S. A., et al., Variation of risks of breast and ovarian cancer associated with different germline mutations of the BRCA2 gene. *Nat Genet*, **1997**, *15(1)*, 103–5.

76 SHATTUCK-EIDENS, D., et al., A collaborative survey of 80 mutations in the BRCA1 breast and ovarian cancer susceptibility gene. Implications for presymptomatic testing and screening. *Jama*, **1995**, *273(7)*, 535–41.

77 BRZOVIC, P. S., et al., Binding and recognition in the assembly of an active BRCA1/BARD1 ubiquitin-ligase complex. *Proc Natl Acad Sci USA*, **2003**, *100(10)*, 5646–51.

78 HAAS, A. L. and T. J. SIEPMANN, Pathways of ubiquitin conjugation. *Faseb J*, **1997**, *11(14)*, 1257–68.

79 DEFFENBAUGH, A. E., et al., Release of ubiquitin-charged Cdc34-S – Ub from the RING domain is essential for ubiquitination of the SCF(Cdc4)-bound substrate Sic1. *Cell*, **2003**, *114(5)*, 611–22.

80 HAKLI, M., et al., Transcriptional coregulator SNURF (RNF4) possesses

ubiquitin E3 ligase activity. *FEBS Lett,* **2004,** *560(1–3)*, 56–62.

81 HASHIZUME, R., et al., The RING heterodimer BRCA1-BARD1 is a ubiquitin ligase inactivated by a breast cancer-derived mutation. *J Biol Chem,* **2001,** *276(18)*, 14537–40.

82 THROWER, J. S., et al., Recognition of the polyubiquitin proteolytic signal. *Embo J,* **2000,** *19(1)*, 94–102.

83 NISHIKAWA, H., et al., Mass spectro-metric and mutational analyses reveal Lys-6-linked polyubiquitin chains catalyzed by BRCA1-BARD1 ubiq-uitin ligase. *J Biol Chem,* **2004,** *279(6)*, 3916–24.

84 HOFMANN, R. M. and C. M. PICKART, Noncanonical MMS2-encoded ubiquitin-conjugating enzyme functions in assembly of novel polyubiquitin chains for DNA repair. *Cell,* **1999,** *96(5)*, 645–53.

85 ULRICH, H. D. and S. JENTSCH, Two Ring-finger proteins mediate cooperation between ubiquitin-conjugating enzymes in DNA repair. *Embo J,* **2000,** *19(13)*, 3388–97.

86 KAPLUN, L., et al., Functions of the DNA damage response pathway target Ho endonuclease of yeast for degradation via the ubiquitin-26S proteasome system. *Proc Natl Acad Sci USA,* **2000,** *97(18)*, 10077–82.

87 SMIT, L. and J. BORST, The Cbl family of signal transduction molecules. *Crit Rev Oncog,* **1997,** *8(4)*, 359–79.

88 HOHENSTEIN, P. and R. H. GILES, BRCA1: a scaffold for p53 response? *Trends Genet,* **2003,** *19(9)*, 489–94.

89 JENSEN, D. E., et al., BAP1: a novel ubiquitin hydrolase which binds to the BRCA1 RING finger and enhances BRCA1-mediated cell growth suppression. *Oncogene,* **1998,** *16(9)*, 1097–112.

90 ANDERSON, S. F., et al., BRCA1 protein is linked to the RNA polymerase II holoenzyme complex via RNA helicase A. *Nat Genet,* **1998,** *19(3)*, 254–6.

91 VAN DYCK, E., et al., Binding of double-strand breaks in DNA by human Rad52 protein. *Nature,* **1999,** *398(6729)*, 728–31.

92 SCULLY, R., et al., Association of BRCA1 with Rad51 in mitotic and meiotic cells. *Cell,* **1997,** *88(2)*, 265–75.

93 WANG, Q., et al., BRCA1 binds c-Myc and inhibits its transcriptional and transforming activity in cells. *Oncogene,* **1998,** *17(15)*, 1939–48.

94 JEFFORD, C. E., et al., Nuclear-cytoplasmic translocation of BARD1 is linked to its apoptotic activity. *Oncogene,* **2004,** *23(20)*, 3509–20.

95 MISHIMA, K., et al., ARD 1, a 64-kDa guanine nucleotide-binding protein with a carboxyl-terminal ADP-ribosylation factor domain. *J Biol Chem,* **1993,** *268(12)*, 8801–7.

96 VITALE, N., J. Moss, and M. VAUGHAN, Interaction of the GTP-binding and GTPase-activating domains of ARD1 involves the effector region of the ADP-ribosylation factor domain. *J Biol Chem,* **1997,** *272(7)*, 3897–904.

97 DIDIER, C., et al., RNF5, a Ring-finger protein that regulates cell motility by targeting paxillin ubiquitination and altered localization. *Mol Cell Biol,* **2003,** *23(15)*, 5331–45.

98 HAMPTON, R. Y., R. G. GARDNER, and J. RINE, Role of 26S proteasome and HRD genes in the degradation of 3-hydroxy-3-methylglutaryl-CoA reductase, an integral endoplasmic reticulum membrane protein. *Mol Biol Cell,* **1996,** *7(12)*, 2029–44.

99 KOSTOVA, Z. and D. H. WOLF, For whom the bell tolls: protein quality control of the endoplasmic reticulum and the ubiquitin-proteasome connection. *Embo J,* **2003,** *22(10)*, 2309–17.

100 LIANG, J. S., et al., Overexpression of the tumor autocrine motility factor receptor Gp78, a ubiquitin protein ligase, results in increased ubiqui-tinylation and decreased secretion of apolipoprotein B100 in HepG2 cells. *J Biol Chem,* **2003,** *278(26)*, 23984–8.

101 SAKATA, E., et al., Parkin binds the Rpn10 subunit of 26S proteasomes through its ubiquitin-like domain. *EMBO Rep,* **2003,** *4(3)*, 301–6.

102 SAEKI, Y., et al., Identification of

ubiquitin-like protein-binding subunits of the 26S proteasome. *Biochem Biophys Res Commun*, **2002**, *296(4)*, 813–9.

103 LEGGETT, D. S., et al., Multiple associated proteins regulate proteasome structure and function. *Mol Cell*, **2002**, *10(3)*, 495–507.

104 JIMENEZ, G. S., et al., p53 regulation by post-translational modification and nuclear retention in response to diverse stresses. *Oncogene*, **1999**, *18(53)*, 7656–65.

105 WANG, X., A. ZALCENSTEIN, and M. OREN, Nitric oxide promotes p53 nuclear retention and sensitizes neuroblastoma cells to apoptosis by ionizing radiation. *Cell Death Differ*, **2003**, *10(4)*, 468–76.

106 YANG, Y., C. C. LI, and A. M. WEISSMAN, Regulating the p53 system through ubiquitination. *Oncogene*, **2004**, *23(11)*, 2096–106.

107 THIEN, C. B. and W. Y. LANGDON, Cbl: many adaptations to regulate protein tyrosine kinases. *Nat Rev Mol Cell Biol*, **2001**, *2(4)*, 294–307.

108 SCHEEL, H. and K. HOFMANN, No evidence for PHD fingers as ubiquitin ligases. *Trends Cell Biol*, **2003**, *13(6)*, 285–7; author reply 287–8.

109 FRUH, K., et al., Immune evasion by a novel family of viral PHD/LAP-finger proteins of gamma-2 herpesviruses and poxviruses. *Virus Res*, **2002**, *88(1–2)*, 55–69.

110 BARTEE, E., et al., Downregulation of major histocompatibility complex class I by human ubiquitin ligases related to viral immune evasion proteins. *J Virol*, **2004**, *78(3)*, 1109–20.

111 CONNELL, P., et al., The co-chaperone CHIP regulates protein triage decisions mediated by heat-shock proteins. *Nat Cell Biol*, **2001**, *3(1)*, 93–6.

112 JIANG, J., et al., CHIP is a U-box-dependent E3 ubiquitin ligase: identification of Hsc70 as a target for ubiquitylation. *J Biol Chem*, **2001**, *276(46)*, 42938–44.

113 MEACHAM, G. C., et al., The Hsc70 co-chaperone CHIP targets immature CFTR for proteasomal degradation. *Nat Cell Biol*, **2001**, *3(1)*, 100–5.

114 KOEGL, M., et al., A novel ubiquitination factor, E4, is involved in multi-ubiquitin chain assembly. *Cell*, **1999**, *96(5)*, 635–44.

115 PATTERSON, C., A new gun in town: the U box is a ubiquitin ligase domain. *Sci STKE*, **2002**, *2002(116)*, PE4.

116 MISRA, S., G. J. MILLER, and J. H. HURLEY, Recognizing phosphatidylinositol 3-phosphate. *Cell*, **2001**, *107(5)*, 559–62.

117 HURLEY, J. H. and T. MEYER, Subcellular targeting by membrane lipids. *Curr Opin Cell Biol*, **2001**, *13(2)*, 146–52.

118 HURLEY, J. H. and S. MISRA, Signaling and subcellular targeting by membrane-binding domains. *Annu Rev Biophys Biomol Struct*, **2000**, *29*, 49–79.

119 MISRA, S. and J. H. HURLEY, Crystal structure of a phosphatidylinositol 3-phosphate-specific membrane-targeting motif, the FYVE domain of Vps27p. *Cell*, **1999**, *97(5)*, 657–66.

120 VOM BAUR, E., et al., Differential ligand-dependent interactions between the AF-2 activating domain of nuclear receptors and the putative transcriptional intermediary factors mSUG1 and TIF1. *Embo J*, **1996**, *15(1)*, 110–24.

121 URANO, T., et al., Efp targets 14–3–3 sigma for proteolysis and promotes breast tumour growth. *Nature*, **2002**, *417(6891)*, 871–5.

122 HORIE, K., et al., Estrogen-responsive Ring-finger protein controls breast cancer growth. *J Steroid Biochem Mol Biol*, **2003**, *85(2–5)*, 101–4.

123 FRIDELL, R. A., et al., Identification of a novel human zinc finger protein that specifically interacts with the activation domain of lentiviral Tat proteins. *Virology*, **1995**, *209(2)*, 347–57.

124 TISSOT, C. and N. MECHTI, Molecular cloning of a new interferon-induced factor that represses human immunodeficiency virus type 1 long terminal repeat expression. *J Biol Chem*, **1995**, *270(25)*, 14891–8.

125 REYMOND, A., et al., The tripartite motif family identifies cell compartments. *Embo J*, **2001**, *20(9)*, 2140–51.

126 KALLIJARVI, J., A. E. LEHESJOKI, and M. LIPSANEN-NYMAN, Mulibrey nanism – a novel peroxisomal disorder. *Adv Exp Med Biol*, **2003**, *544*, 31–7.

127 HAMALAINEN, R. H., et al., Novel mutations in the TRIM37 gene in Mulibrey Nanism. *Hum Mutat*, **2004**, *23(5)*, 522.

128 ZAPATA, J. M., et al., A diverse family of proteins containing tumor necrosis factor receptor-associated factor domains. *J Biol Chem*, **2001**, *276(26)*, 24242–52.

129 PERRY, J., et al., FXY2/MID2, a gene related to the X-linked Opitz syndrome gene FXY/MID1, maps to Xq22 and encodes a FNIII domain-containing protein that associates with microtubules. *Genomics*, **1999**, *62(3)*, 385–94.

130 TROCKENBACHER, A., et al., MID1, mutated in Opitz syndrome, encodes an ubiquitin ligase that targets phosphatase 2A for degradation. *Nat Genet*, **2001**, *29(3)*, 287–94.

131 BODINE, S. C., et al., Identification of ubiquitin ligases required for skeletal muscle atrophy. *Science*, **2001**, *294(5547)*, 1704–8.

132 DOUCAS, V., The promyelocytic (PML) nuclear compartment and transcription control. *Biochem Pharmacol*, **2000**, *60(8)*, 1197–201.

133 SALOMONI, P. and P. P. PANDOLFI, The role of PML in tumor suppression. *Cell*, **2002**, *108(2)*, 165–70.

134 LANGLEY, E., et al., Human SIR2 deacetylates p53 and antagonizes PML/p53-induced cellular senescence. *Embo J*, **2002**, *21(10)*, 2383–96.

135 LOURIA-HAYON, I., et al., The Promyelocytic Leukemia Protein Protects p53 from Mdm2-mediated Inhibition and Degradation. *J Biol Chem*, **2003**, *278(35)*, 33134–41.

136 TANG, A. H., et al., PHYL acts to down-regulate TTK88, a transcriptional repressor of neuronal cell fates, by a SINA-dependent mechanism. *Cell*, **1997**, *90(3)*, 459–67.

137 HU, G. and E. R. FEARON, Siah-1 N-terminal RING domain is required for proteolysis function, and C-terminal sequences regulate oligomerization and binding to target proteins. *Mol Cell Biol*, **1999**, *19(1)*, 724–32.

138 MATSUZAWA, S., et al., p53-inducible human homologue of Drosophila seven in absentia (Siah) inhibits cell growth: suppression by BAG-1. *Embo J*, **1998**, *17(10)*, 2736–47.

139 GERMANI, A., et al., hSiah2 is a new Vav binding protein which inhibits Vav-mediated signaling pathways. *Mol Cell Biol*, **1999**, *19(5)*, 3798–807.

140 RELAIX, F., et al., Pw1/Peg3 is a potential cell death mediator and cooperates with Siah1a in p53-mediated apoptosis. *Proc Natl Acad Sci USA*, **2000**, *97(5)*, 2105–10.

141 XIE, Q., et al., SINAT5 promotes ubiquitin-related degradation of NAC1 to attenuate auxin signals. *Nature*, **2002**, *419(6903)*, 167–70.

142 POLEKHINA, G., et al., Siah ubiquitin ligase is structurally related to TRAF and modulates TNF-alpha signaling. *Nat Struct Biol*, **2002**, *9(1)*, 68–75.

143 KEINO-MASU, K., et al., Deleted in Colorectal Cancer (DCC) encodes a netrin receptor. *Cell*, **1996**, *87(2)*, 175–85.

144 BOEHM, J., et al., Regulation of BOB.1/OBF.1 stability by Siah. *Embo J*, **2001**, *20(15)*, 4153–62.

145 HOUSE, C. M., et al., A binding motif for Siah ubiquitin ligase. *Proc Natl Acad Sci USA*, **2003**, *100(6)*, 3101–6.

146 NAGANO, Y., et al., Siah-1 facilitates ubiquitination and degradation of synphilin-1. *J Biol Chem*, **2003**, *278(51)*, 51504–14.

147 FREW, I. J., et al., Generation and analysis of Siah2 mutant mice. *Mol Cell Biol*, **2003**, *23(24)*, 9150–61.

148 LIU, J., et al., Siah-1 mediates a novel beta-catenin degradation pathway linking p53 to the adenomatous polyposis coli protein. *Mol Cell*, **2001**, *7(5)*, 927–36.

149 BRUZZONI-GIOVANELLI, H., et al., Siah-1 inhibits cell growth by altering the mitotic process. *Oncogene*, **1999**, *18(50)*, 7101–9.

150 DICKINS, R. A., et al., The ubiquitin ligase component Siah1a is required for completion of meiosis I in male

mice. *Mol Cell Biol*, **2002**, *22(7)*, 2294–303.

151 HABELHAH, H., et al., Stress-induced decrease in TRAF2 stability is mediated by Siah2. *Embo J*, **2002**, *21(21)*, 5756–65.

152 NAKAYAMA, K., et al., Siah2 regulates stability of prolyl-hydroxylases, controls HIF1a abundance and modulates physiological responses to hypoxia. *Cell*, **2004**, *117(7)*, 941–52.

153 SALVESEN, G. S. and C. S. DUCKETT, IAP proteins: blocking the road to death's door. *Nat Rev Mol Cell Biol*, **2002**, *3(6)*, 401–10.

154 SHI, Y., A conserved tetrapeptide motif: potentiating apoptosis through IAP-binding. *Cell Death Differ*, **2002**, *9(2)*, 93–5.

155 DU, C., et al., Smac, a mitochondrial protein that promotes cytochrome c-dependent caspase activation by eliminating IAP inhibition. *Cell*, **2000**, *102(1)*, 33–42.

156 VERHAGEN, A. M., et al., Identification of DIABLO, a mammalian protein that promotes apoptosis by binding to and antagonizing IAP proteins. *Cell*, **2000**, *102(1)*, 43–53.

157 SUZUKI, Y., et al., A serine protease, HtrA2, is released from the mitochondria and interacts with XIAP, inducing cell death. *Mol Cell*, **2001**, *8(3)*, 613–21.

158 HAY, B. A., D. A. WASSARMAN, and G. M. RUBIN, Drosophila homologs of baculovirus inhibitor of apoptosis proteins function to block cell death. *Cell*, **1995**, *83(7)*, 1253–62.

159 VUCIC, D., W. J. KAISER, and L. K. MILLER, Inhibitor of apoptosis proteins physically interact with and block apoptosis induced by Drosophila proteins HID and GRIM. *Mol Cell Biol*, **1998**, *18(6)*, 3300–9.

160 WANG, S. L., et al., The Drosophila caspase inhibitor DIAP1 is essential for cell survival and is negatively regulated by HID. *Cell*, **1999**, *98(4)*, 453–63.

161 SUZUKI, Y., Y. NAKABAYASHI, and R. TAKAHASHI, Ubiquitin-protein ligase activity of X-linked inhibitor of apoptosis protein promotes proteasomal degradation of caspase-3 and enhances its anti-apoptotic effect in Fas-induced cell death. *Proc Natl Acad Sci USA*, **2001**, *98(15)*, 8662–7.

162 MACFARLANE, M., et al., Proteasome-mediated Degradation of Smac during Apoptosis: XIAP Promotes Smac Ubiquitination in Vitro. *J Biol Chem*, **2002**, *277(39)*, 36611–6.

163 WILSON, R., et al., The DIAP1 RING finger mediates ubiquitination of Dronc and is indispensable for regulating apoptosis. *Nat Cell Biol*, **2002**, *4(6)*, 445–50.

164 ROTHE, M., et al., The TNFR2-TRAF signaling complex contains two novel proteins related to baculoviral inhibitor of apoptosis proteins. *Cell*, **1995**, *83(7)*, 1243–52.

165 ARCH, R. H., R. W. GEDRICH, and C. B. THOMPSON, Tumor necrosis factor receptor-associated factors (TRAFs) – a family of adapter proteins that regulates life and death. *Genes Dev*, **1998**, *12(18)*, 2821–30.

166 CHUNG, J. Y., et al., All TRAFs are not created equal: common and distinct molecular mechanisms of TRAF-mediated signal transduction. *J Cell Sci*, **2002**, *115(Pt 4)*, 679–88.

167 DENG, L., et al., Activation of the IkappaB kinase complex by TRAF6 requires a dimeric ubiquitin-conjugating enzyme complex and a unique polyubiquitin chain. *Cell*, **2000**, *103(2)*, 351–61.

168 WANG, C., et al., TAK1 is a ubiquitin-dependent kinase of MKK and IKK. *Nature*, **2001**, *412(6844)*, 346–51.

169 MORIGUCHI, T., et al., A novel kinase cascade mediated by mitogen-activated protein kinase kinase 6 and MKK3. *J Biol Chem*, **1996**, *271(23)*, 13675–9.

170 WANG, W., et al., Activation of the hematopoietic progenitor kinase-1 (HPK1)-dependent, stress-activated c-Jun N-terminal kinase (JNK) pathway by transforming growth factor beta (TGF-beta)-activated kinase (TAK1), a kinase mediator of TGF beta signal transduction. *J Biol Chem*, **1997**, *272(36)*, 22771–5.

171 BROWN, K. D., B. S. HOSTAGER, and G. A. BISHOP, Regulation of TRAF2 signaling by self-induced degradation. *J Biol Chem*, **2002**, *277(22)*, 19433–8.

172 BROWN, K. D., B. S. HOSTAGER, and G. A. BISHOP, Differential signaling and tumor necrosis factor receptor-associated factor (TRAF) degradation mediated by CD40 and the Epstein-Barr virus oncoprotein latent membrane protein 1 (LMP1). *J Exp Med*, **2001**, *193(8)*, 943–54.

173 LI, X., Y. YANG, and J. D. ASHWELL, TNF-RII and c-IAP1 mediate ubiquitination and degradation of TRAF2. *Nature*, **2002**, *416(6878)*, 345–7.

174 SHI, C. S. and J. H. KEHRL, Tumor necrosis factor (TNF)-induced germinal center kinase-related (GCKR) and stress-activated protein kinase (SAPK) activation depends upon the E2/E3 complex Ubc13-Uev1A/TNF receptor-associated factor 2 (TRAF2). *J Biol Chem*, **2003**, *278(17)*, 15429–34.

175 HABELHAH, H., et al., Ubiquitination and translocation of TRAF2 is required for activation of JNK but not of p38 or NF-kappaB. *Embo J*, **2004**, *23(2)*, 322–32.

176 LANGDON, W. Y., et al., v-cbl, an oncogene from a dual-recombinant murine retrovirus that induces early B-lineage lymphomas. *Proc Natl Acad Sci USA*, **1989**, *86(4)*, 1168–72.

177 JONGEWARD, G. D., T. R. CLANDININ, and P. W. STERNBERG, sli-1, a negative regulator of let-23-mediated signaling in C. elegans. *Genetics*, **1995**, *139(4)*, 1553–66.

178 LANGDON, W. Y., et al., The c-cbl proto-oncogene is preferentially expressed in thymus and testis tissue and encodes a nuclear protein. *J Virol*, **1989**, *63(12)*, 5420–4.

179 KEANE, M. M., et al., Cloning and characterization of cbl-b: a SH3 binding protein with homology to the c-cbl proto-oncogene. *Oncogene*, **1995**, *10(12)*, 2367–77.

180 KEANE, M. M., et al., cbl-3: a new mammalian cbl family protein. *Oncogene*, **1999**, *18(22)*, 3365–75.

181 KIM, M., et al., Molecular cloning and characterization of a novel cbl-family gene, cbl-c. *Gene*, **1999**, *239(1)*, 145–54.

182 BARTKIEWICZ, M., A. HOUGHTON, and R. BARON, Leucine zipper-mediated homodimerization of the adaptor protein c-Cbl. A role in c-Cbl's tyrosine phosphorylation and its association with epidermal growth factor receptor. *J Biol Chem*, **1999**, *274(43)*, 30887–95.

183 DAVIES, G. C., et al., Cbl-b interacts with ubiquitinated proteins; differential functions of the UBA domains of c-Cbl and Cbl-b. *Oncogene*, **2004**, *23*, 7104–7115.

184 WATERMAN, H., et al., The RING finger of c-Cbl mediates desensitization of the epidermal growth factor receptor. *J Biol Chem*, **1999**, *274(32)*, 22151–4.

185 LEVKOWITZ, G., et al., Ubiquitin ligase activity and tyrosine phosphorylation underlie suppression of growth factor signaling by c-Cbl/Sli-1. *Mol Cell*, **1999**, *4(6)*, 1029–40.

186 YOKOUCHI, M., et al., Ligand-induced ubiquitination of the epidermal growth factor receptor involves the interaction of the c-Cbl RING finger and UbcH7. *J Biol Chem*, **1999**, *274(44)*, 31707–12.

187 RAO, N., et al., Negative regulation of Lck by Cbl ubiquitin ligase. *Proc Natl Acad Sci USA*, **2002**, *99(6)*, 3794–9.

188 HOWLETT, C. J. and S. M. ROBBINS, Membrane-anchored Cbl suppresses Hck protein-tyrosine kinase mediated cellular transformation. *Oncogene*, **2002**, *21(11)*, 1707–16.

189 RAO, N., I. DODGE, and H. BAND, The Cbl family of ubiquitin ligases: critical negative regulators of tyrosine kinase signaling in the immune system. *J Leukoc Biol*, **2002**, *71(5)*, 753–63.

190 TZENG, S. R., et al., Solution structure of the human BTK SH3 domain complexed with a proline-rich peptide from p120cbl. *J Biomol NMR*, **2000**, *16(4)*, 303–12.

191 SOUBEYRAN, P., et al., Cbl-CIN85-endophilin complex mediates ligand-induced downregulation of EGF receptors. *Nature*, **2002**, *416(6877)*, 183–7.

192 PETRELLI, A., et al., The endophilin-CIN85-Cbl complex mediates ligand-

dependent downregulation of c-Met. *Nature*, **2002**, *416(6877)*, 187–90.

193 WANG, Y., et al., c-Cbl is transiently tyrosine-phosphorylated, ubiquitin- ated, and membrane-targeted following CSF-1 stimulation of macrophages. *J Biol Chem*, **1996**, *271(1)*, 17–20.

194 LEVKOWITZ, G., et al., c-Cbl/Sli-1 regulates endocytic sorting and ubiquitination of the epidermal growth factor receptor. *Genes Dev*, **1998**, *12(23)*, 3663–74.

195 JEHN, B. M., et al., c-Cbl binding and ubiquitin-dependent lysosomal degradation of membrane-associated Notch1. *J Biol Chem*, **2002**, *277(10)*, 8033–40.

196 DE MELKER, A. A., et al., c-Cbl ubiquitinates the EGF receptor at the plasma membrane and remains receptor associated throughout the endocytic route. *J Cell Sci*, **2001**, *114(Pt 11)*, 2167–78.

197 MOSESSON, Y., et al., Endocytosis of receptor tyrosine kinases is driven by monoubiquitylation, not polyubiquit- ylation. *J Biol Chem*, **2003**, *278(24)*, 21323–6.

198 ETTENBERG, S. A., et al., Cbl-b- dependent coordinated degradation of the epidermal growth factor receptor signaling complex. *J Biol Chem*, **2001**, *276(29)*, 27677–84.

199 ANDONIOU, C. E., et al., The Cbl proto-oncogene product negatively regulates the Src-family tyrosine kinase Fyn by enhancing its degrada- tion. *Mol Cell Biol*, **2000**, *20(3)*, 851–67.

200 KIM, M., et al., Cbl-c suppresses v-Src- induced transformation through ubiquitin-dependent protein degrada- tion. *Oncogene*, **2004**, *23(9)*, 1645–55.

201 FANG, D., et al., Cbl-b, a RING-type E3 ubiquitin ligase, targets phosphatidy- linositol 3-kinase for ubiquitination in T cells. *J Biol Chem*, **2001**, *276(7)*, 4872–8.

202 MAGNIFICO, A., et al., WW domain HECT E3s target Cbl RING finger E3s for proteasomal degradation. *J Biol Chem*, **2003**, *278(44)*, 43169–77.

203 ALWAN, H. A., E. J. VAN ZOELEN, and J. E. VAN LEEUWEN, Ligand-induced lysosomal epidermal growth factor receptor (EGFR) degradation is preceded by proteasome-dependent EGFR de-ubiquitination. *J Biol Chem*, **2003**, *278(37)*, 35781–90.

204 LONGVA, K. E., et al., Ubiquitination and proteasomal activity is required for transport of the EGF receptor to inner membranes of multivesicular bodies. *J Cell Biol*, **2002**, *156(5)*, 843– 54.

205 NARAMURA, M., et al., c-Cbl and Cbl-b regulate T cell responsiveness by promoting ligand-induced TCR down- modulation. *Nat Immunol*, **2002**, *3(12)*, 1192–9.

206 NARAMURA, M., et al., Altered thymic positive selection and intracellular signals in Cbl-deficient mice. *Proc Natl Acad Sci USA*, **1998**, *95(26)*, 15547– 52.

207 BACHMAIER, K., et al., Negative regulation of lymphocyte activation and autoimmunity by the molecular adaptor Cbl-b. *Nature*, **2000**, *403(6766)*, 211–6.

208 WU, L. C., et al., Identification of a RING protein that can interact in vivo with the BRCA1 gene product. *Nat Genet*, **1996**, *14(4)*, 430–40.

209 STAD, R., et al., Mdmx stabilizes p53 and Mdm2 via two distinct mechanisms. *EMBO Rep*, **2001**, *2(11)*, 1029–34.

210 KENTSIS, A., R. E. GORDON, and K. L. BORDEN, Self-assembly properties of a model RING domain. *Proc Natl Acad Sci USA*, **2002**, *99(2)*, 667–72.

211 ITOH, M., et al., Mind bomb is a ubiquitin ligase that is essential for efficient activation of Notch signaling by Delta. *Dev Cell*, **2003**, *4(1)*, 67–82.

212 TAN, N. G., et al., Human homologue of ariadne promotes the ubiquitylation of translation initiation factor 4E homologous protein, 4EHP. *FEBS Lett*, **2003**, *554(3)*, 501–4.

213 NIKOLAEV, A. Y., et al., Parc: a cytoplasmic anchor for p53. *Cell*, **2003**, *112(1)*, 29–40.

214 ITO, T., et al., Dorfin localizes to Lewy bodies and ubiquitylates synphilin-1. *J Biol Chem*, **2003**, *278(31)*, 29106–14.

215 NIWA, J., et al., Dorfin ubiquitylates mutant SOD1 and prevents mutant

SOD1-mediated neurotoxicity. *J Biol Chem*, **2002**, *277(39)*, 36793–8.

216 WATANABE, M., et al., Histological evidence of protein aggregation in mutant SOD1 transgenic mice and in amyotrophic lateral sclerosis neural tissues. *Neurobiol Dis*, **2001**, *8(6)*, 933–41.

217 JOHNSTON, J. A., et al., Formation of high molecular weight complexes of mutant Cu, Zn-superoxide dismutase in a mouse model for familial amyotrophic lateral sclerosis. *Proc Natl Acad Sci USA*, **2000**, *97(23)*, 12571–6.

218 STIEBER, A., J. O. GONATAS, and N. K. GONATAS, Aggregates of mutant protein appear progressively in dendrites, in periaxonal processes of oligodendrocytes, and in neuronal and astrocytic perikarya of mice expressing the SOD1(G93A) mutation of familial amyotrophic lateral sclerosis. *J Neurol Sci*, **2000**, *177(2)*, 114–23.

219 WANG, J., G. XU, and D. R. BORCHELT, High molecular weight complexes of mutant superoxide dismutase 1: age-dependent and tissue-specific accumulation. *Neurobiol Dis*, **2002**, *9(2)*, 139–48.

220 IWAI, K., An ubiquitin ligase recognizing a protein oxidized by iron: implications for the turnover of oxidatively damaged proteins. *J Biochem (Tokyo)*, **2003**, *134(2)*, 175–82.

221 YAMANAKA, K., et al., Identification of the ubiquitin-protein ligase that recognizes oxidized IRP2. *Nat Cell Biol*, **2003**, *5(4)*, 336–40.

222 CAPILI, A. D., et al., Structure of the C-terminal RING Finger from a RING-IBR-RING/TRIAD Motif Reveals a Novel Zinc-binding Domain Distinct from a RING. *J Mol Biol*, **2004**, *340(5)*, 1117–29.

223 ARDLEY, H. C., et al., Features of the parkin/ariadne-like ubiquitin ligase, HHARI, that regulate its interaction with the ubiquitin-conjugating enzyme, Ubch7. *J Biol Chem*, **2001**, *276(22)*, 19640–7.

224 RANKIN, C. A., et al., E3 ubiquitin-protein ligase activity of Parkin is dependent on cooperative interaction

of RING finger (TRIAD) elements. *J Biomed Sci*, **2001**, *8(5)*, 421–9.

225 ZHANG, Y., et al., Parkin functions as an E2-dependent ubiquitin-protein ligase and promotes the degradation of the synaptic vesicle-associated protein, CDCrel-1 [In Process Citation]. *Proc Natl Acad Sci USA*, **2000**, *97(24)*, 13354–9.

226 ITO, K., et al., N-Terminally extended human ubiquitin-conjugating enzymes (E2s) mediate the ubiquitination of RING finger proteins, ARA54 and RNF8. *Eur J Biochem*, **2001**, *268(9)*, 2725–32.

227 SHIMURA, H., et al., Familial Parkinson disease gene product, parkin, is a ubiquitin-protein ligase. *Nat Genet*, **2000**, *25(3)*, 302–5.

228 IMAI, Y., et al., An unfolded putative transmembrane polypeptide, which can lead to endoplasmic reticulum stress, is a substrate of Parkin. *Cell*, **2001**, *105(7)*, 891–902.

229 WEST, A. B. and N. T. MAIDMENT, *Genetics* of parkin-linked disease. *Hum Genet*. **2004**, *114*, 327–336.

230 TSAI, Y. C., et al., Parkin facilitates the elimination of expanded polyglut-amine proteins and leads to preservation of proteasome function. *J Biol Chem*, **2003**, *278(24)*, 22044–55.

231 TERRENI, L., et al., New mutation (R42P) of the parkin gene in the ubiquitinlike domain associated with parkinsonism. *Neurology*, **2001**, *56(4)*, 463–6.

232 DEMAND, J., et al., Cooperation of a ubiquitin domain protein and an E3 ubiquitin ligase during chaperone/proteasome coupling. *Curr Biol*, **2001**, *11(20)*, 1569–77.

233 WIEDERKEHR, T., B. BUKAU, and A. BUCHBERGER, Protein turnover: a CHIP programmed for proteolysis. *Curr Biol*, **2002**, *12(1)*, R26–8.

234 ALBERTI, S., et al., Ubiquitylation of BAG-1 suggests a novel regulatory mechanism during the sorting of chaperone substrates to the proteasome. *J Biol Chem*, **2002**, *277(48)*, 45920–7.

235 HAYASHI, S., et al., An autopsy case

of autosomal-recessive juvenile parkinsonism with a homozygous exon 4 deletion in the parkin gene [In Process Citation]. *Mov Disord*, **2000**, *15(5)*, 884–8.

236 MORI, H., et al., Pathologic and biochemical studies of juvenile parkinsonism linked to chromosome 6q [see comments]. *Neurology*, **1998**, *51(3)*, 890–2.

237 PORTMAN, A. T., et al., The nigrostriatal dopaminergic system in familial early onset parkinsonism with parkin mutations. *Neurology*, **2001**, *56(12)*, 1759–62.

238 TAKAHASHI, H., et al., Familial juvenile parkinsonism: clinical and pathologic study in a family. *Neurology*, **1994**, *44(3 Pt 1)*, 437–41.

239 FARRER, M., et al., Lewy bodies and parkinsonism in families with parkin mutations. *Ann Neurol*, **2001**, *50(3)*, 293–300.

240 CESARI, R., et al., Parkin, a gene implicated in autosomal recessive juvenile parkinsonism, is a candidate tumor suppressor gene on chromosome 6q25-q27. *Proc Natl Acad Sci USA*, **2003**, *100(10)*, 5956–61.

241 DENISON, S. R., et al., Alterations in the common fragile site gene Parkin in ovarian and other cancers. *Oncogene*, **2003**, *22(51)*, 8370–8.

242 DONG, Z., et al., Dopamine-dependent neurodegeneration in rats induced by viral vector-mediated overexpression of the parkin target protein, CDCrel-1. *Proc Natl Acad Sci USA*, **2003**, *100(21)*, 12438–43.

243 PENG, X. R., et al., The septin CDCrel-1 is dispensable for normal development and neurotransmitter release. *Mol Cell Biol*, **2002**, *22(1)*, 378–87.

244 HUYNH, D. P., et al., The autosomal recessive juvenile Parkinson disease gene product, parkin, interacts with and ubiquitinates synaptotagmin XI. *Hum Mol Genet*, **2003**, *12(20)*, 2587–97.

245 FALLON, L., et al., Parkin and CASK/LIN-2 associate via a PDZ-mediated interaction and are co-localized in lipid rafts and postsynaptic densities in brain. *J Biol Chem*, **2002**, *277(1)*, 486–91.

246 SHIMURA, H., et al., Ubiquitination of a new form of alpha-synuclein by parkin from human brain: implications for Parkinson's disease. *Science*, **2001**, *293(5528)*, 263–9.

247 YANG, Y., et al., Parkin suppresses dopaminergic neuron-selective neurotoxicity induced by Pael-R in Drosophila. *Neuron*, **2003**, *37(6)*, 911–24.

248 PETRUCELLI, L., et al., Parkin protects against the toxicity associated with mutant alpha-synuclein: proteasome dysfunction selectively affects catecholaminergic neurons. *Neuron*, **2002**, *36(6)*, 1007–19.

249 CHUNG, K. K., et al., Parkin ubiquitinates the alpha-synuclein-interacting protein, synphilin-1: implications for Lewy-body formation in Parkinson disease. *Nat Med*, **2001**, *7(10)*, 1144–50.

250 O'FARRELL, C., et al., Transfected synphilin-1 forms cytoplasmic inclusions in HEK293 cells. *Brain Res Mol Brain Res*, **2001**, *97(1)*, 94–102.

251 REN, Y., J. ZHAO, and J. FENG, Parkin binds to alpha/beta tubulin and increases their ubiquitination and degradation. *J Neurosci*, **2003**, *23(8)*, 3316–24.

252 ZHAO, J., et al., Parkin is recruited to the centrosome in response to inhibition of proteasomes. *J Cell Sci*, **2003**, *116(Pt 19)*, 4011–9.

253 HUYNH, D. P., et al., Parkin is associated with actin filaments in neuronal and nonneural cells. *Ann Neurol*, **2000**, *48(5)*, 737–44.

254 CORTI, O., et al., The p38 subunit of the aminoacyl-tRNA synthetase complex is a Parkin substrate: linking protein biosynthesis and neurodegeneration. *Hum Mol Genet*, **2003**, *12(12)*, 1427–37.

255 STAROPOLI, J. F., et al., Parkin is a component of an SCF-like ubiquitin ligase complex and protects postmitotic neurons from kainate excitotoxicity. *Neuron*, **2003**, *37(5)*, 735–49.

256 GOLDBERG, M. S., et al., Parkin-deficient mice exhibit nigrostriatal deficits but not loss of dopaminergic neurons. *J Biol Chem*, **2003**, *278(44)*, 43628–35.

257 IMAI, Y., M. SODA, and R. TAKAHASHI, Parkin Suppresses Unfolded Protein Stress-induced Cell Death through Its E3 Ubiquitin-protein Ligase Activity. *J Biol Chem*, **2000**, *275(46)*, 35661–35664.

258 ITIER, J. M., et al., Parkin gene inactivation alters behaviour and dopamine neurotransmission in the mouse. *Hum Mol Genet*, **2003**, *12(18)*, 2277–91.

259 LORENZETTI, D., et al., The neurological mutant quaking(viable) is Parkin deficient. *Mamm Genome*, **2004**, *15(3)*, 210–7.

260 LOCKHART, P. J., C. A. O'FARRELL, and M. J. FARRER, It's a double knock-out! The quaking mouse is a spontaneous deletion of parkin and parkin co-regulated gene (PACRG). *Mov Disord*, **2004**, *19(1)*, 101–4.

261 GREENE, J. C., et al., Mitochondrial pathology and apoptotic muscle degeneration in Drosophila parkin mutants. *Proc Natl Acad Sci USA*, **2003**, *100(7)*, 4078–83.

262 FEANY, M. B. and W. W. BENDER, A Drosophila model of Parkinson's disease. *Nature*, **2000**, *404(6776)*, 394–8.

263 DARIOS, F., et al., Parkin prevents mitochondrial swelling and cytochrome c release in mitochondria-dependent cell death. *Hum Mol Genet*, **2003**, *12(5)*, 517–26.

264 IMAI, Y., et al., CHIP is associated with Parkin, a gene responsible for familial Parkinson's disease, and enhances its ubiquitin ligase activity. *Mol Cell*, **2002**, *10(1)*, 55–67.

265 AGUILERA, M., et al., Ariadne-1: a vital Drosophila gene is required in development and defines a new conserved family of ring-finger proteins. *Genetics*, **2000**, *155(3)*, 1231–44.

266 TAKEUCHI, H., et al., Hsp70 and Hsp40 improve neurite outgrowth and suppress intracytoplasmic aggregate formation in cultured neuronal cells expressing mutant SOD1. *Brain Res*, **2002**, *949(1–2)*, 11–22.

267 MALTZMAN, W. and L. CZYZYK, UV irradiation stimulates levels of p53 cellular tumor antigen in nontransformed mouse cells. *Mol Cell Biol*, **1984**, *4(9)*, 1689–94.

268 BATES, S. and K. H. VOUSDEN, p53 in signaling checkpoint arrest or apoptosis. *Curr Opin Genet Dev*, **1996**, *6(1)*, 12–8.

269 LANE, D. P., Cancer. p53, guardian of the genome. *Nature*, **1992**, *358(6381)*, 15–6.

270 NIGRO, J. M., et al., Mutations in the p53 gene occur in diverse human tumour types. *Nature*, **1989**, *342(6250)*, 705–8.

271 FINLAY, C. A., The mdm-2 oncogene can overcome wild-type p53 suppression of transformed cell growth. *Mol Cell Biol*, **1993**, *13(1)*, 301–6.

272 CAHILLY-SNYDER, L., et al., Molecular analysis and chromosomal mapping of amplified genes isolated from a transformed mouse 3T3 cell line. *Somat Cell Mol Genet*, **1987**, *13(3)*, 235–44.

273 FAKHARZADEH, S. S., S. P. TRUSKO, and D. L. GEORGE, Tumorigenic potential associated with enhanced expression of a gene that is amplified in a mouse tumor cell line. *Embo J*, **1991**, *10(6)*, 1565–9.

274 MOMAND, J., et al., The mdm-2 oncogene product forms a complex with the p53 protein and inhibits p53-mediated transactivation. *Cell*, **1992**, *69(7)*, 1237–45.

275 JUVEN, T., et al., Wild type p53 can mediate sequence-specific transactivation of an internal promoter within the mdm2 gene. *Oncogene*, **1993**, *8(12)*, 3411–6.

276 WU, X., et al., The p53-mdm-2 autoregulatory feedback loop. *Genes Dev*, **1993**, *7(7A)*, 1126–32.

277 MONTES DE OCA LUNA, R., D. S. WAGNER, and G. LOZANO, Rescue of early embryonic lethality in mdm2-deficient mice by deletion of p53. *Nature*, **1995**, *378(6553)*, 203–6.

278 JONES, S. N., et al., Rescue of embryonic lethality in Mdm2-deficient mice by absence of p53. *Nature*, **1995**, *378(6553)*, 206–8.

279 OLINER, J. D., et al., Amplification of a

gene encoding a p53-associated protein in human sarcomas. *Nature*, **1992**, *358(6381)*, 80–3.

280 MOMAND, J., et al., The MDM2 gene amplification database. *Nucleic Acids Res*, **1998**, *26(15)*, 3453–9.

281 THUT, C., J. GOODRICH, and R. TJIAN, Repression of p53-mediated transcription by Mdm2: a dual mechanism. *Genes Dev*, **1997**, *11(15)*, 1974–86.

282 MIRNEZAMI, A. H., et al., Hdm2 recruits a hypoxia-sensitive corepressor to negatively regulate p53-dependent transcription. *Curr Biol*, **2003**, *13(14)*, 1234–9.

283 HAUPT, Y., et al., Mdm2 promotes the rapid degradation of p53. *Nature*, **1997**, *387(6630)*, 296–9.

284 KUBBUTAT, M. H., S. N. JONES, and K. H. VOUSDEN, Regulation of p53 stability by Mdm2. *Nature*, **1997**, *387(6630)*, 299–303.

285 HONDA, R. and H. YASUDA, Activity of MDM2, a ubiquitin ligase, toward p53 or itself is dependent on the RING finger domain of the ligase. *Oncogene*, **2000**, *19(11)*, 1473–6.

286 LEVEILLARD, T., et al., MDM2 expression during mouse embryogenesis and the requirement of p53. *Mech Dev*, **1998**, *74(1–2)*, 189–93.

287 BARTEL, F., H. TAUBERT, and L. C. HARRIS, Alternative and aberrant splicing of MDM2 mRNA in human cancer. *Cancer Cell*, **2002**, *2(1)*, 9–15.

288 KRAUS, A., et al., Expression of alternatively spliced mdm2 transcripts correlates with stabilized wild-type p53 protein in human glioblastoma cells. *Int J Cancer*, **1999**, *80(6)*, 930–4.

289 EVANS, S. C., et al., An alternatively spliced HDM2 product increases p53 activity by inhibiting HDM2. *Oncogene*, **2001**, *20(30)*, 4041–9.

290 DANG, J., et al., The RING domain of Mdm2 can inhibit cell proliferation. *Cancer Res*, **2002**, *62(4)*, 1222–30.

291 STEINMAN, H. A., et al., An Alternative Splice Form of Mdm2 Induces p53-independent Cell Growth and Tumorigenesis. *J Biol Chem*, **2004**, *279(6)*, 4877–86.

292 LAI, Z., et al., Metal and RNA binding

properties of the hdm2 RING finger domain. *Biochemistry*, **1998**, *37(48)*, 17005–15.

293 BODDY, M. N., P. S. FREEMONT, and K. L. BORDEN, The p53-associated protein MDM2 contains a newly characterized zinc-binding domain called the RING finger. *Trends Biochem Sci*, **1994**, *19(5)*, 198–9.

294 LOHRUM, M. A., et al., Identification of a cryptic nucleolar-localization signal in MDM2. *Nat Cell Biol*, **2000**, *2(3)*, 179–81.

295 POYUROVSKY, M. V., et al., Nucleotide binding by the Mdm2 RING domain facilitates Arf-independent Mdm2 nucleolar localization. *Mol Cell*, **2003**, *12(4)*, 875–87.

296 LOHRUM, M. A., et al., C-terminal ubiquitination of p53 contributes to nuclear export. *Mol Cell Biol*, **2001**, *21(24)*, 8521–32.

297 LUO, J., et al., Acetylation of p53 augments its site-specific DNA binding both in vitro and in vivo. *Proc Natl Acad Sci USA*, **2004**, *101(8)*, 2259–64.

298 ZENG, X., et al., MDM2 suppresses p73 function without promoting p73 degradation. *Mol Cell Biol*, **1999**, *19(5)*, 3257–66.

299 BALINT, E. E. and K. H. VOUSDEN, Activation and activities of the p53 tumour suppressor protein. *Br J Cancer*, **2001**, *85(12)*, 1813–23.

300 LAI, Z., et al., Human mdm2 mediates multiple mono-ubiquitination of p53 by a mechanism requiring enzyme isomerization. *J Biol Chem*, **2001**, *276(33)*, 31357–67.

301 LI, M., et al., Mono- versus polyubiquitination: differential control of p53 fate by Mdm2. *Science*, **2003**, *302(5652)*, 1972–5.

302 GROSSMAN, S. R., et al., Polyubiquitination of p53 by a ubiquitin ligase activity of p300. *Science*, **2003**, *300(5617)*, 342–4.

303 LENG, R. P., et al., Pirh2, a p53-induced ubiquitin-protein ligase, promotes p53 degradation. *Cell*, **2003**, *112(6)*, 779–91.

304 FUCHS, S. Y., et al., MEKK1/JNK signaling stabilizes and activates p53.

Proc Natl Acad Sci USA, **1998**, *95(18)*, 10541–6.

305 SHVARTS, A., et al., MDMX: a novel p53-binding protein with some functional properties of MDM2. *Embo J*, **1996**, *15(19)*, 5349–57.

306 PARANT, J., et al., Rescue of embryonic lethality in Mdm4-null mice by loss of Trp53 suggests a nonoverlapping pathway with MDM2 to regulate p53. *Nat Genet*, **2001**, *29(1)*, 92–5.

307 FINCH, R. A., et al., mdmx is a negative regulator of p53 activity in vivo. *Cancer Res*, **2002**, *62(11)*, 3221–5.

308 MIGLIORINI, D., et al., Mdm4 (Mdmx) regulates p53-induced growth arrest and neuronal cell death during early embryonic mouse development. *Mol Cell Biol*, **2002**, *22(15)*, 5527–38.

309 RIEMENSCHNEIDER, M. J., et al., Amplification and overexpression of the MDM4 (MDMX) gene from 1q32 in a subset of malignant gliomas without TP53 mutation or MDM2 amplification. *Cancer Res*, **1999**, *59(24)*, 6091–6.

310 GU, J., et al., Mutual dependence of MDM2 and MDMX in their functional inactivation of p53. *J Biol Chem*, **2002**, *277(22)*, 19251–4.

311 DE GRAAF, P., et al., Hdmx protein stability is regulated by the ubiquitin ligase activity of Mdm2. *J Biol Chem*, **2003**, *278(40)*, 38315–24.

312 KAWAI, H., et al., DNA damage-induced MDMX degradation is mediated by MDM2. *J Biol Chem*, **2003**, *278(46)*, 45946–53.

313 KAWAI, H., D. WIEDERSCHAIN, and Z. M. YUAN, Critical contribution of the MDM2 acidic domain to p53 ubiquitination. *Mol Cell Biol*, **2003**, *23(14)*, 4939–47.

314 LOWE, S. W. and C. J. SHERR, Tumor suppression by Ink4a-Arf: progress and puzzles. *Curr Opin Genet Dev*, **2003**, *13(1)*, 77–83.

315 BOTHNER, B., et al., Defining the molecular basis of Arf and Hdm2 interactions. *J Mol Biol*, **2001**, *314(2)*, 263–77.

316 POMERANTZ, J., et al., The Ink4a tumor suppressor gene product,

p19Arf, interacts with MDM2 and neutralizes MDM2's inhibition of p53. *Cell*, **1998**, *92(6)*, 713–23.

317 ZHANG, Y., Y. XIONG, and W. G. YARBROUGH, ARF promotes MDM2 degradation and stabilizes p53: ARF-INK4a locus deletion impairs both the Rb and p53 tumor suppression pathways. *Cell*, **1998**, *92(6)*, 725–34.

318 STOTT, F. J., et al., The alternative product from the human CDKN2A locus, p14(ARF), participates in a regulatory feedback loop with p53 and MDM2. *Embo J*, **1998**, *17(17)*, 5001–14.

319 KAMIJO, T., et al., Functional and physical interactions of the ARF tumor suppressor with p53 and Mdm2. *Proc Natl Acad Sci USA*, **1998**, *95(14)*, 8292–7.

320 HONDA, R. and H. YASUDA, Association of p19(ARF) with Mdm2 inhibits ubiquitin ligase activity of Mdm2 for tumor suppressor p53. *Embo J*, **1999**, *18(1)*, 22–7.

321 MIDGLEY, C. A., et al., An N-terminal p14ARF peptide blocks Mdm2-dependent ubiquitination in vitro and can activate p53 in vivo. *Oncogene*, **2000**, *19(19)*, 2312–23.

322 PAN, Y. and J. CHEN, MDM2 promotes ubiquitination and degradation of MDMX. *Mol Cell Biol*, **2003**, *23(15)*, 5113–21.

323 TAO, W. and A. J. LEVINE, P19(ARF) stabilizes p53 by blocking nucleo-cytoplasmic shuttling of Mdm2. *Proc Natl Acad Sci USA*, **1999**, *96(12)*, 6937–41.

324 ZHANG, Y. and Y. XIONG, Mutations in human ARF exon 2 disrupt its nucleolar localization and impair its ability to block nuclear export of MDM2 and p53. *Mol Cell*, **1999**, *3(5)*, 579–91.

325 WEBER, J. D., et al., Nucleolar Arf sequesters Mdm2 and activates p53. *Nat Cell Biol*, **1999**, *1(1)*, 20–6.

326 LLANOS, S., et al., Stabilization of p53 by p14ARF without relocation of MDM2 to the nucleolus. *Nat Cell Biol*, **2001**, *3(5)*, 445–52.

327 LIN, A. W. and S. W. LOWE, Oncogenic ras activates the ARF-p53 pathway to suppress epithelial cell transformation. *Proc Natl Acad Sci USA*, **2001**, *98(9)*, 5025–30.

328 KORGAONKAR, C., et al., ARF function does not require p53 stabilization or Mdm2 relocalization. *Mol Cell Biol*, **2002**, *22(1)*, 196–206.

329 XIRODIMAS, D. P., et al., P14ARF promotes accumulation of SUMO-1 conjugated (H)Mdm2. *FEBS Lett*, **2002**, *528(1–3)*, 207–11.

330 LOHRUM, M. A., et al., Regulation of HDM2 activity by the ribosomal protein L11. *Cancer Cell*, **2003**, *3(6)*, 577–87.

331 ZHANG, Y., et al., Ribosomal protein L11 negatively regulates oncoprotein MDM2 and mediates a p53-dependent ribosomal-stress checkpoint pathway. *Mol Cell Biol*, **2003**, *23(23)*, 8902–12.

332 MARECHAL, V., et al., The ribosomal L5 protein is associated with mdm-2 and mdm-2-p53 complexes. *Mol Cell Biol*, **1994**, *14(11)*, 7414–20.

333 XIRODIMAS, D. P., C. W. STEPHEN, and D. P. LANE, Cocompartmentalization of p53 and Mdm2 is a major determinant for Mdm2-mediated degradation of p53. *Exp Cell Res*, **2001**, *270(1)*, 66–77.

334 LI, L., et al., A TSG101/MDM2 regulatory loop modulates MDM2 degradation and MDM2/p53 feedback control. *Proc Natl Acad Sci USA*, **2001**, *98(4)*, 1619–24.

335 RULAND, J., et al., p53 accumulation, defective cell proliferation, and early embryonic lethality in mice lacking tsg101. *Proc Natl Acad Sci USA*, **2001**, *98(4)*, 1859–64.

336 LU, Q., et al., TSG101 interaction with HRS mediates endosomal trafficking and receptor down-regulation. *Proc Natl Acad Sci USA*, **2003**, *100(13)*, 7626–31.

337 WANG, X., et al., p53 Activation by nitric oxide involves down-regulation of Mdm2. *J Biol Chem*, **2002**, *277(18)*, 15697–702.

338 MAYA, R., et al., ATM-dependent phosphorylation of Mdm2 on serine 395: role in p53 activation by DNA damage. *Genes Dev*, **2001**, *15(9)*, 1067–77.

339 GOLDBERG, Z., et al., Tyrosine phosphorylation of Mdm2 by c-Abl: implications for p53 regulation. *Embo J*, **2002**, *21(14)*, 3715–27.

340 MAYO, L. D. and D. B. DONNER, A phosphatidylinositol 3-kinase/Akt pathway promotes translocation of Mdm2 from the cytoplasm to the nucleus. *Proc Natl Acad Sci USA*, **2001**, *98(20)*, 11598–603.

341 ZHOU, B. P., et al., HER-2/neu induces p53 ubiquitination via Akt-mediated MDM2 phosphorylation. *Nat Cell Biol*, **2001**, *3(11)*, 973–82.

342 MAYO, M. W., et al., PTEN blocks tumor necrosis factor-induced NF-kappa B-dependent transcription by inhibiting the transactivation potential of the p65 subunit. *J Biol Chem*, **2002**, *277(13)*, 11116–25.

343 FREEMAN, D. J., et al., PTEN tumor suppressor regulates p53 protein levels and activity through phosphatase-dependent and -independent mechanisms. *Cancer Cell*, **2003**, *3(2)*, 117–30.

344 KIM, H., et al., Merlin neutralizes the inhibitory effect of Mdm2 on p53. *J Biol Chem*, **2004**, *279(9)*, 7812–8.

345 SHENOY, S. K., et al., Regulation of receptor fate by ubiquitination of activated beta 2-adrenergic receptor and beta-arrestin. *Science*, **2001**, *294(5545)*, 1307–13.

346 KINYAMU, H. K. and T. K. ARCHER, Estrogen receptor-dependent proteasomal degradation of the glucocorticoid receptor is coupled to an increase in mdm2 protein expression. *Mol Cell Biol*, **2003**, *23(16)*, 5867–81.

347 JIN, Y., et al., MDM2 mediates p300/CREB-binding protein-associated factor ubiquitination and degradation. *J Biol Chem*, **2004**, *279(19)*, 20035–43.

348 SENGUPTA, S. and B. WASYLYK, Ligand-dependent interaction of the glucocorticoid receptor with p53 enhances their degradation by Hdm2. *Genes Dev*, **2001**, *15(18)*, 2367–80.

349 LEGUBE, G., et al., Tip60 is targeted to proteasome-mediated degradation by Mdm2 and accumulates after UV irradiation. *Embo J*, **2002**, *21(7)*, 1704–12.

350 HIETANEN, S., et al., Activation of p53 in cervical carcinoma cells by small molecules. *Proc Natl Acad Sci USA*, **2000**, *97(15)*, 8501–6.

351 LANE, D. P. and P. M. FISCHER, Turning the key on p53. *Nature*, **2004**, *427(6977)*, 789–90.

352 PICKSLEY, S. M., et al., The p53-MDM2 interaction in a cancer-prone family, and the identification of a novel therapeutic target. *Acta Oncol*, **1996**, *35(4)*, 429–34.

353 VASSILEV, L. T., et al., In vivo activation of the p53 pathway by small-molecule antagonists of MDM2. *Science*, **2004**, *303(5659)*, 844–8.

354 LI, M., et al., Deubiquitination of p53 by HAUSP is an important pathway for p53 stabilization. *Nature*, **2002**, *416(6881)*, 648–53.

355 CUMMINS, J. M. and B. VOGELSTEIN, HAUSP is Required for p53 Destabilization. *Cell Cycle*, **2004**, *3*, 689–692.

356 LI, M., et al., A dynamic role of HAUSP in the p53-Mdm2 pathway. *Mol Cell*, **2004**, *13(6)*, 879–86.

357 NAKIELNY, S., et al., Nup153 is an M9-containing mobile nucleoporin with a novel Ran-binding domain. *Embo J*, **1999**, *18(7)*, 1982–95.

358 YASEEN, N. R. and G. BLOBEL, Two distinct classes of Ran-binding sites on the nucleoporin Nup-358. *Proc Natl Acad Sci USA*, **1999**, *96(10)*, 5516–21.

359 JONES, J. S., S. WEBER, and L. PRAKASH, The Saccharomyces cerevisiae RAD18 gene encodes a protein that contains potential zinc finger domains for nucleic acid binding and a putative nucleotide binding sequence. *Nucleic Acids Res*, **1988**, *16(14B)*, 7119–31.

360 VAN DER LAAN, R., et al., Characterization of mRAD18Sc, a mouse homolog of the yeast postreplication repair gene RAD18. *Genomics*, **2000**, *69(1)*, 86–94.

361 BORDEN, K. L., RING fingers and B-boxes: zinc-binding protein-protein interaction domains. *Biochem Cell Biol*, **1998**, *76(2–3)*, 351–8.

362 REDDY, B. A., M. KLOC, and L. ETKIN, The cloning and characterization of a maternally expressed novel zinc finger nuclear phosphoprotein (xnf7) in Xenopus laevis. *Dev Biol*, **1991**, *148(1)*, 107–16.

363 BONNEAU, R., et al., De novo prediction of three-dimensional structures for major protein families. *J Mol Biol*, **2002**, *322(1)*, 65–78.

364 BONNEAU, R., et al., Functional inferences from blind ab initio protein structure predictions. *J Struct Biol*, **2001**, *134(2–3)*, 186–90.

365 SCHULTZ, J., et al., SMART, a simple modular architecture research tool: identification of signaling domains. *Proc Natl Acad Sci USA*, **1998**, *95(11)*, 5857–64.

366 LETUNIC, I., et al., SMART 4.0: towards genomic data integration. *Nucleic Acids Res*, **2004**, *32 Database issue*, D142–4.

367 XU, R. X., et al., NMR structure of a protein kinase C-gamma phorbol-binding domain and study of protein-lipid micelle interactions. *Biochemistry*, **1997**, *36(35)*, 10709–17.

368 ONO, Y., et al., Phorbol ester binding to protein kinase C requires a cysteine-rich zinc-finger-like sequence. *Proc Natl Acad Sci USA*, **1989**, *86(13)*, 4868–71.

369 SONG, Z., et al., A novel cysteine-rich sequence-specific DNA-binding protein interacts with the conserved X-box motif of the human major histocompatibility complex class II genes via a repeated Cys-His domain and functions as a transcriptional repressor. *J Exp Med*, **1994**, *180(5)*, 1763–74.

370 KWON, Y. T., et al., The mouse and human genes encoding the recognition component of the N-end rule pathway. *Proc Natl Acad Sci USA*, **1998**, *95(14)*, 7898–903.

371 MU, F. T., et al., EEA1, an early endosome-associated protein. EEA1 is a conserved alpha-helical peripheral membrane protein flanked by cysteine "fingers" and contains a calmodulin-binding IQ motif. *J Biol Chem*, **1995**, *270(22)*, 13503–11.

372 TSUKAZAKI, T., et al., SARA, a FYVE domain protein that recruits Smad2 to the TGFbeta receptor. *Cell*, **1998**, *95(6)*, 779–91.

373 SIMONSEN, A., et al., EEA1 links PI(3)K function to Rab5 regulation of endosome fusion. *Nature*, **1998**, *394(6692)*, 494–8.

374 PATKI, V., et al., A functional PtdIns(3)P-binding motif. *Nature*, **1998**, *394(6692)*, 433–4.

375 MORETT, E. and P. BORK, A novel transactivation domain in parkin. *Trends Biochem Sci*, **1999**, *24(6)*, 229–31.

376 WORTHINGTON, M. T., et al., Metal binding properties and secondary structure of the zinc-binding domain of Nup475. *Proc Natl Acad Sci USA*, **1996**, *93(24)*, 13754–9.

377 THOMPSON, M. J., et al., Cloning and characterization of two yeast genes encoding members of the CCCH class of zinc finger proteins: zinc finger-mediated impairment of cell growth. *Gene*, **1996**, *174(2)*, 225–33.

378 WEBB, J. R. and W. R. MCMASTER, Molecular cloning and expression of a Leishmania major gene encoding a single-stranded DNA-binding protein containing nine "CCHC" zinc finger motifs. *J Biol Chem*, **1993**, *268(19)*, 13994–4002.

379 SUMMERS, M. F., et al., High-resolution structure of an HIV zinc fingerlike domain via a new NMR-based distance geometry approach. *Biochemistry*, **1990**, *29(2)*, 329–40.

380 PARRAGA, G., et al., Zinc-dependent structure of a single-finger domain of yeast ADR1. *Science*, **1988**, *241(4872)*, 1489–92.

5
Ubiquitin-conjugating Enzymes

Michael J. Eddins and Cecile M. Pickart

5.1
Introduction

In this chapter we review the biochemical, structural, and biological properties of ubiquitin-conjugating enzymes (also called E2 enzymes). Because length restrictions preclude a comprehensive treatment, we focus on key findings that have revealed important general insights and principles. Throughout the piece we try to point out important unanswered questions concerning the E2 enzyme family.

A few words about nomenclature are necessary. The yeast E2 genes were numbered in the order of their discovery, but the situation is more complicated in mammals. There are currently three naming systems in use for human E2s: one based on protein molecular mass (e.g. $E2_{25K}$ is a 25-kD E2), one based on temporal order of gene cloning (e.g. UbcH10 is specified by the tenth E2 gene cloned in humans), and one based on relationship to yeast E2s (e.g. HR6A is one of two human homologs of yeast Rad6/Ubc2). The second system is the least ambiguous, but also the least informative. In this chapter, we generally name mammalian E2s according to their relationship to yeast E2s. When this is not possible, we use one of the published names.

5.2
Historical Background

Ubiquitin's best-understood function is that of a protein cofactor in an intracellular protein-degradation pathway that terminates with the destruction of ubiquitin-tagged substrates by 26S proteasomes [1]. The discovery in 1980 that this 76-residue protein is conjugated to proteolytic substrates through the formation of a peptide-like bond, and in an ATP-dependent manner, suggested that ubiquitin activation would be part of the conjugation process [2, 3]. A ubiquitin activating enzyme (E1) was soon identified and shown to employ an aminoacyl-tRNA-synthetase-like mechanism [4]. E1 first catalyzes the addition of an adenylate moiety to the carboxyl group of ubiquitin's C-terminal residue, G76. The AMP-bound

Protein Degradation. Edited by J. Mayer, A. Ciechanover, M. Rechsteiner
Copyright © 2005 WILEY-VCH Verlag GmbH & Co. KGaA, Weinheim
ISBN: 3-527-30837-7

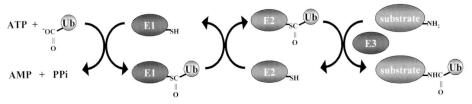

Fig. 5.1. The ubiquitin-conjugation pathway. Steps in ubiquitin activation and substrate modification. E1, ubiquitin activating enzyme; E2, ubiquitin-conjugating enzyme; E3, ubiquitin-protein ligase. Atoms involved in the thiol ester and amide bonds are shown.

ubiquitin is then transferred to a cysteine residue in the E1 active site, concomitant with the formation of a new molecule of ubiquitin adenylate. The thiol-linked ubiquitin is the proximal source of activated ubiquitin for downstream steps.

From a chemical point of view, the E1/ubiquitin thiol ester should be competent to donate ubiquitin to a substrate amino group. In fact, aminoacyl-enzyme thiol esters are used in exactly this way in non-ribosomal polypeptide synthesis, a process that was discovered around the same time as ubiquitin–protein conjugation [5]. In spite of the attractive simplicity of this model, however, biochemical reconstitution studies showed that besides E1 two additional fractions were required to conjugate ubiquitin to a model substrate. They were called ubiquitin carrier protein (E2) and ubiquitin-protein ligase (E3), respectively, since the respective factors seemed to act sequentially [6]. Interestingly, the E2 factor apparently formed a thiol ester with ubiquitin. Based on these results, Hershko and co-workers proposed the "ubiquitin conjugation cascade" (Figure 5.1).

Multiple thiol ester-forming proteins were present in the E2 fraction [6], but only the smallest of them reconstituted substrate ubiquitination catalyzed by the then-known E3 [7]. This result suggested that there could be multiple E2s with distinct functional properties. Confirmation of this hypothesis came with the cloning of the first two E2 genes, *RAD6/UBC2* and *CDC34/UBC3*, which indeed encoded homologous yeast proteins with a signature cysteine-containing active-site motif [8, 9]. The two E2s functioned in distinct biological processes – DNA damage tolerance [8, 10] and cell-cycle control [9] – providing the first hint that ubiquitination might regulate a broad range of biological processes. A family of E2 enzymes naturally suggested that there would also be a family of E3 enzymes. This prediction has since been strikingly confirmed. We now know that specific E2/E3 complexes function to modify specific substrates with ubiquitin.

5.3
What is an E2?

A protein can be identified as an E2 enzyme according to several different criteria. Functionally, the E2 occupies an intermediate position in the conjugation cascade –

that is, it acts between the E1 and the E3 (Figure 5.1). This property accounts for the original name of ubiquitin carrier protein, which drew an analogy to the acyl carrier proteins used in fatty acid biosynthesis and non-ribosomal peptide synthesis [6]. Subsequently, with the recognition that E2 enzymes often play an active role in conjugation, the conjugating enzyme name gained favor.

Mechanistically, the E2 first participates in a thiol ester transfer reaction, in which the activated ubiquitin is moved from the active-site cysteine of E1 to that of the E2 (Figure 5.1). The E2/ubiquitin thiol ester intermediate is strictly required for downstream steps, as shown by ablation of substrate ubiquitination following mutation of the active-site cysteine residues of different E2s (see, for example, Refs. [11, 12]). The ubiquitin is then transferred from the E2 active site to the ε-amino group of the substrate's lysine residue, forming an isopeptide bond. The conjugation site can also be a specific lysine on a previously conjugated ubiquitin, which leads to polyubiquitin chain elongation; chains linked through K48 are the principal signal for targeting substrates to proteasomes [1]. Transfer of ubiquitin to the substrate requires the assistance of the E3 [1, 6]. If this enzyme belongs to the HECT domain family (*H*omologous to *E*6AP *C*-*T*erminus), the ubiquitin is first transferred to an active-site cysteine residue of the E3; if the E3 belongs to the RING domain family (*R*eally *I*nteresting *N*ew *G*ene), ubiquitin is transferred directly to the substrate's amino group (Section 5.6.3). Collectively, these properties constitute the biochemical definition of an E2 enzyme: it is a protein that accepts ubiquitin in thiol ester linkage from E1, and cooperates with an E3 enzyme to deliver this ubiquitin to the substrate.

The functional specialization of individual E2s (Section 5.4) reflects the specificity of interaction of each E2 with its cognate E3(s), in conjunction with the E3's substrate specificity. Therefore an E2 enzyme can also be defined according to the cognate E3(s) with which it interacts. The E3 partners of many of the eleven ubiquitin-dedicated E2s in *Saccharomyces cerevisiae* are conserved in higher organisms (Section 5.4). However, both the E2 and E3 families are much larger in higher organisms than in yeast. Present accounting suggests that there are 50–70 E2s, and hundreds of E3s, in mammals [13, 14].

The amino acids surrounding the thiol ester-forming cysteine residue are particularly highly conserved, but there is sequence similarity throughout the ~150-residue E2 core domain (Figure 5.2). This bioinformatic definition makes it easy to identify E2 genes in sequenced genomes [13, 14]. In fact, many E2s consist of just this core domain (Figure 5.2). The fact that such E2s are often functionally distinct from one another indicates that modest sequence variation within the core domain can be highly significant. Structural biology has begun to shed light on this structure/function correlation (Section 5.6). Other E2s display N- and C-terminal extensions to the core domain (Figure 5.2), which may play a role in E3 and/or substrate specificity (see Refs. [15–19]).

Structural biology provides a final way to define an E2 enzyme. As expected from the strong sequence conservation, the E2 core domain adopts a conserved fold. At the time this article was being prepared, twelve different E2 structures had been deposited in the Protein Data Bank. The average root-mean-square deviation of

```
scUBC1    1 -------MSR-------AKRIMKEIQAVKDDPAAHITLEFVSE------SDIHHLKGTFLGPPGTPYEGGKFVVDIEVPMEYPFKPPKMQFDT  73
atUBC1    1 -------MST-------PARRRIMRDFKRLQQDPPAGISGAPQD-----NNIMLWNAVIFGPDTPWDGGTFKLSLQFSEDYPNKPPTVREFVS  74
scUBC2    1 -------MST-------PARRRIMRDFKRMKEDAPPGVSASPLP-----DNVMVWNAMIIGPADTPYEDGTFRLLLEFDEEYPNKPPHVKFLS  74
hUBC2     1 -------MST-------PARRRIMRDFKRLQEDPPVGVSGAPSE-----NNIMQWNAVIFGPEGTPFEDGTFKLVIEFSEEYPNKPPTVRFLS  74
scUBC4    1 -------MSS-------SKRIAKELSDLERDPPTSCSAGPVG-------DDLYHWQASIMGPADSPYAGGVFFLSIHFPTDYPFKPPKISFTT  72
scUBC13   1 -------MASL------PRRIIKETEKLVSDPVPGITAEPHD-------DNLRYFQVTIEGPEQSPYEDGIFELELYLPDDYPMEAPKVRFLT  73
hUBC13    1 -------MAGL------PRRIIKETQRLLAEPVPGIKAEPDE-------SNARYFHVVIAGPQDSPFEGGTFKLELFLPEEYPMAAPKVRFMT  73
hUBC9     1 -------MSG-------IALSRLAQERKAWRKDHPFGFVAVPTKNPDGTMNLMNWECAIPGKKGTPWEGGLFKLRMLFKDDYPSSPPKCKFEP  79
hUBC7     1 -------MAA-------SRRIMKELEEIRKCGMKNFRNIQVD-------EANLLTWQGLIV-PDNPPYDKGAFRIEINFPAEYPFKPPKITFKT  72
scUBC7    1 -------MSK-------TAQKRLLKELQQLIKDSPPGIVAGPKSE----NNIFIWDCLIQGPPDTPYADGVFNAKLEFPKDYPLSPPKLTFTP  75
hUBC10    1 MASQNRDPAATSVAAARKGAEPSGGAARGPVGKRLQQELMTLMMSGDKGISAFPES-------DNLFKWVGTIHGAAGTVEDLRYKLSLEFPSGYPNAPTVKFLT 100
E2-C      1 MSQQNIDPAANQVRQKERPRDMTTSKERHSVSKRLQQELRTLLMSGDPGITAFPDG-------DNLFKWVATLDGPKDTVYESLKYKLTLEFPSDYPYKPPVKFTT 100

scUBC1   74 KVYHPNISSVTGAICLDILK--------------NAWSPVITIKSALISIQALLSQEPFENDPQDAEVAQHYLRDRESFNKTAALWTRLYASETSNGQKGNVEESDL 165
atUBC1   75 RMFHPNIY-ADGSICLDILQ--------------NQWSPIYDVAAIITSIQSLLCDPNPNSPANSEAARMYSESKREYNRRVRDVVEQSWTAD----------------- 152
scUBC2   75 EMFHPNVY-ANGEICLDILQ--------------NRWTPTYDVASIITSIQSLFNDPNPASPANVEAATLFKDHKSQYVKRVKETVEKSWEDDMDDDDDDDDD------ 165
hUBC2    75 KMFHPNVY-ADGSICLDILQ--------------NRWSPTYDVSSILTSIQSLLDEPNPNSPANSQAAQLYQENKREYEKRVSAIVEQSWNDS---------------- 152
scUBC4   73 KIYHPNIN-ANGNICLDILK--------------DQWSPALTLSKVLLSICSLLTDANPDDPLVPEIAHIYKTDRPKYEATAREWTKKYAV------------------ 148
scUBC13  74 KIYHPNID-RLGRICLDVLK--------------TNWSPALQIRTVLLSIQALLSAPNPNDPLANDVAEDWIKNEQGAKAKAREWTKLYAKKPE--------------- 153
hUBC13   74 KIYHPNVD-KLGRICLDILK--------------DKWSPALQIRTVLLSIQALLSAFNPDDPLANDVAEQWKTNEAQAIETARAWTRLYAMNNI-------------- 152
hUBC9    80 PLFHPNVY-PSGTVCLSILEE-------------DKDWRPAITIKQILLGIQELLNEPNIQDPAQAEAYTIYCQNRVEYEKRVRAQAKKFAPS-------------- 158
hUBC7    73 KIYHPNID-EKGQVCLPVIS--------------AENWKPATKTDQVIQSLIALVNDPQPEHPLRADLAEEYSKDRKKFCKNAEEFTKKYGEKRPVD----------- 154
scUBC7   76 SILHPNIY-PNGEVCISILHSPGDDPNMYELAEERWSPVQSVEKILLSVMSMLSEPNIESGANIDACILWRDNRPEFERQVKLSILKSLGF------------- 165
hUBC10  101 PCYHPNVD-TQGNICLDILK--------------EKWSALYDVRTILLSIQSLLGEPNIDSPLNTHAAELWK-NPTAFKKYLQETYSKQVTSQEP---------- 179
E2-C    101 PCWHPNVD-QSGNICLDILK--------------ENWTASYDVRTILLSLQSLLGEPNASPLNAQAADMWS-NQTEYKKVLHEKY-KTAQSDK----------- 177

scUBC1  166 YGIDHDLIDEFESQGFEKDKIVEVLRRLGVKSLDPNDNNTANRIIEELLK 215
atUBC1  166 ------------------------------------------------- 
scUBC2  166 DDDDEAD------------------------------------------ 172
```

Fig. 5.2. E2 sequence alignments. Sequences of the twelve E2s found in the PDB. The active-site cysteine is colored green, identical residues colored red, and conserved residues colored blue.

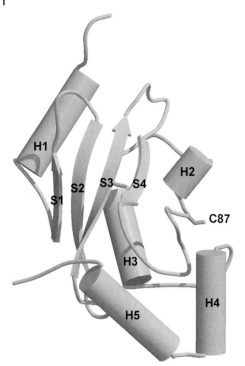

Fig. 5.3. Ubc13 (1JBB). Canonical α/β E2 fold with the active-site cysteine shown in ball-and-stick.

the 150 Cα positions of these structures is less than 2 Å. E2s are α/β proteins containing a central anti-parallel four-stranded β-sheet (S1–S4), four α-helices (H1, H3, H4, H5), and a small 3_{10} helix (H2) (Figure 5.3) [20, 21]. The cysteine is located on an extended loop after β-strand 4 and immediately before the short 3_{10} helix H2. The active-site cysteine sits in a shallow groove composed of residues from the H3–H4 and S4–H2 loops. The canonical α/β E2 fold is highly versatile, allowing E2 enzymes to associate with several different proteins in the ubiquitin conjugation cascade without any perturbation of the E2's tertiary structure (Section 5.6). Residues occupying the face opposite the active site are less conserved than those surrounding the cysteine [21]. Sequence variation in this region contributes to the functional diversity of the E2 family by permitting specific interactions of individual E2s with cognate E3s and (perhaps) substrates.

5.4
Functional Diversity of Ubiquitin-conjugating Enzymes

The functional range of the ubiquitin-conjugating enzyme family is easily appreciated by considering the family members in a single organism. Table 5.1 summa-

Tab. 5.1. E2 enzymes of the yeast *Saccharomyces cerevisiae*.

Gene	Amino acids	Cognate E3	Functions and substrates
UBC1	215 [28]		Short C-terminal tail harbors ubiquitin-associated (UBA) domain [155]
		Unknown	Essential in *ubc4Δubc5Δ* genetic background, suggesting a redundant role with Ubc4/5 in proteasomal turnover of short-lived and abnormal proteins [28]
		Hrd1 [39]	Role in ERAD that is not fulfilled by Ubc4/5 [39, 40]
UBC2 (RAD6)	172 [8]	Ubr1 [156]	Proteasomal degradation of N-end rule [158] substrates, including cohesin fragment [51]
		Ubr1 [52]	Proteasomal degradation of Cup9 transcriptional repressor regulates peptide import
		Rad18 [73, 157]	DNA-damage tolerance [8] via mono-ubiquitination of PCNA [65] (non-proteolytic function)
		Bre1 [67, 69]	Ubiquitination of histone H2B [68] regulates gene transcription and silencing [71] (non-proteolytic function)
UBC3 (CDC34)	295 [9]	SCF E3s	Essential gene; long C-terminal tail; targets diverse substrates for proteasomal degradation [30, 34, 36]; regulation of cell-cycle progression
UBC4	148 [24]	Unknown	Proteasomal degradation of diverse short-lived proteins [24]
		Doa10	Proteasomal degradation of MATα2 transcriptional repressor [49]
		Rsp5	Endocytosis of membrane proteins [60, 61]; protein trafficking (see Ref. [63])
UBC5	148 [24]	See *UBC4*	92% identical to Ubc4; functionally redundant [24]
UBC6	250 [46]		C-terminal tail provides anchoring to ER membrane [46]
		Unknown	Together with Ubc7, proteasomal degradation of some ERAD substrates [47, 48, 159]
		Doa10 [42]	Proteasomal turnover of Ubc6 is Ubc6-, Ubc7-, and Doa10-dependent [42, 50]
		Doa10 [42]	In conjunction with Ubc7, proteasomal turnover of MATα2
UBC7	165 [25]		Localized to ER membrane via Cue1 [37]
		Hrd1 [39]	Role in proteasomal degradation via ERAD [38, 48] confers resistance to cadmium and other ER stresses [25]
		Doa10 [42]	In conjunction with Ubc7, proteasomal turnover of MATα2
UBC8	206 [160]	Unknown	Glucose-induced proteasome degradation of fructose-1,6-bisphosphatase [57]
UBC10	165 [75]	Unknown	Also called Pas2/Pex10. Peroxisome biogenesis [75]; Pex10 is a candidate E3 [78]

Tab. 5.1. (continued)

Gene	Amino acids	Cognate E3	Functions and substrates
UBC11	156 [84]	Unknown	Unknown; similar to clam E2-C (E2-C functions in mitotic cyclin turnover [79], but Ubc11 is dispensable for this process in yeast [84])
UBC13	153		Heterodimerizes with Mms2 (UEV) [72]
		Rad5 [73]	DNA-damage tolerance [72] via polyubiquitination of PCNA [65] (non-proteolytic function)
UBC9	157 [161]		Essential gene; E2 dedicated to Smt3 (SUMO) [163]
		Siz1/2 [162]	Septin modification
UBC12	188 [164]		E2 dedicated to Rub1 (Nedd8)
		SCF E3s	Modification of specific cullin lysine residue activates cullin-based E3s [165]

rizes key properties of the complete set of E2s in the yeast *S. cerevisiae*, including notable structural features, known cognate E3(s) and their key substrates, and biological functions (see also [22, 23]). We cannot give a comprehensive review of E2 functions in higher organisms, but we do comment on some notable instances of functional conservation, expansion, and divergence (see also [23]).

5.4.1
Functions Related to Proteasome Proteolysis

In many cases, the specific function(s) of a given E2 enzyme reflect its role in targeting one or more substrates for degradation by 26S proteasomes. The scope of this function varies considerably between E2 family members, however. At one extreme, the functionally redundant enzymes Ubc4 and Ubc5 are necessary for the turnover of many substrates, as shown by a marked reduction in the rate of turnover of endogenous short-lived and abnormal proteins in a *ubc4Δubc5Δ* yeast strain [24]. The slow growth and stress sensitivity of this strain [24, 25] can also be ascribed to inhibition of proteasomal proteolysis since these phenotypes are characteristic of proteasome mutants [22, 26]. Despite the important role of Ubc4/5 in proteasome degradation, few E3 partners relevant to this function have been identified. One is Ufd4, a HECT-domain E3 that mediates the degradation of linear ubiquitin fusion proteins [27]. A *ufd4Δ* strain grows normally, however, indicating that Ubc4 has other cognate E3s. Rsp5, an essential HECT-domain E3, is one likely candidate since this E3 partners with Ubc4 in other pathways (see below). *UBC1* is essential for viability in the *ubc4Δubc5Δ* strain, suggesting that Ubc1 shares substantial functional overlap with Ubc4/5 in directing substrates to proteasomes for degradation [28].

The Ubc4/5 sub-family of E2s is much larger in mammals, where it includes both constitutively and selectively expressed enzymes. Notable human E2s in this group are UbcH5a/b/c, UbcH7, and UbcH8 (see Ref. [23]). The expansion is likely to reflect the much larger size of the E3 family in higher organisms. However, while the results of *in vitro* conjugation assays and protein–protein interaction studies suggest that certain E3s partner specifically with individual Ubc4/5 subfamily members, the degree of E3 (hence, functional) selectivity in the cellular setting remains quite uncertain (discussed in Refs. [23, 29]). RNA interference studies and mouse knockout models may be helpful in addressing this question in the future.

Ubc3/Cdc34 supports the proteasome-mediated proteolysis of numerous substrates through its role as the specific E2 partner of a large family of multi-subunit RING E3s called SCF E3s (*S*kp-*C*ullin-*F*-box, Section 5.6.3). This role is preserved in higher organisms [30]. Yeast *ubc3* mutants arrest in G1 phase of the cell cycle because they fail to degrade Sic1 [31], an inhibitor of the G1/S transition that is recognized and polyubiquitinated by a specific SCF E3 [32, 33]. Although this is the only essential function of yeast Ubc3 [31], this E2 partners with many other SCF E3s to target diverse substrates for degradation by proteasomes (see Refs. [30, 34–36]). Although studies in yeast suggest that Cdc34 is the main E2 partner of SCF E3s, some SCF E3s seem to partner with Ubc4/5-type E3s (see Refs. [23, 36]). Ubc3 has a long C-terminal tail (Table 5.1), making it the most distinctive yeast E2 in terms of primary structure. A chimeric E2 in which the Ubc3 tail is appended to the core domain of Ubc2 fulfills the essential function of Ubc3 in yeast, suggesting that the tail of Ubc3 is necessary for key interactions with the E3 or Sic1 [15, 16].

Ubc7 is localized to the endoplasmic reticulum (ER) through an interaction with a partner protein, Cue1 [37], and plays a major role in proteasome degradation. Ubc7 acts on misfolded proteins of the ER, which are ejected from this compartment as a prelude to ubiquitination at the cytosolic face of the ER membrane and degradation by cytosolic proteasomes [38]. Ubc7's role in ERAD (***ER A**ssociated **D**egradation*) explains why a *ubc7Δ* strain is conditionally sensitive to agents that induce protein misfolding in the ER [25, 39–41]. Ubc7 frequently partners with Hrd1, an ER-localized RING E3 [39], but some ERAD substrates of Ubc7 seem to be recognized in cooperation with a different ER-localized RING E3, Doa10 [42]. Consistent with Ubc7's prominent role in ERAD, the yeast *UBC7* and *CUE1* genes are induced as part of the *U*nfolded *P*rotein *R*esponse (UPR) and there is a synthetic lethal relationship between certain ERAD and UPR genes [40, 41]. Yeast Ubc1 also plays a significant, but poorly-defined, role in ERAD [39, 40]. Mammalian Ubc7 also functions in ERAD [43–45].

Ubc6 localizes to the ER through its own C-terminal membrane anchor [46]. Although Ubc6 plays a role in ERAD, its function in this process is less conspicuous than that of Ubc7 [43, 47, 48]. Interestingly, Ubc6 and Ubc7 both contribute to the Doa10-dependent degradation of a soluble nuclear protein [42, 49], and Ubc6 is itself rapidly degraded by proteasomes in a manner that depends on its own active-site cysteine, its C-terminal membrane anchor, functional Ubc7, and Doa10 [42, 50]. The purpose of this instability remains mysterious.

Ubc2 functions rather selectively in proteasome proteolysis. In yeast, two specific E3 partners are known, leading to proteasome degradation events that regulate chromosome stability [51], peptide import [52], and homing endonuclease stability [53]. Mammals have two closely-related Ubc2 isoforms, each of which complements most of the functions of the yeast *ubc2Δ* strain [54]. But the mammalian Ubc2 isoforms also have specialized functions – one of them is required for spermatogenesis in the mouse [55] and at least one of them can be inferred to be necessary for cardiovascular development [56].

So far, Ubc8 has been implicated in the regulated turnover of just one substrate, and its E3 partner(s) remain unknown [57]. Interestingly, the closest mammalian relative of yeast Ubc8 is expressed with a restricted tissue specificity and (in some tissues) in a regulated manner [58, 59].

5.4.2
Endocytosis and Trafficking

Just because an E2 functions in proteasome proteolysis does not mean that its functions are limited to this pathway. This is because the E2's functional range is largely determined by the substrate specificity of its E3 partner(s). For example, yeast Ubc4 and Ubc5 play a prominent role in proteasome degradation, but they also cooperate with a HECT E3, Rsp5, to mono-ubiquitinate certain plasma membrane receptors [60–62]. This modification signals receptor endocytosis, leading to degradation in the vacuole (equivalent to the mammalian lysosome). Ubc1 is partially redundant with Ubc4/5 in endocytosis [61, 62], as seen in ubiquitination reactions leading to proteasome degradation (above). Thus, Ubc1, 4, and 5 may be able to substitute for one another in complexes involving many different E3s, probably reflecting the strong conservation of the core domain between Ubc1 and Ubc4/5. Ubc4 and Ubc5 may also act with Rsp5 to regulate protein trafficking downstream of endocytosis (reviewed in Ref. [63]).

5.4.3
Non-proteolytic Functions

As mentioned in Section 5.2, Ubc2 is the defining player in a conserved DNA damage-tolerance pathway [8, 10, 64]. Here Ubc2 partners with a RING E3, Rad18, to modify a DNA polymerase processivity factor with a single ubiquitin [65]. This modification signals error-prone bypass of DNA lesions [66]. Ubc2 partners with a different RING E3 (Table 5.1) to mono-ubiquitinate histone H2B [67–69]. This modification promotes histone methylation, which in turn regulates transcription and silencing [70, 71].

Ubc13 participates in the same DNA damage-tolerance pathway as Ubc2 [72]. Ubc13 collaborates with two enzyme partners (Table 5.1) to modify the DNA polymerase cofactor (above) with a K63-linked polyubiquitin chain, which signals error-free lesion bypass [65, 72, 73]. In higher organisms, Ubc13 also helps to synthesize K63-linked polyubiquitin chains in a second non-proteolytic signaling pathway (see

Ref. [74]). The function and mechanism of Ubc13 are discussed in more detail below (Section 5.7).

5.4.4
E2s of Uncertain Function

Ubc10 is required for the biogenesis of the peroxisome, an oxidative organelle [75]. This E2 plays a role in peroxisomal protein import [76] and is recruited to the peroxisomal membrane through an interaction with a partner protein [77]. Membrane-localized Ubc10 also seems to be spatially proximal to Pex10, which has a RING-like domain [78]. Whether Pex10 is a cognate E3 of Ubc10 remains to be determined, as does the mechanistic role of ubiquitin conjugation in peroxisome biogenesis.

The function of the remaining yeast E2, Ubc11, remains uncertain. Ubc11 is very similar to E2-C, a clam E2 that acts with an essential multi-subunit RING E3, the *a*naphase *p*romoting *c*omplex (APC) or cyclosome, to ubiquitinate mitotic cyclins [79]. This reaction leads to cyclin degradation by proteasomes, which drives exit from mitosis [35, 80]. Mitotic cyclin ubiquitination can be reconstituted *in vitro* with apparent amphibian and fission yeast orthologs of either Ubc11 or Ubc4 [81, 82] and other data implicate both E2s in this process in higher cells [82, 83]. However, mitotic cyclins are efficiently degraded in budding yeast *ubc4Δ* and *ubc11Δ* strains, indicating that other E2 enzymes can support this essential function in *S. cerevisiae* [84].

5.4.5
E2 Enzymes and Disease

There are now several striking examples of disease-related defects in ubiquitin conjugation, but most of them involve E3s rather than E2s. This is not surprising given the paramount role of E3s in substrate selection and the corresponding intensity of research effort that has been focused on E3s. Still, there are several hints that defects at the E2 level of the conjugation cascade can also contribute to disease.

Many viruses subvert the ubiquitin system to evade the host cell's defenses or modulate the cellular environment so as to promote viral replication (see Refs. [85, 86] and Section 5.7). The genome of African swine fever virus encodes an E2 enzyme that is somewhat similar to yeast Ubc3 [87, 88]. This enzyme might alter the activity or specificity of the host cell's conjugation cascade so as to benefit the virus, or it could act on specific viral proteins. *H*erpes *S*implex *V*irus-1 (HSV-1) encodes an E3 enzyme that specifically binds the host cell's Ubc3/Cdc34 enzyme and targets this E2 for ubiquitination and (presumably) degradation – events that may help to stabilize specific cyclins and promote viral replication [89].

A different kind of relationship between an E2 enzyme and disease is exemplified by the finding that the Alzheimer's amyloid-β peptide induces the expression of E2$_{25K}$, a mammalian relative of yeast Ubc1 [90]. E2$_{25K}$ was found to play a major role in amyloid-β-dependent neuronal cell killing. This effect may be related to the E2$_{25K}$-dependent production of aberrant polyubiquitin chains, leading to the inhibition of proteasomes [90, 91]. Other studies showed that the human homolog of yeast Ubc11 is over-expressed in numerous cancer cell lines and primary tumors and that forced over-expression of this E2 in cultured cells can drive proliferation and transformation [92]. Similarly, transformation and chromosomal abnormalities were observed following over-expression of human Ubc2b [93]. Such disease-related over-expression effects could arise in two different ways. The higher E2 concentration could lead to a relaxation of specificity – that is, pairings with non-cognate E3s – leading to inappropriate ubiquitination events. Alternatively, specificity could be maintained, but an inappropriately high flux through the normal E2/E3 pathway could lead to the excessive ubiquitination of cognate substrates.

5.5
E2 Enzymes Dedicated to Ubiquitin-like Proteins (UbLs)

Ubiquitin is just one member of a family of protein modifiers that share a common fold and a common mechanism of isopeptide tagging [70, 94, 95]. Like ubiquitin, individual UbLs are activated at a C-terminal glycine residue by a specific E1 enzyme. Often, the next step is transfer to a specific E2 enzyme. Certain UbL-specific E2 enzymes are so similar to ubiquitin-conjugating enzymes that they were initially thought to be members of the ubiquitin-conjugating enzyme family. This was true of yeast Ubc9 and Ubc12, which are dedicated to Smt3/SUMO and Rub1/Nedd8, respectively (Table 5.1). SUMO modifies numerous cellular proteins and has a broad functional range [94], but the only known target of Nedd8 is a specific lysine residue in one subunit (the cullin) of SCF E3s. Nedd8 modification activates these E3s (see Ref. [70]).

The reader should consult earlier reviews [70, 94, 95] and other chapters in this volume for a detailed discussion of UbL biology and biochemistry. There are two important points for the current discussion. First, the conjugation cascades of UbLs differ from that of ubiquitin chiefly in terms of complexity – there is one conjugating enzyme per UbL, and many fewer E3s. Second, because modifier proteins (including ubiquitin) do not interact strongly with their dedicated E2s (Section 5.6.1), it is believed that E1 enzymes play the major role in matching E2s with the correct modifier protein (see Ref. [96]).

5.6
The Biochemistry of E2 Enzymes

5.6.1
E1 Interaction

An E2 needs to associate with several different proteins in the course of the ubiquitin conjugation cascade, with the first being the E1. Mutational studies conducted with the SUMO-specific E2 Ubc9 suggest that free Ubc9 associates with its free cognate E1 through a surface of Ubc9 that includes the C-terminal residues of α-helix H1 and residues in a loop between β-strands S1 and S2 (Figure 5.3) [97]. This surface of Ubc9 is also important for thiol ester bond formation [97]. Several residues in α-helix H1, particularly the C-terminal residues, are poorly conserved among E2s, and Ubc9 contains a five-residue insertion in the loop between β-strands S1 and S2 (Figure 5.2). Thus, this region of E2s may contribute to specificity for their cognate E1s. Consistent with this idea, the N-terminal helix (H1) of a ubiquitin E2 was found to be important for E2/ubiquitin thiol ester formation [98]. One cautionary note is that Ubc9 displays a substantial affinity for its free E1 [97], whereas ubiquitin E2s bind tightly to their E1 only after it has been loaded with ubiquitin [6, 99, 100]. The structural basis of this effect remains to be determined.

5.6.2
Interactions with Thiol-linked Ubiquitin

As a consequence of interacting with ubiquitin-loaded E1, the E2 accepts the activated ubiquitin at its active-site cysteine residue. This thiol ester complex, although biochemically detectable [6], has not been crystallized because it is labile in comparison to the requirements of structural biology. However, NMR chemical shift perturbations have been used to map the binding surface of ubiquitin onto human Ubc2b [101], yeast Ubc1 [102], and human Ubc13 [103]. All three models map the ubiquitin-binding surface of the E2 to a common area that includes parts of α-helix H3, the loop between α-helices H3 and H4, and residues around the active-site cysteine in the extended S4-H2 loop, including part of the 3_{10} helix H2 (Figures 5.2 and 5.4). The C-terminus of ubiquitin extends around part of the E2 and is constrained in a cleft [102, 103]. Although this contact surface is detectable in the thiol ester, free E2s display a negligible affinity for free ubiquitin [6]. Thus, the covalent E2/ubiquitin bond enables the formation of these non-covalent contacts.

5.6.3
E3 Interactions

After E2/ubiquitin thiol ester formation, the ubiquitin must be transferred to the substrate, which is sometimes another ubiquitin. An E3 is usually required for this reaction *in vitro*, and is always required *in vivo*. There are three known types of E3s: the RING domain, HECT domain, and U-box (UFD2 homology) families. RING and U-box E3s act as bridging factors for E2s and substrates, but HECT E3s use a different mechanism, adding an extra step to the pathway (Section 5.6.3.3).

5.6.3.1 RING E3/E2 Interactions

The small RING domain coordinates two zinc ions in a cross-brace arrangement [104]. The domain is defined by the presence of eight zinc-binding groups (cysteines and histidines) with a conserved spacing, such that the distance between

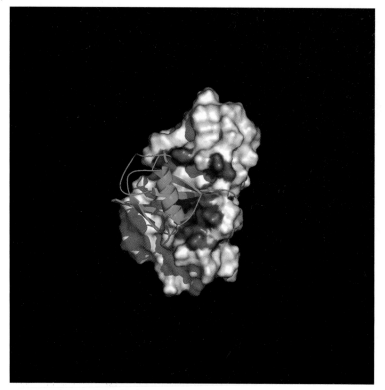

Fig. 5.4. Ubc1/ubiquitin thiol ester complex model (1FXT).
The surface of Ubc1 is shown with residues implicated in
ubiquitin binding colored purple and the active-site cysteine
colored yellow. Ubiquitin is colored green.

the two zincs is conserved at 14 Å [104]. Sequence conservation between RING domains is otherwise minimal. The RING-domain fold consists of a central α-helix and several small β-strands separated by loops with variable lengths [105].

RING E3s can be either single-subunit or multi-subunit enzymes. The crystal structure of UbcH7 complexed to a single-subunit RING E3, c-Cbl, shows that the RING domain is the main site of contact, although there are a few intermolecular hydrogen bonds to a non-RING helix of the E3 [106] (Figure 5.5). The structure of the E2 in the c-Cbl/UbcH7 complex is unchanged relative to free E2 structures. The main basis for the interaction is the packing of several hydrophobic residues of UbcH7, notably F63, into a shallow groove on the RING domain surface. These residues come from the S3–S4 and H2–H3 loops (Figure 5.3). The α-helix H1 of UbcH7 makes the hydrogen-bond contacts to the non-RING α-helix. Interestingly, even though the c-Cbl/UbcH7 structure undoubtedly shows conserved RING/E2 contacts, this complex is catalytically inactive (cited in Ref. [107]). Therefore additional E2/RING contacts may be needed for catalytic competence.

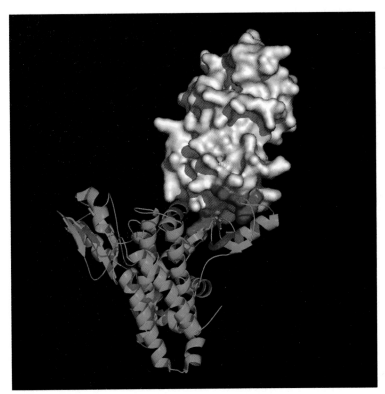

Fig. 5.5. UbcH7/c-Cbl complex (1FBV). The surface of UbcH7 is shown with residues interacting with the c-Cbl RING domain shown in red and the active-site cysteine shown in yellow. c-Cbl is colored green.

The surface of UbcH7 that contacts c-Cbl does not overlap with the E2 surface that contacts ubiquitin (Figures. 5.4 and 5.5; see also Ref. [108]), confirming that the E2/ubiquitin thiol ester can associate with a RING E3. The E2 surface that contacts c-Cbl does, however, overlap the E2 surface implicated in E1/E2 interactions (Section 5.6.1). Thus, the E1 may have to depart from the E2/ubiquitin complex before E2/E3 interactions can take place.

The closest approach of a RING-domain residue to the active-site cysteine of UbcH7 is about 15 Å, arguing against a role for RING E3s in chemical catalysis [106]. Instead, RING E3s have been proposed to facilitate ubiquitination by inducing physical proximity of the E2/ubiquitin thiol ester and the substrate [23, 30, 106, 109]. Catalysis would result from the increased local concentrations of the two reactants (discussed further below).

RING/E2 interactions have also been studied with BRCA1. This E3 is unique in that it must heterodimerize with a second RING-domain protein, BARD1, in order to display maximal E3 activity [110]. Even though the heterodimer interface leaves both RING domains available for interaction [110], UbcH5c binds exclusively to the BRCA1 RING [107]. The interacting surface of UbcH5c is homologous to the surface of UbcH7 that contacts the c-Cbl RING domain in that several residues of UbcH5c pack into a cleft on the BRCA1 RING domain. But UbcH5c also makes several contacts with the C-terminus of the BRCA1 RING domain and with a non-RING region of the heterodimer [107]. These extra contacts are not observed when UbcH7 binds to the BRCA1/BARD1 complex [107]. Because the BRCA1/BARD1/UbcH7 complex is inactive, the extra contacts observed with UbcH5c may help to create a competent E2/E3 complex.

Despite the greater complexity of multi-subunit RING E3s, a common theme is evident – all SCF E3s, as well as several other types of cullin-based E3s, utilize a common RING-domain subunit, the small protein Rbx1 (reviewed in Refs. [30, 111]). Four subunits compose the minimal SCF E3 ligase complex: a cullin scaffold, Rbx1, an adaptor protein (Skp1), and a substrate-binding subunit that connects to the adaptor through a conserved domain called the F-box (see Ref. [30]). The cullin acts as a scaffold, with Rbx1 binding to one end to form a cullin/Rbx1 subcomplex that recruits the E2 and, in many cases, displays a substrate-independent ubiquitin-ligase activity (see Refs. [23, 30]).

The crystal structure of the mammalian SCF^{Skp2} (Skp1/Cul1/F-box^{Skp2}/Rbx1) E3 ligase shows a remarkably rigid, elongated complex [112]. The Cul1 scaffold contains three cullin-repeat motifs that span ~110 Å, with Rbx1 binding to a discrete C-terminal α/β domain. The Skp1/F-box^{Skp2} complex binds to the opposite (N-terminal end) of the cullin. Rbx1 displays a hydrophobic groove, as seen previously in the c-Cbl RING domain [106, 112]. In c-Cbl, this groove provides an interaction surface for UbcH7 and it is reasonable to assume a similar mode of interaction in the case of Rbx1. Interestingly, the site where Nedd8 modifies the cullin is close to where the E2 binds, consistent with data which suggest that neddylation modulates E2 binding or activity [113, 114].

A model of the full SCF/E2 complex [112] shows that the end of Skp2 which binds the substrate is pointed toward the Rbx1-bound E2, with a 50-Å gap between the two. Models based on two other SCF structures show similar distances between the F-box protein and the E2 [109, 115]. Whether this gap can be bridged by the bound substrate is currently unclear. It has been suggested that the E2 may bind to Rbx1 somewhat differently than UbcH7 is observed to bind in the c-Cbl RING/UbcH7 complex, but it is not obvious that this can lead to a 20 Å movement of the E2 toward the bound substrate as suggested [109].

One could also imagine that the bound substrate and E2 "meet" through conformational changes of the SCF complex. However, the rigid separation produced by the Cul1 scaffold seems to be important for activity – introducing a flexible linker into the center of Cul1 produced a protein that could still bind an E2, but did not catalyze substrate ubiquitination [112]. An interesting study established a positive correlation between the rate of dissociation of the Ubc3/ubiquitin intermediate from the RING domain and the rate of Sic1 ubiquitination catalyzed by SCF^{Cdc4} [116]. The authors proposed that the role of the RING domain is to bring the charged E2 into the vicinity of the SCF-bound substrate, but that release of the charged E2 is important to bridge the gap and enable multiple substrate lysines to be targeted. However, although these mechanisms may place the substrate's lysine residue in the general vicinity of the E2's active site, it is unclear that they can establish an effective orientation of the lysine residue and the thiol ester bond. Since there is no known consensus site for ubiquitination [23, 117, 118], it is unlikely that specific molecular contacts in the vicinity of the substrate's lysine residue are used to position this attacking group. Overall, it remains unclear how the substrate's lysine residue approachs the E2 active site.

5.6.3.2 U-box E3/E2 Interactions

The U-box family of E3s bind E2s through the small U-box domain [119]. Some U-box E3s do not seem to have their own cognate substrates, but instead promote polyubiquitination of the substrates of other E3s [120]. Other U-box E3s have defined cognate substrates and behave in a canonical manner [121, 122].

An NMR structure [123] confirmed an earlier prediction [124] that the U-box domain has a RING-domain-like fold. Remarkably, the U-box domain uses hydrogen-bonding networks in place of zinc coordination to support the characteristic cross-brace arrangement. These interactions stabilize a globular fold consisting of a central α-helix surrounded by several β-strands, which are separated by loops of variable length [123]. There is a shallow groove in the surface located in a position homologous to the E2-interacting surface of RING domains. Mutational studies have linked E3 ligase activity to some of the residues in the surface groove [123, 125]. Since these mutations do not disrupt the U-box fold, they are likely to abrogate E2 binding. Although it is likely that E2s bind similarly to the U-box and RING domains, no E2/U-box structure has been reported so far.

5.6.3.3 HECT E3/E2 Interactions

HECT-domain E3s are defined by the presence of a domain of $\sim$350 residues that is homologous to the C-terminus of the founding family member, E6AP (E6 *A*ssociated *P*rotein [126]). E6AP is known for its role in binding the E6 protein of oncogenic human papilloma viruses and targeting the p53 tumor suppressor for ubiquitination and degradation [127]. HECT-domain E3s possess an active-site cysteine residue positioned $\sim$35 residues upstream of the C-terminus; a thiol ester with ubiquitin is formed at this site and is required for substrate ubiquitination [128].

The crystal structure of an E6AP/UbcH7 complex showed that the HECT domain is L-shaped, with a large, mostly α-helical, N-terminal lobe and a small C-terminal lobe with an α/β structure [108] (Figure 5.6). UbcH7 binds to the end of the N-terminal lobe and somewhat parallel to the C-terminal lobe, forming an overall U-shaped complex. UbcH7 binds in a large hydrophobic groove in the N-terminal lobe [108]. As seen in other E2/E3 structures, neither the E2 enzyme nor the HECT domain changes its overall fold upon binding. UbcH7 contacts its binding groove with residues from the S3–S4 loop and the H2–H3 loop (Figure 5.3). A few contacts are also made with the C-terminal portion of α-helix H1.

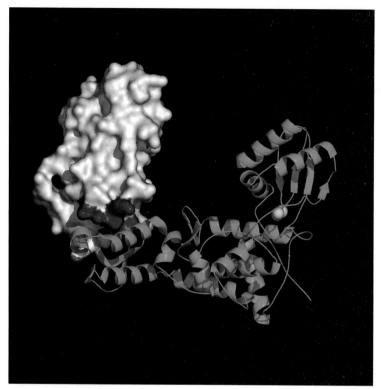

Fig. 5.6. UbcH7/E6AP (1C4Z). The surface of UbcH7 is shown with residues interacting with the E6AP HECT domain shown in blue and the active-site cysteine shown in yellow. E6AP is colored green with the active-site cysteine between the N- and C-lobe shown in yellow.

Remarkably, these are the same two loops and helix that bind the RING domain in the c-Cbl/UbcH7 structure [106]. That two different E3s contact a largely similar surface on UbcH7 (Figures 5.5 and 5.6) can be explained through the nature of key side-chain contacts. The S3–S4 loop seems to provide most of the specificity, as it contains the F63 residue that is present in all E2s that are known to bind both HECT E3s and c-Cbl [106, 108]. In c-Cbl, the main contacts for F63 are isoleucine and tryptophan residues located in the RING groove [106]. Both F63 and its contact site in c-Cbl are seen to vary in other E2/RING E3 pairs, suggesting that interactions between these three specific residues are needed, but that the nature of the contact can vary [106]. In other words, a different E2 could bind to a different E3 with a similar geometry, but through different types of side-chain contacts. UbcH7-F63 makes specific contacts with six E6AP residues, so this interaction is likely to be important for all HECT/E2 pairs [108]. The H2–H3 loop that makes the other major contacts with the HECT domain is part of the more variable E2 surface (Section 5.3). The specific contacts made between the H2–H3 loop of UbcH7 and the E6AP HECT domain could be used to correctly predict the E2 preferences of E6AP and Rsp5 [108]. Thus, residues in the S3–S4 and H2–H3 loops play an important role in determining the specificity of E2/E3 interactions.

Positioned near the bend between the two lobes of the HECT domain is its active-site cysteine [108]. This residue is 41 Å away from the active-site cysteine of UbcH7 (Figure 5.6), suggesting that a large conformational change is needed to bring about transfer of ubiquitin from E2 to E3. Such a mechanism was confirmed in the crystal structure of another HECT domain [129]. The WWP1-HECT domain resembles the E6AP HECT domain in having two lobes, but their relative positions differ. In the WWP1-HECT domain the two lobes form an inverted T-shape as opposed to the L-shape seen in the E6AP HECT structure [108, 129]. This conformational change can be brought about by a rotation around three residues in a hinge loop connecting the two lobes. Modeling in the E2 in a position homologous to that seen in the E6AP/UbcH7 structure, the distance between the two active-site cysteines decreases to 16 Å [129]. With additional rotation around the hinge loop the WWP1 active-site cysteine can be brought within 5 Å of the E2's active-site cysteine. Mutational studies suggested that the flexibility of this hinge loop is indeed important for ligase activity [129].

This flexibility requirement points to possible models for ubiquitin transfer and polyubiquitin chain elongation. One possibility is that the HECT domain adds ubiquitin to target substrates one at a time. This would imply that the E3 changes specificity – from recognizing the substrate to recognizing the ubiquitin – following transfer of the first ubiquitin to the substrate. A different possibility is that the chain is built up by the HECT E3 first, and then transferred as a unit to the substrate. This would require two active sites, one to hold the growing polyubiquitin chain, and the other to hold the next ubiquitin to be added. HECT E3/E2 complexes would satisfy this condition. RING E3/E2 complexes cannot, and thus would have to utilize another mechanism, presumably building on their rigid architecture. In an attractive model [129], the HECT cysteine could hold the first ubiquitin (and later the growing chain), while the C-lobe could rotate to position the first ubiquitin's target lysine near the thiol ester bond of the bound E2/ubiquitin intermediate. Subsequent rounds of ubiquitin addition to the chain terminus would require the C-lobe to keep rotating, ultimately ending in steric problems for the chain which could favor its transfer to the substrate.

A third general model for polyubiquitin chain extension is that the initiation and elongation phases of the reaction involve different E2 enzymes, different E3 enzymes, or different E2/E3 complexes. The modification of a substrate with a non-canonical polyubiquitin chain follows the third model. The Rad6/Rad18 complex ligates the first ubiquitin, while the Mms2/Ubc13/Rad5 complex extends the chain [65, 73]. The extension of K48-linked chains from ubiquitin fusion proteins seems to involve the sequential action of two different E3s with the same E2 [120]. In another possible example, two E2s (orthologs of Ubc11 and Ubc4) have been suggested to act sequentially with the APC in the polyubiquitination of mitotic cyclins in fission yeast [82].

5.6.4
E2/Substrate Interactions

With HECT domain E3s, all of the chemistry of isopeptide bond formation occurs at the E3 active site. With RING and U-box E3s, however, the E2 participates directly in this chemical reaction, so the substrate's lysine must closely approach the E2's active site. A crystal structure of the SUMO E2 Ubc9, complexed with a large fragment of RanGAP1 (an efficient sumoylation substrate), reveals the specificity of this interaction [130]. Unlike ubiquitination, sumoylation is site-specific. The target lysine for sumoylation lies within a tetrapeptide sequence motif Ψ-K-X-D/E, where Ψ is a hydrophobic residue, K is the target lysine, and X is any residue. Ubc9 makes specific interactions with each of these consensus-motif residues in a manner that places the lysine ε-amino group within 3.5 Å of the Ubc9 active-site cysteine [130]. The lysine approaches the cysteine from what is expected to be its unencumbered (by SUMO) side [102]. The interacting surface on Ubc9 involves α-helix H4, the loop preceding it, and the extended S4–H2 loop, including the active-site cysteine [130] (Figure 5.3). This surface does not overlap with the presumptive binding surface for SUMO. This mode of interaction is unlikely to hold with ubiquitin E2s, since no general consensus site for ubiquitination is known.

A model for ubiquitin E2/substrate interactions has also been proposed for the special case in which the substrate is ubiquitin [131]. The crystal structure of the Mms2/Ubc13 complex led to the modeling of an E2/UEV/ubiquitin (donor)/ ubiquitin (acceptor) model. As discussed in Section 5.7, UEV (*U*biquitin *E2 V*ariant) proteins such as Mms2 are homologous to E2s, but lack the active-site cysteine residue. Known UEV/Ubc13 complexes act as E2 enzymes specialized for the synthesis of K63-linked polyubiquitin chains [72, 132]. In the model [131], Ubc13 is bound to the donor ubiquitin through a thiol ester bond in a manner that agrees well with inferences from NMR analysis of the Ubc1/ubiquitin thiol ester [102] (Figure 5.7). The position of the non-covalently bound acceptor ubiquitin is determined by the orientation of Mms2 on Ubc13 (Figure 5.7). The acceptor ubiquitin has its K63 side chain placed to enter the active site of Ubc13 to form a diubiquitin conjugate. The model suggests that K63 of ubiquitin is selected as the conjugation site through steric exclusion of other lysines, as determined by an interaction between Mms2 and a region of ubiquitin that is distant from K63 [131]. Recent NMR studies have confirmed and refined this model [103]. Thus, the substrate lysine is presented to the active-site cysteine through an indirect mechanism, in contrast to the Ubc9/RanGAP1 example in which the E2 interacts directly with the lysine residue itself [130]. Unlike most ubiquitination reactions, the modification of ubiquitin itself is often site-specific. The Mms2/Ubc13/ubiquitin model can help to explain this phenomenon.

5.6.5
E2 Catalysis Mechanism

Chemical catalytic mechanisms in the ubiquitin conjugation cascade have proved difficult to decipher. The reactions leading to E2/ubiquitin thiol ester and isopep-

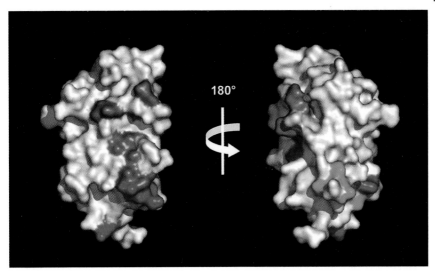

Fig. 5.7. Ubc13 interaction surface. The interacting surfaces have been mapped onto Ubc13. The active-site cysteine is shown in yellow. Colored surfaces contact: covalently bound ubiquitin (purple); RING domains (red); E1 (presumptive, green); acceptor ubiquitin involved in K63-linked polyubiqutin-chain synthesis (blue).

tide bond formation would be facilitated by electrostatic stabilization of the oxyanion and deprotonation of the attacking amino group (isopeptide bond formation) by a general base [133, 134]. However, while the sequence conservation around the E2 active site is very high (Figure 5.2), all E2 structures show a lack of candidate catalytic residues close to the cysteine (see Refs. [23, 135]). Although catalytic residues could be contributed by other enzymes in the cascade or by the E2 backbone, structural data argue strongly against a chemical catalytic contribution where it may be needed most – in reactions involving RING and U-box domain E3s.

Recent studies [136] addressed the role of a strictly conserved asparagine positioned just upstream of the active-site cysteine (N79 in Ubc1 numbering, Figure 5.2). In existing E2 structures the asparagine is hydrogen-bonded to the backbone or a side chain. It is distant from the E3 contact surface and, as expected, it is dispensable for E2 binding to RING domain E3s. However, the asparagine is critical for E2-catalyzed and RING E3/E2-catalyzed ubiquitin conjugation reactions. The similar effect of asparagine mutation on the two types of conjugation reactions is reasonable given that E2s do not experience structural perturbations upon binding to E3s (above). The data suggest that an intrinsic catalytic role of the asparagine side chain is brought into play through RING-mediated recruitment of the catalytically competent E2/ubiquitin thiol ester. The asparagine is dispensable for upstream and downstream thiol transfer reactions, suggesting that catalytic residues for these reactions may be located in the E1 and HECT E3 active sites.

A specific proposal for the role of the asparagine was developed in a model which breaks the hydrogen bonds to the backbone and rotates the asparagine to-

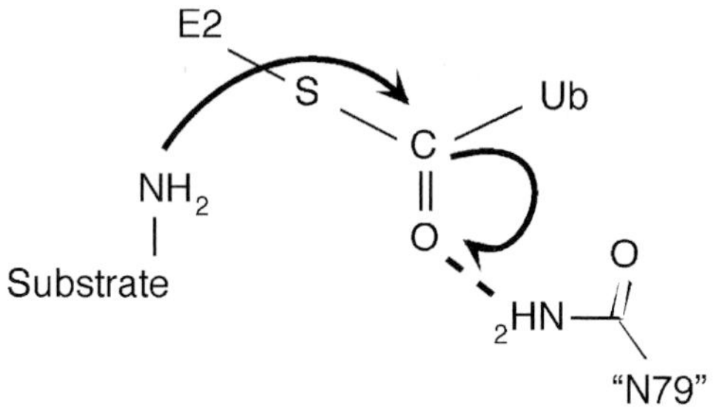

Fig. 5.8. Model for catalytic role of E2 active-site asparagine. The side chain of the asparagine in the conserved "HPN" motif (Figure 5.2) stabilizes the oxyanion that forms when the substrate's lysine attacks the E2/ubiquitin thiol ester bond. N79 is numbering for Ubc1 (Figure 5.2).

ward the active-site cysteine [136]. Molecular modeling suggested that the asparagine can be positioned to donate a hydrogen bond to the oxyanion (Figure 5.8) [136]. Many cysteine proteases, including deubiquitinating enzymes, use an amide side chain in this manner [134, 137–139]. Structural studies of a deubiquitinating enzyme have shown that the entry of ubiquitin into the active site causes a histidine and an asparagine to shift their positions so that the histidine becomes the general base and the asparagine provides the oxyanion hole [137]. Similarly, ubiquitin binding in the E2 active site could be a trigger that repositions the asparagine. So far, no general base is evident in E2s, but this group may not be needed due to the lability of the thiol ester bond.

5.7
Functional Diversification of the E2 Fold

Increasing evidence suggests that evolution has used (and is using) the E2 fold for new purposes. In one apparent example of functional expansion, E2 core domains have been observed to be embedded within much larger polypeptide chains [140, 141]. The functional properties of these massive E2s remain poorly characterized, and it is likely that more of them will be discovered. But the clearest case of functional diversification is provided by the UEV proteins. UEVs are related to E2s in their primary, secondary, and tertiary structures, but they lack an active-site cysteine residue and therefore cannot function as canonical E2s [142]. Nonetheless they play several different roles in ubiquitin-dependent pathways.

Mms2 and its close (mammalian) relative Uev1a form heterodimers with Ubc13 [72, 132]. Each complex plays a key role in the synthesis of K63-linked chains, which act as non-proteolytic signals in different cellular pathways. The Mms2/Ubc13 complex participates in the *UBC2/RAD6*-dependent DNA damage tolerance pathway by polyubiquitinating the DNA polymerase processivity factor called PCNA

(*P*roliferating *C*ell *N*uclear *A*ntigen) [65, 72]. To be activated for this pathway, PCNA is first mono-ubiquitinated by the Rad6/Rad18 complex, and then modified with a K63-linked polyubiquitin chain by the Mms2/Ubc13/Rad5 complex (Rad18 and Rad5 are RING E3s) [65, 73]. The Mms2/Ubc13 complex has a core ubiquitin polymerization activity [72]. Rad5 might stimulate this activity [132] or target the Uev/E2 complex to PCNA [73], or both.

The related human Uev1a/Ubc13 complex is involved in NFκB signal transduction [132]. It plays an intermediate role in the signaling cascade that starts with a proinflammatory cytokine signal and culminates in the nuclear translocation of the active NFκB transcription factor. In this pathway the Ubc13/Uev1a complex modifies a RING E3, Traf6, with K63-linked polyubiquitin chains [132]. This modification is linked to Traf6 oligomerization. It instigates a cascade of kinase reactions ultimately cause the ubiquitination and degradation of NFκB's inhibitory partner, IκBα [74, 143].

The crystal structure of Mms2 has been solved alone and in complex with Ubc13 [131, 144]. The overall fold is similar to that in E2s, containing a central four-stranded anti-parallel β-sheet surrounded by α-helices. Differences include the absence of the C-terminal α-helix H5 in the shorter Mms2 protein. The helical N-terminus of Mms2 is also extended compared to Ubc13, and this region plays the major role in heterodimer formation. Two ubiquitins can bind to the heterodimer (Section 5.6.4) and the surfaces they contact do not overlap with the surface contacted by Rad5 [131, 145].

Ubiquitin plays a crucial role in a protein-trafficking pathway that delivers specific cargo proteins to regions of the late endosome membrane that invaginate into the lumen, thereby targeting these proteins to the vacuole/lysosome (reviewed in Ref. [146]). A different UEV protein, called Tsg101 in humans (*T*umor *S*usceptibility *G*ene) and Vps23 in yeast (*V*acuolar *P*rotein *S*orting), is part of a large complex that plays a critical role in the sorting step. Cargo proteins are selected based on their conjugation to mono-ubiquitin; the specific role of the UEV protein is to bind the cargo-linked mono-ubiquitin moiety [147]. HIV-1 and certain other viruses subvert this function of Tsg101 in order to bud from the plasma membrane [148–150]. Mechanistically, Tsg101 is recruited to the virus budding sites by binding to a tetrapeptide "PTAP" motif in the late domain of viral proteins such as HIV-1-GAG. Tsg101 is essential for virus budding from the plasma membrane [148], so it is possible that the endocytic budding machinery is hijacked to the plasma membrane via the Tsg101/GAG interaction [85].

The solution structure of the Tsg101 UEV domain has been solved alone and in complex with a PTAP-containing peptide [151, 152]. Human Tsg101 contains 390 amino acids, with the UEV domain located at the N-terminus [152]. The UEV domain is the minimal region needed to bind HIV-1 GAG, and is also the domain involved in mono-ubiquitin recognition and binding [153]. The overall fold of the UEV domain is similar to that of E2s. One notable difference is the presence of an extra N-terminal α-helix on Tsg101 [152]. The other major difference is the absence in Tsg101 UEV of the two C-terminal α-helices of the E2s – a truncation that was also seen in the Mms2 structures. This truncation appears to be a special trait of UEV proteins [154]. In Tsg101 UEV, the absence of the C-terminal helices helps to create the binding site for the PTAP peptide [151].

When aligning the structures of a canonical E2, Mms2, and the Tsg101 UEV domain, the hydrophobic core and the region surrounding the vestigial active site are quite similar, but the Tsg101 UEV domain differs from Mms2 and canonical E2s in the positions of the first two β-strands [152]. In Tsg101 they are elongated and shifted toward the N-terminus, forming a β-hairpin that extends 11 residues outside the main body of the domain [152]. This loop is important for ubiquitin binding by Tsg101 [152]. As determined by chemical shift mapping and mutagenesis studies, the Tsg101/ubiquitin binding interface involves the bottom half of the four-stranded β-sheet, including the β-hairpin (loop S1–S2). The binding interface for ubiquitin on Tsg101 is distinct from the surface that Mms2 uses to position the acceptor ubiquitin within the Ubc13/Mms2 complex [131, 144]. Thus, two different UEVs bind ubiquitin in two different ways. The structural biology and biochemistry of UEVs illustrates how modest changes to an E2-like module can create new, functionally important interaction sites. The UEV domain is just one of a growing set of small domains that can endow other protein domains with ubiquitin-binding capability (reviewed in Ref. [63]). Such binding elements are likely to play important roles in transducing ubiquitin signals in diverse cellular pathways.

5.8
Concluding Remarks

We have emphasized the biochemical properties of E2s, particularly interactions with other factors in the conjugation cascade, because these properties are central to the biological actions of E2s. We have tried to give a flavor of the "creativity and economy" [103] with which E2s have evolved to maximize the interaction potential of a relatively small and conserved surface (Figure 5.7). Owing to the large scope of the relevant literature and the limited length of this chapter, we have not done full justice to the biological breadth of the E2 enzyme family. For example, we have focused on yeast and mammalian enzymes, but ubiquitin conjugation is increasingly being studied in other model organisms, including flies, worms, and plants. These systems offer powerful tools to address outstanding questions about ubiquitin-dependent pathways in general and E2 enzymes in particular. What are some of those questions? Significant uncertainties remain concerning E2 catalysis and mechanism, as discussed in Section 5.6. Another important question has been largely ignored in this review – exactly why are there so many E2s? One appealing model is that the identity of the E2 can modulate the substrate specificity of the E3, but experimental evidence for this model remains sparse. Another possibility is that the E2 has little or no influence on substrate choice, but rather helps to control the flux of activated ubiquitin to its cognate E3. In view of the remarkable developments in ubiquitin biology over the last decade, we should be prepared for both interesting and unexpected answers to these (and other) questions.

References

1 HERSHKO, A. and CIECHANOVER, A. The ubiquitin system. *Annu. Rev. Biochem.* **1998**, *67*, 425–79.

2 CIECHANOVER, A., HELLER, H., ELIAS, S., HAAS, A. L., and HERSHKO, A. ATP-dependent conjugation of reticulocyte proteins with the polypeptide required for protein degradation. *Proc. Natl. Acad. Sci. USA* **1980**, *77*, 1365–68.

3 HERSHKO, A., CIECHANOVER, A., HELLER, H., HAAS, A. L., and ROSE, I. A. Proposed role of ATP in protein breakdown: conjugation of proteins with multiple chains of the polypeptide of ATP-dependent proteolysis. *Proc. Natl. Acad. Sci. USA* **1980**, *77*, 1783–86.

4 HAAS, A. L., WARMS, J. V. B., HERSHKO, A., and ROSE, I. A. Ubiquitin-activating enzyme. Mechanism and role in protein-ubiquitin conjugation. *J. Biol. Chem.* **1982**, *257*, 2543–48.

5 CANE, D. E. and WALSH, C. T. The parallel and convergent universes of polyketide synthases and nonribosomal peptide synthetases. *Chem. Biol.* **1999**, *6*, R319–R25.

6 HERSHKO, A., HELLER, H., ELIAS, S., and CIECHANOVER, A. Components of ubiquitin-protein ligase system. *J. Biol. Chem.* **1983**, *258*, 8206–14.

7 PICKART, C. M. and ROSE, I. A. Functional heterogeneity of ubiquitin carrier proteins. *J. Biol. Chem.* **1985**, *260*, 1573–81.

8 JENTSCH, S., McGRATH, J. P., and VARSHAVSKY, A. The yeast DNA repair gene RAD6 encodes a ubiquitin-conjugating enzyme. *Nature* **1987**, *329*, 131–34.

9 GOEBL, M. G., YOCHEM, J., JENTSCH, S., McGRATH, J. P., VARSHAVSKY, A., and BYERS, B. The yeast cell cycle gene CDC34 encodes a ubiquitin-conjugating enzyme. *Science* **1988**, *241*, 1331–35.

10 REYNOLDS, P., WEBER, S., and PRAKASH, L. RAD6 gene of Saccharomyces cerevisiae encodes a protein containing a tract of 13 consecutive aspartates. *Proc. Natl. Acad. Sci. USA* **1985**, *82*, 168–72.

11 SUNG, P., PRAKASH, S., and PRAKASH, L. Stable ester conjugate between the Saccharomyces cerevisiae RAD6 protein and ubiquitin has no biological activity. *J. Mol. Biol.* **1991**, *221*, 745–49.

12 BANERJEE, A., DESHAIES, R. J., and CHAU, V. Characterization of a dominant negative mutant of the cell cycle ubiquitin-conjugating enzyme Cdc34. *J. Biol. Chem.* **1995**, *270*, 26209–15.

13 SEMPLE, C. A. The comparative proteomics of ubiquitination in mouse. *Genome Res* **2003**, *13*, 1389–94.

14 WONG, B. R., PARLATI, F., QUK, K., DEMO, S., PRAY, T., HUANG, J., PAYAN, D. G., and BENNETT, M. K. Drug discovery in the ubiquitin regulatory pathway. *Drug Discov. Today* **2003**, *8*, 746–54.

15 KOLMAN, C. J., TOTH, J., and GONDA, D. K. Identification of a portable determinant of cell cycle function within the carboxyl-terminal domain of the yeast CDC34 (UBC3) ubiquitin conjugating (E2) enzyme. *EMBO J.* **1992**, *11*, 3081–90.

16 SILVER, E. T., GWOZD, T. J., PTAK, C., GOEBL, M. G., and ELLISON, M. J. A chimeric ubiquitin conjugating enzyme that combines the cell cycle properties of CDC34 (UBC3) and the DNA repair properties of RAD6 (UBC2): implications for the structure, function, and evolution of the E2s. *EMBO J.* **1992**, *11*, 3091–98.

17 PTAK, C., PRENDERGAST, J. A., HODGINS, R., KAY, C. M., CHAU, V., and ELLISON, M. J. Functional and physical characterization of the cell cycle ubiquitin-conjugating enzyme CDC34 (UBC3). *J. Biol. Chem.* **1994**, *269*, 26539–45.

18 HODGINS, R., GWOZD, C., ARNASON, T., CUMMINGS, M., and ELLISON, M. J. The tail of a ubiquitin-conjugating enzyme redirects multi-ubiquitin chain synthesis from the lysine 48-

linked configuration to a novel nonlysine-linked form. *J. Biol. Chem.* **1996**, *271*, 28766–71.

19 MATHIAS, N., STEUSSY, C. N., and GOEBL, M. G. An essential domain within Cdc34p is required for binding to a complex containing Cdc4p and Cdc53p in *Saccharomyces cerevisiae*. *J. Biol. Chem.* **1998**, *273*, 4040–45.

20 COOK, W. J., JEFFREY, L. C., SULLIVAN, M. L., and VIERSTRA, R. D. Three-dimensional structure of a ubiquitin-conjugating enzyme (E2). *J. Biol. Chem.* **1992**, *267*, 15116–21.

21 COOK, W. J., JEFFREY, L. C., XU, Y., and CHAU, V. Tertiary structures of class I ubiquitin-conjugating enzymes are highly conserved: crystal structure of yeast Ubc4. *Biochemistry* **1993**, *32*, 13809–17.

22 HOCHSTRASSER, M. Ubiquitin-dependent protein degradation. *Annu. Rev. Genet.* **1996**, *30*, 405–39.

23 PICKART, C. M. Mechanisms underlying ubiquitination. *Annu. Rev. Biochem.* **2001**, *70*, 503–33.

24 SEUFERT, W. and JENTSCH, S. Ubiquitin-conjugating enzymes UBC4 and UBC5 mediate selective degradation of short-lived and abnormal proteins. *EMBO J.* **1990**, *9*, 543–50.

25 JUNGMANN, J., REINS, H.-A., SCHOBERT, C., and JENTSCH, S. Resistance to cadmium mediated by ubiquitin-dependent proteolysis. *Nature* **1993**, *361*, 369–71.

26 HEINEMEYER, W., KLEINSCHMIDT, J. A., SAIDOWSKY, J., ESCHER, C., and WOLF, D. H. Proteinase yscE, the yeast proteasome/multicatalytic-multifunctional proteinase: mutants unravel its function in stress induced proteolysis and uncover its necessity for cell survival. *EMBO J.* **1991**, *10*, 555–62.

27 JOHNSON, E. S., MA, P. C., OTA, I. M., and VARSHAVSKY, A. A proteolytic pathway that recognizes ubiquitin as a degradation signal. *J. Biol. Chem.* **1995**, *270*, 17442–56.

28 SEUFERT, W., McGRATH, J. P., and JENTSCH, S. UBC1 encodes a novel member of an essential subfamily of yeast ubiquitin-conjugating enzymes

involved in protein degradation. *EMBO J.* **1990**, *9*, 4535–41.

29 GONEN, H., BERCOVICH, B., ORIAN, A., CARRANO, A., TAKIZAWA, C., YAMANAKA, K., PAGANO, M., IWAI, K., and CIECHANOVER, A. Identification of the ubiquitin carrier proteins, E2s, involved in signal-induced conjugation and subsequent degradation of IκBα. *J. Biol. Chem.* **1999**, *274*, 14823–30.

30 DESHAIES, R. J. SCF and cullin/RING H2-based ubiquitin ligases. *Annu. Rev. Cell Devel. Biol.* **1999**, *15*, 435–67.

31 SCHWOB, E., BOHM, T., MENDENHALL, M. D., and NASMYTH, K. The B-type cyclin kinase inhibitor p40SIC1 controls the G1 to S transition in S. cerevisiae. *Cell* **1994**, *79*, 233–44.

32 FELDMAN, R. M. R., CORRELL, C. C., KAPLAN, K. B., and DESHAIES, R. J. A complex of Cdc4p, Skp1p, and Cdc53p/cullin catalyzes ubiquitination of the phosphorylated CDK inhibitor Sic1p. *Cell* **1997**, *91*, 221–30.

33 SKOWYRA, D., CRAIG, K. L., TYERS, M., ELLEDGE, S. J., and HARPER, J. W. F-box proteins are receptors that recruit phosphorylated substrates to the SCF ubiquitin-ligase complex. *Cell* **1997**, *91*, 209–19.

34 PATTON, E. E., WILLEMS, A. R., SA, D., KURAS, L., THOMAS, D., CRAIG, K. L., and TYERS, M. Cdc53 is a scaffold protein for multiple Cdc34/Skp1/F-box protein complexes that regulate cell division and methionine biosynthesis in yeast. *Genes Dev.* **1998**, *12*, 692–705.

35 JACKSON, P. K., ELDRIDGE, A. G., FREED, E., FURSTENTHAL, L., HSU, J. Y., KAISER, B. K., and REIMANN, J. D. R. The lore of the RINGs: substrate recognition and catalysis by ubiquitin ligases. *Trends Cell Biol.* **2000**, *10*, 429–39.

36 KUS, B., CALDON, C. E., ANDORN-BROZA, R., and EDWARDS, A. M. Functional interactions of 13 yeast SCF complexes with a set of yeast E2 enzymes in vitro. *Proteins: Structure, Function, and Bioinformatics* **2004**, *54*, 455–67.

37 BIEDERER, T., VOLKWEIN, C., and SOMMER, T. Role of Cue1p in ubiquitination and degradation at the ER surface. *Science* **1997**, *278*, 1806–09.

38 KOSTOVA, Z. and WOLF, D. H. For whom the bell tolls: protein quality control of the endoplasmic reticulum and the ubiquitin-proteasome connection. *EMBO J.* **2003**, *22*, 2309–17.

39 BAYS, N. W., GARDNER, R. G., SEELIG, L. P., JOAZEIRO, C. A., and HAMPTON, R. Y. Hrd1p/Der3p is a membrane-anchored ubiquitin ligase required for ER-associated degradation. *Nature Cell Biol.* **2000**, *3*, 24–29.

40 FRIEDLANDER, R., JAROSCH, E., URBAN, J., VOLKWEIN, C., and SOMMER, T. A regulatory link between ER-associated protein degradation and the unfolded-protein response. *Nature Cell Biol.* **2000**, *2*, 379–84.

41 TRAVERS, K. J., PATIL, C. K., WODICKA, L., LOCKHART, D. J., WEISSMAN, J. S., and WALTER, P. Functional and genomic analyses reveal an essential coordination between the unfolded protein response and ER-associated degradation. *Cell* **2000**, *101*, 249–58.

42 SWANSON, R., LOCHER, M., and HOCHSTRASSER, M. A conserved ubiquitin ligase of the nuclear envelope/endoplasmic reticulum that functions in both ER-associated and Matα2 repressor degradation. *Genes Dev.* **2001**, *15*, 2660–74.

43 TIWARI, S. and WEISSMAN, A. M. Endoplasmic reticulum (ER)-associated degradation of T cell receptor subunits. *J. Biol. Chem.* **2001**, *276*, 16193–200.

44 WEBSTER, J. M., TIWARI, S., WEISSMAN, A. M., and WOJCIKIEWICZ, R. J. H. Inositol 1,4,5-trisphosphate receptor ubiquitination is mediated by mammalian Ubc7, a component of the endoplasmic reticulum-associated degradation pathway, and is inhibited by chelation of intracellular Zn^{2+}. *J. Biol. Chem.* **2003**, *278*, 38238–46.

45 FANG, S., FERRONE, M., YANG, C., JENSEN, J. P., TIWARI, S., and WEISSMAN, A. M. The tumor autocrine motility factor receptor, gp78, is a ubiquitin protein ligase implicated in degradation from the endoplasmic reticulum. *Proc. Natl. Acad. Sci. USA* **2001**, *98*, 14422–27.

46 SOMMER, T. and JENTSCH, S. A protein translocation defect linked to ubiquitin conjugation at the endoplasmic reticulum. *Nature* **1993**, *365*, 176–79.

47 BIEDERER, T., VOLKWEIN, C., and SOMMER, T. Degradation of subunits of the Sec61p complex, an integral component of the ER membrane, by the ubiquitin-proteasome pathway. *EMBO J.* **1996**, *15*, 2069–76.

48 HILLER, M. M., FINGER, A., SCHWEIGER, M., and WOLF, D. H. ER degradation of a misfolded luminal protein by the cytosolic ubiuqitin-proteasome pathway. *Science* **1996**, *273*, 1725–28.

49 CHEN, P., JOHNSON, P., SOMMER, T., JENTSCH, S., and HOCHSTRASSER, M. Multiple ubiquitin-conjugating enzymes participate in the in vivo degradation of the yeast MATα2 repressor. *Cell* **1993**, *74*, 357–69.

50 WALTER, J., URBAN, J., VOLKWEIN, C., and SOMMER, T. Sec61p-independent degradation of the tail-anchored ER membrane protein Ubc6p. *EMBO J.* **2001**, *20*, 3124–31.

51 RAO, H., UHLMANN, F., NASMYTH, K., and VARSHAVSKY, A. Degradation of a cohesin subunit by the N-end rule pathway is essential for chromosome stability. *Nature* **2001**, *410*, 955–59.

52 BYRD, C., TURNER, G. C., and VARSHAVSKY, A. The N-end rule pathway controls the import of peptides through degradation of a transcriptional repressor. *EMBO J.* **1998**, *17*, 269–77.

53 KAPLUN, L., IVANTSIV, Y., KORNITZER, D., and RAVEH, D. Functions of the DNA damage response pathway target Ho endonuclease of yeast for degradation via the ubiquitin-26S proteasome pathway. *Proc. Natl. Acad. Sci. USA* **2000**, *97*, 10077–82.

54 KOKEN, M. H. M., HOOGERBRUGGE, J. W., JASPERS-DEKKER, I., DE WIT, J., WILLEMSEN, R., ROEST, H. P., GROOTEGOED, J. A., and HOEIJMAKERS, J. H. J. Expression of the ubiquitin-conjugating DNA repair enzymes HHR6A and B suggests a role in spermatogenesis and chromatin modification. *Dev. Biol.* **1996**, *173*, 119–32.

55 ROEST, H. P., VAN KLAVEREN, J., DE WIT, J., VAN GURP, C. G., KOKEN, M. H. M., VERMEY, M., VAN ROIJEN, J. H., HOOGERBRUGGE, J. W., VREEBERG, M. T. M., BAARENDS, W. M., BOOTSMA, D., GROOTEGOED, J. A., and HOEIJMAKERS, J. H. J. Inactivation of the HR6B ubiquitin-conjugating DNA repair enzyme in mice causes male sterility associated with chromatin modification. *Cell* **1996**, *86*, 799–810.

56 KWON, Y. T., KASHINA, A. S., DAVYDOV, I. V., HU, R.-G., AN, J. Y., SEO, J. W., DU, F., and VARSHAVSKY, A. An essential role of N-terminal arginylation in cardovascular development. *Science* **2002**, *297*, 96–99.

57 SCHULE, T., ROSE, M., ENTIAN, K.-D., THUMM, M., and WOLF, D. H. Ubc8p functions in catabolite degradation of fructose-1,6-bisphosphatase in yeast. *EMBO J.* **2000**, *19*, 2161–67.

58 LI, Y.-P., LECKER, S. H., CHEN, Y., WADDELL, I. D., GOLDBERG, A. L., and REID, M. B. TNF-α increases ubiquitin-conjugating activity in skeletal muscle by up-regulating UbcH2/E2$_{20K}$. *FASEB J.* **2003**, *17*, 1048–57.

59 WEFES, I., MASTRANDREA, L. D., HALDEMAN, M., KOURY, S. T., TAMBURLIN, J., PICKART, C. M., and FINLEY, D. Induction of ubiquitin-conjugating enzymes during terminal erythroid differentiation. *Proc. Natl. Acad. Sci. USA* **1995**, *92*, 4982–86.

60 GITAN, R. S., SHABABI, M., KRAMER, M., and EIDE, D. J. A cytosolic domain of the yeast Zrt1 zinc transporter is required for its post-translational inactivation in response to zinc and cadmium. *J. Biol. Chem.* **2003**, *278*, 39558–64.

61 HICKE, L. and RIEZMAN, H. Ubiquitination of a yeast plasma membrane receptor signals its ligand-stimulated endocytosis. *Cell* **1996**, *84*, 277–87.

62 DUNN, R. and HICKE, L. Domains of the Rsp5 ubiquitin-protein ligase required for receptor-mediated and fluid-phase endocytosis. *Mol. Biol. Cell* **2001**, *12*, 421–35.

63 HICKE, L. and DUNN, R. Regulation of membrane protein transport by ubiquitin and ubiquitin-binding proteins. *Annu. Rev. Cell Devel. Biol.* **2003**, *19*, 141–72.

64 ULRICH, H. D. Degradation or maintenance: actions of the ubiquitin system on eukaryotic chromatin. *Eukaryotic Cell* **2002**, *1*, 1–10.

65 HOEGE, C., PFANDER, B., MOLDOVAN, G.-L., PYROWOLAKIS, G., and JENTSCH, S. *RAD6*-dependent DNA repair is linked to modification of PCNA by ubiquitin and SUMO. *Nature* **2002**, *419*, 135–41.

66 STELTER, P. and ULRICH, H. D. Control of spontaneous and damage-induced mutagenesis by SUMO and ubiquitin conjugation. *Nature* **2003**, *425*, 188–91.

67 HWANG, W. W., VENKATASUBRAHMANYAM, S., IANCULESCU, A. G., TONG, A., BOONE, C., and MADHANI, H. D. A conserved RING finger protein required for histone H2B monoubiquitination and cell size control. *Mol. Cell* **2003**, *11*, 261–66.

68 ROBZYK, K., RECHT, J., and OSLEY, M. A. Rad6-dependent ubiquitination of histone H2B in yeast. *Science* **2000**, *287*, 501–04.

69 WOOD, A., KROGAN, N. J., DOVER, J., SCHNEIDER, J., HEIDT, J., BOATENG, M. A., DEAN, K., GOLSHANI, A., ZHANG, Y., GREENBLATT, J. F., JOHNSTON, M., and SHILATIFARD, A. Bre1, an E3 ubiquitin ligase required for recruitment and substrate selection of Rad6 at a promoter. *Mol. Cell* **2003**, *11*, 267–74.

70 SCHWARTZ, D. C. and HOCHSTRASSER, M. A superfamily of protein tags: ubiquitin, SUMO and related modifiers. *Trends Biochem. Sci.* **2003**, *28*, 321–28.

71 SUN, Z.-W. and ALLIS, C. D. Ubiquitination of histone H2B regulates H3 methylation and gene silencing in yeast. *Nature* **2002**, *418*, 104–08.

72 HOFMANN, R. M. and PICKART, C. M.

Noncanonical *MMS2*-encoded ubiquitin-conjugating enzyme functions in assembly of novel polyubiquitin chains for DNA repair. *Cell* 1999, *96*, 645–53.

73 ULRICH, H. D. and JENTSCH, S. Two RING finger proteins mediate cooperation between ubiquitin-conjugating enzymes in DNA repair. *EMBO J.* 2000, *19*, 3388–97.

74 SUN, L. and CHEN, Z. J. The novel functions of ubiquitination in signaling. *Curr. Opin. Cell Biol.* 2004, *16*, 119–26.

75 WIEBEL, F. F. and KUNAU, W. H. The Pas2 protein essential for peroxisome biogenesis is related to ubiquitin-conjugating enzymes. *Nature* 1992, *359*, 73–76.

76 VAN DER KLEI, I. J., HILBRANDS, R. E., KIEL, J. A. K. W., RASMUSSEN, S. W., CREGG, J. M., and VEENHUIS, M. The ubiquitin-conjugating enzyme Pex4p of *Hansenula polymorpha* is required for efficient functioning of the PTS1 import machinery. *EMBO J.* 1998, *17*, 3608–18.

77 KOLLER, A., SNYDER, W. B., FABER, K. N., WENZEL, T. J., RANGELL, L., KELLER, G. A., and SUBRAMANI, S. Pex22p of *Pichia pastoris*, essential for peroxisomal matrix protein import, anchors the ubiquitin-conjugating enzyme, Pex4p, on the peroxisomal membrane. *J. Cell Biol.* 1999, *146*, 99–112.

78 ECKERT, J. H. and JOHNSSON, N. Pex10p links the ubiquitin conjugating enzyme Pex4p to the protein import machinery of the peroxisome. *J. Cell Sci.* 2003, *116*, 3623–34.

79 ARISTARKHOV, A., EYTAN, E., MOGHE, A., ADMON, A., HERSHKO, A., and RUDERMAN, J. V. E2-C a cyclin-selective ubiquitin carrier protein required for the destruction of mitotic cyclins. *Proc. Natl. Acad. Sci. USA* 1996, *93*, 4294–99.

80 PETERS, J. M. The anaphase-promoting complex: proteolysis in mitosis and beyond. *Mol. Cell* 2002, *9*, 931–43.

81 YU, H., KING, R. W., PETERS, J.-M., and KIRSCHNER, M. W. Identification of a novel ubiquitin-conjugating enzyme involved in mitotic cyclin degradation. *Curr. Biol.* 1996, *6*, 455–66.

82 SEINO, H., KISHI, T., NISHITANI, H., and YAMAO, F. Two ubiquitin-conjugating enzymes, UbcP1/Ubc4 and UbcP4/Ubc11, have distinct functions for ubiquitination of mitotic cyclin. *Mol. Cell. Biol.* 2003, *23*, 3497–505.

83 TOWNSLEY, F. M., ARISTARKHOV, A., BECK, S., HERSHKO, A., and RUDERMAN, J. V. Dominant-negative cyclin-selective ubiquitin carrier protein E2-C/UbcH10 blocks cells in metaphase. *Proc. Natl. Acad. Sci. USA* 1997, *94*, 2362–67.

84 TOWNSLEY, F. M. and RUDERMAN, J. V. Functional analysis of the *Saccharomyces cerevisiae UBC11* gene. *Yeast* 1998, *14*, 747–57.

85 PORNILLOS, O., GARRUS, J. E., and SUNDQUIST, W. I. Mechanisms of enveloped RNA virus budding. *Trends Cell Biol.* 2002, *12*, 569–79.

86 FURMAN, M. H. and PLOEGH, H. L. Lessons from viral manipulation of protein disposal pathways. *J. Clin. Invest.* 2002, *110*, 875–79.

87 HINGAMP, P. M., ARNOLD, J. E., MAYER, R. J., and DIXON, L. K. A ubiquitin conjugating enzyme encoded by African swine fever virus. *EMBO. J.* 1992, *11*, 361–66.

88 RODRIGUEZ, J. M., SALAS, M. L., and VINUELA, E. Genes homologous to ubiquitin-conjugating proteins and eukaryotic transcription factor SII in African swine fever virus. *Virology* 1992, *186*, 40–52.

89 HAGGLUND, R. and ROIZMAN, B. Characterization of the novel E3 ubiquitin ligase encoded in exon 3 of herpes simplex virus-1-infected cell protein 0. *Proc. Natl. Acad. Sci. USA* 2002, *99*, 7889–94.

90 SONG, S., KIM, S.-Y., HONG, Y.-M., JO, D.-G., LEE, J.-Y., SHIM, S. M., CHUNG, C.-W., SEO, S. J., YOO, Y. J., KOH, J.-Y., LEE, M. C., YATES, A. J., ICHIJO, H., and JUNG, Y.-K. Essential role of E2–25K/Hip-2 in mediating amyloid-*β* toxicity. *Mol. Cell* 2003, *12*, 553–63.

91 LAM, Y. A., PICKART, C. M., ALBAN, A.,

LANDON, M., JAMIESON, C., RAMAGE, R., MAYER, R. J., and LAYFIELD, R. Inhibition of the ubiquitin-proteasome system in Alzheimer's disease. *Proc. Natl. Acad. Sci. USA* **2000**, *97*, 9902–06.

92 OKAMOTO, Y., OZAKI, T., MIYAZAKI, K., AOYAMA, M., MIYAZAKI, M., and NAKAGAWARA, A. UbcH10 is the cancer-related E2 ubiquitin-conjugating enzyme. *Cancer Res.* **2003**, *63*, 4167–73.

93 SHEKHAR, M. P. V., LYAKHOVICH, A., VISSCHER, D. W., HENG, H., and KONDRAT, N. Rad6 over-expression induces multinucleation, centrosom amplification, abnormal mitosis, aneuploidy, and transformation. *Cancer Res.* **2002**, *62*, 2115–24.

94 MULLER, S., HOEGE, C., PYROWOLAKIS, G., and JENTSCH, S. SUMO, ubiquitin's mysterious cousin. *Nature Rev. Mol. Cell Biol.* **2001**, *2*, 202–10.

95 HOCHSTRASSER, M. Evolution and function of ubiquitin-like protein-conjugation systems. *Nature Cell Biol.* **2000**, *2*, E153–E57.

96 WALDEN, H., PODGORSKI, M. S., and SCHULMAN, B. A. Insights into the ubiquitin transfer cascade from the structure of the E1 for NEDD8. *Nature* **2003**, *422*, 330–34.

97 BENCSATH, K. P., PODGORSKI, M. S., PAGALA, V. R., SLAUGHTER, C. A., and SCHULMAN, B. A. Identification of a multifunctional binding site on Ubc9p required for Smt3p conjugation. *J. Biol. Chem.* **2002**, *277*, 47938–45.

98 SULLIVAN, M. L. and VIERSTRA, R. D. Cloning of a 16-kDa ubiquitin carrier protein from wheat and Arabidopsis thaliana. Identification of functional domains by in vitro mutagenesis. *J. Biol. Chem.* **1991**, *266*, 23878–85.

99 HAAS, A. L., BRIGHT, P. M., and JACKSON, V. E. Functional diversity among putative *E2* isozymes in the mechanism of ubiquitin-histone ligation. *J. Biol. Chem.* **1988**, *163*, 13268–75.

100 SIEPMANN, T. J., BOHNSACK, R. N., TOKGOZ, Z., BABOSHINA, O., and HAAS, A. L. Protein interactions within the N-end rule ubiquitin ligation pathway. *J. Biol. Chem.* **2003**, *278*, 9448–57.

101 MIURA, T., KLAUS, W., GSELL, B., MIYAMOTO, C., and SENN, H. Characterization of the binding interface between ubiquitin and class I human ubiquitin-conjugating enzyme 2b by multidimensional heteronuclear NMR spectroscopy in solution. *J. Mol. Biol.* **1999**, *290*, 213–28.

102 HAMILTON, K. S., ELLISON, M. J., BARBER, K. R., WILLIAMS, R. S., HUZIL, J. T., MCKENNA, S., PTAK, C., GLOVER, M., and SHAW, G. S. Structure of a conjugating enzyme-ubiquitin thiolester intermediate reveals a novel role for the ubiquitin tail. *Structure* **2001**, *9*, 897–904.

103 MCKENNA, S., MORAES, T., PASTUSHOK, L., PTAK, C., XIAO, W., SPYRACOPOULOS, L., and ELLISON, M. J. An NMR-based model of the ubiquitin-bound human ubiquitin conjugation complex Mms2.Ubc13. The structural basis for lysine 63 chain catalysis. *J. Biol. Chem.* **2003**, *278*, 13151–58.

104 BORDEN, K. L. RING domains: master builders of molecular scaffolds? *J. Mol. Biol.* **2000**, *295*, 1103–12.

105 BORDEN, K. L. RING fingers and B-boxes: zinc-binding protein-protein interaction domains. *Biochem. Cell Biol.* **1998**, *76*, 351–58.

106 ZHENG, N., WANG, P., JEFFREY, P. D., and PAVLETICH, N. P. Structure of a c-Cbl-UbcH7 complex: RING domain function in ubiquitin-protein ligases. *Cell* **2000**, *102*, 533–39.

107 BRZOVIC, P. S., KEEFFE, J. R., NISHIKAWA, H., MAYAMOTO, K., FOX, D., FUKUDA, M., OHTA, T., and KLEVIT, R. Binding and recognition in the assembly of an active BRCA1-BARD1 ubiquitin ligase complex. *Proc. Natl. Acad. Sci. USA* **2003**, *100*, 5646–51.

108 HUANG, L., KINNUCAN, E., WANG, G., BEAUDENON, S., HOWLEY, P. M., HUIBREGTSE, J. M., and PAVLETICH, N. P. Structure of an E6AP-UbcH7 complex: insights into ubiquitination by the E2-E3 enzyme cascade. *Science* **1999**, *286*, 1321–26.

109 Wu, G., Xu, G., Schulman, B. A., Jeffrey, P. D., Harper, J. W., and Pavletich, N. P. Structure of a β-TrCP1-Skp1-β-catenin complex: destruction motif binding and lysine specificity of the SCF$^{β\text{-}TrCP1}$ ubiquitin ligase. *Mol. Cell* **2003**, *11*, 1145–456.

110 Brzovic, P. S., Rajagopal, P., Hoyt, D. W., King, M.-C., and Klevit, R. E. Structure of a BRCA1-BARD1 heterodimeric RING-RING complex. *Nature Struct. Biol.* **2001**, *8*, 833–37.

111 Conaway, R. C. and Conaway, J. W. The von Hippel-Lindau tumor suppressor complex and regulation of hypoxia-inducible transcription. *Adv. Cancer Res.* **2002**, *85*, 1–12.

112 Zheng, N., Schulman, B. A., Song, L., Miller, J. J., Jeffrey, P. D., Wang, P., Chu, C., Koepp, D. M., Elledge, S. J., Pagano, M., Conaway, R. C., Conaway, J. W., Harper, J. W., and Pavletich, N. P. Structure of the Cul1-Rbx1-Skp1-F box^{Skp2} SCF ubiquitin ligase complex. *Nature* **2002**, *416*, 703–09.

113 Kawakami, T., Chiba, T., Suzuki, T., Iwai, K., Yamanaka, K., Minato, N., Suzuki, H., Shimbara, N., Hidaka, Y., Osaka, F., Omata, M., and Tanaka, K. NEDD8 recruits E2-ubiquitin to SCF E3 ligase. *EMBO J.* **2001**, *20*, 4003–12.

114 Wu, K., Chen, A., Tan, P., and Pan, Z.-Q. The Nedd8-conjugated ROC1-CUL1 core ubiquitin ligase utilizes Nedd8 charged surface residues for efficient polyubiquitin chain assembly catalyzed by Cdc34. *J. Biol. Chem.* **2002**, *277*, 516–27.

115 Orlicky, S., Tang, X., Willems, A., Tyers, M., and Sicheri, F. Structural basis for phosphodependent substrate selection and orientation by the SDF^{Cdc4} ubiquitin ligase. *Cell* **2003**, *112*, 243–56.

116 Deffenbaugh, A. E., Scaglione, K. M., Buranda, T., Sklar, L. A., and Skowyra, D. Release of ubiquitin-charged Cdc34-S∼Ub from the RING domain is essential for ubiquitination of the SCF^{Cdc4}-bound substrate Sic1. *Cell* **2003**, *114*, 611–22.

117 Petroski, M. D. and Deshaies, R. J. Context of multiubiquitin chain attachment influences the rate of Sic1 degradation. *Mol. Cell* **2003**, *11*, 1435–44.

118 Peng, J., Schwartz, D., Elias, J. E., Thoreen, C. C., Cheng, D., Marsischky, G., Roelofs, J., Finley, D., and Gygi, S. P. A proteomics approach to understanding protein ubiquitination. *Nature Biotechnol.* **2003**, *21*, 921–26.

119 Pringa, E., Martinez-Noel, G., Muller, U., and Harbers, K. Interaction of the ring finger-related U-box motif of a nuclear dot protein with ubiquitin-conjugating enzymes. *J. Biol. Chem.* **2001**, *276*, 19617–23.

120 Koegl, M., Hoppe, T., Schlenker, S., Ulrich, H. D., Mayer, T. U., and Jentsch, S. A novel ubiquitination factor, E4, is involved in multiubiquitin chain assembly. *Cell* **1999**, *96*, 635–44.

121 Jiang, J., Ballinger, C. A., Wu, Y., Dai, Q., Cyr, D. M., Hohfeld, J., and Patterson, C. CHIP is a U-box-dependent E3 ubiquitin ligase: identification of Hsc70 as a target for ubiquitylation. *J. Biol. Chem.* **2001**, *276*, 42938–44.

122 Murata, S., Minami, Y., Minami, M., Chiba, T., and Tanaka, K. CHIP is a chaperone-dependent E3 ligase that ubiquitylates unfolded protein. *EMBO Rep.* **2001**, *2*, 1133–8.

123 Ohi, M. D., Vander Kooi, C. W., Rosenberg, J. A., Chazin, W. J., and Gould, K. L. Structural insights into the U-box, a domain associated with multi-ubiquitination. *Nature Struct. Biol.* **2003**, *10*, 250–55.

124 Aravind, L. and Koonin, E. V. The U box is a modified RING finger – a common domain in ubiquitination. *Curr. Biol.* **2000**, *10*, R132–R24.

125 Hatakeyama, S., Yada, M., Matsumoto, M., Ishida, N., and Nakayama, K. I. U box proteins as a new family of ubiquitin-protein ligases. *J. Biol. Chem.* **2001**, *276*, 33111–20.

126 Huibregtse, J. M., Scheffner, M., Beaudenon, S., and Howley, P. M. A family of proteins structurally and functionally related to the E6-AP ubiquitin-protein ligase. *Proc. Natl. Acad. Sci. USA* **1995**, *92*, 2563–67.

127 HUIBREGTSE, J. M., SCHEFFNER, M., and HOWLEY, P. M. A cellular protein mediates association of p53 with the E6 oncoprotein of human papillomavirus types 16 or 18. *EMBO J.* **1991**, *10*, 4129–35.

128 SCHEFFNER, M., NUBER, U., and HUIBREGTSE, J. M. Protein ubiquitination involving an E1-E2-E3 enzyme ubiquitin thioester cascade. *Nature* **1995**, *373*, 81–83.

129 VERDECIA, M. A., JOAZEIRO, C. A. P., WELLS, N. J., FERRER, J.-L., BOWMAN, M. E., HUNTER, T., and NOEL, J. P. Conformational flexibility underlies ubiquitin ligation mediated by the WWP1 HECT domain E3 ligase. *Mol. Cell* **2003**, *11*, 249–59.

130 BERNIER-VILLAMOR, V., SAMPSON, D. A., MATUNIS, M. J., and LIMA, C. D. Structural basis for E2-mediated SUMO conjugation revealed by a complex between ubiquitin-conjugating enzyme Ubc9 and RanGAP1. *Cell* **2002**, *108*, 345–56.

131 VANDEMARK, A. P., HOFMANN, R. M., TSUI, C., PICKART, C. M., and WOLBERGER, C. Molecular insights into polyubiquitin chain assembly: crystal structure of the Mms2/Ubc13 heterodimer. *Cell* **2001**, *105*, 711–20.

132 DENG, L., WANG, C., SPENCER, E., YANG, L., BRAUN, A., YOU, J., SLAUGHTER, C., PICKART, C., and CHEN, Z. J. Activation of the IκB kinase complex by TRAF6 requires a dimeric ubiquitin-conjugating enzyme complex and a unique polyubiquitin chain. *Cell* **2000**, *103*, 351–61.

133 CARTER, P. and WELLS, J. A. Dissecting the catalytic triad of a serine protease. *Nature* **1988**, *332*, 564–68.

134 RAWLINGS, N. D. and BARRETT, A. J. Families of cysteine peptidases. *Meth. Enzymol.* **1994**, *244*, 461–86.

135 TONG, H., HATEBOER, G., PERRAKIS, A., BERNARDS, R., and SIXMA, T. K. Crystal structure of murine/human Ubc9 provides insight into the variability of the ubiquitin-conjugating system. *J. Biol. Chem.* **1997**, *272*, 21381–87.

136 WU, P.-Y., HANLON, M., EDDINS, M., TSUI, C., ROGERS, R., JENSEN, J. P., MATUNIS, M. J., WEISSMAN, A. M., WOLBERGER, C., and PICKART, C. M. A conserved catalytic residue in the E2 enzyme family. *EMBO J.* **2003**, *22*, 5241–50.

137 HU, M., LI, P., LI, M., LI, W., YAO, T., WU, J.-W., GU, W., COHEN, R. E., and SHI, Y. Crystal structure of a UBP-family deubiquitinating enzyme in isolation and in complex with ubiquitin aldehyde. *Cell* **2002**, *111*, 1041–54.

138 MOSSESSOVA, E. and LIMA, C. D. Ulp1-SUMO crystal structure and genetic analysis reveal conserved interactions and a regulatory element essential for cell growth in yeast. *Mol. Cell* **2000**, *5*, 865–76.

139 JOHNSTON, S. C., RIDDLE, S. M., COHEN, R. E., and HILL, C. P. Structural basis for the specificity of ubiquitin C-terminal hydrolases. *EMBO J.* **1999**, *18*, 3877–87.

140 BERLETH, E. S. and PICKART, C. M. Mechanism of ubiquitin conjugating enzyme E2–230K: catalysis involving a thiol relay? *Biochemistry* **1995**, *35*, 1664–71.

141 HAUSER, H.-P., BARDROFF, M., PYROWOLAKIS, G., and JENTSCH, S. A giant ubiquitin-conjugating enzyme related to IAP apoptosis inhibitors. *J. Cell Biol.* **1998**, *141*, 1415–22.

142 SANCHO, E., VILA, M. R., SANCHEZ-PULIDO, L., LOZANO, J. J., PACIUCCI, R., NADAL, M., FOX, M., HARVEY, C., BERCOVICH, B., LOUKILI, N., CIECHANOVER, A., LIN, S. L., SANZ, F., ESTIVILL, X., VALENCIA, A., and THOMSON, T. M. Role of UEV-1, an inactive variant of the E2 ubiquitin-conjugating enzymes, in in vitro differentiation and cell cycle behavior of HT-29-M6 intestinal mucosecretory cells. *Mol. Cell Biol.* **1998**, *18*, 576–89.

143 WANG, C., DENG, L., HONG, M., AKKARAJU, G. R., INOUE, J.-I., and CHEN, Z. J. TAK1 is a ubiquitin-dependent kinase of MKK and IKK. *Nature* **2001**, *412*, 346–51.

144 MORAES, T. F., EDWARDS, R. A., MCKENNA, S., PASTUSHOK, L., XIAO,

W., GLOVER, J. N., and ELLISON, M. J. Crystal structure of the human ubiquitin conjugating enzyme complex, hMms2-hUbc13. *Nature Struct. Biol.* **2001**, *8*, 669–73.

145 ULRICH, H. D. Protein-protein interactions in an E2-RING finger complex: implications for ubiquitin-dependent DNA damage repair. *J. Biol. Chem.* **2003**, *278*, 7051–58.

146 KATZMANN, D. J., ODORIZZI, G., and EMR, S. D. Receptor downregulation and multivesticular-body sorting. *Nature Rev. Mol. Cell Biol.* **2002**, *3*, 893–905.

147 KATZMANN, D. J., BABST, M., and EMR, S. D. Ubiquitin-dependent sorting into the multivesicular body pathway requires the function of a conserved endosomal protein sorting complex, ESCRT-I. *Cell* **2001**, *106*, 145–55.

148 GARRUS, J. E., von SCHWEDLER, U. K., PORNILLOS, O. W., MORHAM, S. G., ZAVITZ, K. H., WANG, H. E., WETTSTEIN, D. A., STRAY, K. M., COTE, M., RICH, R. L., MYSZKA, D. G., and SUNDQUIST, W. I. Tsg101 and the vaculolar protein sorting pathway are essential for HIV-1 budding. *Cell* **2001**, *107*, 55–65.

149 MARTIN-SERRANO, J., ZANG, T., and BIENIASZ, P. D. HIV-1 and Ebola virus encode small peptide motifs that recruit Tsg101 to sites of particle assembly to facilitate egress. *Nature Med.* **2001**, *7*, 1313–19.

150 CARTER, C. A. Tsg101: HIV-1's ticket to ride. *Trends Microbiol.* **2002**, *10*, 203–05.

151 PORNILLOS, W., ALAM, S. L., DAVIS, D. R., and SUNDQUIST, W. I. Structure of the Tsg101 UEV domain in complex with the PTAP motif of the HIV-1 p6 protein. *Nature Struct. Biol.* **2002**, *9*, 812–17.

152 PORNILLOS, O., ALAM, S. L., RICH, R. L., MYSZKA, D. G., DAVIS, D. R., and SUNDQUIST, W. I. Structure and functional interactions of the Tsg101 UEV domain. *EMBO J.* **2002**, *21*, 2397–406.

153 GOFF, A., EHRLICH, L. S., COHEN, S. N., and CARTER, C. A. Tsg101 control of human immunodeficiency virus type 1 Gag trafficking and release. *J. Virol.* **2003**, *77*, 9173–82.

154 KOONIN, E. V. and ABAGYAN, R. A. TSG101 may be the prototype of a class of dominant negative ubiquitin regulators. *Nature Genet.* **1997**, *16*, 330–31.

155 HOFMANN, K. and BUCHER, P. The UBA domain: a sequence motif present in multiple enzyme classes of the ubiquitination pathway. *Trends Biochem. Sci.* **1996**, *21*, 172–73.

156 DOHMEN, R. J., MADURA, K., BARTEL, B., and VARSHAVSKY, A. The N-end rule is mediated by the UBC2(RAD6) ubiquitin-conjugating enzyme. *Proc. Natl. Acad. Sci. USA* **1991**, *88*, 7351–55.

157 BAILLY, V., LAMB, J., SUNG, P., PRAKASH, S., and PRAKASH, L. Specific complex formation between yeast RAD6 and RAD18 proteins: a potential mechanism for targeting RAD6 ubiquitin-conjugating activity to DNA damage sites. *Genes Dev.* **1994**, *8*, 811–20.

158 VARSHAVSKY, A. The N-end rule pathway of protein degradation. *Genes Cells* **1997**, *2*, 13–28.

159 BRAUN, S., MATUSCHEWSKI, K., RAPE, M., THOMS, S., and JENTSCH, S. Role of the ubiquitin-selective CDC48$_{UFD1/NPL4}$ chaperone (segregase) in ERAD of OLE1 and other substrates. *EMBO J.* **2002**, *21*, 615–21.

160 QIN, S., NAKAJIMA, B., NOMURA, M., and ARFIN, S. M. Cloning and characterization of a *Saccharomyces cerevisiae* gene encoding a new member of the ubiquitin-conjugating protein family. *J. Biol. Chem.* **1991**, *266*, 15549–54.

161 SEUFERT, W., FUTCHER, B., and JENTSCH, S. Role of a ubiquitin-conjugating enzyme in degradation of S- and M-phase cyclins. *Nature* **1995**, *373*, 78–81.

162 JOHNSON, E. S. and GUPTA, A. A. An E3-like factor that promotes SUMO conjugation to the yeast septins. *Cell* **2001**, *106*, 735–44.

163 JOHNSON, E. S. and BLOBEL, G. Ubc9p is the conjugating enzyme for the

ubiquitin-like protein Smt3p. *J. Biol. Chem.* **1997**, *272*, 26799–802.

164 LIAKOPOULOS, D., DOENGES, G., MATUSCHEWSKI, K., and JENTSCH, S. A novel protein modification pathway related to the ubiquitin system. *EMBO J.* **1998**, *17*, 2208–14.

165 LAMMER, D., MATHIAS, N., LAPLAZA, J. M., JIANG, W., LIU, Y., CALLIS, J., GOEBL, M., and ESTELLE, M. Modification of yeast Cdc53p by the ubiquitin-related protein Rub1p affects function of the SCF ^{Cdc4} complex. *Genes Dev.* **1998**, *12*, 914–26.

6
The SCF Ubiquitin E3 Ligase

Leigh Ann Higa and Hui Zhang

6.1
Introduction

One of the most effective ways to activate or inactivate a biological process rapidly is to specifically eliminate through proteolysis the critical proteins that regulate or participate in the process. Eukaryotic cells utilize ubiquitin-dependent proteolysis to regulate responses to diverse signals during development and metabolism [1, 2]. With more than 30 000 genes encoded in the human genome, selective degradation of a particular protein in response to a regulatory signal poses a great challenge to the cell. The ubiquitin-dependent proteolysis pathway ensures that each protein is degraded in a temporal and spatially regulated fashion in response to such diverse signals or environmental cues [2]. In this system, the doomed protein is specifically modified by ubiquitin, a small peptide consisting of 76 amino acid residues [1]. The enzymatic cascade is set in motion when ubiquitin is first activated by an activating enzyme, E1, at the expense of ATP. The activated ubiquitin, which is covalently linked to the E1 enzyme by a thioester bond, is transferred to a member of a family of ubiquitin E2-conjugating enzymes. Last but not least, the doomed protein substrate is recognized by an ubiquitin E3 ligase, which often aids in ubiquitin transfer from E2 to substrate. Polyubiquitinated proteins are then degraded by the 26S proteasome. Since E3 ligases define the substrate specificity, studies suggest that intricate and fascinating mechanisms specify a large number of ubiquitin E3 ligases for the selective and timely elimination of a particular substrate through ubiquitin-dependent proteolysis [1–3]. In this chapter, we will focus on the function and regulation of the SCF (**S**kp1, **C**ul1/Cdc53, **F**-box proteins) family of ubiquitin E3 ligases. Unlike the HECT-domain E3 ligases, which consist of a single polypeptide, the SCF E3 ligase is composed of multiple protein subunits. This multiprotein complex regulates many important biological processes such as the cell cycle, transcription, and inflammation response. In addition, SCF is subject to regulation at various levels by complex signaling processes, and some of the regulatory mechanisms are exclusive to this class of E3 ligases. Accordingly, alteration of the function and regulation of the SCF ubiquitin E3 ligase has been associated with human diseases such as cancer.

Protein Degradation. Edited by J. Mayer, A. Ciechanover, M. Rechsteiner
Copyright © 2005 WILEY-VCH Verlag GmbH & Co. KGaA, Weinheim
ISBN: 3-527-30837-7

6.2
Discovery of the SCF Complex

One of the largest ubiquitin E3 ligase families, SCF ubiquitin E3 ligases are assembled from Skp1, Cul1/Cdc53 an F-box protein, and Roc1 (also called Rbx1 or Hrt1) [3]. Skp1 and the F-box protein Skp2 (*S*-phase *k*inase associated *p*rotein 1 and 2), were initially identified during analysis of the cyclin A/CDK2 complex [4]. Skp2 expression was found to be highly elevated in many cancer cells and is required for G1 cells to enter S phase. However, Skp1 and Skp2 can form a complex independent of cyclin A/CDK2, suggesting that this binary complex may have a cell-cycle function independent of the cyclin A/CDK2 kinase activity.

Yeast Skp1 was isolated as a high copy suppressor of yeast cdc4 temperature-sensitive mutant at restrictive temperature, and as a protein that interacts with human cyclin F, a protein that can also suppress the cell-cycle defects of cdc4 mutant when it is expressed in high copy in yeast [5]. Skp1 also directly interacts with the yeast Cdc4 protein, which encodes eight WD40 repeats (WD repeats), and cyclin F. Since Skp1 also directly binds to Skp2 which contains seven *l*eucine-*r*ich *r*epeats (LRR), these observations suggest that Skp2, Cdc4, and cyclin F may share a common mechanism for Skp1 binding. Indeed, sequence alignment of all three proteins indicates that they possess a relatively conserved 40–45 amino acid motif which mediates the binding of Skp1 [5] (Figure 6.1). This motif had been previously identified in some WD repeat-containing proteins but its significance was unknown [6]. This motif is therefore called the F-box, after the cognate region in cyclin F, and is present in a wide variety of otherwise unrelated proteins [5, 7–9]. Accordingly, the proteins that contain this motif are called F-box proteins [5]. The function of Skp1 was further revealed by earlier observations that yeast Cdc4, Cdc53, and Cdc34 temperature-sensitive mutants all fail to perform yeast Start-related events (G1 progression into S phase, nuclear DNA replication, and spindle formation) and accumulate yeast CDK inhibitor $p40^{Sic1}$ at the restrictive temperature [10–12]. Since Cdc34 encodes an ubiquitin E2-conjugating enzyme [10], Cdc34, Cdc4, and Cdc53 are likely to act in concert to regulate the G1/S transition by controlling the ubiquitin-dependent proteolysis of $p40^{Sic1}$. Certain Skp1 mutants also accumulate $p40^{Sic1}$ and expression of Skp1 in cdc4 mutants is sufficient to suppress the accumulation of $p40^{Sic1}$ in cdc4 mutants at restrictive temperature [5]. These observations suggested that Skp1 is involved in the Cdc4-, Cdc53-, and Cdc34-mediated ubiquitin-dependent proteolysis of $p40^{Sic1}$.

Cullin-1 (Cul1) was originally isolated from *Caenorhabditis elegans* as a negative regulator of cell proliferation during development [13]. Loss of Cul1 (or *lin-19*) in

```
Skp2:   D S L P D E L L L G I F S C L C L P E L L K V S G V C K R W Y R L A S D - E S L W  (a.a.98-137)
Cdc4:   T S L P F E I S L K I F N Y L Q F E D I I N S L G V S Q N W N K I I R K S T S L W  (a.a.276-316)
CycF:   L S L P E D V L F H I L K W L S V E D I L A V R A V H S Q L K D L V D N H A S V W  (a.a.33-73)
```

Fig. 6.1. The F-box motif in human Skp2, budding yeast Cdc4, and human cyclin F (CycF). The conserved amino acids are highlighted.

C. elegans causes hyperplasia in all tissues. In the proliferating cells, the progression from G1 to S phase is accelerated. The normal developmentally programmed mitotic arrests are overridden, with additional cell divisions that produce abnormally small cells. It was found that Cul1 belongs to a conserved family of cullins that share extensive homology [13]. The cloning of yeast Cdc53 revealed that it is an ortholog of Cul1. Biochemical analyses suggested that Cdc53, Cdc4, and Cdc34 form a protein complex [11]. These studies laid the foundation for the more detailed studies of SCF ubiquitin E3 ligase and related cullin-containing ubiquitin E3 ligases.

6.3
The Components of the SCF Complex

The essential components of the SCF ubiquitin E3 ligase include Skp1, Cul-1/ Cdc53, one of many F-box proteins, and the RING-H2-finger protein Roc1 (Rbx1 or Hrt1) (Figure 6.2). Although initial studies did not reveal the presence of a fourth component of the SCF complex [14, 15], later work showed that a RING-H2-finger protein, Roc1, is an essential subunit of the SCF complex [3]. The SCF complex thus contains three invariable components (Roc1, Cul1, and Skp1) which provide a core structure to which one of the many substrate-specific subunits (F-box proteins) binds. The Roc1–Cul1–Skp1 core also independently interacts with the ubiquitin E2-conjugating enzyme to couple ubiquitin transfer to the substrates [3]. One of the F-box proteins binds directly to a specific substrate and such interaction facilitates the polyubiquitination of the substrate by ubiquitin

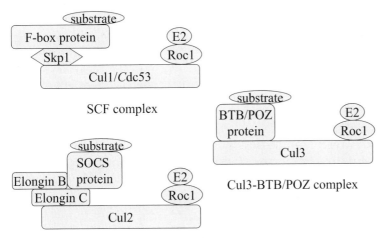

Fig. 6.2. The structures of SCF, Cul2–Elongin B-C, and Cul3–BTB/POZ ubiquitin E3 ligase complexes with the bound substrates and E2 enzymes.

transfer from the ubiquitin-charged E2. Since there are multiple F-box proteins [3], this mechanism illustrates how the same core complex can control the abundance of a diverse spectrum of substrates. Related E3 ligases built with a conserved cullin as a core protein employ similar strategies to extend the substrate specificity.

6.3.1
Roc1/Rbx1/Hrt1

The RING-H2 protein, Roc1 (also called Rbx1 and Hrt1, but Roc1 hereafter) was identified through its interaction with either mammalian Cul1, Cul2, or yeast Cdc53 [3]. It was found that addition of Roc1 stimulates the polyubiquitination activity of SCF complexes. Roc1 and its homologs are highly conserved in evolution. Roc1 contains a cysteine-containing and zinc-binding RING-finger domain in its C-terminal half that is distinct from but related to other RING-finger E3 ligases such as c-Cbl. A close homolog, Roc2/Rbx2/Hrt2 (also called Sag), also exists [3]. Both Roc1 and Roc2 can bind to Cul1 and related cullins (Cul2–5, Cul7, and Cul8) through a highly conserved C-terminal region, the cullin homology domain [3, 16, 17]. A more distant Roc1 homolog, Apc11, is a component of the *A*naphase *p*romoting *c*omplex/*C*yclosome (Apc/C) [3], an ubiquitin E3 ligase complex that regulates mitosis. Apc11 binds to a distant cullin-related protein, Apc2, in the Apc/C complex [18]. Genetic studies confirmed the essential role of Roc1 in the SCF E3 ligase complex. Roc1 also binds to ubiquitin E2 ligases such as Cdc34, and thus serves as the link between the E2 (Cdc34) conjugating enzyme and Cul1/Cdc53. The RING-H2 domain of Roc1 has been shown to be required for the E2-binding and ubiquitin-ligation reaction. However, although biochemical studies suggest that Roc1 and Roc2 share the same biochemical properties of cullin binding and act as a link between the cullin E3 ligases and the E2 enzymes, the physiological roles of Roc1a (a Roc1 ortholog) and Roc1b (a Roc2 ortholog) appear to differ in *Drosophila melanogaster* [19]. Drosophila Roc1a is required for cell proliferation, and cells lacking Roc1a fail to proliferate during development. However, expression of Roc1b under the control of Roc1a promoter does not rescue the Roc1a-deficient phenotype. In addition, Roc1a deficiency causes the differential accumulation of SCF substrates in *Drosophila*. While F-box protein Slimb/*β*-Trcp is required for both the proteolytic degradation of Armadillo/*β*-Trcp (Arm) and the proteolytic processing of the Cubitus interruptus (Ci) [20], Roc1a null mutants only accumulate Ci but not Arm. These studies suggest that an additional mechanism may exist to distinguish between Roc1a and Roc1b and various SCF substrates. *Drosophila* encode a third Roc1-like protein, Roc2, but its function in the cullin ubiquitin E3 ligase is not clear.

6.3.2
Cullin-1 (Cul1)

In the SCF complex, Cul1 forms the core scaffold that associates with Roc1 at the extreme C-terminal region [3]. At its the amino terminal region, Cul1 interacts with Skp1. The Roc1-binding domain exhibits the highest conservation among cullins and was initially identified as the cullin homology domain [3, 13]. The conservation of this domain is consistent with the fact that Roc1 can bind to almost all cullins (Cul1–5, Cul7, and Cul8) and this binding couples cullin E3 ligases to the ubiquitin E2-conjugating enzymes.

The N-terminal region for Skp1 binding is conserved between orthologs of Cul1 from different species, and also displays significant homology in the equivalent regions among other cullins such as Cul2, Cul3, Cul4, and Cul5 [21]. In Cul2 and Cul5 this region has been shown to interact with Elongin C [22, 23], a protein that shares considerable homology with Skp1. In another parallel, the N-terminal region of Cul3 has been shown to interact with BTB/POZ proteins that display a similar three-dimensional crystal structure to that of Skp1 [24].

The crystal structure of Cul1 has been resolved [21] and found to resemble a long stalk connecting two protein-interaction domains at either end of the stalk. The globular domain of 360 amino acids at the C-terminal region of Cul1 forms a complex with Roc1. In this region, a four-helix bundle and an α/β domain form two winged-helix domains. This creates a V-shaped groove in which the C-terminal half of Roc1, containing the RING-finger domain, is situated. This Cul1 region spans the cullin homology domain which is highly conserved between all cullins, being present even in the more distantly related cullin homolog Apc2, a subunit of the Apc/C ubiquitin E3 ligase. This sequence conservation suggests that other cullins also use the same strategy to interact with RING-H2 proteins such as Roc1 [21]. Overall, the globular domain of Cul1/Roc1 generates a surface area for interaction with the E2 enzyme [21]. Although deletion mutant analysis of Cul1 suggests that the first 219 amino acids are required to bind Skp1 [3], the crystal structure of the N-terminal region of the Cul1 protein reveals that Cul1 forms three cullin repeats (about 120 amino acids each) in an arch-like shape [21]. The first N-terminal repeat forms the domain for Skp1 interaction while the other two repeats form a long stalk between the Skp1/F-box protein-binding domain and the Roc1 interaction domain. In the crystal structure of Cul1, it does not seem that the F-box proteins such as Skp2 bind or interact with Cul1 [21]. This is consistent with the biochemical analysis that F-box proteins require Skp1 for binding to Cul1 [5].

6.3.3
Skp1

Skp1 serves as an adaptor protein that provides a molecular link between Cul1/Roc1 and the F-box proteins [4, 5]. The Skp1 protein contains two separate protein-interaction domains that are conserved among its family members between species [21]. The N-terminal region of Skp1 (~1–70 a.a.) interacts with Cul1 while the C-terminal half (100–163 a.a.) binds the F-box proteins [21]. The use of Skp1 as an adaptor to link the core ubiquitin E3 ligase components of Cul1/Roc1 with numerous and diverse substrate-targeting subunits, the F-box proteins, represents a strategy to specifically target many proteins for ubiquitination

[3]. The role of Skp1 is to bring the substrate-targeting subunit, the F-box protein, into proximity with the Cul1/Roc1/E2 complex to promote ubiquitin transfer from the E2-ubiquitin to the F-box protein-bound substrates.

The crystal structure of Skp1 reveals that it contains a BTB/POZ-like domain at the N-terminal region [25]. It has been found that BTB/POZ proteins bind to Cul3 and act as the substrate-targeting subunits for Cul3 E3 ligase-mediated ubiquitin-dependent proteolysis [24]. Thus the similarity between Skp1 and BTB/POZ proteins is significant. The structure of Skp1 also confirmed its similarity with Elongin C, a Skp1-like protein that interacts with Cul2. Both Skp1 and Elongin C also share structural resemblance to the potassium channel tetramerization domain, which also belongs to the BTB/POZ superfamily [25, 26]. Thus the BTB/POZ-like structures determine the interaction between cullins and their adaptor proteins such as Skp1, Elongin C, and other BTB/POZ-containing proteins.

Interestingly, although in mammals and single-cell organisms such as yeast there is only one conserved Skp1 homolog, other multi-cellular organisms encode multiple Skp1-like proteins. In *Arabidopsis thaliana*, at least 19 Asks (*Arabidopsis Sk*p1-like) are predicted and genetic studies suggest that Asks1 is part of SCF[Tir] and SCF[Coi1] complexes that regulate the responses to the plant hormones auxin and jasmonate, respectively [27, 28]. It also regulates vegetative and flower development and male meiosis. Other Asks exhibit different abilities to interact with F-box proteins. In addition, seven Skp1 homologs have been identified in *Drosophila melanogaster* and at least 21 *Sk*p1-*r*elated proteins (Skrs) have been identified in *C. elegans* [29–31]. In *C. elegans*, while loss of Skr7, 8, 9, and 10 results in slow growth and morphological abnormalities, Skr1 and Skr2 are essential for embryonic development. The presence of such large families of Skp1-related proteins in these organisms suggest that selective expression of these Skp1-related proteins during development or in a particular tissue may represent an additional level of regulation for their protein substrates.

In addition to targeting substrate proteins for degradation, Skp1 has been associated with certain activities that remain to be further characterized. For example, Emi1 (also called FBX5), an F-box protein that binds to Skp1 [32], contains a zinc-binding region near its carboxy terminus that is separate from its F-box region. This zinc-finger domain is required for binding to Cdc20 or Cdh1, substrate-targeting subunits of Apc/C ubiquitin E3 ligase. The binding of Emi1 to Cdc20 or Cdh1 inhibits Apc activity and thus regulates mitosis [32]. Furthermore, yeast Skp1 also binds to the kinetochore Cbf3 complex and is essential for the yeast kinetochore/centromere function in G2 [3]. However, the precise roles of Skp1 in these biological processes still remain unclear.

6.3.4
F-box Proteins

F-box proteins serve as the substrate-targeting subunit of the SCF ubiquitin E3 ligase [5]. They are structurally diverse but they all contain a relatively conserved signature motif of about 45–50 amino acids [5]. This motif, the F-box, was initially identified among human Skp2, yeast Cdc4, and human cyclin F, which all bind to Skp1 [5]. F-box proteins also contain a separate protein–protein interaction domain that mediates the binding to various substrates [5, 14, 15]. The binding of F-box proteins to their selected substrates usually targets the respective substrate for polyubiquitination and subsequent proteolysis through the 26S proteasome. However, F-box proteins can also mediate the processing of certain protein precursors to their cleavage products through limited ubiquitin-dependent proteolysis [20, 33].

The existence of a large repertoire of F-box proteins means that SCF E3 ligase is one of the largest E3 ligase families (other large E3 families such as the Cul2- and Cul3-containing ubiquitin E3 ligases are also related to the SCF E3 ligase) [2, 24]. In the yeast *Saccharomyces cerevisiae*, at least 11 F-box proteins that contain conserved F-box domains have been identified [7]. In *Drosophila* at least 22 F-box proteins exist, while more than 326 F-box proteins are predicted in the genome of *C. elegans* [34]. In human and mouse, the presence of at least 38 conserved F-box proteins has been reported [4, 5, 8, 9]. However, many F-box proteins may contain a less-canonical F-box motif [35]. In such cases, identification of the candidate F-box proteins in the protein databases using the standard sequence search algorithms is quite difficult. The classification of such a protein as a member of the F-box family relies on confirming its association with the other components of the SCF complex and its activity towards a particular protein substrate [35]. The prototypical F-box proteins such as Skp2, Cdc4, or β-Trcp have been relatively well studied. These studies clearly indicate that F-box proteins act as the substrate-targeting subunit of the SCF ubiquitin E3 ligases [3].

In addition to the F-box motif, many conserved F-box proteins contain either *l*eucine-*r*ich *r*epeats (LRR) such as Skp2 or yeast Grr1, or WD40 repeats, which are present in Cdc4 or β-Trcp [5]. In human and mouse, seven F-box proteins contain WD40 repeats (Fbws) while 10 F-box proteins have LRR repeats (Fbls) at their C-terminal domain [8, 9]. However, a large number of F-box proteins contain other protein–protein interaction modules or unknown domains (Fbxs). The LRR or WD-40 repeats of F-box proteins have been shown to mediate the interaction between F-box proteins and their respective substrates through phosphorylated serines or threonines [34]. The differential binding specificities of protein–protein interaction domains found in F-box proteins confers the substrate specificity of the SCF ubiquitin E3 ligase.

6.4
E2-conjugating Enzymes for the SCF E3 Ligases

The function of the SCF E3 ligase complex is to facilitate ubiquitin transfer from the E2-conjugating enzymes to the protein substrates. Although more than two dozen E2s exist, genetic studies suggest the Cdc34 E2 conjugation enzyme is especially involved in regulating SCF substrates [12]. These observations are further strengthened by the association between Cdc34 and components of SCF complexes. From yeast to human, this highly conserved E2 is also required for the *in vitro* polyubiquitination reactions for the substrates of SCF E3 ligases [14, 15, 36, 37]. However, other E2 enzymes, such as human UbcH5, can also function *in vitro* for polyubiquitination of certain substrates of the SCF complexes with similar [38], if not greater, efficiency than Cdc34.

6.5
Substrates and Substrate Recognition

Both genetic and biochemical analyses suggest that the SCF E3 ligase targets a wide spectrum of important proteins for ubiquitin-dependent proteolysis (see Table 6.1 for examples). A common feature of the physiological substrates of various SCF

Tab. 6.1. F-box proteins.

Protein	Species	Substrates	Function of substrates
Skp2	*H. sapiens*	p27	CDK inhibitor
		p21	CDK inhibitor
		Rb2/p130	Rb-related protein, CDK inhibitor
		Orc1	Component of origin recognition complex
Beta-Trcp	*H. sapiens*	β-catenin	Transcription factor
		IκB	Inhibitor of transcription
		CD4	HIV Vpu target, surface receptor
		Emi1	Inhibitor of anaphase-promoting complex/cyclosome
		CDC25A	Phosphatase, positive regulator of Cdks
hCdc4/Fbw7	*H. sapiens*	Cyclin E	G1 cyclin
		Notch	Receptor
		Presenilin 1	Familial Alzheimer's disease gene
Tome-1	*X. laevis*	Wee1	CDK inhibitory kinase
Slimb (beta-Trcp homologue)	*D. melanogaster*	Armadillo	β-catenin homolog
		Cubitus interruptus	Transcription factor
Archipelago (Ago) (hCdc4/Fbw homologue)	*D. melanogaster*	Cyclin E	G1 cyclin
Cdc4	*S. cerevisiae*	Cdc6	Replication initiation protein
		Sic1	Cdk inhibitor
		Far1	Cdk inhibitor
		Gcn4	Transcription repressor
Grr1	*S. cerevisiae*	Cln1	G1 cyclins
		Cln2	G1 cyclin
Met30	*S. cerevisiae*	Met4	Transcription factor
		Swe1	Wee1-like kinase
Pop1/Pop2	*S. pombe*	Rum	Cdk inhibitor
Tir1	*A. thaliana*	AXR2/IAA7	Auxin response
		AXR3/IAA17	Auxin reponse
Ebf1/Ebf2	*A. thiana*	EIN3	Transcription factor in ethylene response

ligases such as CDK inhibitors p40^{Sic1} or p27^{Kip1}, β-catenin, or IκB shows a requirement for phosphorylation of substrates on either serine or threonine for SCF-mediated ubiquitin-dependent proteolysis [2, 3]. The WD40 and LRR repeats in the F-box proteins bind phosphorylated substrates independent of the F-box, which interacts with Skp1. The binding of various F-box proteins to phosphorylated serines or threonines within a particular substrate appears to be quite specific [36–39], suggesting that the phosphorylation-mediated binding of F-box proteins is dependent on the peptide sequences surrounding the phosphorylation site of protein substrates. The binding of F-box proteins to the phosphorylated substrates promotes the interaction of Skp1 and its associated Cul1/Roc1 with the substrates, facilitating ubiquitin transfer from the E2-conjugating enzymes to the substrates [3]. Subsequent polyubiquitination of the substrates is coupled to their proteolysis by the 26S proteasome.

However, studies also suggest that many F-box proteins can bind and target a number of proteins for polyubiquitination in a phosphorylation- and sequence-dependent manner [34, 36, 37, 40–43]. Furthermore, evidence also suggests that some F-box proteins can selectively bind to several sites containing phosphorylated serines/threonines within a single protein substrate [44, 45]. These observations raise the question of how substrate specificity is defined. In several cases, additional mechanisms appear to be involved in specifying substrate selection by the F-box proteins. Several well characterized examples of SCF substrates will be presented to highlight our current understanding of the substrate-specificity of the SCF ubiquitin E3 ligase.

6.5.1
Skp2 and Its Substrates

One of the best characterized SCF complexes is SCF^{SKP2}. Skp2 was originally isolated by its highly elevated expression in many cancer cell lines and by its association with Skp1 and cyclin A/Cdk2/Cks1 [4]. Skp2 was also found to be critical in regulating the progression of mammalian G1 cells into S phase. One of the critical G1 cell-cycle regulators is p27^{Kip1}. In the cell cycle, p27 protein levels are regulated by ubiquitin-dependent proteolysis, being high in G0 and early G1 and low in late G1 and throughout S phase [46]. The high G1 level of p27 is required for preventing the premature activation of cyclin E/CDK2 or other G1- or S-specific cyclin/Cdks. In late G1, p27 is rapidly proteolyzed through ubiquitin-dependent degradation [46], promoting the release of active cyclin E/Cdk2 kinase and consequently the S phase entry. p27 degradation requires a single phosphorylation site at threonine 187 located at the C-terminal end of the protein [47]. Cyclin E-associated kinases can phosphorylate p27 at this critical site *in vitro* and in transfection systems. Conversion of threonine 187 to non-phosphorylatable alanine in p27 greatly stabilizes this protein. The F-box protein Skp2 specifically binds to the phosphorylated threonine 187 in p27 and targets p27 for polyubiquitination and subsequent proteolysis while other F-box proteins such as β-Trcp do not [36, 37]. Deletion of the F-box region in Skp2 promotes the interaction between Skp2 and the phosphorylated p27 but causes stabilization of p27 *in vivo* [37]. A particularly unique requirement for substrate binding and recognition by Skp2 involves an accessory protein, p9^{Cks1} [48, 49]. Cks1 was previously identified as a Cdk-binding and Suc1-like protein, and initially isolated as a suppressor of certain Cdc28/Cdc2 mutants in yeast [50]. SCF^{Skp2}-mediated p27 polyubiquitination requires Cks1, and the activity associated with p27 polyubiquitination is independent of Cdk binding but depends on its direct and specific interaction with Skp2. A close Cks1 homolog, Cks2, cannot substitute for Cks1 in this reaction. The polyubiquitination of p27 has been reconstituted with purified Skp2, Skp1, Cul1, Roc1, Cks1, cyclin E/CDK2, Cdc34, and E1 in the presence of ATP and ubiquitin [36, 49]. Overexpression of Skp2 is sufficient to induce p27 down-regulation and in some cases, induces S phase [51]. Strikingly, genetically engineered Skp2 knockout mice and Cks1 deficient mice share similar phenotypes [49, 52]. Mouse embryonic fibroblasts derived from Skp2$^{-/-}$ and Cks1$^{-/-}$ mice both accumulate p27 and its binding partner cyclin E [49, 52]. Thus it appears that Cks1 is specifically evolved in mammals to facilitate Skp2-mediated substrate polyubiquitination. Interestingly, Skp2 was isolated as a protein complex that contains cyclin A/Cdk2/Cks1 [4]. One possible role for cyclin A/CDK2 is recruitment of Cks1 into the proximity of Skp2 to facilitate Skp2 binding to phosphorylated p27. Alternatively, cyclin A/CDK2 can also bind and phosphorylate p27 at threonine 187 to promote Skp2 binding.

Although p27 remains the critical target of Skp2 in late G1, additional polyubiquitinated substrates of Skp2 have been identified. These include the retinoblastoma-related protein Rb2/p130 [41], Cdk inhibitor p21 [40], and other proteins. Characterization of Skp2 binding sites in these proteins reveals that while phosphorylation of serine 130 in p21 is required for Skp2 binding and polyubiquitination, phosphorylation of serine 672 in p130 is essential for the interaction with Skp2 and p130 polyubiquitination. One common feature among these characterized Skp2 substrates is the presence of minimum serine/threonine followed immediately by a proline residue (S or T/P) in the Skp2-binding motifs. However, it is still not clear how Skp2 selects its binding site after the phosphorylation of serines or threonines in these and other substrates.

6.5.2
β-Trcp and Its Substrates

The substrate consensus sequences are best illustrated in the case of β-Trcp (also called Fbw1, FWD1, and Hos in vertebrates, and Slimb in Drosophila), an F-box protein that contains seven WD40 repeats at its C-terminal region [20]. Initial genetic evidence in *Drosophila* suggests that the Drosophila β-Trcp homolog Slimb regulates proliferation and axis formation during development through the Wingless/Wnt and Hedgehog signaling pathways [20]. Genetic mosaic analysis of Slimb *Drosophila* mutants indicates that the slimb defect causes the abnormal accumulation of Armadillo, the Drosophila homolog of β-catenin, a transcription factor involved in the Wingless/Wnt pathway. In human, β-catenin is the target of the human tumor suppressor protein *a*denomatous *p*olyposis *c*oli (APC) which is often mutated in familial colorectal cancers [53]. In the absence of wingless signaling, the cytoplasmic β-catenin is unstable and is degraded by ubiquitin-dependent proteolysis. However, an active Wingless signaling pathway stabilizes β-catenin, which is subsequently transported from the cytoplasm into the nucleus to activate Wingless transcription programs. β-catenin is destabilized by phosphorylation on two conserved serines (serines 33 and 37, underlined in the sequence of DSGIHS) catalyzed by the glycogen synthase kinase-3b (GSK-3b) and casein kinase Iε (CKIε), through binding of the scaffold protein Axin and APC [53]. Phosphorylation of these two conserved serines in this N-terminal region of β-catenin triggers its ubiquitin-dependent proteolysis mediated by SCF^{β-Trcp} [39]. In addition to β-catenin, β-Trcp also binds to two phosphorylated serines (serines 32 and 36, DSGLDS) of IκB [38, 39], an inhibitor of NFκB, and regulates the NFκB-mediated inflammatory and other responses. Initial studies on the binding sites of β-catenin, IκB, and HIV-1 protein Vpu (another β-Trcp-binding protein when it is phosphorylated at two serines on DSGNES) [54], suggest that dual phosphorylation of serines within a consensus sequence of DSGXXS is sufficient for β-Trcp binding. This binding triggers the polyubiquitination of β-catenin and IκB. In addition to Armadillo, slimb mutation in *Drosophila* also causes the abnormal accumulation of Cubitus interruptus (Ci), producing phenotypes that resemble the ectopic activation of the Hedgehog signaling pathway [20]. In the absence of Hedgehog signal, full length Ci (p155) is processed by the proteasome in a β-Trcp-dependent manner to generate a smaller p55 form of Ci, which acts as a repressor for Hedgehog-regulated transcription. Mammalian β-Trcp exhibits similar processing activity towards NFκB1 [33]. However, studies on the phosphorylation-dependent processing of NFκB1 demonstrate that it occurs when the serines in the DSGXXXS motif of these proteins are phosphorylated [33]. The extra amino acid between the two phosphorylated serines suggests a certain tolerance by β-Trcp. More strikingly, characterization of the β-Trcp-mediated polyubiquitination of Cdc25A in response to DNA damage indicates that higher tolerance of the spacer between the dual phosphorylated serines exists [42, 43]. In this case, β-Trcp binds and targets Cdc25A for polyubiquitination once the two serines in DSGXXXXS are phosphorylated. The tolerance of two additional amino acid residues in the spacer region between the two phosphorylated serines suggests that β-Trcp is substantially flexible in its binding to substrates within the consensus sequence. Intriguingly, Cdc25A degradation is triggered after its phosphorylation at serine 76 by the DNA damage checkpoint kinase CHK1 in response to DNA damage. This phosphorylation, which precedes the serine 79 and serine 82 utilized for phosphorylation-dependent binding of β-Trcp in the DSGXXXXS motif, is essential for the DNA-damage-induced Cdc25A proteolysis by the SCF^{β-Trcp} E3 ligase. The mechanism by which the phosphorylated serine 76 triggers the β-Trcp-mediated Cdc25A polyubiquitination is not clear. These findings suggest that a more complicated regulation exists for the polyubiquitination of Cdc25 by the SCF^{β-Trcp} E3 ligase.

6.5.3
Yeast Cdc4 and Its Substrates

The substrate recognition mechanisms discussed above suggest that phosphorylation at a particular site (or sites) is sufficient to bind the F-box proteins Skp2 or β-

Trcp to their respective substrates. These studies also suggest certain flexibilities in the binding of these F-box proteins to their substrate consensus sites. The yeast F-box Cdc4, which contains eight WD40 repeats, has been implicated in mediating the ubiquitin-dependent degradation of Cdk inhibitor p40^{Sic1} and Far1, replication initiation protein Cdc6, and transcription repressor GCN4. Characterization of the yeast SCF^{Cdc4}-mediated polyubiquitination of yeast CDK inhibitor p40^{Sic1} provides additional insights into the mechanism of the interaction between the phosphory-lated substrates and F-box proteins.

It has been established that phosphorylation of Sic1 is absolutely required for SCF^{Cdc4}-mediated polyubiquitination [3, 45]. Sic1 is phosphorylated by the Cln/Cdc28 kinase at the minimum consensus sequences of serine/threonine followed immediately by a proline (S/TP) [3]. In addition, the Cln/Cdc28 kinase preferen-tially phosphorylates the S/TP site containing basic amino acid residues (arginines or lysines) [3]. Initial characterization of Sic1 phosphorylation sites reveals that at least nine such sites exist for Cln/Cdc28 phosphorylation and subsequent Cdc4 binding [3, 45]. Since Skp2 binds to p27 only when threonine 187 of p27 is phos-phorylated [36, 37], a systematic characterization of Sic1 phosphorylation sites was conducted to determine which one of these nine sites is critical for the binding of yeast Cdc4 F-box protein [45]. Initially, all of the potential serine/threonine phos-phorylation sites were removed by site-directed mutagenesis of the Sic1 protein. This mutant is extremely stable and cannot be degraded by SCF^{Cdc4} *in vivo*. Sys-tematic re-addition of serine/threonine phosphorylation sites to the Sic1 mutant protein suggests that while addition of any single serine or threonine is not suffi-cient to trigger its degradation, re-addition of at least six phosphorylation sites of the potential nine serines/threonines restores the Sic1 sensitivity to SCF^{Cdc4}. This differs from the polyubiquitination of the substrates of SCF^{Skp2}, in which a single phosphorylation constitutes the binding site for F-box protein [45]. However, the requirement of multiple phosphorylation sites is not unique to yeast Sic1 and its F-box protein Cdc4. The ubiquitin-dependent proteolysis of yeast Cln2, which is mediated through the F-box protein Grr1 encoding seven LRRs also depends on the phosphorylation of a cluster of at least four serines/threonines in the Cln2 pro-tein [44]. The requirement of at least six phosphorylation sites in Sic1 suggests that these sites may cooperate to allow the multiply phosphorylated Sic1 to bind to Cdc4 and raises the question of how Cdc4 can count the number of phosphoryla-tions to properly target Sic1 for polyubiquitination.

It turns out that not all the phosphorylated sites are created equal. The mecha-nism for binding cooperativity by multiple phosphorylation sites in Sic1 is demon-strated in part by the observation that a high affinity phosphorylated Cdc4-binding site on human cyclin E (threonine 380) is sufficient to destabilize Sic1 that lacks all original nine phosphorylation sites through Cdc4 [44]. Thus it is unlikely that the eight WD40 repeats of Cdc4 contain six or more phosphorylation binding modules for the binding of multiply phosphorylated Sic1 to Cdc4. Rather, it appears that a single phosphorylation site is necessary for Sic1 binding to Cdc4.

The co-crystal structure of yeast Cdc4 and its phosphorylated substrates, as well as that of β-Trcp and its phosphopeptide substrate derived from β-catenin, have been resolved [55, 56]. The *β*-Trcp and Cdc4 proteins contain either seven or eight WD40 repeats, respectively, which correspond to the formation of seven- or eight-blade *β*-propeller structures. The phosphopeptide substrates lie across the top surface, in alignment with the active E2-binding site in the Roc1-binding domain located at the C-terminal end of Cul1. All seven or eight *β*-propeller blades of *β*-Trcp or Cdc4 interact with the phosphorylated peptide substrates. One significant feature of the co-crystal structure of Cdc4 with its respective substrate phosphopeptides is that there is only one phosphorylation-binding site on the surface of the WD40 propeller repeats in Cdc4 [55]. This phosphorylation site is reminiscent of the *β*-Trcp site which binds the phosphorylated serine 37 in *β*-catenin [56]. Consistent with the genetic and biochemical characterization, the structure of Cdc4 does not suggest that it can contain six phospho-binding modules for Sic1 [55].

How then can we explain the observed cooperativity of Sic1 phosphorylation in binding Cdc4? A model was proposed to explain this cooperativity [45, 55]. It hypothesizes that while each single phosphorylation site in Sic1 may constitute a suboptimal binding site for Cdc4, the binding and subsequent release of each phosphorylated site will somehow increase the local concentration of Sic1 near Cdc4 [45, 55].

The presence of the multiply phosphorylated suboptimal sites in Sic1 should accelerate binding and dissociation cycles of Sic1 within the WD40-repeat domain. In proximity to Cdc4, this should elevate the effective concentration of this form of Sic1 above its simple diffusional rate. In terms of Cdc4 binding, this process should favor multiply phosphorylated Sic1 over those containing fewer phosphates. Biologically, Sic1 prevents premature entry into S phase in yeast by inhibiting the S phase cyclin/Cdk kinase, Clb5/6/Cdc28. Thus, multiple phosphorylations of Sic1 may require a higher level of G1 Cln/Cdc28 and promote a shaper transition of G1 to S phase. Consistent with this possibility, conversion of the positively charged arginine or lysine residues downstream of the serine/threonine-proline (S/TP) sites to neutral amino acids (such as alanine) in Sic1 reduces the number of phosphorylated sites in Sic1 required for the binding of Cdc4. Such changes may convert the suboptimal Cdc4-binding sites to high affinity ones for the binding and ubiquitination of Sic1 by SCF[Cdc4]. While this possibility may explain the Sic1 and Cdc4 interaction, it also provides an interesting model for substrate recognition and selection by other F-box proteins in which a degenerate phosphorylation consensus site is present in the substrates.

6.6
Structure of the SCF E3 Ligase Complex

Elucidation of the structure of the SCF ubiquitin E3 ligase complex and their substrates should help resolve certain issues regarding the mechanism by which SCF E3 ligase promotes ubiquitin transfer from the E2 enzyme to the protein substrate. Recently, the structures of Cul1/Roc1 in complex with Skp1 and Skp2, as well as that of Skp1/*β*-Trcp and Cdc4, have been reported [21, 25, 55, 56]. These studies suggest a rigid structure for the SCF complexes and thus provide an insight into the mechanism by which the SCF E3 ligase modulates the polyubiquitination of the protein substrates.

The overall shape of SCF[Skp2] consists of an elongated structure with Cul1 as the scaffold protein [21]. The structure of Cul1 displays a long stalk connecting protein interaction domains at either end of the stalk. While the C-terminal region of Cul1 forms a complex with the RING-H2 protein Roc1, the extreme and opposite N-terminal region of the Cul1 protein forms the domain for Skp1 interaction. The other two cullin repeats connecting these two functional domains of Cul1 adopt an arch-like shape. One surprise is that there is a substantial space of about 50 Å between the Skp1/F-box protein-binding domain and the Roc1 interaction domain, which recruits the active E2. In addition, a prominent feature of the SCF structure is the rigidity of the Cul1 stalk. This rigidity appears to be required to separate the substrate-binding domain of Skp1 from the E2-binding domain of Roc1. Attempts to alter the distance or the rigidity of Cul1 by incorporating a more flexible swivel in the connecting Cul1 stalk results in loss of SCF ubiquitin E3 ligase activity towards its physiological substrates. This rigid structure has also been observed in c-Cbl RING-finger protein which represents an independent E3 ligase family [57].

The structures of Skp2/Skp1, and Skp1/β-Trcp containing the phospho-β-catenin substrate peptide, and Skp1/yeast Cdc4/phospho-substrate peptides have been reported [21, 25, 55, 56]. While the conserved F-box regions of Skp2, β-Trcp, and Cdc4 interact with Skp1, additional linker repeats between the F-box and the LRR or WD40 repeats also support the interaction with Skp1. In addition, the C-terminal tail of Skp2 also folds back into the linker repeats between the F-box and the linker region to provide additional interaction with the C-terminal region of Skp1. These conformations suggest that the Skp1 and Skp2 interaction, as well as Skp1 binding to β-Trcp and Cdc4, is relatively rigid. This rigid structure may position the LRR or WD40 domains of these F-box proteins to orient the substrates in a particular direction towards the E2 site by Cul1/Roc1 interface. Thus the structural rigidity of the SCF E3 ligases and the existence of a substantial distant gap between the substrate-binding domain and the E2/RING-H2 domain may be a common feature required for the polyubiquitination of the substrates. However, the distance between the F-box-bound substrates and the Cul1/RING-H2 protein-bound E2 poses a structural limit for their direct interaction and thus the ubiquitin-transfer reaction from E2 to substrates.

The SCF E3 ligase, unlike other E3 ligases such as the HECT E3 ligase, does not appear to form a covalent thioester bond between ubiquitin and the E3 ligase [2]. This may suggest that the SCF E3 ligase could use a different mechanism to drive the ubiquitin-transfer reaction. Interestingly, it has been found that although the Cdc34 E2-conjugating enzyme binds to Roc1, which in turn binds to Cul1, the covalent linkage between ubiquitin and Cdc34 leads to an increased dissociation of the ubiquitin-charged Cdc34 from the Roc1/Cul1 interaction [58]. These observations suggest that Cdc34 may be constantly dissociated from the Roc1/Cul1-binding domain in a cyclic fashion during the ubiquitin-transfer reaction to extend the elongating ubiquitin chain on the substrate [58]. Thus the function of Roc1/Cul1 is probably to bring Cdc34 into the vicinity of the SCF-bound substrates, which lie on the top surface of the β-propeller repeats of WD40 or LRR in line with the E2 active site [21, 55, 56]. The formation of a thioester bond between ubiquitin and Cdc34 facilitates Cdc4 release from the Roc1/Cul1 interaction. Once the ubiquitin-charged Cdc34 is released into the region surrounding the substrate, the resulting increase in the effective concentration of ubiquitin-Cdc34 near the substrates promotes ubiquitin transfer from Cdc34 to the substrates. This model appears to provide a mechanistic explanation for the distance between the F-box protein-bound substrates and the Roc1/Cul1-bound E2, and the requirement for the rigid structure of SCF complexes. In addition, it also helps to address how the E2 protein can cope with the increasing distance between the elongating polyubiquitin chain and the fixed positions of the Roc1/Cul1-bound E2 and the F-box-bound substrates in bringing about the ubiquitin-transfer reaction. This cyclic association and dissociation of the E2 enzyme may also help interpret some observations for the effects of Cop9-signalosome complex (Csn) [59]. Although biochemically Csn plays an inhibitory role towards SCF through deneddylation of Cul1 and deubiquitination of the ubiquitinated substrates [59], the loss of Csn often produces accumulation of the SCF substrates. Since one function of Cul1 neddylation is thought to increase the binding of E2 to Roc1, loss of Csn *in vivo* may affect the cyclic binding of E2 to SCF complex during the polyubiquitination reaction of SCF substrates [59].

The rigid structure of the SCF complexes may also underlie the observed selectivity of the lysine residues in the substrates [21, 56]. It was found that only a subset of lysines in the SCF substrates such as p27 or Sic1 can be efficiently polyubiquitinated by SCF-E2 enzymes [60, 61]. Conversion of these lysines to arginines stabilizes these proteins even though they still contain other lysines. A single polyubiquitination chain on one of these critical lysines appears to be sufficient for substrate degradation by the 26S proteasome [60]. It thus appears that only those lysines in the SCF substrates that are in sight of or in the vicinity of the E2 enzyme may be used as ubiquitination receptors during the SCF-mediated ubiquitin-transfer reaction. The rigidity of the SCF complex is likely to contribute to such a restriction on the use of lysines in the substrates.

However, it is still possible that other mechanisms may exist to bridge the gap between substrate and E2 in the SCF-mediated ubiquitin-transfer reaction. For example, reports suggest that SCF may form higher order structures to facilitate the degradation of protein substrates. The *S. pombe* F-box proteins Pop1 and Pop2 have been shown to form heterodimers, and evidence suggests that these interactions may be important for the degradation of their *in vivo* substrates [62].

6.7
Regulation of SCF Activity

Several mechanisms have been shown to regulate the activity of the SCF complex. The expression of F-box proteins such as Skp2 is regulated by cell-cycle-dependent transcription [4]. The expression of Skp2 is high in late G1, S, and G2/M phase but low in the early G1 phase. In addition, Skp2 expression appears to be regulated by cell attachment and by the Pten/PI-3 kinase signaling pathway in certain cells [63, 64]. Tome-1, an F-box protein that triggers mitosis by targeting the mitotic inhibitory kinase Wee1 for proteolysis, is regulated by ubiquitin-dependent degradation through the Apc/C-Cdh1 E3 ligase in the G1 cells [35].

The expression of Cul1 has also been reported to increase in cycling cells after growth factor stimulation as compared with that of G0 cells [65]. In *C. elegans* and other organisms where multiple Skp1-related genes exist, the expression of individual Skp1-related genes is also regulated in a development- or cell-specific manner [27]. The regulated expression of these genes may play an important role in controlling the temporal functions of these SCF complexes in cell-cycle, development, and tissue specificity.

Another level of control is mediated through the control of F-box protein stabilities by the SCF complex using an auto-ubiquitination mechanism. Deletion of the F-box motif of various F-box proteins such as yeast Cdc4 or β-Trcp abolishes the interaction between Skp1/Cul1 and the F-box proteins [66]. Consequently the F-box proteins become more stable. This regulation may provide a means to recycle the components of SCF complexes between different F-box proteins. In addition, the levels of a particular F-box protein may be in part regulated by the balance between autoubiquitination and substrate-specific ubiquitination and thus could be sensitive to the presence of the substrates.

Cul1, together with other cullin family members, is uniquely regulated by a covalent modification with an ubiquitin-like protein, Nedd8 (Rub1 in yeast), at the conserved lysine 720 or equivalent sites in other cullins [59]. Neddylation appears to be important for the activity of SCF complexes and is essential for many organisms. The neddylation of Cul1 depends on the neddylation-activating E1 (APPBP1 and UBA3) and its specific conjugating E2 enzyme (UBC12). It was found that Csn, which shares substantial homology with the lid subcomplex of the 26S proteasome, exhibits an intrinsic peptidase activity towards neddylated Cul1 and other cullins [67]. Csn binds to cullins and this binding promotes the deneddylation of Cul1 and other cullins [67]. Csn may also play a role in recruiting the deubiquitination enzymes to reverse the ubiquitination of the SCF substrates [68]. One function of neddylation may be associated with the recruitment of E2 to the Cul1/Roc1 complex [68]. This possibility is also consistent with the structural determination that the neddylation receptor lysine 720 of Cul1 lies within the same surface as the Roc1-binding site in the C-terminal domain of Cul1 [68]. Binding of Roc1 to Cul1 or Cul2 can promote the neddylation of these cullins in plants and animals [68]. The binding of Roc1 to Cul1 may provide an open configuration for the neddylation of lysine 720 in Cul1, which lies close to the Roc1-binding site.

In addition, studies suggest that the association between Skp1/F-box proteins and Cul1/Roc1 is highly regulated. In particular, reports suggest that the binding of a Cul1-binding protein, Cand1/Tip120, to Cul1 or Cul1/Roc1 complexes causes the dissociation of Skp1/F-box proteins from Cul1. Cand1 (cullin-associated nedd8-dissociated protein 1) or Tip120 was isolated as a protein that binds to the Cul1/Roc1 complex only when Cul1 is not neddylated [69, 70]. Both *in vitro* and *in vivo*, there is a dynamic equilibrium between Cand1, Cul1 neddylation/deneddylation, and Skp1/F box protein binding. Assembly of the SCF complex is in part regulated by Cand1 and Cul1 neddylation. In the absence of Cul1 neddylation, Cand1 binds to the Cul1/Roc1 complex and such an interaction dissociates Skp1/F-box proteins from the Cul1/Roc1 complex. Neddylation of Cul1 promotes Cand1 dissociation, and facilitates the incorporation of Skp1/F-box proteins and SCF complex assembly. However, it remains unclear whether Cand1 is required for each cycle of ubiquitin transfer by the SCF and E2 enzymes to the substrate. Although Csn acts as an inhibitor of Cul1 and related cullin E3 complexes, loss of Csn activity often results in the accumulation, rather than enhanced degradation, of the SCF substrates [59]. The accumulation of SCF substrates in CSN mutants suggests that neddylation of Cul1 may be required for the repeated cycles of ubiquitin transfer *in vivo*.

6.8
The SCF Complex and the Related Cullin-containing Ubiquitin E3 Ligase

The SCF ubiquitin E3 ligase serves as the prototype of many related ubiquitin E3 ligases containing one of the cullin family members (Figure 6.2). So far, eight highly conserved cullins (Cul1–8) are found from yeast to human [3, 16, 17, 71]. All cullin-containing E3 ligases appear to bind one of the RING-H2 proteins, Roc1 or Roc2, and are subject to modification and regulation by the Nedd8 pathway. These multiprotein complexes probably represent the largest branch of the ubiquitin E3 ligases owing to their utilization of distinct substrate-targeting subunits.

Among these E3 ligases, Cul2 and Cul5 bind to Elongin C, an Skp1-like protein that is highly conserved between yeast and human [22, 23]. Elongin C also binds to Elongin B, an ubiquitin-like protein that is absent in SCF complexes. The Cul2/ElonginB/ElonginC complex interacts with the *v*on *H*ippel–Lindau tumor suppressor (VHL), a substrate-targeting subunit that regulates the stability of the hypoxia-inducible transcription factor HIFα in response to oxygen levels [23, 72]. Mutation of VHL is associated with many renal cell carcinomas and these mutations affect the VHL activity as the substrate-targeting unit of the Cul2/ElonginB/C complex. VHL also belongs to the large family of SOCS box proteins which are candidate substrate-targeting subunits of the Cul2/ElonginB/C complex that regulates signal transduction and many other biological processes [2]. For example, studies have shown that Cul2/Elongin C maintains germ cell fate in *C. elegans* by selectively targeting the germ cell-specific zinc-finger proteins for ubiquitin-dependent proteolysis in the soma cell but not in the germ cell [73].

Studies indicate that Cul3 binds to and employs BTB/POZ proteins as substrate-targeting subunits [24]. Cul3 has been shown to regulate mammalian embryonic cyclin E levels and also meiosis in *C. elegans*. A large number of BTB/POZ proteins exist. In human, more than 200 BTB/POZ proteins have been identified while more than 100 BTB/POZ proteins exist in *Drosophila* and *C. elegans*. The association of the BTB/POZ protein with Cul3 also has structural relevance to the SCF and

Cul2/Elongin B/C E3 complexes, since both Skp1 and Elongin C show the BTB/POZ-like protein fold. Thus the presence of the large family of cullin-containing ubiquitin E3 ligases suggests their profound regulatory roles in various important biological processes.

6.9
Perspectives

The SCF complex represents one of the largest ubiquitin E3 ligase families. The diversity of the F-box proteins allows the involvement of SCF in regulating various biological processes. So far, only a small number of F-box proteins have been characterized; a large body of work remains to further identify the substrates and regulation of other F-box proteins at cellular and organismal levels. In addition, the SCF E3 ligase represents the prototype of an extended family of E3 ligases that contain cullins. These cullin-containing E3 ligases, with at least hundreds of subunit-targeting subunits, are involved in a spectrum of biological events encompassing cell cycle, cell fate, and various signaling pathways. Alterations of many F-box proteins and SCF-regulated pathways are also associated with human diseases [74]. Understanding the function and regulation of this ubiquitin-dependent proteolysis mechanism should provide new insight into the treatment of human diseases such as cancer.

References

1 HERSHKO, A. and A. CIECHANOVER, The ubiquitin system. *Annu Rev Biochem*, **1998**, *67*, 425–79.

2 PICKART, C. M., Mechanisms underlying ubiquitination. *Annu Rev Biochem*, **2001**, *70*, 503–33.

3 DESHAIES, R. J., SCF and Cullin/RING H2-based ubiquitin ligases. *Annu Rev Cell Dev Biol*, **1999**, *15*, 435–67.

4 ZHANG, H., et al., p19Skp1 and p45Skp2 are essential elements of the cyclin A-CDK2 S phase kinase. *Cell*, **1995**, *82(6)*, 915–25.

5 BAI, C., et al., SKP1 connects cell cycle regulators to the ubiquitin proteolysis machinery through a novel motif, the F-box. *Cell*, **1996**, *86(2)*, 263–74.

6 KUMAR, A. and J. V. PAIETTA, The sulfur controller-2 negative regulatory gene of Neurospora crassa encodes a protein with β-transducin repeats. *Proc Natl Acad Sci USA*, **1995**, *92(8)*, 3343–7.

7 PATTON, E. E., A. R. WILLEMS, and M. TYERS, Combinatorial control in ubiquitin-dependent proteolysis: don't Skp the F-box hypothesis. *Trends Genet.*, **1998**, *14*, 236–43.

8 CENCIARELLI, C., et al., Identification of a family of human F-box proteins. *Curr Biol*, **1999**, *9(20)*, 1177–9.

9 WINSTON, J. T., et al., A family of mammalian F-box proteins. *Curr Biol*, **1999**, *9(20)*, 1180–2.

10 GOEBL, M. G., et al., The yeast cell cycle gene CDC34 encodes a ubiquitin-conjugating enzyme. *Science* **1988**, *241(4871)*, 1331–5.

11 MATHIAS, N., et al., Cdc53p acts in concert with Cdc4p and Cdc34p to control the G1-to-S-phase transition and identifies a conserved family of proteins. *Mol Cell Biol*, **1996**, *16(12)*, 6634–43.

12 SCHWOB, E., et al., The B-type cyclin kinase inhibitor p40SIC1 controls the

G1 to S transition in S. cerevisiae. *Cell*, **1994**, *79(2)*, 233–44.

13 KIPREOS, E. T., et al., cul-1 is required for cell cycle exit in C. elegans and identifies a novel gene family. *Cell*, **1996**, *85*, 829–839.

14 FELDMAN, R. M., et al., A complex of Cdc4p, Skp1p, and Cdc53p/cullin catalyzes ubiquitination of the phosphorylated CDK inhibitor Sic1p. *Cell*, **1997**, *91*, 221–30.

15 SKOWYRA, D., et al., F-box proteins are receptors that recruit phosphorylated substrates to the SCF ubiquitin-ligase complex. *Cell*, **1997**, *91*, 209–19.

16 DIAS, D. C., et al., CUL7: A DOC domain-containing cullin selectively binds Skp1.Fbx29 to form an SCF-like complex. *Proc Natl Acad Sci USA*, **2002**, *99(26)*, 16601–6.

17 ARAI, T., et al., Targeted disruption of p185/Cul7 gene results in abnormal vascular morphogenesis. *Proc Natl Acad Sci USA*, **2003**, *100(17)*, 9855–60.

18 PAGE, A. M. and P. HIETER, The anaphase-promoting complex: new subunits and regulators. *Annu Rev Biochem*, **1999**, *68*, 583–609.

19 NOUREDDINE, M. A., et al., Drosophila Roc1a encodes a RING-H2 protein with a unique function in processing the Hh signal transducer Ci by the SCF E3 ubiquitin ligase. *Dev Cell*, **2002**, *2(6)*, 757–70.

20 JIANG, J. and G. STRUHL, Regulation of the Hedgehog and Wingless signalling pathways by the F-box/WD40-repeat protein Slimb. *Nature*, **1998**, *391(6666)*, 493–6.

21 ZHENG, N., et al., Structure of the Cul1-Rbx1-Skp1-F boxSkp2 SCF ubiquitin ligase complex. *Nature*, **2002**, *416(6882)*, 703–9.

22 PAUSE, A., et al., The von Hippel-Lindau tumor-suppressor gene product forms a stable complex with human CUL-2, a member of the Cdc53 family of proteins. *Proc Natl Acad Sci USA*, **1997**, *94(6)*, 2156–61.

23 QUERIDO, E., et al., Degradation of p53 by adenovirus E4orf6 and E1B55K proteins occurs via a novel mechanism involving a Cullin-containing complex. *Genes Dev*, **2001**, *15(23)*, 3104–17.

24 KREK, W., BTB proteins as henchmen of Cul3-based ubiquitin ligases. *Nat Cell Biol*, **2003**, *5(11)*, 950–1.

25 SCHULMAN, B. A., et al., Insights into SCF ubiquitin ligases from the structure of the Skp1-Skp2 complex. *Nature*, **2000**, *408(6810)*, 381–6.

26 STEBBINS, C. E., W. G. KAELIN, JR., and N. P. PAVLETICH, Structure of the VHL-ElonginC-ElonginB complex: implications for VHL tumor suppressor function. *Science*, **1999**, *284(5413)*, 455–61.

27 MARROCCO, K., et al., The Arabidopsis SKP1-like genes present a spectrum of expression profiles. *Plant Mol Biol*, **2003**, *52(4)*, 715–27.

28 ZHAO, D., et al., Members of the Arabidopsis-SKP1-like gene family exhibit a variety of expression patterns and may play diverse roles in Arabidopsis. *Plant Physiol*, **2003**, *133(1)*, 203–17.

29 NAYAK, S., et al., The Caenorhabditis elegans Skp1-related gene family: diverse functions in cell proliferation, morphogenesis, and meiosis. *Curr Biol*, **2002**, *12(4)*, 277–87.

30 YAMANAKA, A., et al., Multiple Skp1-related proteins in Caenorhabditis elegans: diverse patterns of interaction with Cullins and F-box proteins. *Curr Biol*, **2002**, *12(4)*, 267–75.

31 WING, J. P., et al., Drosophila Morgue is an F box/ubiquitin conjugase domain protein important for grim-reaper mediated apoptosis. *Nat Cell Biol*, **2002**, *4(6)*, 451–6.

32 REIMANN, J. D., et al., Emi1 is a mitotic regulator that interacts with Cdc20 and inhibits the anaphase promoting complex. *Cell*, **2001**, *105(5)*, 645–55.

33 CIECHANOVER, A., et al., Mechanisms of ubiquitin-mediated, limited processing of the NF-κB1 precursor protein p105. *Biochimie*, **2001**, *83(3–4)*, 341–9.

34 KIPREOS, E. T. and M. PAGANO, The F-box protein family. *Genome Biol*, **2000**, *1(5)*, REVIEWS3002.

35 AYAD, N. G., et al., Tome-1, a trigger

of mitotic entry, is degraded during G1 via the APC. *Cell*, **2003**, *113(1)*, 101–13.

36 Tsvetkov, L. M., et al., p27(Kip1) ubiquitination and degradation is regulated by the SCF(Skp2) complex through phosphorylated Thr187 in p27. *Curr Biol*, **1999**, *9(12)*, 661–4.

37 Carrano, A. C., et al., SKP2 is required for ubiquitin-mediated degradation of the CDK inhibitor p27. *Nat Cell Biol*, **1999**, *1(4)*, 193–9.

38 Yaron, A., et al., Identification of the receptor component of the IκBalpha-ubiquitin ligase. *Nature*, **1998**, *396(6711)*, 590–4.

39 Winston, J. T., et al., The SCFβ-TRCP-ubiquitin ligase complex associates specifically with phosphorylated destruction motifs in IκBalpha and β-catenin and stimulates IκBalpha ubiquitination in vitro [published erratum appears in *Genes Dev* **1999** Apr 15; *13(8)*: 1050]. *Genes Dev*, **1999**, *13(3)*, 270–83.

40 Bornstein, G., et al., Role of SCFSkp2 ubiquitin ligase in the degradation of p21Cip1 during S-phase. *J Biol Chem*, **2003**, *278(28)*, p. 25752–7.

41 Tedesco, D., J. Lukas, and S. I. Reed, The pRb-related protein p130 is regulated by phosphorylation-dependent proteolysis via the protein-ubiquitin ligase SCF(Skp2). *Genes Dev*, **2002**, *16(22)*, 2946–57.

42 Busino, L., et al., Degradation of Cdc25A by β-TrCP during S phase and in response to DNA damage. *Nature*, **2003**, *426(6962)*, 87–91.

43 Jin, J., et al., SCF{beta}-TRCP links Chk1 signaling to degradation of the Cdc25A protein phosphatase. *Genes Dev*, **2003**, *17(24)*, 3062–3074.

44 Berset, C., et al., Transferable domain in the G(1) cyclin Cln2 sufficient to switch degradation of Sic1 from the E3 ubiquitin ligase SCF(Cdc4) to SCF(Grr1). *Mol Cell Biol*, **2002**, *22(13)*, 4463–76.

45 Nash, P., et al., Multisite phosphorylation of a CDK inhibitor sets a threshold for the onset of DNA replication. *Nature*, **2001**, *414(6863)*, 514–21.

46 Pagano, M., et al., Role of the ubiquitin-proteasome pathway in regulating abundance of the cyclin-dependent kinase inhibitor p27. *Science*, **1995**, *269(5224)*, 682–5.

47 Sheaff, R. J., et al., Cyclin E-CDK2 is a regulator of p27Kip1. *Genes Dev*, **1997**, *11*, 1464–78.

48 Ganoth, D., et al., The cell-cycle regulatory protein Cks1 is required for SCF(Skp2)-mediated ubiquitinylation of p27. *Nat Cell Biol*, **2001**, *3(3)*, 321–4.

49 Spruck, C., et al., A CDK-Independent Function of Mammalian Cks1. Targeting of SCF(Skp2) to the CDK Inhibitor p27(Kip1). *Mol Cell*, **2001**, *7(3)*, 639–650.

50 Tang, Y. and S. I. Reed, The Cdk-associated protein Cks1 functions both in G1 and G2 in Saccharomyces cerevisiae. *Genes Dev*, **1993**, *7(5)*, 822–32.

51 Sutterluty, H., et al., p45SKP2 promotes p27Kip1 degradation and induces S phase in quiescent cells. *Nat Cell Biol*, **1999**, *1(4)*, 207–14.

52 Nakayama, K., et al., Targeted disruption of Skp2 results in accumulation of cyclin E and p27(Kip1), polyploidy and centrosome overduplication. *Embo J*, **2000**, *19(9)*, 2069–81.

53 Maniatis, T., A ubiquitin ligase complex essential for the NF-B, Wnt/Wingless, and Hedgehog signaling pathways. *Genes & Dev*, **1999**, *13*, 505–510.

54 Margottin, F., et al., A novel human WD protein, h-beta TrCp, that interacts with HIV-1 Vpu connects CD4 to the ER degradation pathway through an F-box motif. *Mol Cell*, **1998**, *1(4)*, 565–74.

55 Orlicky, S., et al., Structural basis for phosphodependent substrate selection and orientation by the SCFCdc4 ubiquitin ligase. *Cell*, **2003**, *112(2)*, 243–56.

56 Wu, G., et al., Structure of a β-TrCP1-Skp1-β-catenin complex: destruction motif binding and lysine specificity of the SCF(β-TrCP1) ubiquitin ligase. *Mol Cell*, **2003**, *11(6)*, 1445–56.

57 Zheng, N., et al., Structure of a c-Cbl-UbcH7 complex: RING domain function in ubiquitin-protein ligases. *Cell*, **2000**, *102(4)*, 533–9.

58 DEFFENBAUGH, A. E., et al., Release of ubiquitin-charged Cdc34-S-Ub from the RING domain is essential for ubiquitination of the SCF(Cdc4)-bound substrate Sic1. *Cell*, **2003**, *114(5)*, 611–22.

59 COPE, G. A. and R. J. DESHAIES, COP9 signalosome: a multifunctional regulator of SCF and other cullin-based ubiquitin ligases. *Cell*, **2003**, *114(6)*, 663–71.

60 PETROSKI, M. D. and R. J. DESHAIES, Context of multiubiquitin chain attachment influences the rate of Sic1 degradation. *Mol Cell*, **2003**, *11(6)*, 1435–44.

61 SHIRANE, M., et al., Down-regulation of p27(Kip1) by two mechanisms, ubiquitin-mediated degradation and proteolytic processing. *J Biol Chem*, **1999**, *274(20)*, 13886–93.

62 WOLF, D. A., F. McKEON, and P. K. JACKSON, F-box/WD-repeat proteins pop1p and Sud1p/Pop2p form complexes that bind and direct the proteolysis of cdc18p. *Curr Biol*, **1999**, *9(7)*, 373–6.

63 MAMILLAPALLI, R., et al., PTEN regulates the ubiquitin-dependent degradation of the CDK inhibitor p27(KIP1) through the ubiquitin E3 ligase SCF(SKP2). *Curr Biol*, **2001**, *11(4)*, 263–7.

64 CARRANO, A. C. and M. PAGANO, Role of the F-box protein Skp2 in adhesion-dependent cell cycle progression. *J Cell Biol*, **2001**, *153(7)*, 1381–90.

65 O'HAGAN, R. C., et al., Myc-enhanced expression of Cul1 promotes ubiquitin-dependent proteolysis and cell cycle progression. *Genes Dev*, **2000**, *14(17)*, 2185–91.

66 ZHOU, P. and P. M. HOWLEY, Ubiquitination and degradation of the substrate recognition subunits of SCF ubiquitin-protein ligases. *Mol Cell*, **1998**, *2(5)*, 571–80.

67 LYAPINA, S., et al., Promotion of NEDD-CUL1 conjugate cleavage by COP9 signalosome. *Science*, **2001**, *292(5520)*, 1382–5.

68 WOLF, D. A., C. ZHOU, and S. WEE, The COP9 signalosome: an assembly and maintenance platform for cullin ubiquitin ligases? *Nat Cell Biol*, **2003**, *5(12)*, 1029–33.

69 ZHENG, J., et al., CAND1 binds to unneddylated CUL1 and regulates the formation of SCF ubiquitin E3 ligase complex. *Mol Cell*, **2002**, *10(6)*, 1519–26.

70 LIU, J., et al., NEDD8 modification of CUL1 dissociates p120(CAND1), an inhibitor of CUL1-SKP1 binding and SCF ligases. *Mol Cell*, **2002**, *10(6)*, 1511–8.

71 MICHEL, J. J., J. F. McCARVILLE, and Y. XIONG, A role for Saccharomyces cerevisiae Cul8 ubiquitin ligase in proper anaphase progression. *J Biol Chem*, **2003**, *278(25)*, 22828–37.

72 OHH, M., et al., Ubiquitination of hypoxia-inducible factor requires direct binding to the β-domain of the von Hippel-Lindau protein. *Nat Cell Biol*, **2000**, *2(7)*, 423–7.

73 DeRENZO, C., K. J. REESE, and G. SEYDOUX, Exclusion of germ plasm proteins from somatic lineages by cullin-dependent degradation. *Nature*, **2003**, *424(6949)*, 685–9.

74 PAGANO, M. and R. BENMAAMAR, When protein destruction runs amok, malignancy is on the loose. *Cancer Cell*, **2003**, *4(4)*, 251–6.

7

The Structural Biology of Ubiquitin–Protein Ligases

Ning Zheng and Nikola P. Pavletich

7.1
Introduction

Ubiquitination, the conjugation of ubiquitin to proteins, is a major post-translational modification mechanism that regulates a broad spectrum of biological functions [1]. One of the major functional roles of ubiquitination is to control the turnover rate of the substrate proteins, whereby the substrates are targeted to the 26S proteasome and proteolytically degraded. By adjusting the abundance of key proteins in cellular pathways, ubiquitination can switch many regulatory circuits to different states. Ubiquitination has also been found to regulate proteins through processes other than targeting proteins to the proteasome. These include endocytotic pathways, where ubiquitination serves as a sorting signal, and the DNA-damage response, where ubiquitination has a poorly understood but essential role [2].

The fundamental step of protein ubiquitination involves the formation of an amide bond between the ubiquitin C-terminus and the ε-amino group of a substrate lysine residue. Variations in the way additional ubiquitin molecules are conjugated to the substrate-linked ubiquitin confer different cellular functionalities on the modifier. To target proteins for proteasome-dependent degradation, a poly-ubiquitin chain is assembled on the substrates through the successive conjugation of ubiquitin to the K48 residue of the previous ubiquitin. A similar ubiquitin chain assembled through conjugation to the K63 residue, or the absence of any ubiquitin chain extension, on the other hand, lead to non-proteolytic signaling events.

The conjugation of ubiquitin to proteins is mediated by ubiquitin–protein ligases, which function at the last step of a three-enzyme reaction (Figure 7.1) [1, 3]. In the first reaction, the ubiquitin-activating enzyme, E1, activates free ubiquitin by utilizing ATP to form a high-energy thioester bond between the ubiquitin C-terminus and an E1 cysteine residue. The activated ubiquitin is then transferred to the ubiquitin-conjugating enzyme, E2, which forms a similar thioester linkage between its own active-site cysteine and ubiquitin. The final step of protein ubiquitination is catalyzed by the ubiquitin–protein ligases, E3, which bind both an E2 and a protein substrate and promote the ubiquitin transfer from the E2 to the sub-

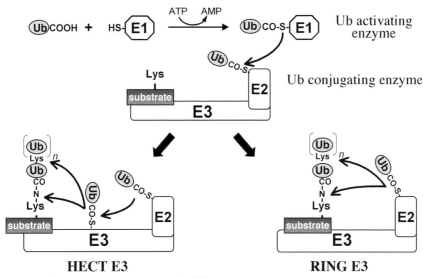

HECT E3   **RING E3**

Fig. 7.1. The E1–E2–E3 enzyme cascade of ubiquitin conjugation.

strate. The importance of the ubiquitin–protein ligases in ubiquitination is underscored by their roles in both determining the specificity of the modification and catalyzing the ubiquitin-transfer reaction. Coupled to various cellular signaling events, ubiquitin ligases ensure that the ubiquitination process is temporally controlled and tightly regulated with a high degree of substrate specificity. Not surprisingly, ubiquitin–protein ligases serve as the key regulators in many cellular pathways. Abnormal ubiquitin–protein ligase activity has been implicated in numerous human diseases such as cancer and neurological disorders [4].

Although the first E3 ligase activity was described in the early 1980s [5], it is only more recently that the E3s have emerged as a large superfamily of proteins and protein complexes. Central to this was the realization that the RING domain is a common motif in many E3s (reviewed in Ref. [6]). The rapidly growing number of ubiquitin ligases and the increasing recognition of their biological importance in recent years have been followed by several structural studies of E3s and of E3 complexes. In this chapter, we will review the recent advances in the structural biology of ubiquitin–protein ligases, focusing on their general architecture, their substrate recognition and E2-binding activities, and, importantly, the mechanistic insights into E3-catalyzed protein ubiquitination provided by these structures.

7.2
The Two Major Classes of Ubiquitin–Protein Ligases

The E3 ubiquitin–protein ligases represent a large and diverse family of proteins and protein complexes [3]. The human genome alone is estimated to code for hundreds of E3s, whereas the number of E2s is estimated at around thirty, and there is only one identified E1. The diversity of the E3 ubiquitin–protein ligases is reflected in their early loose definition as proteins or protein complexes that are required, in addition to the E1 and E2 activities, for the ubiquitination of a substrate. Today, we know that most E3s carry out three functions. They bind the substrate, thus conferring substrate specificity to the ubiquitination pathway, they bind a cognate E2, and they promote the ubiquitination of the substrate [1]. Identification and the realization of a common structural motif, namely the RING domain, in many otherwise divergent ubiquitin ligases have greatly facilitated the classification of known E3s [3, 6]. To date, all characterized ubiquitin ligases can be grouped into two major classes: the HECT (*H*omologous to *E*6AP *C*-*T*erminus) class and the RING/RING-like class. These two classes of E3s contain different signature domains and mediate substrate ubiquitination in functionally distinct ways.

HECT E3s share a conserved ∼40-kDa C-terminal catalytic HECT domain, preceded by divergent N-terminal domains that bind different protein substrates [3]. To mediate ubiquitination, HECT E3s first form a thioester intermediate between their active-site cysteine and the ubiquitin C-terminus and then transfer ubiquitin to the substrate (Figure 7.1) [7]. The RING class E3s do not form such a thioester intermediate with ubiquitin. Instead, they promote the direct transfer of ubiquitin from the E2 to the substrate. RING E3s are structurally diverse, containing from one to over ten subunits, yet they all have a RING domain in common [6, 8]. The RING domain is a ∼60-amino acid structural domain stabilized by two to three zinc atoms. In most cases that have been studied, the RING domain has the main E2-binding activity. The U-box, which is structurally related to the RING domain but lacks the zinc ligands, has recently also been shown to assemble into RING-like E3s [9].

Despite the difference in sequence conservation, structure, and the way they mediate ubiquitination, all ubiquitin ligases functionally share two common activities. They bind the substrate, conferring specificity to the reaction, and they also bind a cognate E2 (Figure 7.1). In regulating protein stability, the ability of E3s to interact with the substrate protein is often governed by the phosphorylation or other post-translational modification of the substrate. E3s studied to date are highly specific for either the Ubc2 or Ubc4 class of E2s, but generally appear to be less specific for individual E2s within each class, at least *in vitro* [10].

7.3
Mechanistic Questions About E3 Function

One of the central mechanistic questions regarding ubiquitination has been whether the reaction utilizes general acid/base catalysis, possibly in a manner analogous to the catalysis of peptide-bond cleavage. For example, an acidic catalytic residue could deprotonate the substrate lysine and make it a better nucleophile for attacking the ubiquitin thioester bond. In addition, a basic catalytic residue could polarize the thioester bond making the carbonyl carbon a better electrophile, and it could also stabilize the likely tetrahedral intermediate resulting from the nucleophilic attack. This is still an unanswered question for the HECT E3s.

In the case of RING E3s, three distinct mechanisms can be envisioned. One mechanism is recruitment and positioning. In principle, the RING E3 could promote ubiquitination by simply increasing the effective concentration of the entire substrate protein around the E2. Such an effect could be stereochemically more precise. For example, the E3 could increase the effective concentration of a portion of the substrate that includes the ubiquitination-site lysine at the E2 active site, or position and orient the lysine ε-amino group next to the ubiquitin-E2 thioester bond optimally for nucleophilic attack. The second possible mechanism is that the RING E3 could provide amino acids that act as acid/base catalysts at the active site of the E2, perhaps in a manner analogous to the *G*TPase *a*ctivating *p*roteins (GAPs) [11]. The third possibility is that E3-binding could cause significant conformational changes of the E2 to activate the enzyme for ubiquitin transfer. This was initially proposed based on the observation that RING domains of ubiquitin ligases can catalyze the polymerization of ubiquitin in a substrate-independent manner [12, 13]. As will be discussed later, the results of structural studies favor the model that the E3 raises the effective concentration of the lysine-containing substrate portion as the likely mechanism of catalysis. This, however, does not exclude the possibility that ubiquitination requires acid/base catalysis. Catalytic groups can, in principle, be provided by the E2. Although the vicinity of the E2 active-site cysteine is devoid of solvent-exposed polar or charged residues that are also conserved [10], a catalytic role could be played by E2 backbone groups, or by conserved residues elsewhere on the E2 that are brought into the active site through a hypothetical conformational change [14]. An alternative possibility is that the reaction is substrate catalyzed, with ubiquitin providing catalytic residues.

Another question regarding the mechanism of ubiquitin transfer is how the specificity for the ubiquitinated lysine is achieved. Until now, no sequence of motifs has been found that dictates which lysine on a protein is ubiquitinated. Therefore, it is possible that any lysine residue on a substrate could be modified. This is supported by studies showing that mutation of a large fraction of all lysines on a protein is necessary to reduce ubiquitination of certain proteins to a detectable extent (see, for example, Refs. [15, 16]). In other cases, however, ubiquitination of substrate proteins showed clear lysine specificity. Early work by Alexander Varshavsky and colleagues indicated that the polyubiquitin chain was conjugated to two specific lysine residues in the 1045 amino acid model substrate, β-galactosidase fusion protein [17]. More recently, lysine specificity for ubiquitination has been demonstrated in several physiological ubiquitination substrates. One of the best-characterized cases is IκBα, which gets ubiquitinated only at two adjacent lysines out of a total of 51 lysines. Mutation of both IκBα lysines completely abolishes its ubiquitination and degradation [18, 19]. The lysine specificity of ubiquitination in some proteins could be functionally required for several reasons. First, like protein phosphorylation, the ubiquitination of certain proteins is reversible with de-ubiquitination being mediated by protein-specific ubiquitin hydrolases [20]. Modification of a specific lysine residue might be favored by the system to gain precise control. Second, ubiquitination is one of many post-translational modifications of proteins that occur at lysine side chains. These include conjugation of other ubiquitin-like proteins, acetylation, and methylation [21]. Competition among these modifications at a specific lysine residue might be utilized as a mechanism to integrate signals from different pathways. For example, the same two lysines of IκBα that get ubiquitinated can also be modified by the ubiquitin-like protein SUMO (*s*mall *u*biquitin-related *mo*difier), and this has been shown to block the ubiquitination and destruction of IκBα [22]. Site-specific acetylation of the p53 tumor suppressor and the E2F-1 transcription factor has also been shown to block their ubiquitination [23, 24].

Ubiquitin–protein ligases promote not only the attachment of ubiquitin to the protein substrates but also the extension of the ubiquitin chain. What determines the choice between mono- *vs.* polyubiquitination is not well understood. It is possible that certain E3s catalyze only mono-ubiquitination. Alternatively, factors other than E3s might be responsible for the attachment of a single ubiquitin. For example, ubiquitin-binding accessory proteins have been suggested to block extension of the ubiquitin chain [25], whereas E3-associated ubiquitin hydrolase could trim down the polyubiquitin chain. The identification and characterization of *u*biquitin *E2 v*ariant proteins (UEVs) have provided an explanation for the assembly of K63-linked polyubiquitin chains [26, 27]. As discussed later, UEVs can be considered as special E3s, with the ubiquitin chain as their substrates.

Unlike many other enzymatic reactions, ubiquitination involves multiple enzymes and proteinaceous reactants. Although the structures of some individual components of the enzymatic system, such as the E2 and the RING fold, had been long known [28, 29], basic questions such as how the ubiquitin ligases recognize their substrates and recruit their cognate E2s, and how the E3 ligases couple ligand binding to ubiquitin transfer, could be best answered when the structures of several E3-substrate and E3–E2 complexes became available. These structures also set the framework for addressing questions about the mechanism of ubiquitin transfer.

7.4
The E6AP HECT Domain in Complex With UbcH7

First identified as a protein associated with the *h*uman *p*apilloma *v*irus (HPV) E6 [30], E6AP has been shown to be the cellular E3 that cooperates with the viral E6 protein to ubiquitinate p53 in HPV-infected cells [31]. It has since become the prototypical HECT E3, and in fact the E3–ubiquitin thioester intermediate was first discovered using E6AP [7]. E6AP was independently identified as the gene mutated in *A*ngelman *S*yndrome (AS), an inherited developmental syndrome characterized by severe motor dysfunction and mental retardation [32]. The endogenous substrates of E6AP involved in AS remain unknown.

Similar to other HECT E3s, E6AP consists of a ∼40-kDa C-terminal HECT domain and an N-terminal region containing sequences involved in binding E6-p53, and presumably the endogenous substrate(s). Distinct from many other HECT E3s, which have protein–protein interaction motifs such as WW domains, E6AP has no recognizable motifs in the N-terminal region [7]. E6AP also lacks the C2 domain, which is found in some HECT E3s that act on membrane-bound substrates. Nevertheless, biochemical studies have demonstrated that the E6AP HECT domain has the following activities that are likely to be common to all HECT E3s. The E6AP HECT binds a cognate E2 and accepts ubiquitin from the E2, forming a ubiquitin–thioester intermediate with its active-site cysteine. It then transfers ubiquitin to the ε-amino groups of lysine side chains on the substrate by catalyzing the amide bond formation, and subsequently transfers additional ubiquitin molecules to the growing end of the polyubiquitin chain (Figure 7.1) [33, 34].

The Ubc4 but not the Ubc2 family of E2s has been shown to function with E6AP and several other HECT E3s. E6AP appears to have preference for specific E2s within the Ubc4 family, and this has been shown to be due to specificity in the HECT domain–E2 interactions [33, 35–37]. In a yeast-two-hybrid assay, the closely related UbcH7 and UbcH8 E2s, but not the UbcH5, bind to E6AP. When tested *in vitro*, UbcH7 and UbcH8 showed the highest rates of ubiquitin–thioester intermediate formation with E6AP, whereas UbcH5 supports E6AP–ubiquitin–thioester formation at lower rates [37].

The crystal structure of the E6AP HECT domain bound to UbcH7 was the first solved structure of an E3 catalytic domain and of an E3–E2 complex. While the structure revealed the fold of the HECT domain, the details of the HECT catalytic site, the E3–E2 interactions, and the basis of specificity of E6AP for its cognate E2, it also raised many new questions. The structure showed that the HECT domain consists of two loosely packed lobes (Figure 7.2) [38], with an elongated N-terminal one (N-lobe) interacting with the E2, and a globular C-terminal one (C-lobe) bearing the active-site cysteine (Cys820). The C-lobe packs at one end of the N-lobe, forming an overall L-shape with the N-lobe being the base. The active-site cysteine of the HECT E3 is found near the junction of the two lobes, where a shallow and broad cleft is formed at the interface. The E2 active-site cysteine has an open line-of-sight to the E3 cysteine, the two being separated by ∼40 Å (Figure 7.2).

The large distance between the E2 and E3 active sites in the complex was unexpected, as trans-thioesterification is thought to proceed through an associative mechanism. For the HECT domain to take the ubiquitin from the E2, the E3 active cysteine has to be in close vicinity to the E2's active site. It has therefore been hypothesized that the HECT domain must undergo a significant conformational change to accept ubiquitin from the E2 [38]. Support for this model came from a subsequent report of the structure of the HECT domain from WWP1, determined by Joseph Noel and coworkers [39]. Although the structures of the individual N and C lobes of the two HECT domains are very similar, the relative orientation and position of the two lobes is very different in the two structures. Instead of packing at one end of the N-lobe, the C-lobe of the WWP1 HECT is interacting with the middle part of the N-lobe, at a position spatially related to that of the E6AP HECT C-lobe by a ∼100° rotation around the hinge loop connecting the lobes. Under this structural arrangement, the two active-site cysteines are brought

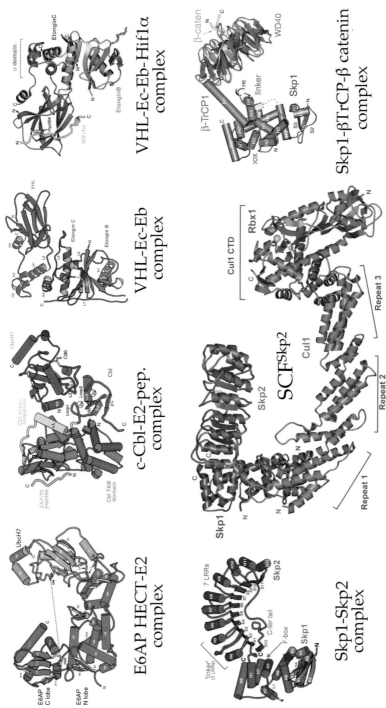

Fig. 7.2. Crystal structures of ubiquitin ligases discussed in this chapter. The structures are colored as following: The HECT N lobe is purple, C lobe green, and the E2 blue, the c-Cbl TKB domain is blue, and the E2 blue, the c-Cbl TKB domain purple, the phosphopeptide orange, and the E2 blue. In the VHL complexes, VHL is red, ElonginC blue, ElonginB green and Hif1α orange. In the Skp1–Skp2, SCF^{Skp2} and Skp1–βTrCP-β catenin complexes, Skp1 is blue, the F box protein red, Cul1 green, Rbx1 purple, and β-catenin orange.

much closer to each other, although the two are still 16 Å apart (Figure 7.3). Thus, the HECT must undergo an additional conformational change during catalysis [39]. In principle, the conformations observed in the two crystal structures could be induced by crystal packing and are not biologically relevant. However, the junction of the two lobes in both structures is lined with highly conserved residues from both sides in spite of the loosely packing interface (Figure 7.3), suggesting that this part of the structure, its potential conformational changes, and the specific structural configuration observed in the crystal are functionally important [38]. Moreover, conserved residues from both the N- and C-lobes, including a subset of those found in E6AP, are also juxtaposed at the active site cleft in the WWP1 HECT domain, again implying that the conformation seen in the WWP1 HECT crystal is functionally relevant. It is thus more likely that the conformations of the HECT domain seen in these two structures represent different steps of the ubiquitin-transfer reaction. Although the number of other steps involved remains to be determined, the large movement of the C-lobe relative to the N-lobe around the hinge loop is very likely required during the transfer of ubiquitin from the E2 to the HECT E3. Indeed, mutations that restrain the flexibility of the hinge loop between the two lobes caused significant decrease of the ubiquitin-transfer activity of the WWP1 HECT [39]. So far, neither the substrate-binding domain nor the substrate of any HECT E3s has been successfully co-crystallized with the HECT domain. It remains unclear how the HECT and its active site are orientated relatively to the substrate. It is conceivable that the movement of the HECT C-lobe might also be involved in transferring the ubiquitin it has taken from the E2 to the substrate and/or the growing polyubiquitin chain.

In many enzymes, conserved residues near the active site often participate in the catalytic reaction. To investigate the possible involvement of acid/base catalysis, mutagenesis studies of the E6AP active site have been carried out. Although mutations of several polar or charged residues reduced the efficiency of ubiquitin transfer, a residue that would be consistent with a role in deprotonating the substrate lysine has not been found [38]. Intriguingly, unlike the RING domain, the HECT domain of all known HECT E3s is always found at the C-terminus of the polypeptide. The extreme C-terminus of HECT is located near the active-site cysteine. A frameshift mutation of E6AP resulting in the extension of the C-terminus by 16 amino acids has been found in AS cases. Together, these lines of evidence raise the possibility that the C-terminal carboxylic acid group might participate in the reaction, especially in deprotonating the substrate lysine residue attacking the ubiquitin–thioester bond.

In the crystal of the E6AP-UbcH7 complex, the E3-bound E2 has an α/β structure very similar to the structures of other E2s crystallized by themselves, arguing against the model that E3s activate the E2 by causing conformational changes. The E6AP E3 interacts with one end of the overall elongated UbcH7 through a shallow hydrophobic groove on the N-lobe of its HECT domain (Figure 7.3). The major E2–E3 contacts are made by side chains from two E2 loops, termed L1 and L2 loops, while a portion of the N-terminal α-helix of the E2 is involved in minor intermolecular contacts (Figure 7.3). The structure indicates that the primary determinants of

(A)

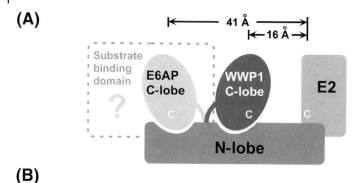

(B)

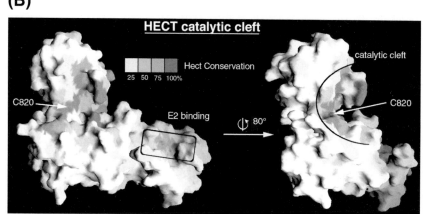

(C)

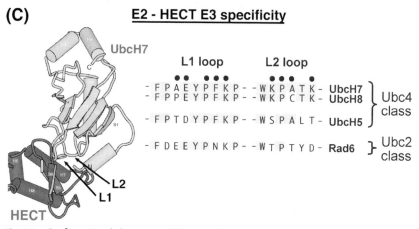

Fig. 7.3. Conformational changes and E2-binding specificity of the HECT domain. (A) Schematic diagram of the structures of E6AP and WWP1 HECT domains with their N-lobes superimposed indicating the potential large domain movement of the HECT E3. (B) Surface representation of the E6AP HECT domain, showing that conserved HECT domain residues map to the catalytic cleft defined by the active-site cysteine (Cys820). (C) Interactions of UbcH7 with a hydrophobic groove of the E6AP HECT domain through the E2's L1 and L2 loops. Alignment of L1 and L2 loop sequences from representative E2s shows that E6AP contacting residues (indicated by dots) are mostly conserved in the E2s that function with E6AP.

E6AP's specificity for UbcH7 are in the L1 and L2 loops of the E2. The hallmark of the interface is an E2 L1 loop phenylalanine (Phe63), which inserts its side chain into the center of the hydrophobic HECT groove. This phenylalanine residue is highly conserved in the Ubc4 but not the Ubc2 E2 subfamily, explaining the specificity of E6AP for the Ubc4 class of E2s. The preference of E6AP for UbcH7/UbcH8 over UbcH5 within the Ubc4 class can be explained by the additional contacts made by the L1 and L2 loops. For example, two L2-loop lysines that contact E6AP are conserved between UbcH7 and UbcH8 but not in UbcH5 (Figure 7.3).

7.5
The c-Cbl–UbcH7 Complex

The product of the proto-oncogene *c-Cbl* negatively regulates activated *r*eceptor *t*yrosine *k*inases (RTKs) such as PDGFR, EGFR, CSF-1R, and Met, by promoting their ubiquitination and subsequent degradation [40]. Although it was initially thought that c-Cbl mediates the polyubiquitination of the RTKs, recent studies have suggested that mono-ubiquitination of RTKs is sufficient for their internalization and degradation [41, 42]. Therefore, one of the functional roles of c-Cbl in RTK down-regulation is likely to mediate their mono-ubiquitination, possibly at multiple sites, providing a signal for their endocytosis and degradation in the lysosome. Consistent with this hypothesis, it has been shown that in addition to ubiquitinating the activated RTKs, c-Cbl also recruits and mono-ubiquitinates adapter proteins involved in endocytosis [43–45].

c-Cbl is a member of the Cbl protein family that is found in most multicellular organisms. In mammals, the Cbl family consists of two additional members (Cbl-b and Cbl-3), whereas in invertebrates such as *Drosophila melangogaster* and *Caenorhabditis elegans*, only one member (D-Cbl and SLI-1, respectively) has been identified [46]. All these Cbl proteins share three highly conserved structural domains, including an N-terminal SH2-containing *t*yrosine *k*inase-*b*inding (TKB) domain, a RING domain, and a ∼40-residue short linker region connecting the two. Two recognizable sequence motifs, including a proline-rich region and a ubiquitin-associated (UBA) domain, are also found in some but not all Cbl family members. c-Cbl recognizes activated RTKs by binding a phosphotyrosine sequence motif through its TKB domain, whereas it binds an E2 through its conserved RING domain [12, 47]. The structure of the TKB domain bound to a phosphopeptide derived from the non-receptor tyrosine kinase ZAP-70 has been reported by Michael Eck's group [48]. In that structure, the TKB domain is formed by three interacting sub-domains comprising a four-helix (4H) bundle, a calcium-binding EF hand, and a variant SH2 domain. Building on this work, we subsequently determined the structure of a nearly intact c-Cbl including both the TKB and RING domains bound to the same ZAP-70 phosphopeptide as well as the UbcH7 E2 [49].

The c-Cbl–E2–ZAP70 peptide complex adapts a compact structure with multiple inter- and intra-molecular interfaces (Figure 7.2) [49]. The RING domain is anchored on the TKB domain through extensive interactions with the 4H bundle, while the linker forms an ordered loop and an α-helix, which packs closely with the TKB domain next to the RING. As expected, the E2 predominantly interacts with the c-Cbl RING domain, but it is also in contact with the c-Cbl linker helix (Figure 7.2). The E2 active-site cysteine is located on the side of the complex opposite where the phosphorylated substrate peptide binds.

The structure of the complex provides several insights into the mechanism by which the RING E3 mediates ubiquitination. First, the structure helped to rule out the model of acid/base catalysis, as there were no c-Cbl residues in the vicinity of the E2 active-site cysteine (the closest c-Cbl residue is 15 Å away). Second, as there was no significant conformational change within the UbcH7 E2 upon binding to the Cbl protein, the structure helped to rule out the possibility that the RING E3 activates the E2 by causing conformational changes. Therefore, the mechanism underlying the catalysis could only be plausibly explained by the general recruitment model [49]. Intriguingly, it has been reported that mutations in the linker region inactivate c-Cbl and render it tumorigenic in mice, presumably through a dominant negative effect. In particular, one of the mutations mapping to a tyrosine (Y371) at the linker–TKB interface had been shown to abolish c-Cbl function without qualitatively affecting binding to either RTKs or E2 [12]. This suggests that the Cbl E3 may do more than just recruit the protein substrate to the E2 [49]. In addition to bringing together the substrate and the E2, the precise relative position and orientation of the substrate-binding and E2-binding domains might be important to the functions of the RING E3. In fact, this is consistent with the rigid appearance of the c-Cbl structure, where all functional domains are packed closely to each other. Since the binding site of the substrate phosphopeptide is located far from the E2, the precise positioning of the rest of the substrate would require additional contacting surface from the E3. Strikingly, the molecular surface of c-Cbl revealed a surface channel lined with conserved residues (Figure 7.4). The channel runs from the substrate peptide-binding site to the general vicinity of the E2 active site, suggesting that it might be involved in directing the substrate polypeptide chain towards the E2. Taken together, these structural features of the E3–E2 complex suggest that RING E3s may serve to position and orient the substrate optimally for ubiquitin transfer [49]. Furthermore, the c-Cbl E3 may provide steric and distance restraints that determine which substrate lysine(s), relative to the c-Cbl binding phosphotyrosine epitope, will be ubiquitinated at the highest rates [49].

The structure of the c-Cbl–E2 complex revealed that the c-Cbl RING domain binds to essentially the same structural elements of the E2 as the E6AP HECT domain does (Figure 7.4). These include both the L1 and L2 specificity loops and the N-terminal E2 helix. Similar to the HECT domain, the RING domain forms a shallow hydrophobic groove on its surface, accommodating the L1 and L2 loops. The same E2 L1 loop phenylalanine (Phe63) inserts into the center of the groove. Although with a different side, the N-terminal E2 helix, which packs against the HECT domain, also interacts with the linker helix of the RING (Figure 7.4). The similarities of E2–E3 interactions between the two cases are even more striking considering that the E2-binding grooves of the RING and HECT domains are structurally unrelated. Therefore, it is likely that most E2–E3 interactions will occur the

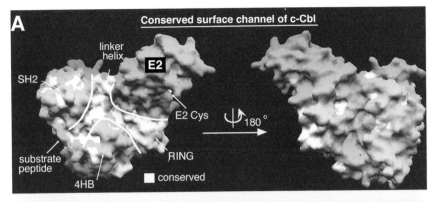

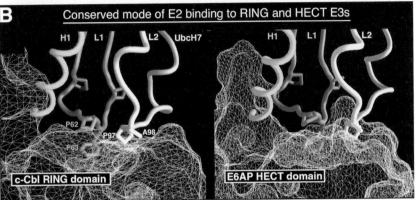

Fig. 7.4. Structural features of the c-Cbl RING E3 for substrate and E2 binding. (A) A conserved surface channel is found at one side of c-Cbl running from the peptide-binding site to the vicinity of the E2 active site. (B) The c-Cbl RING domain recognizes the same structural elements of the E2 as the E6AP HECT domain does despite the completely different folds of the two E3s (surfaces represented as a white net).

same way and that the L1 and L2 loops of the E2 are the principal determinants of E2–E3 specificity [49].

7.6
The SCF E3 Superfamily

The SCF and SCF-like complexes are multi-subunit RING-type E3s that represent the largest E3 family known to date. This superfamily of E3s are involved in regulating cell-cycle progression, signal transduction pathways, transcriptional control, and multiple aspects of cell growth and development (reviewed in Ref. [50]). All members of this E3 superfamily contain two basic components, a member of the cullin protein family and a RING-domain protein. The cullin subunit serves as the

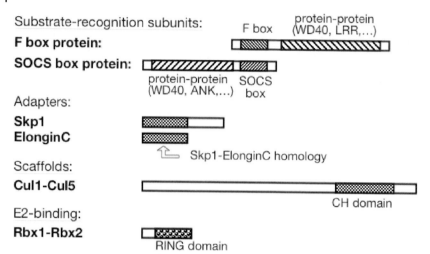

Fig. 7.5. Subunit families of the SCF and SCF-like E3s.

scaffold of the E3 complex, whereas the RING-domain protein recruits the E2. In the human genome, five cullins (Cul1–Cul5) and two cullin-interacting RING proteins (Rbx1 and Rbx2) have been identified (Figure 7.5). In addition, three more genes have been found to share sequence homology to a part of the cullins, termed the *C*ullin-*H*omology (CH) domain (APC2, KIAA0708, and KIAA0076). Among these three, APC2 has been shown to interact with an Rbx-homologous protein, APC11, together forming the core of the *a*naphase-*p*romoting *c*omplex (APC), an important E3 complex regulating the cell cycle [51–53]. KIAA0076 has been shown to interact with Rbx1 (also called Roc1 and Hrt1) and renamed as Cul7 [54].

The prototype of this E3 superfamily is the SCF complex, whose scaffold and RING subunits are Cul1 and Rbx1, respectively. An SCF complex also contains two more components, an adapter protein Skp1, which interacts with the scaffold subunit, and an interchangeable substrate-binding subunit, termed the F-box protein (Figure 7.5). A similar SCF-like complex is formed based on Cul2, in which Elongin C serves as the adapter protein, whereas the substrate-binding subunit is one of the SOCS-box proteins. Recent studies have revealed a new family of SCF-like complexes built on Cul3 [55, 56]. In these complexes, the adapter and substrate-binding functions are combined in a single polypeptide, which is a member of the emerging BTB protein family.

The F-box protein family is the largest substrate-recognition subunit family. It enables the eukaryotic cells to use the SCF E3 machinery to ubiquitinate a large number of diverse protein substrates. So far, over 70 F-box proteins have been identified in the human genome [57, 58]. F-box proteins all share an ~40-amino acid F-box motif, which is usually followed by a C-terminal protein–protein interaction domain such as the WD40 repeats β-propeller (Fbw subfamily) and *l*eucine-*r*ich *r*epeats (LRRs; Fbl subfamily; Figure 7.5) [59, 60]. F-box proteins interact with the Skp1 adapter protein through their F-box motif to assemble with the rest of the SCF complex. F-box proteins play a central role in the phosphorylation-controlled destruction of regulatory proteins. For example, SCF^{Skp2} (superscript denotes the F-box protein) recognizes p27^{Kip1} only after the latter has been phosphorylated during the G1–S transition [61, 62]; SCF^{Fbw7} binds only phosphorylated CyclinE [63–65]; and SCF$^{\beta TrCP}$ recognizes a doubly-phosphorylated destruction motif sequence in the β-catenin and IκBα proteins [66].

The SOCS-box protein family plays a similar role in the Cul2-based E3 complexes to the F-box proteins in the SCF complex [60, 67]. The adapter protein ElonginC shares limited sequence homology over ∼115 amino acids with Skp1 and also requires an obligate partner ElonginB for its function (Figure 7.5) [67]. SOCS-box proteins bind ElonginC through their common SOCS-box motif and bind substrates through their N-terminal protein–protein interaction domains such as ankyrin repeats (ASB subfamily) and WD40 repeats (WSB subfamily) (Figure 7.5). Initially identified as suppressors of cytokine signaling, the SOCS-box protein family has expanded to over 40 members in the human genome [68]. Whether all SOCS-box proteins can assemble into E3 complexes remains to be tested, but a large body of studies has demonstrated that the SOCS box protein VHL mediates the ubiquitination of Hif1α in oxygen-response pathways [69]. Furthermore, SOCS-1 and SOCS-3 have been shown to ubiquitinate the *i*nsulin *r*eceptor *s*ubstrate 1 and 2 (IRS1 and IRS2) in response to insulin stimulation [70].

Crystal structures of a number of sub-complexes of the SCF and SCF-like E3s have been reported. These include the VHL–ElonginC–ElonginB complex bound to a Hifα peptide, Skp1–Skp2, Skp1–β-TrCP bound to a β-catenin peptide, Skp1–Cdc4 bound to a consensus peptide, and Cul1–Rbx1–Skp1–F-box^{Skp2}. Together, these structures have not only revealed the general architecture of the SCF complex and delineated the structural and functional roles of each subunit, but also shed light on how these multisubunit RING E3 complexes mediate substrate ubiquitination.

7.6.1
The VHL–ElonginC–ElonginB–Hif1 α Complex

The VHL gene was first identified by positional cloning as a tumor suppressor gene whose germline mutation is associated with the rare inherited von Hippel–Lindau cancer predisposition syndrome [71]. The disorder is characterized by tumors of the central nervous system, kidney, retina, pancreas, and adrenal gland. Soon after, VHL was found to bind the ElonginC and ElonginB proteins [72], which were known previously as factors that stimulated the transcriptional elongation factor ElonginA *in vitro*. Although the association of VHL with ElonginC and ElonginB suggested that VHL may function in transcription elongation, the subsequent detection of Cul2 in the same complex, together with the sequence homology between Skp1 and ElonginC, indicated that these four proteins form an SCF-like ubiquitin ligase complex [73].

To date, the best-characterized substrate of the VHL–Elongin*B*–Elongin*C*–

*C*ul2–*R*bx1 (VBC-CR) E3 complex is the α-subunit of the heterodimeric *h*ypoxia-*i*nducible *f*actors (HIFs). HIFs are transcription factors that play a central role in the cellular response to low oxygen levels (hypoxia) by activating the expression of genes involved in angiogenesis, erythropoiesis, and energy metabolism [75]. VHL recognizes the *o*xygen-dependent *d*egradation *d*omain (ODD) of Hif1α only when a proline residue (Pro564) of the ODD is hydroxylated [75, 76]. This proline hydroxylation modification is carried out by a family of recently identified *H*IF *p*rolyl *h*ydroxylases (HPHs) only in the presence of oxygen [77, 78]. Under normal conditions (normaxia), where there is enough oxygen delivered, Hif1α is constantly synthesized, hydroxylated, ubiquitinated, and degraded. Under hypoxic conditions, however, HPHs fail to hydroxylate Pro564 of Hif1α owing to the lack of oxygen, which allows Hif1α to escape from VBC-CR-mediated ubiquitination and be stabilized. This explains why VHL-associated tumors often have constitutively high levels of Hif1α and is associated with the development of highly vascularized tumors.

The crystal structure of the VHL–ElonginC–ElonginB complex reveals that VHL has two structural domains, an N-terminal β-domain rich in β-sheet and a C-terminal α-domain adopting a three-helix cluster structure [67]. The α-domain, where the SOCS box motif is found, interacts with ElonginC, whereas the β-domain is not involved in any intermolecular contacts (Figure 7.2). The α- and β-domains are connected by two linkers and an extensive interface with a network of hydrogen bonds, indicating that the two are rigidly connected. Several residues mutated in VHL tumors are found at the interface, including the most frequently mutated one (Arg167). This suggests that the relative arrangement of the substrate- and ElonginC-binding domains is important for VHL function, a structural feature also observed in the c-Cbl structure [67].

Mapping of the tumor-derived VHL mutations on the VHL structure revealed two patches of solvent-exposed residues (Figure 7.6). One patch is located on the portion of the α-domain involved in ElonginC binding, confirming the role ElonginC binding plays in the tumor suppressor function of VHL. The second patch,

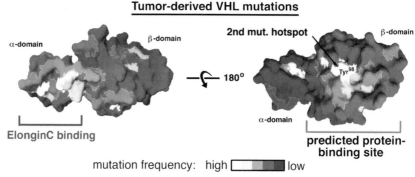

Tumor-derived VHL mutations

mutation frequency: high [   ] low

Fig. 7.6. Mis-sense mutations derived from VHL tumors map either to hydrophobic core-residues of the α- and β-domains, or to two clusters of surface residues. One of these clusters corresponds to the ElonginC-binding site, while the other maps to the β-domain, suggesting the presence of a protein-binding site.

which includes the second most frequently mutated VHL residue (Tyr98), is mapped onto the β-domain, strongly suggesting that this domain has a protein-binding site [67]. This prediction was confirmed by the crystal structure of the VBC complex bound to a hydroxyproline-containing Hif1α peptide reported later (Figure 7.2) [79]. The VBC–Hif1α structure showed that the Hif1α peptide binds the β-domain of VHL in an extended conformation. The hydroxyproline inserts into a gap in the VHL hydrophobic core, precisely at the protein-binding site predicted by tumor-derived mutations (Figure 7.6). The 4-hydroxyl group of the hydroxyproline is recognized by a pair of buried serine and histidine residues of VHL [79].

ElonginC adopts an α/β structure similar to the BTP/POZ fold [67]. ElonginB does not interact with VHL, and appears to have a structural role in stabilizing ElonginC. Based on the sequence homology between ElonginC and Skp1, it was expected that Skp1 would also have a similar BTP/POZ fold structure, except for a ~40-residue C-terminal Skp1 region extending beyond the homology region [67]. Since the functional relationship between the F-box and Skp1 is highly analogous to that between the SOCS-box and ElonginC, even before the structure of F-box has been determined it had been predicted that the F-box might have a similar structure to that of the SOCS-box. This was supported by threading analysis showing that among a library of 1925 folds, the fold most consistent with the F-box sequence was the three-helix cluster structure of the VHL α-domain [67]. The structural similarity between the SOCS box and F-box proteins became even more obvious when the Skp1–Skp2 structure became available.

7.6.2
Skp1–Skp2 Complex

One of the best-studied F-box proteins is Skp2, which contains an LRR substrate-binding domain. Skp2 is involved in regulating the G1–S transition in mammalian cells by controlling the levels of the cyclin-dependent kinase inhibitor p27^{Kip1} [80]. In quiescent cells, p27^{Kip1} binds and inhibits Cdk2/cyclinA/E kinase complexes, whose activities are necessary for DNA replication in S phase. Upon stimulation by mitogenic signals, p27^{Kip1} is phosphorylated and targeted to the SCF^{Skp2} for ubiquitination and subsequent degradation. Since p27^{Kip1} functions as a tumor suppressor, Skp2 can be classified as an oncogene. In fact, Skp2 is over-expressed in many tumors and has transforming capacity [81]. Unlike canonical F-box proteins, however, Skp2 is essential but not sufficient for recruiting p27 to the ligase complex. It has been shown that an additional subunit, the 79-amino acid Cks1 protein, is also required for p27 ubiquitination by the SCF^{Skp2} E3 [61, 62]. Other studies have expanded the substrates of SCF^{Skp2} to include the pRb family member p130 and c-Myc [82–84]. The recognition and ubiquitination of p130 by SCF^{Skp2}, however, does not require Cks1.

The crystal structure of a nearly full length Skp1 bound to Skp2 without an N-terminal 100-residue region of unknown function has been determined. Confirming the prediction from the VHL structure, the structure of the complex showed

F box and SOCS box structurally and functionally related

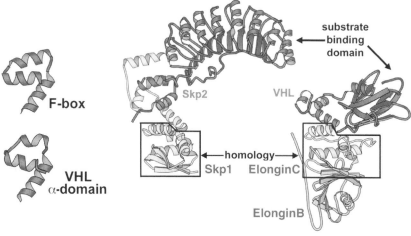

Fig. 7.7. Comparison of the Skp2–Skp1 and VHL–ElonginC–ElonginB complexes reveals the structural similarity between the F-box and SOCS-box (VHL α-domain) and similarity of the assembly of the intact Skp2–Skp1 and VHL–ElonginC–ElonginB complexes.

that the F-box has a three-helix cluster structure very similar to the SOCS box domain of VHL [60] (Figures 7.2 and 7.7). Different from the SOCS box of VHL, which interacts predominantly with the BTB/POZ fold of ElonginC, the F-box of Skp2 mainly interacts with the C-terminal extension of Skp1. This makes the relative orientation between the F-box and the Skp1 BTB/POZ fold different from that of the SOCS-box and the ElonginC BTB/POZ fold [60]. However, when the BTB/POZ folds of Skp1 and ElonginC are aligned, the substrate-binding LRR domain of Skp1 and the β-domain of VHL end up pointing in the same direction (Figure 7.7). This common structural arrangement is functionally relevant as the BTB/POZ fold is the cullin-binding domain of these adapter proteins, as discussed later.

Another striking similarity between the Skp1–Skp2 and ElonginC–VHL complexes is the rigid structural coupling between the adapter-binding motif and the substrate-binding domain in both cases. In the Skp1–Skp2 structure, the linker sequence connecting the F-box and LRRs adopts the structure of three non-canonical LRRs, which together with seven predicted LRRs form a single structural domain packing directly against the F-box motif. In addition, a portion of Skp1 also participates in the F-box–LRR interface (Figure 7.7). The functional importance of this rigid coupling has been further demonstrated by mutation studies showing that altering an Skp1 residue involved in this interface rendered a non-functional mutant in a yeast complement assay without abolishing its Skp2-binding activity [60].

The Skp1–Skp2 structure thus recapitulated the rigidity of the intermolecular and inter-domain interfaces recurrently observed with the RING E3 structures. Together with mutational analyses in all cases including the c-Cbl and VHL, this common structural feature strongly suggests that the rigid structural architecture of these RING E3 components is functionally important. Therefore, the ubiquitin ligase activity of the RING E3s is best explained by the model that RING E3s catalyze ubiquitination by optimally positioning the substrate for ubiquitin transfer. These concepts were further confirmed by the structure of the SCF^{Skp2} complex.

7.6.3
SCF^{Skp2} Structure

A nearly complete picture of an SCF complex became available when two crystal structures were reported, one of the full length Cul1 bound to full length Rbx1, and the other of the quaternary complex consisting of Cul1, Rbx1, Skp1, and the F-box domain of Skp2 (hereafter referred to SCF^{Skp2}). SCF^{Skp2} has an overall highly elongated structure with Rbx1 and Skp1 segregated to opposite ends (Figure 7.2) [85]. This structural arrangement is organized by Cul1, which interacts with both Rbx1 and Skp1 and serves as the scaffold of the complex. Cul1 consists of an ~400-amino acid N-terminal helical region (hereafter NTD) that adopts a long stalk-like structure, and a ~350-amino acid C-terminal globular α/β domain (hereafter CTD). The two domains pack across an extensive interface that is invariant in the two different crystals [85].

The Cul1 NTD comprises three novel helical repeats, each is formed by five α-helices. The three repeats pack consecutively in a regular manner with extensive hydrophobic interfaces, overall adopting an arc-shaped structure with an ~110-Å span. There is little difference in the NTD in the two crystal structures. One side of the first repeat at the tip of the NTD stalk binds Skp1, interacting with its BTB/POZ fold. The Cul1 NTD residues that contact Skp1 are highly conserved in Cul1 orthologs but not in the cullin paralogs, Cul2 through Cul5, explaining the specificity of Skp1 to Cul1 but not the other cullins. Interestingly, when the sequences of the orthologs of individual cullins are compared, the residues that correspond to the Skp1-interacting residues of Cul1 are evolutionarily conserved and distinct among different cullins (Figure 7.8). This led to the prediction that each cullin would recruit a different protein-binding partner similar to the way that Cul1 binds Skp1. Since Cul2 is known to bind the ElonginC adapter, it is conceivable that the binding of Cul2 to the BTB/POZ domain of ElonginC should structurally mimic the Cul1–Skp1 binding (Figure 7.8). In agreement with the prediction, studies have shown that Cul3 interacts with a family of proteins containing both a BTB/POZ domain and a protein–protein interaction domain [55, 56]. Therefore, interaction between the BTB/POZ fold and the cullin NTD is likely to be a common mechanism involved in the recruitment of adapter/substrate-binding partners by most if not all cullins. (No BTB/POZ fold has been found in any of the protein subunits of the more recently found Cul4-containing E3 ligase complex [86].)

The major function of the Cul1 CTD is to recruit the RING-finger protein Rbx1. The CTD binds Rbx1 through two distinct interfaces (Figure 7.2). One involves an

Other cullins may have an adapter/protein binding site at their NTD tip

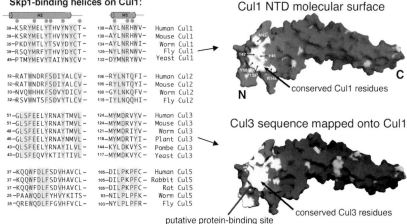

Skp1-binding helices on Cul1:

39-KSRYMELYTHVYNYCT-	138-AYLNRHWV-	Human Cul1
39-KSRYMELYTHVYNYCT-	138-AYLNRHWV-	Mouse Cul1
35-PKDYMTLYTSVYDYCT-	134-AYLNRHWI-	Worm Cul1
39-RSQYMRFYTHVYDYCT-	129-NYLNRNWV-	Fly Cul1
45-PTMYMEVYTAIYNYCV-	132-DYMNRYWV-	Yeast Cul1
32-RATWNDRFSDIYALCV-	106-RYLNTQFI-	Human Cul2
32-RATWNDRFSDIYALCV-	106-RYLNTQYI-	Mouse Cul2
33-NVQWHHKFSDVYDICV-	108-GYLNKQFV-	Worm Cul2
32-RSVWNTSFSDVYTLCV-	115-IYLNQQHI-	Fly Cul2
51-GLSFEELYRNAYTMVL-	124-MYMDRVYV-	Human Cul3
51-GLSFEELYRNAYTMVL-	124-MYMDRVYV-	Mouse Cul3
48-GLSFEELYRNAYTMVL-	121-MYMDRIYV-	Worm Cul3
46-GLSFEELYRNAYNMVL-	118-MYMDRTYI-	Plant Cul3
43-QLSFEELYRNAYILVL-	144-KYLDKVYS-	Pombe Cul3
43-DLSFEQVYKTIYTIVL-	117-MYMDKVYC-	Yeast Cul3
37-KQQWFDLFSDVHAVCL-	105-DILPKPFC-	Human Cul5
37-KQQWFDLFSDVHAVCL-	105-DILPKPFC-	Rabbit Cul5
37-KQQWFDLFSDVHAVCL-	105-DILPKPFC-	Rat Cul5
25-PAAWQDLFYHVYKITS-	93-NILPLPFK-	Worm Cul5
35-QREWQDLFFGVHFVCL-	103-NYLPLPFR-	Fly Cul5

Cul1 NTD molecular surface

conserved Cul1 residues

Cul3 sequence mapped onto Cul1

conserved Cul3 residues

putative protein-binding site

Fig. 7.8. Analysis of the Skp1-binding site of Cul1 and its comparison with other cullins indicate that other cullins all have a protein-binding site at their NTD tip. Left, sequence alignment of orthologs of individual cullins, showing ortholog-restricted conservation of residues that correspond to the Skp1-interacting residues of Cul1 (indicated by dots). Right, surface representation of the Cul1 NTD (top) and Cul3 NTD model (bottom) highlighting surface residues invariant in the respective orthologs.

intermolecular β-sheet, where a five-stranded β-sheet is formed by four β-strands from Cul1 and one central strand from Rbx1. Immediately following the Rbx1 β-strand, the Rbx1 RING domain is embedded in a wide Cul1 CTD cleft, where the second Cul1–Rbx1 interface is formed. The unusual intermolecular β-sheet between Cul1 and Rbx1 likely accounts for the extremely high affinity between the two proteins, as it involves both an extensive hydrogen bond network and hydrophobic packing [53]. The functional reason for this bipartite interacting mode between the two proteins remains to be clarified.

The Rbx1 RING domain is unique among the RING motifs as it contains three coordinated zinc ions instead of two. A 20-residue insertion sequence of Rbx1 provides the extra zinc-binding site. The major part of the Rbx1 RING, however, adopts a structure highly similar to the canonical RING motif, with a hydrophobic groove on its surface analogous to the c-Cbl E2-binding groove. Mutagenesis studies have confirmed that this groove is involved in E2 binding by Rbx1, whereas the insertion region is not essential for its E3 ligase activity [85].

In the cell, a sub-population of all cullins is covalently modified by the ubiquitin-like protein, Nedd8/Rub1, on a specific lysine residue at their C-terminal domain [87]. To date, cullins are the only known substrates of Nedd8 modification (neddylation). It is generally believed that neddylation regulates the cullin-based ubiquitin ligases, although the precise mechanism remains elusive. Nedd8 is essential for cell viability in fission yeast and for early development of *C. elegans*, *Arabidopsis*, and mice [88–91]. In *in vitro* systems, neddylation of Cul1 in the SCF complexes has been shown to enhance their E3 activities toward natural substrates [92–94]. The crystal structure of the SCF complex reveals that the target lysine residue in cullins for neddylation is in close vicinity to the Rbx1 RING domain, suggesting that Nedd8 might be able to modulate the functions of Rbx1 and/or its associated E2 [85]. Other studies have identified a cellular protein, Cand1, which can bind the Cul1–Rbx1 complex and inhibits its association with Skp1 and F-box proteins [95, 96]. It has been shown that neddylation of Cul1 is able to release the inhibition and promote the assembly of the SCF. The structural basis of the Cand1–Nedd8-mediated assembly and disassembly of the SCF complex remains to be determined. Noticeably, the Cul1 lysine residue targeted for neddylation is surrounded by a number of highly conserved and surface-exposed cullin residues, which could potentially interact with Nedd8 or Cand1.

Consistent with the structural observations of c-Cbl, VHL, and Skp1–Skp2, the SCF^{Skp2} structure suggests that the whole SCF ubiquitin ligase is a rigid assembly. The functional importance of the rigidity of the SCF has been further tested with a Cul1 mutant, whose NTD and CTD were engineered to be linked by a flexible linker sequence (Figure 7.9) [85]. This mutant was made in two steps. First, the hydrophobic packing interface between the NTD and the CTD was disrupted by mutating two residues on the NTD and three on the CTD to polar amino acids. Second, the two domains were connected with 12- or 18-residue polar linkers. The

Rigidity of Cul1 scaffold important for SCF function

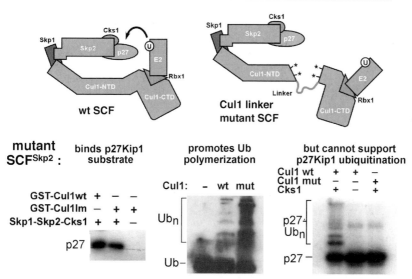

Fig. 7.9. The rigidity of the Cul1 scaffold is important for SCF E3 activity. Top, schematic diagram of the flexible linker Cul1 mutant. Bottom, the SCF containing the Cul1 linker mutant retains the ability to bind phosphoryl- ated p27^{Kip1} and to promote ubiquitin poly-merization, but fails to ubiquitinate p27^{Kip1} in an *in vitro* assay reconstituted with purified components.

mutations of the NTD and CTD made the two otherwise insoluble domains completely soluble when expressed separately in *Escherichia coli* (the CTD requires co-expression with Rbx1), indicating that they retain their structural integrity. In fact, the two slightly altered domains also remain functional as the flexibly linked Cul1 mutants are able to bind phosphorylated p27^{Kip1} in an Skp1–Skp2–Cks1-dependent manner and also to promote the substrate-independent polymerization of ubiquitin. Therefore, the substrate- and E2-binding activities of the engineered Cul1 are intact (Figure 7.9). Strikingly, the flexible Cul1 mutants fail to ubiquitinate p27^{Kip1} *in vitro*, indicating that the rigidity of the Cul1 scaffold is indeed crucial for its E3 activity (Figure 7.9) [85].

These findings, together with the structural and mutational analyses of the c-Cbl, VHL and Skp1–Skp2 complexes, supported the model that the RING E3s catalyze ubiquitination by optimally positioning the substrate for E2-mediated ubiquitin transfer. What is the extent of this positioning effect? It could in principle range from just raising the effective concentration of the lysine-containing substrate segment to the precise positioning of the ε-amino group of the lysine side chain at the E2 active site [85]. Answers to this question require a model of a RING E3 bound to both a substrate and an E2. While an SCF–E2 complex model can be constructed based on the c-Cbl–E2 structure and the similarity between Rbx1 and c-Cbl RING domains, structural and biochemical studies of Skp1–βTrCP1 bound to a substrate peptide helped provide structural insights into the substrate-E3 relationship.

7.6.4
Skp1–βTrCP1–β-Catenin Peptide and Skp1–Cdc4–CyclinE Peptide Complexes

βTrCP1 is a WD40-containing F-box protein conserved from *C. elegans* to humans. The two best-characterized substrates of the SCF^{βTrCP1} complex are β-catenin and IκBα [97, 98]. These SCF^{βTrCP1} substrates contain the DSGΦXS destruction motif (Φ representing a hydrophobic and X any amino acid), which is recognized by βTrCP1 when both serine residues in the motif are phosphorylated. In the Wnt signaling pathway, the destruction motif of β-catenin (Ser33 and Ser37) is constitutively phosphorylated by the GSK3β–APC-Axin complex in the absence of extracellular signals, resulting in SCF^{βTrCP1}-mediated polyubiquitination of β-catenin. The subsequent degradation of β-catenin maintains a relatively low homeostatic level of the protein. In response to the Wnt signals, phosphorylation of β-catenin is inhibited, leading to its stabilization, translocation to the nucleus, and transcriptional activation of proliferation-associated genes [99]. Loss of β-catenin ubiquitination, through mutations either in the destruction motif of β-catenin or in upstream components of the pathway, is among the most commonly observed alterations in colon cancer [100]. SCF^{βTrCP1}-mediated ubiquitination of IκBα functions in the opposite manner in the NF-κB signal transduction pathways. IκBα normally inhibits the NF-κB transcription activator by binding and sequestering it in the cytoplasm. Stimuli from extracellular signals or viral infection activate signaling pathways that lead to the phosphorylation of the two serines in IκBα and its ubiquitination.

Upon degradation of IκBα, NF-κB is released and translocates into the nucleus, where it will activate transcription [101].

Cdc4 is another conserved WD40-containing F-box protein, which plays an important role in regulation of cell cycle. In yeast, SCF^{Cdc4} is responsible for ubiquitinating the Cdk inhibitor Sic1 [102], whereas in higher eukaryotes, the same SCF complex (SCF$^{hCdc4/Fbw7/Ago}$) recognizes Cyclin E and promotes its degradation [63–65]. Mutations and alterations of hCDC4 have been found in numerous cancer cell lines and have been associated with abnormally high levels of Cyclin E.

The crystal structures of both an Skp1–βTrCP1 complex bound to a 26-amino acid human β-catenin peptide that contains the doubly phosphorylated destruction motif and a yeast Skp1–Cdc4 complex bound to a phosphorylated Cyclin E model peptide have been determined [66, 103]. The Skp1–βTrCP1 complex has a relatively elongated structure, with Skp1 and the β-catenin peptide located at opposite ends (Figure 7.2). The seven WD40 repeats of βTrCP1 form a torus-like structure that is characteristic of this fold (commonly referred to as a β-propeller). As seen in other β-propeller structures, the βTrCP1 β-propeller has a narrow channel running through the middle of the torus-like structure, presenting the β-catenin peptide-binding site. Of the 26 β-catenin residues in the crystals, only an 11-residue segment (residues 30 to 40) centered on the doubly phosphorylated destruction motif (residues 32 to 37), is ordered in the structure. The phosphoserine, aspartic acid, and hydrophobic residues of the destruction motif are recognized directly by contacts from βTrCP1. The structure of the Skp1–Cdc4 complex resembles closely that of Skp1–βTrCP1, except that the β-propeller of Cdc4 contains eight instead of seven structural repeats [103]. The substrate peptide is also specifically recognized by the conserved residues lining up the channel in the middle of the Cdc4 propeller.

The rigid structural coupling of the F-box and substrate-binding domains was once again revealed in both structures. The βTrCP1 F-box is linked to the WD40 domain through an ∼65-amino acid α-helical domain (linker domain), which interacts extensively with both the F-box and with one face of the WD40 β-propeller (Figure 7.2). A similar structured helical linker is also found in the structure of yeast Cdc4. It has been further shown that mutations in the linker helix of Cdc4 designed to affect the rigid coupling between the F-box and WD40 domains disrupted Cdc4 function in vivo [103].

7.6.5
Model of the SCF in Complex With E2 and Substrates

With the handful of known structures of several SCF sub-complexes and c-Cbl–E2, models of the SCF bound to E2 and substrate have been constructed (Figure 7.10). A model of the SCF^{Skp2}–E2 complex was built by superimposing Cul1–Rbx1–Skp1–F-box^{Skp2} on Skp1–Skp2 and docking the UbcH7 E2 onto the Rbx1 RING domain based on the way it binds c-Cbl. In addition, superimposing the Skp1–F-box portions of the Skp1–βTrCP1 and Skp1–Cdc4 complex with the SCF^{Skp2}–E2 model has made available the models of the SCF^{βTrCP1}–β-catenin–E2 and the

SCF - substrate - E2 models

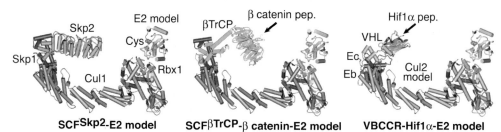

Fig. 7.10. Models of the SCF^{Skp2} complex (left), the SCF$^{\beta TrCP1}$–β-catenin peptide complex (middle), and the VBC–CR–Hif1α peptide complex (right) bound to E2. The E2 active site cysteine and the β-catenin and Hif1α peptides are labeled.

SCF^{Cdc4}–CyclinE–E2 complexes [66, 85, 103]. In all cases, no intermolecular collision was found in the final models. The substrate-binding domains of all three F-box proteins are positioned on the same side of the SCF complex as the E2. In addition, these domains are all oriented toward the E2 active site. Remarkably, the positions of the WD40 domain in the SCF$^{\beta TrCP1}$ and SCF^{Cdc4} models are strikingly similar to the position of the LRR domain in the SCF^{Skp2} structure. In the superimposition of all three SCFs, roughly half of the WD40 domain structure overlaps the LRR domain [103]. Finally, in the SCF$^{\beta TrCP1}$–β-catenin–E2 complex, the β-catenin peptide faces the E2, with both the N- and C-termini of the peptide pointing toward the E2 active-site cysteine (Figure 7.10). Based on the structural similarity between the recognition of F-box by Skp1 and SOCS-box by ElonginC, a model of the VBC–Hif1α peptide complex bound to a model of the Cul2–Rbx1 complex has also been constructed by superimposing the structurally conserved portions of Skp1 and ElonginC from the Cul1–Rbx1–Skp1–Skp2 and VHL–ElonginB–ElonginC–Hif1α peptide complexes (Figure 7.10) [79]. In this model, the position of the VHL β domain relative to the rest of the complex is again very similar to the βTrCP1 WD40 and Skp2 LRR domains. Even though the VHL β-domain does not extend as far towards the E2 as the latter two, the N-terminus of the Hif1α peptide is in clear sight of the E2 active site.

Although the substrate-binding domains are aligned with the E2 in these models, a large gap is always found in between. For example, the N- and C-terminal ends of the structured portions of the β-catenin peptide are ∼50 Å away from the E2 active-site cysteine (Figure 7.10), while the Hif1α peptide is ∼75 Å away from the E2 cysteine. It should be noticed that these peptide–E2 distances have considerable uncertainty as the E2 is modeled onto the Cul1–Rbx1 complex based on the c-Cbl–E2 structure and the similarity between the RING domains of Rbx1 and c-Cbl. In addition to the RING domain, c-Cbl also uses the linker helix to bind the E2. Yet a counterpart to this is not apparent in the SCF. Pivoting about the E2–Rbx1 interface of the model could affect the final position of the E2 active site, which is ~20 Å away from the E2–Rbx1 contacting point. For example, a 45° tilt of the E2 can result in an estimated $+/- \sim 20$-Å error in the position of the E2 active site [66]. Further structural studies of the SCF complex bound to an E2 will be necessary to clarify this uncertainty.

Even with the uncertainty in E2 active-site position, the models have suggested that there would be no E3 residues near the E2 active site, in agreement with the observations made in the c-Cbl–E2 structure. This again ruled out the possibility that the SCF E3 provides acid/base catalysis and the possibility that the SCF positions the ε-amino group of the lysine at the E2 active site [66]. The only plausible mechanism left accounting for the catalysis mediated by the SCF in substrate ubiquitination is that the E3 complex helps increase the effective concentration of a portion of the substrate that contains the physiological ubiquitination-site lysine at the E2 active site. This model made the testable prediction that the distance between the destruction motif and the ubiquitinated lysine is a determinant of the ubiquitination efficiency.

7.6.6
Mechanism of RING E3-mediated Catalysis

The prediction of the effective concentration model has been tested in an *in vitro* system reconstituted with E1, E2 (UbcH5), and SCF$^{\beta TrCP1}$, all purified to >90% homogeneity [66]. To be able to measure the ubiquitination of the substrate lysine independently of the ubiquitination of lysine(s) on ubiquitin, a ubiquitin mutant that lacks lysines and thus does not form polyubiquitin chains was used. The natural substrates of the SCF$^{\beta TrCP1}$ E3, β-catenin and IκBα were chosen for the detailed analyses. To make the correlations between functional and structural observations straightforward, the biochemical studies have focused on β-catenin, the substrate co-crystallized with Skp1–βTrCP1.

Although the ubiquitination site(s) of β-catenin was previously unknown, two lysines of IκBα located ten and nine residues upstream of the destruction motif had been shown to be necessary and sufficient for ubiquitination and degradation [18, 19]. By analogy, a β-catenin lysine (Lys19) 13 residues upstream of the destruction motif was predicted to be the site where β-catenin is ubiquitinated. This was confirmed by the *in vitro* ubiquitination assay, in which a 26-amino acid β-catenin peptide that contains Lys19 and the doubly phosphorylated destruction motif was ubiquitinated in an SCF$^{\beta TrCP1}$-dependent manner to an overall level comparable to an IκBα substrate peptide (Figure 7.11). These isolated β-catenin and IκBα peptides should accurately reflect the context of these destruction motifs in their respective full-length proteins, since Lys19 and the destruction motif of β-catenin are both in a 133-residue N-terminal region that has been previously shown to have a disordered structure by proteolytic digestion analysis [104]. The destruction motif of IκBα similarly resides outside the structured ankyrin-repeat domain.

With the system established, a series of mutant β-catenin peptides where the spacing between Lys19 and the destruction motif was increased or decreased in four-residue steps (wt−8, wt−4, wt+4 and wt+8 peptides in Figure 7.11) was

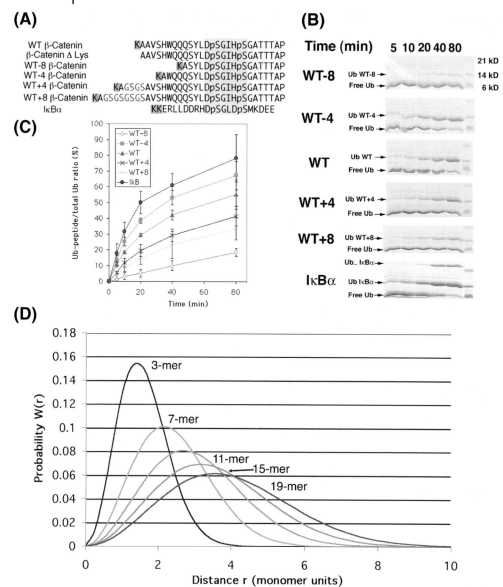

(A)

WT β-Catenin	KAAVSHWQQQSYLDpSGIHpSGATTTAP
β-Catenin Δ Lys	AAVSHWQQQSYLDpSGIHpSGATTTAP
WT-8 β-Catenin	KASYLDpSGIHpSGATTTAP
WT-4 β-Catenin	KAWQQQSYLDpSGIHpSGATTTAP
WT+4 β-Catenin	KAGSGSAVSHWQQQSYLDpSGIHpSGATTTAP
WT+8 β-Catenin	KAGSGSGSGSAVSHWQQQSYLDpSGIHpSGATTTAP
IκBα	KKERLLDDRHDpSGLDpSMKDEE

Fig. 7.11. The rate of ubiquitination by the SCF$^{\beta\text{TrCP1}}$ is dependent on the lysine-destruction motif spacing. (A) Sequences of the wild type and mutant β-catenin and IκBα peptides, with the destruction motif and ubiquitinated lysine(s) highlighted. (B) Time courses of ubiquitination of the wild-type and mutant peptides, visualized by Coomassie staining. (C) The reaction yields plotted with error bars from four experiments. (D) The radial distribution functions ($W(r)$) calculated for random unperturbed polymers of length 3, 7, 11, 15, and 19 monomer units [66]. These correspond to the number of residues between the lysine and the first destruction motif residue ordered in the structure, for the wt−8, wt−4, wild type, wt+4 and wt+8 peptides. The distance plotted on the x-axis is expressed in dimensionless units as a multiple of the monomer unit length (r/l).

tested for *in vitro* ubiquitination. The deleted residues are unlikely to be involved in βTrCP1 binding as they are present in the crystallized β-catenin peptide but are disordered in the structure. This is confirmed by experiments showing that the mutant peptides have affinities for βTrCP1 undistinguishable from that of the wild-type peptide [66].

Strikingly, changing the Lys19-destruction motif spacing by four to eight residues had two- to three-fold effects on the rate of ubiquitination. Increasing the spacing reduced ubiquitination, with the wt+8 peptide being ubiquitinated at an ~three-fold lower rate. Surprisingly, reducing the spacing by four residues increased the ubiquitination rate slightly but consistently, although reducing it further by eight residues (wt−8) resulted in a very poor substrate with only trace amounts of product at the longest reaction times. These findings indicated that SCF^{βTrCP1} substrates have an optimal destruction-motif lysine spacing of 9 to 13 residues [66]. Interestingly, the 9-residue lysine-destruction-motif spacing of the wt−4 mutant is comparable to the 9 and 10-residue spacing found in IκBα (for Lys20 and Lys19, respectively). IκBα is ubiquitinated at a rate closer to the wt−4 peptide than the wt β-catenin peptide (compare early time points in Figure 7.11, when the majority of the IκBα peptide is mono-ubiquitinated), suggesting that the optimal lysine-destruction-motif spacing is closer to 9 than to 13 residues. If the spacing is an important determinant, it is expected to be conserved in the β-catenin and IκBα orthologs and paralogs. Indeed, all of these proteins contain a lysine located 9 to 14 residues upstream of the destruction motif (Table 1).

The apparent effect of the spacing between the lysine residue and the destruction motif on the ubiquitination rates is intuitively consistent with the model that the SCF catalyzes ubiquitination by increasing the effective concentration of specific lysine(s) at the E2 active site [85]. According to this model, a specific SCF will catalyze ubiquitination maximally when the distance between its substrate-binding site and the E2 active site, a parameter likely unique to a particular F-box protein, closely matches the spatial distance between the substrate's SCF-binding motif and its lysine residue(s). Unstructured polypeptides do not have a single, well-defined length in solution, but rather they sample a distribution of lengths, each associated with a probability. Assuming that the reaction rate is proportional

Tab. 7.1.   Closest lysine upstream of destruction motif.

Substrate	Residues
IκBα (human)	9, 10
IκBβ (human)	9
IκBε (human)	11
Cactus (fly IκB)	10, 12
β-catenin (human)	13
Armadillo (fly β-catenin)	14
Bar-1 (worm β-catenin)	10
Plakoglobin (human)	11

to the probability that the lysine side chain will hit the E2 active site, the effective concentration model predicts that the relative ubiquitination rates of our wild-type and mutant peptides would be directly proportional to the relative probabilities that the lysine-destruction-motif distances match the optimal distance.

Except for very short polypeptides, it is not yet possible to calculate *ab initio* the length distribution for a given polypeptide sequence, especially as the length distribution also depends on the nature of the amino acids and the distribution of charged and hydrophobic groups. However, approaches based on random polymer theory with empirical corrections have been shown to agree qualitatively with experimental data [105]. Therefore, polymer theory can be used to analyze and compare the length distribution and its associated probability of a polypeptide. Figure 7.11 shows the length distributions for the polypeptide segments between the lysines and the destruction motifs of the wild-type and mutant *β*-catenin peptides. Assuming that the optimal distance corresponds to that of the wt−4 peptide, which had the highest ubiquitination rate in the *in vitro* ubiquitination assay, the probability that the wt+8 peptide will sample this distance is 2.5-fold lower. This is comparable to the 3-fold lower ubiquitination rate of the wt+8 peptide relative to the wt−4 peptide. These experimental results are thus consistent with the predictions of the effective concentration model. Overall, these structural and biochemical studies have provided a solid body of evidence supporting the model that the SCF E3, and likely other RING E3s, catalyze ubiquitination by increasing the effective concentration of specific substrate lysine(s) at the E2 active site [66].

7.7
The Mms2–Ubc13 Complex

The topology of the polyubiquitin chains determines the nature of the signals they encode. Whereas K48-linked polyubiquitin chains signal for proteasome-dependent degradation, K63-linked chains serves as non-proteolytic signals in cellular processes such as DNA repair and IKK activation. A family of *u*biquitin *E*2 *v*ariant proteins (UEVs) has been identified as mediating the assembly of K63 chains. These E2 variants share sequence homology with the canonical E2s, yet they all lack the active-site cysteine [26]. Yeast *MMS2* gene product, Mms2, is a UEV protein, which functions in the *RAD6/RAD18* post-replication DNA-damage repair pathway [27]. Mms2 performs its biological roles together with a specific E2, yeast Ubc13. It has been shown that Mms2 and Ubc13 are able to form a stable complex and catalyze the assembly of K63 polyubiquitin chains in the presence of E1 only *in vitro*. This suggests that the function of Mms2 might be equivalent to that of an E3, except that the substrate is ubiquitin itself.

The crystal structure of the Mms2–Ubc13 complex has been determined [106]. It shows that the two proteins both adopt the canonical *α/β* E2 fold and together form an asymmetric heterodimer (Figure 7.12). In the crystal, one end of Mms2 packs against one side of Ubc13, resulting in an overall T-shaped dimer structure. At the junction of the two proteins, a hydrophobic channel is formed, leading toward

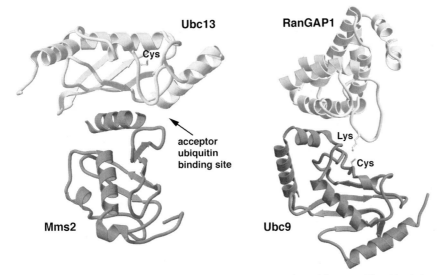

Fig. 7.12. Crystal structures of the Mms2/Ubc13 (left) and the RanGAP1–Ubc9 (right) complexes. The proposed binding site for the acceptor ubiquitin in the Mms2/Ubc13 structure is indicated by arrow. The side chains of the substrate lysine and the Ubc9 active cysteine in the RanGAP1–Ubc9 structure are shown.

the Ubc13 active-site cysteine. Mutations of the surface residues of this channel impaired K63 ubiquitin-chain assembly without disrupting the heterodimerization of the two proteins and thioester formation on Ubc13 [106]. Together, these results suggest that the acceptor ubiquitin likely interacts with this channel during Mms2–Ubc13-mediated K63 chain assembly. In conjunction with previous NMR studies of the ubiquitin–E2 interaction [107], another surface cleft was identified on the other side of Ubc13 as the donor ubiquitin binding site [106].

As repeatedly seen in all the RING E3 ligase structures, there is no residue from Mms2 close enough to the Ubc13 active-site cysteine to mediate acid/base catalysis [106]. Neither does Mms2-binding lead to any significant conformational changes that might activate the Ubc13 E2. Instead, the structure of the Mms2–Ubc13 complex strongly suggests that Mms2 promotes K63 ubiquitin-chain assembly by binding to the Ubc13 E2, forming a ubiquitin-binding site, and positioning the acceptor ubiquitin in an orientation to present its Lys63 residue toward the E2 active site, a mechanism similar to the one derived above for the RING E3s.

7.8
The RanGAP1–Ubc9 Complex

Among the several ubiquitin-like modifiers, SUMO is highly conserved from yeast to human and has been shown to regulate a variety of cellular functions such as nucleocytoplasmic transport, signal transduction, and transcriptional control

[108]. Protein modification by SUMO (sumoylation) requires a similar E1–E2–E3 system as for ubiquitination, although the SUMO E2 Ubc9 can specifically conjugate SUMO to most substrates in the absence of an E3. This is possible partly because the target lysine site in most SUMO-modified proteins is within a consensus sequence motif ΨKxD/E, which is directly recognized by Ubc9. A report of the crystal structure of a RanGAP1–Ubc9 complex marked the first structure of an E2-substrate complex in the ubiquitin and ubiquitin-like conjugation systems [109]. RanGAP1 is the best documented substrate for sumoylation. The structure of RanGAP1–Ubc9 revealed that Ubc9 binds the substrate by both interacting with the consensus sumoylation motif and keeping extensive contacts with two nearby RanGAP1 helices (Figure 7.12). The sumoylation motif of RanGAP1, LKSE, adopts an extended conformation, interacting closely with a rather flat Ubc9 surface with the lysine residue positioned immediately next to the E2 cysteine. Importantly, no residues in the E2 or the substrate surrounding the E2 cysteine and the substrate lysine appear to play a role in deprotonating the lysine, indicating that the SUMO transfer reaction does not involve acid/base catalysis [109]. This was further confirmed by structure-based mutagenesis studies. In agreement with the model described above for the mechanism by which RING E3s mediates ubiquitin transfer, the SUMO E2-substrate structure has implications that SUMO conjugation on the substrate is also catalyzed by proper orientation of the lysine residue toward the E2 active-site cysteine.

7.9
Summary and Perspective

Ubiquitin–protein ligases play a central role in conferring the specificity of protein ubiquitination and promoting ubiquitin transfer from E2 to the substrates. Structural studies of a series of prototypical ubiquitin–protein ligases have significantly advanced our understanding of how these enzymes carry out their biological functions. For the HECT E3s, whose E3 activities involve the formation of a thioester intermediate, structural analyses have revealed a large domain movement within the catalytic domain. For the much larger RING class E3 family, all structural studies support the model that this family of ubiquitin ligases promote protein ubiquitination by optimally orienting and positioning the substrate relative to the E2 active site to raise the effective concentration of the specific lysine(s) for ubiquitin attachment. Further studies will be needed to address the questions of how the extension of the ubiquitin chain is mediated by the E3s and how the E3s are regulated by cellular factors.

Acknowledgments

We are grateful to members of Nikola Paveltich's laboratory whose work, ideas, and enthusiasm for science have made this chapter possible. In particular, we would like to thank Charles E. Stebbins, Lan Huang, Brenda A. Schulman, Jung-Hyun Min and Geng Wu whose work is described here, and Philip D. Jeffrey, whose ability to solve difficult X-ray crystallography problems played a key role in several of the studies. This work was supported by the NIH, the Howard Hughes Medical Institute, and the Leukemia and Lymphoma Society.

References

1 HERSHKO, A. and CIECHANOVER, A. The ubiquitin system. *Annu Rev Biochem* **1998**, *67*, 425–79.

2 HICKE, L. Protein regulation by monoubiquitin. *Nat Rev Mol Cell Biol* **2001**, *2*, 195–201.

3 PICKART, C. M. Mechanisms underlying ubiquitination. *Annu Rev Biochem* **2001**, *70*, 503–33.

4 CIECHANOVER, A. The ubiquitin proteolytic system and pathogenesis of human diseases: a novel platform for mechanism-based drug targeting. *Biochem Soc Trans* **2003**, *31*, 474–81.

5 HERSHKO, A., HELLER, H., ELIAS, S. and CIECHANOVER, A. Components of ubiquitin–protein ligase system. Resolution, affinity purification, and role in protein breakdown. *J Biol Chem* **1983**, *258*, 8206–14.

6 JOAZEIRO, C. A. and WEISSMAN, A. M. RING finger proteins: mediators of ubiquitin ligase activity. *Cell* **2000**, *102*, 549–52.

7 SCHEFFNER, M., NUBER, U. and HUIBREGTSE, J. M. Protein ubiquitination involving an E1–E2–E3 enzyme ubiquitin thioester cascade. *Nature* **1995**, *373*, 81–3.

8 JACKSON, P. K. et al. The lore of the RINGs: substrate recognition and catalysis by ubiquitin ligases. *Trends Cell Biol* **2000**, *10*, 429–39.

9 CYR, D. M., HOHFELD, J. and PATTERSON, C. Protein quality control: U-box-containing E3 ubiquitin ligases join the fold. *Trends Biochem Sci* **2002**, *27*, 368–75.

10 HAAS, A. L. and SIEPMANN, T. J. Pathways of ubiquitin conjugation. *Faseb J* **1997**, *11*, 1257–68.

11 SCHEFFZEK, K., AHMADIAN, M. R. and WITTINGHOFER, A. GTPase-activating proteins: helping hands to complement an active site. *Trends Biochem Sci* **1998**, *23*, 257–62.

12 JOAZEIRO, C. A. et al. The tyrosine kinase negative regulator c-Cbl as a RING-type, E2-dependent ubiquitin–protein ligase. *Science* **1999**, *286*, 309–12.

13 LORICK, K. L. et al. RING fingers mediate ubiquitin-conjugating enzyme (E2)-dependent ubiquitination. *Proc Natl Acad Sci USA* **1999**, *96*, 11364–9.

14 WU, P. Y. et al. A conserved catalytic residue in the ubiquitin-conjugating enzyme family. *Embo J* **2003**, *22*, 5241–50.

15 KORNITZER, D., RABOY, B., KULKA, R. G. and FINK, G. R. Regulated degradation of the transcription factor Gcn4. *Embo J* **1994**, *13*, 6021–30.

16 TREIER, M., STASZEWSKI, L. M. and BOHMANN, D. Ubiquitin-dependent c-Jun degradation in vivo is mediated by the delta domain. *Cell* **1994**, *78*, 787–98.

17 CHAU, V. et al. A multiubiquitin chain is confined to specific lysine in a targeted short-lived protein. *Science* **1989**, *243*, 1576–83.

18 BALDI, L., BROWN, K., FRANZOSO, G. and SIEBENLIST, U. Critical role for lysines 21 and 22 in signal-induced, ubiquitin-mediated proteolysis of I kappa B-alpha. *J Biol Chem* **1996**, *271*, 376–9.

19 SCHERER, D. C., BROCKMAN, J. A., CHEN, Z., MANIATIS, T. and BALLARD, D. W. Signal-induced degradation of I kappa B alpha requires site-specific ubiquitination. *Proc Natl Acad Sci USA* **1995**, *92*, 11259–63.

20 WILKINSON, K. D. Ubiquitination and deubiquitination: targeting of proteins

for degradation by the proteasome. *Semin Cell Dev Biol* **2000**, *11*, 141–8.

21 FREIMAN, R. N. and TJIAN, R. Regulating the regulators: lysine modifications make their mark. *Cell* **2003**, *112*, 11–7.

22 DESTERRO, J. M., RODRIGUEZ, M. S. and HAY, R. T. SUMO-1 modification of IkappaBalpha inhibits NF-kappaB activation. *Mol Cell* **1998**, *2*, 233–9.

23 LI, M., LUO, J., BROOKS, C. L. and GU, W. Acetylation of p53 inhibits its ubiquitination by Mdm2. *J Biol Chem* **2002**, *277*, 50607–11.

24 MARTINEZ-BALBAS, M. A., BAUER, U. M., NIELSEN, S. J., BREHM, A. and KOUZARIDES, T. Regulation of E2F1 activity by acetylation. *Embo J* **2000**, *19*, 662–71.

25 DI FIORE, P. P., POLO, S. and HOFMANN, K. When ubiquitin meets ubiquitin receptors: a signalling connection. *Nat Rev Mol Cell Biol* **2003**, *4*, 491–7.

26 SANCHO, E. et al. Role of UEV-1, an inactive variant of the E2 ubiquitin-conjugating enzymes, in in vitro differentiation and cell cycle behavior of HT-29-M6 intestinal mucosecretory cells. *Mol Cell Biol* **1998**, *18*, 576–89.

27 HOFMANN, R. M. and PICKART, C. M. Noncanonical MMS2-encoded ubiquitin-conjugating enzyme functions in assembly of novel polyubiquitin chains for DNA repair. *Cell* **1999**, *96*, 645–53.

28 COOK, W. J., JEFFREY, L. C., SULLIVAN, M. L. and VIERSTRA, R. D. Three-dimensional structure of a ubiquitin-conjugating enzyme (E2). *J Biol Chem* **1992**, *267*, 15116–21.

29 SAURIN, A. J., BORDEN, K. L., BODDY, M. N. and FREEMONT, P. S. Does this have a familiar RING? *Trends Biochem Sci* **1996**, *21*, 208–14.

30 SCHEFFNER, M., WERNESS, B. A., HUIBREGTSE, J. M., LEVINE, A. J. and HOWLEY, P. M. The E6 oncoprotein encoded by human papillomavirus types 16 and 18 promotes the degradation of p53. *Cell* **1990**, *63*, 1129–36.

31 HUIBREGTSE, J. M., SCHEFFNER, M. and HOWLEY, P. M. A cellular protein mediates association of p53 with the E6 oncoprotein of human papillomavirus types 16 or 18. *Embo J* **1991**, *10*, 4129–35.

32 JIANG, Y., LEV-LEHMAN, E., BRESSLER, J., TSAI, T. F. and BEAUDET, A. L. Genetics of Angelman syndrome. *Am J Hum Genet* **1999**, *65*, 1–6.

33 SCHWARZ, S. E., ROSA, J. L. and SCHEFFNER, M. Characterization of human hect domain family members and their interaction with UbcH5 and UbcH7. *J Biol Chem* **1998**, *273*, 12148–54.

34 WANG, G., YANG, J. and HUIBREGTSE, J. M. Functional domains of the Rsp5 ubiquitin–protein ligase. *Mol Cell Biol* **1999**, *19*, 342–52.

35 SCHEFFNER, M., HUIBREGTSE, J. M. and HOWLEY, P. M. Identification of a human ubiquitin-conjugating enzyme that mediates the E6-AP-dependent ubiquitination of p53. *Proc Natl Acad Sci USA* **1994**, *91*, 8797–801.

36 NUBER, U., SCHWARZ, S., KAISER, P., SCHNEIDER, R. and SCHEFFNER, M. Cloning of human ubiquitin-conjugating enzymes UbcH6 and UbcH7 (E2-F1) and characterization of their interaction with E6-AP and RSP5. *J Biol Chem* **1996**, *271*, 2795–800.

37 KUMAR, S., KAO, W. H. and HOWLEY, P. M. Physical interaction between specific E2 and Hect E3 enzymes determines functional cooperativity. *J Biol Chem* **1997**, *272*, 13548–54.

38 HUANG, L. et al. Structure of an E6AP-UbcH7 complex: insights into ubiquitination by the E2–E3 enzyme cascade. *Science* **1999**, *286*, 1321–6.

39 VERDECIA, M. A. et al. Conformational flexibility underlies ubiquitin ligation mediated by the WWP1 HECT domain E3 ligase. *Mol Cell* **2003**, *11*, 249–59.

40 THIEN, C. B. and LANGDON, W. Y. Cbl: many adaptations to regulate protein tyrosine kinases. *Nat Rev Mol Cell Biol* **2001**, *2*, 294–307.

41 HAGLUND, K., DI FIORE, P. P. and DIKIC, I. Distinct monoubiquitin signals in receptor endocytosis. *Trends Biochem Sci* **2003**, *28*, 598–603.

42 MOSESSON, Y. et al. Endocytsosis of receptor tyrosine kinases is driven by mono-, not poly-ubiquitylation. *J Biol Chem* **2003**.

43 PETRELLI, A. et al. The endophilin-CIN85-Cbl complex mediates ligand-dependent downregulation of c-Met. *Nature* **2002**, *416*, 187–90.

44 SOUBEYRAN, P., KOWANETZ, K., SZYMKIEWICZ, I., LANGDON, W. Y. and DIKIC, I. Cbl-CIN85-endophilin complex mediates ligand-induced downregulation of EGF receptors. *Nature* **2002**, *416*, 183–7.

45 HAGLUND, K. et al. Multiple mono-ubiquitination of RTKs is sufficient for their endocytosis and degradation. *Nat Cell Biol* **2003**, *5*, 461–6.

46 THIEN, C. B. and LANGDON, W. Y. c-Cbl: a regulator of T cell receptor-mediated signalling. *Immunol Cell Biol* **1998**, *76*, 473–82.

47 LEVKOWITZ, G. et al. Ubiquitin ligase activity and tyrosine phosphorylation underlie suppression of growth factor signaling by c-Cbl/Sli-1. *Mol Cell* **1999**, *4*, 1029–40.

48 MENG, W., SAWASDIKOSOL, S., BURAKOFF, S. J. and ECK, M. J. Structure of the amino-terminal domain of Cbl complexed to its binding site on ZAP-70 kinase. *Nature* **1999**, *398*, 84–90.

49 ZHENG, N., WANG, P., JEFFREY, P. D. and PAVLETICH, N. P. Structure of a c-Cbl-UbcH7 complex: RING domain function in ubiquitin–protein ligases. *Cell* **2000**, *102*, 533–9.

50 DESHAIES, R. J. SCF and Cullin/Ring H2-based ubiquitin ligases. *Annu Rev Cell Dev Biol* **1999**, *15*, 435–67.

51 ZACHARIAE, W. et al. Mass spectrometric analysis of the anaphase-promoting complex from yeast: identification of a subunit related to cullins. *Science* **1998**, *279*, 1216–9.

52 YU, H. et al. Identification of a cullin homology region in a subunit of the anaphase-promoting complex. *Science* **1998**, *279*, 1219–22.

53 OHTA, T., MICHEL, J. J., SCHOTTELIUS, A. J. and XIONG, Y. ROC1, a homolog of APC11, represents a family of cullin partners with an associated ubiquitin ligase activity. *Mol Cell* **1999**, *3*, 535–41.

54 DIAS, D. C., DOLIOS, G., WANG, R. and PAN, Z. Q. CUL7: A DOC domain-containing cullin selectively binds Skp1.Fbx29 to form an SCF-like complex. *Proc Natl Acad Sci USA* **2002**, *99*, 16601–6.

55 PINTARD, L. et al. The BTB protein MEL-26 is a substrate-specific adaptor of the CUL-3 ubiquitin-ligase. *Nature* **2003**, *425*, 311–6.

56 XU, L. et al. BTB proteins are substrate-specific adaptors in an SCF-like modular ubiquitin ligase containing CUL-3. *Nature* **2003**, *425*, 316–21.

57 KIPREOS, E. T. and PAGANO, M. The F-box protein family. *Genome Biol* **2000**, *1*, REVIEWS3002.

58 WINSTON, J. T., KOEPP, D. M., ZHU, C., ELLEDGE, S. J. and HARPER, J. W. A family of mammalian F-box proteins. *Curr Biol* **1999**, *9*, 1180–2.

59 BAI, C. et al. SKP1 connects cell cycle regulators to the ubiquitin proteolysis machinery through a novel motif, the F-box. *Cell* **1996**, *86*, 263–74.

60 SCHULMAN, B. A. et al. Insights into SCF ubiquitin ligases from the structure of the Skp1–Skp2 complex. *Nature* **2000**, *408*, 381–6.

61 GANOTH, D. et al. The cell-cycle regulatory protein Cks1 is required for SCF(Skp2)-mediated ubiquitinylation of p27. *Nat Cell Biol* **2001**, *3*, 321–4.

62 SPRUCK, C. et al. A CDK-independent function of mammalian Cks1: targeting of SCF(Skp2) to the CDK inhibitor p27Kip1. *Mol Cell* **2001**, *7*, 639–50.

63 KOEPP, D. M. et al. Phosphorylation-dependent ubiquitination of cyclin E by the SCFFbw7 ubiquitin ligase. *Science* **2001**, *294*, 173–7.

64 MOBERG, K. H., BELL, D. W., WAHRER, D. C., HABER, D. A. and HARIHARAN, I. K. Archipelago regulates Cyclin E levels in Drosophila and is mutated in human cancer cell lines. *Nature* **2001**, *413*, 311–6.

65 STROHMAIER, H. et al. Human F-box protein hCdc4 targets cyclin E for proteolysis and is mutated in a breast cancer cell line. *Nature* **2001**, *413*, 316–22.

66 WU, G. et al. Structure of a beta-TrCP1-Skp1-beta-catenin complex: destruction motif binding and lysine specificity of the SCF(beta-TrCP1)

ubiquitin ligase. *Mol Cell* **2003**, *11*, 1445–56.

67 STEBBINS, C. E., KAELIN, W. G., JR. and PAVLETICH, N. P. Structure of the VHL–ElonginC–ElonginB complex: implications for VHL tumor suppressor function. *Science* **1999**, *284*, 455–61.

68 KILE, B. T. et al. The SOCS box: a tale of destruction and degradation. *Trends Biochem Sci* **2002**, *27*, 235–41.

69 IVAN, M. et al. HIFalpha targeted for VHL-mediated destruction by proline hydroxylation: implications for O_2 sensing. *Science* **2001**, *292*, 464–8.

70 RUI, L., YUAN, M., FRANTZ, D., SHOELSON, S. and WHITE, M. F. SOCS-1 and SOCS-3 block insulin signaling by ubiquitin-mediated degradation of IRS1 and IRS2. *J Biol Chem* **2002**, *277*, 42394–8.

71 LATIF, F. et al. Identification of the von Hippel-Lindau disease tumor suppressor gene. *Science* **1993**, *260*, 1317–20.

72 DUAN, D. R. et al. Inhibition of transcription elongation by the VHL tumor suppressor protein. *Science* **1995**, *269*, 1402–6.

73 PAUSE, A. et al. The von Hippel-Lindau tumor-suppressor gene product forms a stable complex with human CUL-2, a member of the Cdc53 family of proteins. *Proc Natl Acad Sci USA* **1997**, *94*, 2156–61.

74 LONERGAN, K. M. et al. Regulation of hypoxia-inducible mRNAs by the von Hippel-Lindau tumor suppressor protein requires binding to complexes containing elongins B/C and Cul2. *Mol Cell Biol* **1998**, *18*, 732–41.

75 IVAN, M. and KAELIN, W. G., JR. The von Hippel-Lindau tumor suppressor protein. *Curr Opin Genet Dev* **2001**, *11*, 27–34.

76 JAAKKOLA, P. et al. Targeting of HIF-alpha to the von Hippel-Lindau ubiquitylation complex by O_2-regulated prolyl hydroxylation. *Science* **2001**, *292*, 468–72.

77 BRUICK, R. K. and McKNIGHT, S. L. A conserved family of prolyl-4-hydroxylases that modify HIF. *Science* **2001**, *294*, 1337–40.

78 EPSTEIN, A. C. et al. C. elegans EGL-9 and mammalian homologs define a family of dioxygenases that regulate HIF by prolyl hydroxylation. *Cell* **2001**, *107*, 43–54.

79 MIN, J. H. et al. Structure of an HIF-1alpha-pVHL complex: hydroxyproline recognition in signaling. *Science* **2002**, *296*, 1886–9.

80 NAKAYAMA, K. I., HATAKEYAMA, S. and NAKAYAMA, K. Regulation of the cell cycle at the G1–S transition by proteolysis of cyclin E and p27Kip1. *Biochem Biophys Res Commun* **2001**, *282*, 853–60.

81 GSTAIGER, M. et al. Skp2 is oncogenic and overexpressed in human cancers. *Proc Natl Acad Sci USA* **2001**, *98*, 5043–8.

82 TEDESCO, D., LUKAS, J. and REED, S. I. The pRb-related protein p130 is regulated by phosphorylation-dependent proteolysis via the protein-ubiquitin ligase SCF(Skp2). *Genes Dev* **2002**, *16*, 2946–57.

83 KIM, S. Y., HERBST, A., TWORKOWSKI, K. A., SALGHETTI, S. E. and TANSEY, W. P. Skp2 regulates Myc protein stability and activity. *Mol Cell* **2003**, *11*, 1177–88.

84 VON DER LEHR, N. et al. The F-box protein Skp2 participates in c-Myc proteosomal degradation and acts as a cofactor for c-Myc-regulated transcription. *Mol Cell* **2003**, *11*, 1189–200.

85 ZHENG, N. et al. Structure of the Cul1-Rbx1-Skp1-F-boxSkp2 SCF ubiquitin ligase complex. *Nature* **2002**, *416*, 703–9.

86 GROISMAN, R. et al. The Ubiquitin Ligase Activity in the DDB2 and CSA Complexes Is Differentially Regulated by the COP9 Signalosome in Response to DNA Damage. *Cell* **2003**, *113*, 357–67.

87 HORI, T. et al. Covalent modification of all members of human cullin family proteins by NEDD8. *Oncogene* **1999**, *18*, 6829–34.

88 DHARMASIRI, S., DHARMASIRI, N., HELLMANN, H. and ESTELLE, M. The RUB/Nedd8 conjugation pathway is required for early development in Arabidopsis. *Embo J* **2003**, *22*, 1762–70.

89 Kurz, T. et al. Cytoskeletal regulation by the Nedd8 ubiquitin-like protein modification pathway. *Science* **2002**, *295*, 1294–8.

90 Osaka, F. et al. Covalent modifier NEDD8 is essential for SCF ubiquitin-ligase in fission yeast. *Embo J* **2000**, *19*, 3475–84.

91 Tateishi, K., Omata, M., Tanaka, K. and Chiba, T. The NEDD8 system is essential for cell cycle progression and morphogenetic pathway in mice. *J Cell Biol* **2001**, *155*, 571–9.

92 Read, M. A. et al. Nedd8 modification of cul-1 activates SCF(beta(TrCP))-dependent ubiquitination of Ikappa-Balpha. *Mol Cell Biol* **2000**, *20*, 2326–33.

93 Wu, K., Chen, A. and Pan, Z. Q. Conjugation of Nedd8 to CUL1 enhances the ability of the ROC1-CUL1 complex to promote ubiquitin polymerization. *J Biol Chem* **2000**, *275*, 32317–24.

94 Morimoto, M., Nishida, T., Honda, R. and Yasuda, H. Modification of cullin-1 by ubiquitin-like protein Nedd8 enhances the activity of SCF(skp2) toward p27(kip1). *Biochem Biophys Res Commun* **2000**, *270*, 1093–6.

95 Zheng, J. et al. CAND1 binds to unneddylated CUL1 and regulates the formation of SCF ubiquitin E3 ligase complex. *Mol Cell* **2002**, *10*, 1519–26.

96 Liu, J., Furukawa, M., Matsumoto, T. and Xiong, Y. NEDD8 modification of CUL1 dissociates p120(CAND1), an inhibitor of CUL1-SKP1 binding and SCF ligases. *Mol Cell* **2002**, *10*, 1511–8.

97 Hart, M. et al. The F-box protein beta-TrCP associates with phosphorylated beta-catenin and regulates its activity in the cell. *Curr Biol* **1999**, *9*, 207–10.

98 Li, Q. and Verma, I. M. NF-kappaB regulation in the immune system. *Nat Rev Immunol* **2002**, *2*, 725–34.

99 Polakis, P. Wnt signaling and cancer. *Genes Dev* **2000**, *14*, 1837–51.

100 Polakis, P. The oncogenic activation of beta-catenin. *Curr Opin Genet Dev* **1999**, *9*, 15–21.

101 Chen, Z. et al. Signal-induced site-specific phosphorylation targets I kappa B alpha to the ubiquitin-proteasome pathway. *Genes Dev* **1995**, *9*, 1586–97.

102 Feldman, R. M., Correll, C. C., Kaplan, K. B. and Deshaies, R. J. A complex of Cdc4p, Skp1p, and Cdc53p/cullin catalyzes ubiquitination of the phosphorylated CDK inhibitor Sic1p. *Cell* **1997**, *91*, 221–30.

103 Orlicky, S., Tang, X., Willems, A., Tyers, M. and Sicheri, F. Structural basis for phosphodependent substrate selection and orientation by the SCFCdc4 ubiquitin ligase. *Cell* **2003**, *112*, 243–56.

104 Huber, A. H., Nelson, W. J. and Weis, W. I. Three-dimensional structure of the armadillo repeat region of beta-catenin. *Cell* **1997**, *90*, 871–82.

105 Haas, E., Wilchek, M., Katchalski-Katzir, E. and Steinberg, I. Z. Distribution of end-to-end distances of oligopeptides in solution as estimated by energy transfer. *Proc Natl Acad Sci USA* **1975**, *72*, 1807–11.

106 VanDemark, A. P., Hofmann, R. M., Tsui, C., Pickart, C. M. and Wolberger, C. Molecular insights into polyubiquitin chain assembly: crystal structure of the Mms2/Ubc13 heterodimer. *Cell* **2001**, *105*, 711–20.

107 Miura, T., Klaus, W., Gsell, B., Miyamoto, C. and Senn, H. Characterization of the binding interface between ubiquitin and class I human ubiquitin-conjugating enzyme 2b by multidimensional heteronuclear NMR spectroscopy in solution. *J Mol Biol* **1999**, *290*, 213–28.

108 Muller, S., Hoege, C., Pyrowolakis, G. and Jentsch, S. SUMO, ubiquitin's mysterious cousin. *Nat Rev Mol Cell Biol* **2001**, *2*, 202–10.

109 Bernier-Villamor, V., Sampson, D. A., Matunis, M. J. and Lima, C. D. Structural basis for E2-mediated SUMO conjugation revealed by a complex between ubiquitin-conjugating enzyme Ubc9 and RanGAP1. *Cell* **2002**, *108*, 345–56.

8
The Deubiquitinating Enzymes

Nathaniel S. Russell and Keith D. Wilkinson

8.1
Introduction

In the mid-1970s ubiquitin was found to be a covalent modifier of proteins [1]. At the time, it was quite surprising to find a protein that covalently modified another protein. Since then, the reversible covalent modification of proteins by other proteins is known to be commonplace and ubiquitin is used to covalently modify hundreds of proteins, often for the purpose of targeting them to the proteasome for degradation.

Protein degradation through the ubiquitin–proteasome system is facilitated by covalently linking ubiquitin to the ε-amino group of a lysine of a substrate protein through its C-terminal glycine [2]. A polyubiquitin (polyUb) chain is formed by linking subsequent ubiquitins to the lysine 48 residue of the preceding ubiquitin in the chain. A chain of four ubiquitins is sufficient for the targeted protein to be recognized and degraded by the proteasome [3]. Conjugation of ubiquitin to other proteins is catalyzed by a three-enzyme cascade [4]. Conjugation begins by activation of ubiquitin by an E1, or Ub-activating enzyme, forming a high-energy thiol ester bond in an ATP-dependent reaction. The ubiquitin is transferred to an E2, or Ub-conjugating enzyme, which then usually pairs with an E3, or Ub-ligase enzyme, to conjugate the ubiquitin to a specific target protein.

The usefulness of ubiquitin conjugation is not limited to the ubiquitin–proteasome pathway. Mono- and polyubiquitin are used as signals in various pathways including endocytosis, DNA repair, apoptosis, and transcriptional regulation [5–8]. Polyubiquitin chains can be formed using lysine residues other than K48, the linkage required for proteasomal degradation [9–11]. In addition, there are a number of other ubiquitin-like proteins that also behave as signaling molecules although they are not involved directly in proteasomal degradation. This group includes SUMO (*s*mall *u*biquitin-related *mo*difier), Nedd8 (*n*eural precursor cell *e*xpressed, *d*evelopmentally *d*own-regulated *8*), ISG15 (*i*nterferon-*s*timulated *g*ene *15*), and others [12–14]. These proteins are conjugated to substrates in a similar manner to ubiquitin, using an E1, E2, and E3 cascade of enzymes specific for the particular ubiquitin-like protein involved [15–17].

Protein Degradation. Edited by J. Mayer, A. Ciechanover, M. Rechsteiner
Copyright © 2005 WILEY-VCH Verlag GmbH & Co. KGaA, Weinheim
ISBN: 3-527-30837-7

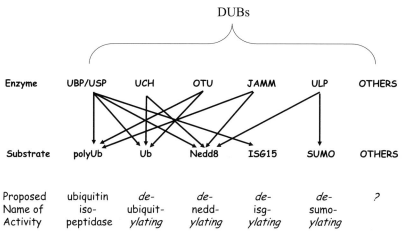

Fig. 8.1. DUB families and substrate specificity. DUBs can be classified by genetic relationships (family) or by substrate specificities (activity). Arrows point towards the substrates that members of each family can process in a physiologically relevant way. Each family is capable of processing multiple substrates and each activity can be catalyzed by members of more than one family.

Soon after it was shown that ubiquitin is conjugated to proteins, it was determined that this was a reversible process and deubiquitinating enzymes, or DUBs, could remove ubiquitin from ubiquitinated proteins [18, 19]. As the genes for ubiquitin and ubiquitin-like proteins were identified it became clear that all ubiquitin family members were synthesized as proproteins and processed to reveal the C-terminal glycylglycine of the active proteins [20]. Based on this information, DUBs were defined as proteases that cleave at the C-terminus of ubiquitin or ubiquitin-like proteins to reverse conjugation to target proteins and also process the proproteins.

Over 100 DUBs have been identified (see Table 8.1 for a list of the DUBs whose roles are known or suspected) and they are used to regulate ubiquitin and ubiquitin-like protein metabolism. Since cells utilize a combination of mono-ubiquitin, polyubiquitin, and ubiquitin-like proteins for a multitude of reasons and conjugate them to thousands of proteins, a regulatory system has evolved that is exceedingly complex and must be exquisitely regulated (see Figure 8.1). DUBs help regulate this system by processing proubiquitin into a mature form, recycling free polyubiquitin chains into monomeric Ub, assisting the degradation of proteasomal substrates, and regulating the ubiquitination levels of proteins in cellular pathways other than proteolysis. Thus, DUBs play crucial roles in determining the cellular fates of many proteins and regulating cellular function.

The study of DUBs has moved at a rapid rate since their initial discovery in the early 1980s. Yet despite all the progress, the total number of DUBs and the substrate specificity of most DUBs are still undetermined. The discovery of novel DUB families including the JAMM isopeptidases and OTU DUBs has highlighted that there may be still more unidentified DUBs. Because of the large number of

Tab. 8.1. Physiological roles of DUBs revealed by deletion or knockdown experiments.

DUB	Organism	Deletion/knockdown phenotype	Functional role
Ubiquitin C-terminal hydrolases (UCHs)			
UCH-L1	mouse	gracile axonal dystrophy	predominant neuronal UCH [112]
UCH-L3	mouse	no detectable phenotype	undetermined neuronal function [113]
UCH37	human	unknown	edits polyubiquitin chains at proteasome [91]
BAP1	human	unknown	tumor suppressor? [22]
YUH1	S. cerevisiae	cannot process proRUB1	processes proRUB1, Ub-adducts [83]
Ubiquitin specific processing proteases (UBPs)			
UBP1	S. cerevisiae	no phenotype detected	undetermined [97]
UBP2	S. cerevisiae	no phenotype detected	undetermined [96]
UBP3/ USP10	S. cerevisiae, human	polyubiquitin accumulation	transcriptional silencing inhibitor, regulates membrane transport [66, 118]
UBP6	S. cerevisiae	low levels of free ubiquitin	processes polyubiquitin chains at proteasome [94]
UBP8	S. cerevisiae	increase in ubiquitinated histone H2B	transcriptional regulator [88]
UBP14/IsoT	S. cerevisiae, human	increased polyubiquitin levels, proteasome defects	recycles free polyubiquitin to mono-ubiquitin [33, 86]
UBP16	S. cerevisiae	no phenotype detected	undetermined function at mitochondria [70]
DOA4	S. cerevisiae	Ub-depletion, defective Ub recycling	recycles Ub and polyubiquitin adducts [71, 84]
Unp/USP4	mouse, human	unknown	undetermined
USP7	human	indirect p53 activation	regulates p53 ubiquitination [36]
UBPy/USP8	human	increase in protein ubiquitination	cell-growth regulation [119]
USP14	mouse (UBP6 homolog)	ataxia	regulating synaptic activity plus proteasome [109]
USP21	human	unknown	process Ub and Nedd8 conjugates, growth regulator? [32]
USP25	human	unknown	over-expression has possible role in Down's Syndrome [120]
UBP41	human	unknown	promotes apoptosis [6]
UBP43	mouse	accumulation of Isg15 conjugates	processes Isg15, regulates Isg15 conjugate levels [31]
CYLD	human	cylindromatosis	regulates K63 polyubiquitination of substrates in NF-κB pathway [7, 103, 104]
fat facets	Drosophila	defective germ cell specification, eye formation	regulates specific developmental processes [74, 75]
DUB1, 2, 2A	mouse	unknown	cytokine specific growth regulators

Tab. 8.1. *(continued)*

DUB	Organism	Deletion/knockdown phenotype	Functional role
VDU1	human	unknown	regulation of Ub-proteasome pathway? [64]
JAMM Isopeptidases			
Rpn11	Yeast, human	lethal	processes polyubiquitin chains at proteasome [46, 47]
Csn5	Yeast, Drosophila	defects in SCF E3s in yeast, lethal in Drosophila	regulation of cullin neddylation levels [45]
OTU DUBs			
cezanne	human	unknown	negative regulation of NF-κB pathway [43]
A20	mouse	severe inflammation, premature death	negative regulation of NF-κB pathway [44, 121]
otubain1	human	unknown	editing DUB? [42, 122]
otubain2	human	unknown	undetermined [42]
VCIP 135	human	unknown	membrane fusion after mitosis [72]
Ubiquitin-like proteases (ULPs)			
Ulp1	S. cerevisiae	lethal	regulates cell cycle progression [39]
Ulp2	S. cerevisiae	increased SUMO conjugates, DNA repair defective	desumoylating enzyme [123]
Den1/SENP8	mouse	unknown	deneddylates cullins [26, 37]
SENP6	human	unknown	involved in reproductive function? [124]
Others			
Apg4B	mouse	unknown	processes autophagy-related UbLs [116]
ataxin-3	human	unknown	processes polyubiquitin chains? [125]

The organism listed for each DUB refers either to where it was discovered or where the work characterizing the deletion strain and function was performed. DUBs with multiple organism identifiers are either highly similar in sequence in each organism or functional homologs. An unknown deletion phenotype indicates that a deletion, knockout, or knockdown of a particular DUB has yet to be generated. No detectable phenotype indicates that a deletion strain has been made, but no phenotypes were observed.

potential DUB substrates and the exquisite specificity that some individual DUBs exhibit (see Figure 8.1) study of these enzymes has often been challenging. A burgeoning amount of structural data and recent technical advances are being used to address this challenge. The goal of this chapter is to highlight recent developments in the DUB field by giving an overview of DUB families, including DUBs that act on ubiquitin-like proteins, to discuss how DUBs achieve their specificity, and to show how the physiological roles of DUBs and their substrates are being elucidated.

8.2
Structure and *In Vitro* Specificity of DUB Families

8.2.1
Ubiquitin C-terminal Hydrolases (UCH)

The first class of DUBs discovered, the *u*biquitin *C*-terminal *h*ydrolases (UCHs), is a relatively small class with only four members in humans and one in budding yeast. UCHs are cysteine proteases related to the papain family of cysteine proteases. Most UCHs consist entirely of a catalytic core that has a molecular mass of about 25 kDa, although Bap1 and UCH37 have C-terminal extensions [21, 22]. All UCHs have a highly conserved catalytic triad consisting of the active-site cysteine, histidine, and aspartate residues that are absolutely required for function [23].

In vitro studies have determined that UCHs have significant activity in removing small adducts from the C-terminus of ubiquitin, including short peptides, ethyl ester groups, and amides [24]. They are also very efficient at cotranslationally processing the primary gene products (proubiquitin or Ub-ribosomal subunit fusions) to expose the C-terminal gly–gly motif required for conjugation of ubiquitin and ubiquitin-like proteins to substrates. However, UCHs are unable to cleave the isopeptide bond between ubiquitins in a polyubiquitin chain or to act on ubiquitin conjugated to a folded protein. They are similarly inefficient in acting upon small peptide substrates based on the sequence of the ubiquitin C-terminus [25]. As described below, the binding of ubiquitin is required for optimal UCH activity. Nedd8, a closely related ubiquitin-like protein, is also a substrate for human UCH-L3, albeit with three orders of magnitude less efficiency than ubiquitin [26].

The crystal structures of human and yeast UCHs have been solved, the latter in complex with the inhibitor ubiquitin aldehyde [27, 28]. The UCH fold is closely related to that of the papain family of cysteine proteases. Ubiquitin is bound in a cleft on a surface that is highly conserved in all UCHs. NMR studies on the binding of ubiquitin to human UCH-L3 show a similar mode of interaction and define three regions on the surface of ubiquitin involved in this binding [29]. As noted below and in Figure 8.2, the same surface of the ubiquitin fold is also involved in binding to USP7 and ULP1. This is remarkable as these DUBs are from different families and not significantly homologous in sequence or structure.

A second feature of UCHs is the presence of a "blocking loop" spanning the active site and limiting the size of substrates that can be accommodated. Yuh1 and other UCH DUBs contain a mobile, ~20-residue loop that is disordered in the unliganded protein, but becomes ordered upon substrate binding. The loop passes directly over the active site and the leaving group attached to the gly–gly at the ubiquitin C-terminus has to pass directly through this loop in order to access the catalytic cysteine. The maximum diameter of this loop was calculated to be ~15 Å, too small for any folded substrate save a single helix [27]. The loop thus allows small substrates to be efficiently cleaved, but excludes larger Ub–protein conjugates. This loop explains, at least in part, the preference of UCHs for small or disordered leaving groups.

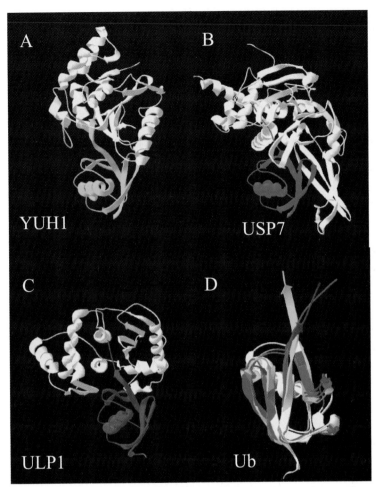

Fig. 8.2. Substrate binding by DUBs revealed by X-ray crystal structures. In the ribbon diagrams, the DUB is represented in white and the substrate in color. The ubiquitin (yellow or green) or SUMO (red) substrates are shown in the same orientation to highlight the similarity of substrate binding by different DUB classes. (A) Ubiquitin (yellow) bound to YUH1. (B) Ubiquitin (green) bound to USP7. (C) SUMO (red) bound to ULP1. (D) Superimposition of substrates from A–C. The regions of each substrate that are within 3.5 Å of the DUB when bound are highlighted in color to demonstrate the conserved regions that are recognized by the different DUB classes.

8.2.2
Ubiquitin-specific Processing Proteases (UBP/USP)

The ubiquitin specific processing proteases (referred to as UBPs in yeast and USPs in human and mouse) were the second class of DUBs discovered. Catalytically, the UBPs are very similar to the UCHs in that they also utilize the catalytic triad of an active-site cysteine and a conserved histidine and aspartate. The UBP catalytic core of about 400 amino acids contains blocks of conserved sequences (Cys and His boxes) around these catalytic residues [23]. The UBPs are generally larger and more variable in size than the UCH class, ranging from 50 to 300 kDa. N-terminal extensions to the catalytic core account for most of the increased size although a few UBPs have C-terminal extensions. These N-terminal extensions are highly divergent in sequence, unlike the conserved regions of the catalytic core. The sequence and size variations of these extensions are thought to aid in determining UBP localization and substrate specificity.

There are 16 UBPs in yeast and more than 50 USPs identified in humans, making the UBP/USP family much larger than the UCH family [30]. UBPs also process a wider variety of substrates than UCH DUBs, including proubiquitin, free polyubiquitin chains of various linkages, and mono-or polyubiquitin conjugated to target proteins in vitro and in vivo. In addition, some family members can act on ubiquitin-like proteins. UBP43 has been demonstrated to act on ISG15 while USP21 cleaves conjugated Nedd8 [31, 32].

The diversity of the UBPs and breadth of substrates they act upon, makes them useful in a wide variety of cellular pathways and locations. UBPs regulate apoptosis, DNA repair, endocytosis, and transcription in addition to the ubiquitin–proteasome pathway (see below). The same diversity presents a challenge in determining the specificity of UBPs and with the exception of Isopeptidase T (UBP14/USP5), there have been few quantitative studies of in vitro specificity [33, 34]. In general, specificity has been described with qualitative "yes or no" assays that are not particularly useful in suggesting in vivo roles.

The structure of one UBP catalytic domain has been solved, that of USP7 complexed to ubiquitin aldehyde [35]. The data may be applicable to the way in which other UBPs function because the catalytic core of many UBPs is highly conserved. USP7 (also called HAUSP) is a human ubiquitin-specific protease that regulates the turnover of p53 [36]. USP7 consists of four structural domains; an N-terminal domain known to bind p53 and EBNA1, a catalytic domain, and two C-terminal domains. The 40-kDa catalytic domain exhibits a three-part architecture comprising Fingers, Palm, and Thumb (see Figure 8.2). The leaving ubiquitin moiety is specifically coordinated by the Fingers, with its C-terminus placed in a deep cleft between the Palm and Thumb where the catalytic residues are located. The domains form a pocket ideal for binding ubiquitin. Residues in the structure important for the above functions are conserved amongst UBPs, indicating that many UBPs may utilize the Fingers, Palm, and Thumb architecture to bind and cleave ubiquitinated substrates.

Another interesting structural observation is that water molecules cushion ubiquitin in the binding pocket. This is necessary because the binding surfaces of ubiquitin are uncharged, and the USP7 binding pocket is made up of predominantly acidic amino acid residues. These water molecules form extensive networks of hydrogen bonds with the bound ubiquitin and USP7. It is possible that they contribute to USP7's substrate specificity by allowing the protein to provide for relatively weak binding of ubiquitin and forcing itself to interact with the target protein to achieve specificity. This seems to be borne out by the fact that ubiquitin does not form a tightly bound complex with USP7 [35].

8.2.3
Ubiquitin-like Specific Proteases (ULP)

The *u*biquitin-*l*ike specific *p*roteases (ULPs) are a third class of DUB first thought to act only on SUMO-related ubiquitin-like proteins. There are two yeast ULPs and seven human ULPs (also called sentrin specific proteases, or SENPs). Further analysis determined that ULPs have little or no activity on ubiquitin substrates, but one (SENP8) acts on Nedd8 [26, 37, 38]. Despite acting on non-ubiquitin substrates, ULPs are still classified as DUBs because the function and mechanism of catalysis is so similar to those of the DUBs that act on ubiquitin. ULPs lack significant sequence homology to other DUBs and are more closely related to viral protein-processing proteases [39].

In addition to the lack of sequence homology, ULPs have little structural homology to other DUB classes except in the active site. The structure of ULP1 (see Figure 8.2) in complex with the C-terminal aldehyde of yeast SUMO (SMT3) illustrates that, like most other DUBs, ULPs are thiol proteases, utilizing a conserved catalytic triad consisting of an active-site cysteine, histidine, and aspartate [40]. Also, they require a gly–gly motif at the C-terminus of their UbL substrate for tight binding. The SUMO binding pocket of ULP1 recognizes SUMO through a number of polar and charged-residue interactions, including multiple salt bridges that are not present in the USP7 ubiquitin-binding site, and does not utilize water molecules or a "blocking loop".

8.2.4
OTU DUBs

A class of DUBs only identified since 2002 is the OTU (*o*varian *tu*mor protein) DUB class. The OTU domain was originally identified in an ovarian tumor protein from *Drosophila melanogaster*, and over 100 proteins from organisms ranging from bacteria to humans are annotated as having an OTU domain. The members of this protein superfamily were annotated as cysteine proteases, but no specific function had been demonstrated for any of these proteins. The first hint of a role for OTU proteins in the ubiquitin pathway was afforded by the observation that an OTU-domain-containing protein, HSPC263, reacted with ubiquitin vinyl sulfone (an active-site-directed irreversible inhibitor of DUBs) [41].

Then two groups almost simultaneously discovered that several OTU-containing proteins have DUB activity. Two human OTU DUBs were identified by purification with Ub-aldehyde (a reversible DUB inhibitor) affinity resin [42]. These proteins, named otubain1 and 2 (OTU-domain Ub-aldehyde binding protein) have a mass of approximately 35 kDa and are able to cleave polyubiquitin chains *in vitro*. However, the cleavage mechanism and their true substrates *in vivo* have yet to be determined. The other OTU DUB found was Cezanne, a 100-kDa protein that is similar to the A20 negative regulator of NF-κB [43]. Like A20, Cezanne plays a role in regulating NF-κB signaling pathways and has general DUB activity [44]. These OTU DUBs have highly conserved catalytic cysteine and histidine residues, implying that they utilize a catalytic triad to catalyze cleavage of polyubiquitin. It is unclear if most proteins containing OTU domains are DUBs, as analysis of the OTU family for DUB activity is only just beginning.

8.2.5
JAMM Isopeptidases

JAMM isopeptidases also constitute a recently identified class of DUBs. The members of this interesting class of DUBs were the first non-cysteine protease DUBs identified. Two JAMM isopeptidases have been confirmed as DUBs: Rpn11, which acts on ubiquitin conjugates, and Csn5, which acts on Nedd8 conjugates [45–47]. A number of other eukaryotic proteins have been annotated as containing the JAMM motif, but whether they have DUB activity has yet to be determined. Instead of cysteine proteases, they are metalloproteases belonging to a family of proteins that contain the Jab1/Csn5 and MPN domains [48]. Their activity depends on the JAMM motif (EX_nHS/THX_7SXXD) in the JAMM domain. The two histidines and an aspartic acid act as ligands to bind a metal ion, presumably zinc although this has not been proven, to achieve catalysis through polarization of a bound water molecule. A glutamic acid serves as a general acid–base catalyst. The crystal structure of a JAMM metalloprotease from *Archaeoglobolus fulgidus* bacteria has been recently been solved, but no structures of a JAMM isopeptidase with DUB activity are yet available [49, 50].

8.3
DUB Specificity

Why are there so many DUBs and how do they achieve specificity? The numerous DUBs identified to date suggest that DUBs have specifically evolved to act on distinct cellular substrates rather than to have general deubiquitinating activity (see Figure 8.1). We can ask what common features of these enzymes define them as DUBs and what differences allow specific DUBs to act on mono- *vs.* polyubiquitin? How have they evolved to cleave only ISG15 or SUMO-modified substrates, for instance? A body of data has been accumulated that at least partially answers these questions.

8.3.1
Recognition of the Ub-like Domain

All DUBs appear to recognize the body of the ubiquitin fold. UCH-L3, for example, makes contact with three regions of ubiquitin; residues 6–12, 41–48, and 69–74 [29]. These surfaces are highly conserved in Nedd8, but divergent in ISG15 and SUMO. Correspondingly, UCH-L3 can cleave ubiquitin and Nedd8 adducts but not those of the other ubiquitin-like proteins [26].

The same regions appear to be important for interactions of ubiquitin with many other DUBs (see Figure 8.2) and Ub-binding proteins. Importantly, all ubiquitin-binding domains examined utilize these same surfaces in binding ubiquitin.

Recognition of a ubiquitin domain can be accomplished by *ub*iquitin-*a*ssociated domains (UBA), which are present in many proteins, including some DUBs, and interact with polyubiquitin up to 1000-fold better than mono-ubiquitin [51]. However, other binding domains such as UIM (*u*biquitin-*in*teracting *m*otif) and CUE (*c*oupling of *u*biquitin conjugation to *E*R degradation) domains utilized in endocytic pathways prefer binding mono-ubiquitin [52–54].

Polyubiquitin-binding proteins recognize a subset of this binding surface of ubiquitin, often described as the hydrophobic patch. It is a group of three amino acids, Leu8, Ile44, and Val70, which are oriented in the ubiquitin molecule to form a small hydrophobic patch [55]. Polyubiquitin chains incorporating ubiquitins with mutations at residues 8 and 44 were unable to be disassembled by DUBs present in the 19S subunit of the proteasome [56]. In addition to providing a recognition site for DUBs, the patch is also important in determining the quaternary structure of polyubiquitin, another feature utilized by DUBs in substrate recognition. One UbL protein, ISG15, consists of a fusion of two ubiquitin domains. The crude mimicking of an Ub-dimer could potentially contribute to its specific recognition by deISGylating enzymes.

Polyubiquitin chains linked through all seven lysines in ubiquitin have been detected *in vivo*, and these poorly characterized forms of non-K48-linked polyubiquitin are also likely to have significant roles in the cell [57]. Polyubiquitin chains that are linked through different lysines are expected to be different enough in structure that individual DUBs could distinguish between them. K63-linked polyubiquitin is a well characterized alternative linkage and unlike K48-linked polyubiquitin, is not involved in proteolytic degradation [58, 59]. Structural data confirms the idea that these two types of polyubiquitin can have different structures [60, 61]. The structures of these dimers were solved by NMR analysis and they were found to have quite different conformations. Indicative of this, non-hydrolyzable ubiquitin-dimer analogs containing different linkages have markedly different effectiveness when used to inhibit the enzymatic activity of Isopeptidase T [62]. Isopeptidase T binds and cleaves polyubiquitin linked through at least four of the seven possible chain linkages found *in vivo*, although the catalytic efficiency of these activities is not known [10]. It is interesting to speculate that Isopeptidase T utilizes its two UBA domains to regulate binding of different polyubiquitin substrates. Mutational analysis of the UBA domains and structural data are needed to determine if this is the case and whether it is applicable to other DUBs as well.

8.3.2
Recognition of the Gly–Gly Linkage

The central feature that defines all DUBs is that they recognize and act at the C-terminus of the ubiquitin or ubiquitin-like domain. All mature ubiquitin and ubiquitin-like proteins have a C-terminal gly–gly motif and DUB cleavage releases leaving groups attached to the carboxyl group of the C-terminal glycine. With the exception of the JAMM metalloproteases, DUB catalysis starts with the nucleophilic attack of the catalytic cysteine on the carbonyl carbon of the scissile bond to form the tetrahedral intermediate. This is converted to an acyl-enzyme intermediate by expelling the C-terminal leaving group. Attack by a water molecule allows regeneration of the free thiol on the catalytic cysteine and releases free ubiquitin. The JAMM isopeptidases appear to use a classical metalloprotease mechanism [50].

DUB structures have evolved to recognize this C-terminal glycylglycine with exquisite specificity. Analysis of ubiquitin-fusion proteins lacking the gly–gly motif has clearly shown that they are not cleaved efficiently by DUBs [63]. All DUBs exclude larger amino acids at the C-terminus of the ubiquitin domain by having a deep cleft in their respective structures that is only large enough to hold two glycines. The narrowest region of USP7's catalytic cleft sterically excludes amino acids with any type of side chain, enforcing specificity for ubiquitin conjugates [35]. However, the end of the cleft is open, which allows USP7 to act on large ubiquitin conjugates like its substrate, ubiquitinated p53. ULP1 uses a similar type of cleft to recognize the gly–gly motif except that it uses a tryptophan residue to restrict access to the catalytic site when a substrate is bound to the enzyme [40]. UCHs have a similarly constrained cleft and also use the previously described "blocking loop" to assist in specifically recognizing the C-terminus of ubiquitin.

8.3.3
Recognition of the Leaving Group

In principle, DUBs might also recognize the leaving group to which ubiquitin is attached. In fact, such a mechanism seems likely as several DUBs have little affinity for ubiquitin and several have been shown to bind the un-ubiquitinated target protein (see Table 8.2). Interactions between DUBs and putative substrates have been shown for the mammalian DUBs VDU1, USP11, and UBPy, as well as UBP3 from yeast and fat facets from *Drosophila* [64–68]. In other cases, DUB-binding proteins may serve as scaffolds or adaptors that localize DUBs (discussed below).

8.3.4
Substrate-induced Conformational Changes

DUBs are not general hydrolases for cleaving after a gly–gly sequence even though they recognize the gly–gly motif at the C-terminus of ubiquitin and ubiquitin-like proteins. What is so special about these particular gly–gly sequences that DUBs will only recognize and act on them and not others? The answer comes from the fact that DUBs interact with the rest of the ubiquitin or ubiquitin-like substrate, and this interaction causes conformational changes in the DUB that are necessary to achieve catalysis. These changes result in rapid and efficient cleavage of only the particular substrate that the DUB is equipped to bind. It also explains why peptides with a gly–gly in them are not susceptible to cleavage by DUBs as they are lacking the substrate-binding domains that cause the DUB conformational change required for cleavage. The different DUB classes utilize a number of conformational

Tab. 8.2. Identification of DUBs and DUB-binding partners through physical and genetic interaction screens.

DUB	Affinity	Characterized by			Interaction partner(s)
		MS	Yeast two-hybrid	Synthetic lethal	
UCHs					
UCH37			X		S14, UIP1 [21]
UCH-L3			X		Nedd8 [126]
UCH-L1			X		JAB1, p27 [127]
UBPs					
DOA4				X	SLA1, SLA2 [128]
UBP3	X		X	X	SIR4, Bre5, Stu1 [66, 118, 129]
UBP6		X			19S proteasome [130]
UBP8		X			SAGA, SLIK acetyl transferases [88]
USP5	X				ubiquitin [131]
USP7			X		ataxin [132]
USP11			X		RanBPM [67]
CYLD		X			NEMO [7]
fat facets					Vasa [75]
UBPy			X		CDC25(Mm) [133]
OTU DUBs					
cezanne			X		ubiquitin [43]
otubain1 and 2	X				ubiquitin aldehyde [42]
VCIP 135	X				VCP/P47 [72]
JAMM Isopeptidases					
Rpn11				X	UBP6 [92]
Others					
ataxin 3			X		RAD23, HHR23A, HHR23B [134]

changes that are induced upon substrate binding to assist in promoting efficient cleavage.

UCH DUBs have been the most thoroughly analyzed. Comparison of the ubiquitin–UCH complex with unliganded UCH shows two significant conformational differences that contribute to keeping the unliganded enzyme in an inactive state. First, the previously described "blocking loop" becomes ordered as it interacts with ubiquitin. Invariant residues form hydrogen bonds with the ubiquitin substrate and other UCH residues, indicating that the loop has functional importance during substrate binding [27]. Second, the side chain of L9 in UCH-L3 intrudes into the substrate-binding cleft, occluding the catalytic cysteine and preventing binding of peptide substrates [29]. When ubiquitin binds to the UCH-L3, an interaction between ubiquitin and UCH-L3 repositions L9, allowing access to the active site cleft. Thus, the energy of ubiquitin binding is required to activate UCH-L3, allowing its cleavage. This type of selectivity (where ubiquitin binding is required for activity) may be necessary to prevent deleterious cleavage of other protein substrates by UCHs.

A similar situation was observed when the crystal structure of USP7 was solved in the absence of substrate [35]. The catalytic cysteine of the unliganded protein is not in an orientation that would allow catalysis to take place. The histidine residue needed to interact with the active-site cysteine is too distant for a catalytic-triad mechanism to function. Binding of ubiquitin aldehyde induces a significant conformational change that realigns the catalytic triad residues so the hydrogen bonding required for catalysis can take place. Thus, like UCH DUBs, the unliganded protease is inactive and only becomes catalytically active when it is binding substrates.

ULP1 also uses conformational changes to "clamp down" on the gly–gly motif when a SUMO substrate is bound. Trp448 lies directly above the active site and interacts with the SUMO C-terminus by Van der Waals interactions, sandwiching the gly–gly motif between Trp448 and the active-site cysteine when SUMO binds [40]. Despite the various methods utilized, all DUBs require a conformational change triggered by binding of a specific substrate to catalyze cleavage. These required conformational changes are driven by the energy of interaction between the DUB and the body of the ubiquitin domain.

8.4
Localization of DUBs

While many DUBs are cytoplasmic, localization of DUBs is also known to be important in regulating DUB specificity. The localization of ULP1, for example, is important in determining its substrates. The N-terminal domain of ULP1 is known to localize the enzyme to the nuclear envelope, and truncation mutations lacking this domain remain in the cytoplasm [69]. When the truncated protein is expressed in ΔULP1 yeast strains, the cells grow at wild-type levels, and the truncated protein is able to cleave SUMO substrates *in vitro*. However, analysis of ΔULP1 cells expressing this truncation shows an accumulation of SUMO conjugates. Apparently, the localization of ULP1 to the nuclear envelope is necessary in order for it to act on specific nuclear-envelope-localized substrates. The localization helps constrain ULP1 isopeptidase activity so ULP1 does not inappropriately act on cytoplasmic substrates. Other examples of DUB activity regulated by localization include UBP6, which is fully active only when bound to the proteasome (see below) and UBP16 residence on the outer membrane of the mitochondria, although its function there is undetermined [70].

Other DUBs have been found to associate with membranes and regulate membrane-associated cellular processes, although they appear not to be membrane anchored like UBP16. The ability of DOA4 to remove ubiquitin from membrane-bound endocytic substrates promotes their degradation in the vacuole or lysosome [71]. DUBs are also important for membrane fusion events as shown by the fact that an OTU domain DUB, VCIP135 (*V*CP/p47 complex-*i*nteracting *p*rotein of

135 kd), is necessary for p97–p47-mediated Golgi cisternae reassembly after mitosis [72]. Also, a neuronal DUB, synUSP, was found to localize to post-synaptic lipid rafts (membrane microdomains involved in membrane trafficking and signal transduction) [73]. However, its function at that location has yet to be characterized.

A well-studied example of a tissue-specific DUB activity is fat facets, a UBP originally found in *Drosophila* [74]. It is important in eye development and germ-cell specification and is active only in specific cell types during certain stages of development [65, 75]. The lack of fat facets results in defective posterior patterning, germ-cell specification, and eye formation. Fat facets activity is required to prevent the inappropriate degradation of vasa and liquid facets. In this case, the role of the DUB appears to be defined by the restricted expression of its known substrates.

Temporal regulation of DUB expression also appears important. D'Andrea and colleagues first described a small family of DUBs that are induced as immediate early gene products of cytokine stimulation [76]. Different cytokines were shown to induce different DUBs and the expression of these enzymes was short-lived [77]. It appears that these DUBs may be involved in down-regulating cytokine receptors, perhaps by removing the ubiquitin involved in sorting of the receptor at the early endosome. Likewise, UBP43, the short-lived processing protein for ISG15, is present at very low levels in normal cells and highly expressed upon interferon induction [78].

8.5
Probable Physiological Roles for DUBs

8.5.1
Proprotein Processing

One important function of DUBs is the processing of ubiquitin or ubiquitin-like proteins to their mature forms. Ubiquitin is expressed in cells as either linear polyubiquitin or N-terminally fused to certain ribosomal proteins [79, 80]. These gene products are processed by DUBs to separate the ubiquitin into monomers and expose the gly–gly motif at the C-terminus. Many DUBs process linear polyubiquitin or Ub-fusion proteins *in vitro*, but this processing appears to take place cotranslationally *in vivo* and is extremely rapid. This makes analysis difficult and leaves unanswered the question of which DUBs actually perform this function *in vivo*. Multiple DUBs may be able to perform this processing at a physiologically relevant level since DUB deletions rarely shows processing defects [81].

Ubiquitin-like proteins are also expressed as proproteins with a short C-terminal extension of a few amino acids that must be removed to make the UbL available for conjugation to target proteins. All ULPs have been shown to metabolize their respective proprotein to an active form *in vitro* although again it is unclear which ULPs are responsible for this activity *in vivo*. An exception to the confusion is the finding that RUB1 (the yeast homolog of Nedd8) is processed by YUH1 in *Saccha-*

romyces cerevisiae. Conjugation of RUB1 to Cdc53 is required for efficient assembly of certain SCF (**s**kp1, **c**ullin, **F**-box) E3 ubiquitin ligases [82]. Yuh1 deletion strains do not process Rub1 or modify Cdc53 with Rub1 [83]. Modification of Cdc53 by Rub1 could be reconstituted in a ΔYuh1 strain by expressing a mature Rub1 construct lacking the C-terminal asparagine normally removed by processing. This demonstrated that Yuh1 processes RUB1 proprotein into the mature form *in vivo*. It is not known which DUB performs the processing of proNedd8.

8.5.2
Salvage Pathways: Recovering Mono-ubiquitin Adducts and Recycling Polyubiquitin

It has been speculated that without UCH function, all ubiquitin in the cell would be conjugated with glutathione or other cellular amines and therefore unavailable for conjugation. This would quickly result in the cessation of the ubiquitin–proteasome system function and cell death due to lack of active ubiquitin to conjugate to substrates. The effectiveness of UCH DUBs in liberating ubiquitin from other small adducts makes them likely candidates to act on these particular adducts. In addition, the cell must regenerate mono-ubiquitin from polyubiquitin and various mono-ubiquitinated proteins to maintain levels of mono-ubiquitin for conjugation. Doa4 appears to remove small peptides attached to mono- and di-ubiquitin intermediates resulting from proteasomal degradation as well as removing ubiquitin from proteins targeted for endocytosis [84, 85]. Loss of Doa4 function in yeast results in depleted levels of mono-ubiquitin and increased cell death during stationary phase.

Another function for DUBs is regenerating free ubiquitin from unanchored polyubiquitin chains removed from proteasome substrates or proteins targeted for other pathways. Polyubiquitin inhibits the proteasome and lowers the amount of free ubiquitin available for conjugation to proteins. Thus, these chains need to be processed to mono-ubiquitin to prevent polyubiquitin accumulation and inhibition of the proteasome. This type of DUB activity has been well characterized *in vivo* and *in vitro* and Isopeptidase T appears to be the DUB that is responsible for the majority of this activity. Deletion of UBP14 in yeast is not lethal, although large amounts of polyubiquitin build up in the cell and proteasome function is impaired [86]. Isopeptidase T seems to serve as a general DUB for regenerating mono-ubiquitin as it cleaves polyubiquitin containing various linkages [59].

8.5.3
Regulation of Mono-ubiquitination

DUBs have increasingly been found to be important in regulating the ubiquitination level of proteins not targeted to the proteasome for degradation. Some DUBs are active participants in the regulation of mono-ubiquitin (or mono-UbL) conjugation and others can regulate the conjugation of multiple types of ubiquitin or UbLs to a single substrate. For instance, deneddylating enzymes may regulate the neddylation of cullin proteins both by processing proNedd8 and by removing Nedd8 from neddylated cullins [26, 38]. As a component of the SCF E3 ligase complexes, cullins require neddylation in order for their cognate E3 ligase to be efficiently assembled [82]. Regulation of this modification indirectly regulates the ubiquitination of a subset of proteins. Defects in deneddylation could lead to inappropriate ubiquitination of substrates owing to inappropriate recruitment of E2s to the SCF E3s [87].

Regulating mono-ubiquitination of proteins by DUBs is important in histone modification where ubiquitination is thought to modulate chromatin structure and transcriptional activity. Normally, about 10% of the histone core octomers contain ubiquitinated histones and the ubiquitin is removed at mitosis by DUB activity. UBP8 has been demonstrated to regulate the ubiquitination of histone H2B, which is important in transcriptional activation of many genes [88].

Many cell-surface receptors are ubiquitinated upon internalization and the ubiquitin is removed by DUBs at the early endosome. Properly sorted receptors are then shuttled to the lysosome for degradation. In the absence of Doa4, the ubiquitin is not removed upon sorting and instead is co-degraded in the vacuole, resulting in ubiquitin depletion [84]. Another DUB, UBP3 assists Golgi-ER retrograde transport by deubiquitinating B′-COP, thus preventing its degradation [66]. Mono-ubiquitinated B′-COP cannot be assembled into the COP1 complex without UBP3/Bre5 complex DUB activity. Disruption of the complex in ΔBre5 strains reduces the efficiency of Golgi-ER transport and facilitates the polyubiquitination and degradation of B′-COP by the proteasome.

One fascinating observation is that PCNA (proliferating cell nuclear antigen) can be modified by multiple forms of ubiquitin, demonstrating that DUBs with different specificities can act at the same location on a specific substrate. PCNA can be modified by mono-ubiquitin, 63-linked polyubiquitin, or SUMO at K164 [89]. Modification of PCNA by mono- or polyubiquitin determines whether it is utilized in translesion synthesis or error-free DNA repair, respectively. SUMO modification prevents PCNA function in DNA repair and instead promotes DNA replication. It is probable that multiple DUBs, as yet unidentified, are required to regulate PCNA modification.

8.5.4
Processing of Proteasome-bound Polyubiquitin

DUBs play a crucial role in regulating the function of the proteasome. For a long time it was unclear what happens to polyubiquitin conjugated to a proteasome substrate when that substrate is at the proteasome ready for degradation. Was the conjugated polyubiquitin processed by the proteasome and degraded or was it removed by a DUB and released from the proteasome? The small 13-Å diameter entrance to the 20S catalytic core of the proteasome requires all substrates to be fed through as unfolded polypeptides [90]. A branched polypeptide such as a ubiquitin–protein conjugate apparently has difficulty fitting through the pore, greatly reducing proteasome efficiency [47]. Thus, it seemed likely that DUBs must remove polyubiquitin from proteasome substrates before they enter the 20S catalytic core of the proteasome.

To date, three DUBs are known to perform this function and all are components of the 19S lid of the mammalian proteasome. The first described was UCH37, although its exact function is still unclear [91]. UCH37 is thought to be an editing DUB that assists in clearing the proteasome of ubiquitinated proteins. UCH37 slowly cleaves one ubiquitin at a time from the distal end of the polyubiquitin chain. If chain trimming is faster than the degradation process, loss of the polyubiquitin signal could result in partial degradation or release of proteins from the proteasome. UCH37 activity could also be necessary to recover proteasomes that are having difficulty degrading ubiquitinated proteins.

UCH37 is only found in higher eukaryotes, but the other two proteasome-bound DUBs, RPN11 and UBP6 (USP14), are found in all eukaryotes. RPN11 and UBP6 remove polyubiquitin from substrates that are committed to degradation by the proteasome [92]. The mechanisms for this, and exactly what role each DUB plays in removing ubiquitin, are not fully understood. Interestingly, the Rpn11 DUB activity was first detected over 10 years ago when 26S proteasome DUB activity was inhibited by *o*-phenanthroline, a metal chelator [93]. The metalloprotease DUB activity was not identified until recently [46]. Rpn11 is thought to remove polyubiquitin chains from proteasome substrates before they are degraded, allowing the unfolded substrate to enter the pore of the 20S subunit of the proteasome. It has been proposed that Rpn11 removes most of the polyubiquitin chain attached to a proteasome substrate and then UBP6 acts to remove the remaining one or two ubiquitin residues.

Despite the lack of mechanistic understanding, the DUBs are clearly required for efficient proteasomal degradation to take place. The Rpn11 deletion is lethal in yeast and temperature-sensitive mutants show massive accumulation of polyubiquitin conjugates [46, 47]. UBP6 is approximately 300 times more active when it is associated with the proteasome than in its purified form [94]. The UBP6 deletion is not lethal in yeast, but a large decrease in the cellular pool of mono-ubiquitin occurs, indicating that ubiquitin is fed into the proteasome and degraded rather than being released from the proteasome and recycled [95].

8.6
Finding Substrates and Roles for DUBs

Surprisingly, little is known about the *in vivo* substrate specificity of DUBs. Difficulty in defining the substrate specificity of individual DUBs often arises from a lack of observable phenotypes in deletion strains. Deletion studies in yeast where up to 4 of the 17 DUBs have been deleted in a single strain have not produced significant phenotypes [96]. It is unclear if this is due to the subtle nature of the phenotypes or if the remaining DUBs compensate for the missing ones. However, several tactics have been fruitful in defining the physiological roles of DUBs. First, definition of *in vitro* specificities can be useful in focusing genetic screens. For example, the first UBPs were cloned and analyzed after it was discovered that they could cleave ubiquitin-fusion proteins [96, 97]. Second, directed screening of deletions or knockdown studies to identify roles for DUBs have also been successful (see Table 8.1 for DUB-deletion phenotypes). Study of DUB deletions, including UCH-L1, UCH-L3, and USP14 (see below), in the mouse have demonstrated their importance in neuronal function. Third, potential roles for DUBs have also been identified by physical and genetic interaction screens. Table 8.2 shows in more detail the interaction screens that have been used in discovering DUBs and characterizing their *in vivo* roles by identifying novel binding partners. For example, Cezanne was suggested to be a DUB after two-hybrid studies demonstrated its interaction with ubiquitin and UBP6 was identified as a component of the 19S subunit of the proteasome by mass spectrometry.

8.7
Roles of DUBs Revealed in Disease

8.7.1
NF-κB Pathway

NF-κB is a transcription factor that can be activated by a number of cellular signals, including stress, inflammation (via tumor necrosis factor) and antigen receptors among others [98, 99]. After receptor stimulation, a cascade ensues that results in the release of NF-κB from its inhibitor IκB. Released NF-κB translocates to the nucleus and activates transcription of a number of genes. Ubiquitin metabolism plays a significant regulatory role in the NF-κB pathway. For NF-κB release from IκB and nuclear translocation to take place, IκB is phosphorylated by IκB kinases, resulting in K48-linked polyubiquitination and proteasomal degradation of IκB [100]. A number of other proteins involved in this pathway such as NEMO, IKKγ, and TRAF6 have K63-linked polyubiquitin chains conjugated to them [101, 102]. It is not clear what purpose the K63-linked chains serve, but they appear to be a regulatory component of the NF-κB pathway.

Most of the DUB activity characterized in the NF-κB pathway appears to act on K63-linked polyubiquitin, suggesting that modulation of K63-linked polyubiquitination by DUBs is important for control of the NF-κB pathway. CYLD, a tumor suppressor gene, has been confirmed as a DUB [7, 103, 104]. Loss of CYLD function leads to cylindromatosis, a syndrome characterized by large benign tumors on the face and neck. This is one of the few examples where a defective DUB has been defined as the direct cause of a specific disease. Preferred *in vivo* substrates of CYLD are believed to be 63-linked polyubiquitin–protein conjugates of NEMO, TRAF6, and TRAF2 components of the NF-κB pathway, but the exact *in vivo* regulatory role of CYLD is still unknown.

OTU family DUBs such as Cezanne and A20 also play significant roles as negative regulators of the NF-κB pathway [43, 44]. A20 can cleave K48- and K63-linked polyubiquitin chains *in vitro* while Cezanne has only been tested on K48-linked chains. Although these DUBs are known to be part of the NF-κB pathway, their *in vivo* substrates are unknown. It is also unclear as to how these DUBs negatively regulate the NF-κB pathway.

8.7.2
Neural Function

DUBs, specifically UCHs, appear to play significant roles in neurodegenerative diseases such as Parkinson's, Alzheimer's, Huntington's, and others [105]. A mutant form of UCH-L1 with reduced enzymatic activity has been found in a small family of Parkinson's patients and the S18Y allele of UCH-L1 has been associated with a reduced risk of sporadic Parkinson's disease [106, 107]. Many of the inclusion bodies found in patients with Parkinson's are known to contain high amounts of UCH-L1, ubiquitin, and ubiquitinated proteins, as determined by immunostaining [108]. This suggests that defects in some DUBs or their regulation can cause significant harm to the neuronal system, resulting in disease.

DUBs have also been implicated in the formation of other neural inclusion bodies. In addition to the case for their involvement in Parkinson's disease it has been shown that the mutation of USP14 (the mammalian homolog of yeast UBP6) results in Ataxia in the mouse [109]. Many neurological diseases, including Ataxia, result in damaged or mutated proteins aggregating as polyubiquitinated forms at the *m*icrotubule *o*rganizing *c*enter (MTOC) to form inclusions called aggresomes [110]. An adapter, the tubulin deacetylase HDAC6 (*h*istone *dea*cetylase 6), has recently been shown to bind these polyubiquitinated proteins and tether them to the microtubules where they are then transported to the MTOC [111]. The classic Lewy Body of Parkinson's disease has all the hallmarks of such an aggresome. The formation of an aggresome is thought to be protective and in its absence the aggregated proteins can trigger apoptosis. Thus, the dynamics of ubiquitination and aggregate formation are important responses to this type of cellular stress and several DUBs can modulate this process.

Deletion of UCH-L1 and/or UCH-L3 in mice has demonstrated that they are both involved in neuronal regulation, but have separate functions. The GAD (*gra*cile *a*xonal *d*ystrophy) mouse has been shown to lack UCH-L1, the predominant neuronal UCH [112]. These mice show a unique neuronal "dying back" phenotype that results in paralysis of the limbs due to death of nerves originating in the gracile nucleus. Mice lacking UCH-L3 have no obvious abnormalities or defects [113]. However, the double deletion mouse shows more severe defects including reduced weight, a more severe gracile axonal dystrophy than the L1 deletion, and earlier lethality caused by a loss of the ability to swallow resulting in starvation [114]. This demonstrates that the two DUBs are not redundant and have separate neuronal functions.

8.8
New Tools for DUB Analysis

Despite all the DUB structures and substrates previously described, in most cases the *in vivo* substrate for a particular DUB is unknown. Structural and localization data can provide clues to determine *in vivo* DUB specificity, especially if one knows what ubiquitin or ubiquitin-like protein it acts upon. Genomic databases have helped, but many annotated DUBs have never been tested for DUB activity and some DUBs thought to act on one type of substrate (based on their homology) are found to act on another when tested. The characterization of hundreds of potential DUBs is a daunting task and *in vivo* characterization is even more difficult. To make headway, novel tools are needed to conclusively identify potential DUBs and their substrates to help direct appropriate *in vivo* studies.

8.8.1
Active-site-directed Irreversible Inhibitors and Substrates

The most promising tools developed for this sort of analysis are active-site-directed irreversible inhibitors of DUBs. These inhibitors are ubiquitin or ubiquitin-like proteins chemically modified at the C-terminus by an electrophilic moiety such as a Michael acceptor or alkyl halide. The modified ubiquitin can be incubated with a purified DUB or a cell lysate containing DUB activity. *Ub*iquitin *v*inyl *sul*fone (UbVS) is one such irreversible inhibitor because the vinyl sulfone moiety reacts with the active-site cysteine of the DUB, forming a thioether linkage. The covalent adduct is stable and can be detected in a variety of ways. Labeling of DUBs is specific, as only a DUB active-site cysteine will efficiently react with the vinyl sulfone moiety.

To create these inhibitors, an N-terminally tagged ubiquitin or ubiquitin-like protein (lacking the C-terminal glycine) is expressed using the intein expression system (New England Biolabs). Briefly, in this system a fusion protein consisting of a ubiquitin or a ubiquitin-like protein lacking the C-terminal glycine, an intein linker, and a *c*hitin-*b*inding *d*omain (CBD) is expressed in *E. coli*. Clarified cell lysate is incubated with chitin beads to bind the ubiquitin-fusion protein. The ubiquitin is then cleaved from the CBD and intein linker by adding mercaptoethanesulfonic acid (MESNA). After MESNA elution, the resulting truncated ubiquitin C-terminal thioester is reacted with glycine vinyl methyl sulfone to create the Ub or UbL vinyl sulfone derivative. The N-terminal tag on the ubiquitin molecule allows analysis of DUB labeling by immunoprecipitation and Western blotting.

This labeling has been used with success in yeast where 6 of the 17 known DUBs were labeled with UbVS [115]. Incomplete labeling likely results from DUBs that do not act on mono-ubiquitin or where the UbVS could not access the active site. The labeling has also been used with great success in mammalian cell lysates to identify novel ubiquitin DUBs [41]. A novel deneddylating enzyme and a novel DUB that acts on autophagy-related UbL proteins have also been identified using vinyl sulfone labeled probes [26, 37, 116].

This ubiquitin intein system can also be utilized to make a DUB substrate rather than inhibitors by attaching a C-terminal fluorescent tag such as 7-amidomethylcoumarin (AMC) instead of vinyl sulfone. DUBs cleave the ubiquitin derivative and release the fluorescent tag, a process that can be followed fluorometrically. Fluorometric assays can then be used to determine a particular DUB's preferred substrate or to quantitate DUB activity in crude lysates. AMC substrates have turned out to be excellent tools for identifying the substrates of individual DUBs. Den1, for example, was shown to cleave Nedd8–AMC 60 000-fold faster than it cleaves ubiquitin–AMC, and the ratio was even higher when compared to SUMO–AMC [26]. Clearly, these reagents are powerful tools for identifying novel DUBs and identifying potential DUB substrates.

8.8.2
Non-hydrolyzable Polyubiquitin Analogs

Other modified ubiquitin reagents that are useful in analyzing DUBs are non-hydrolyzable polyubiquitin analogs. These analogs are polyubiquitin chains where the ubiquitins are linked by cross-linking reagents. To synthesize them, one ubiquitin is mutated to cysteine at the C-terminal glycine and another has cysteine introduced at a particular lysine residue. These ubiquitins can then be linked through their cysteine residues with a bifunctional thiol reagent such as dichloroacetone (DCA). As the native ubiquitin sequence contains no cysteines, the ubiquitins will only be linked through the introduced cysteine residues. The result is a ubiquitin dimer analog that mimics physiological dimers. The isopeptide bond is replaced by a DCA linkage, but the ubiquitin subunits retain the appropriate spatial orientation. Thus, DUBs should bind these dimers, but will be unable to cleave them because they cannot hydrolyze the DCA linkage. To make longer polyubiquitin chains, cysteine residues or sulfhydryl groups must be introduced at the desired lysine and the C-terminal glycine on the same ubiquitin molecule.

These chain analogs have been used to characterize DUBs in two fashions. First, they can be used as inhibitors of DUBs [62]. Cleavage of Ub–AMC by Isopeptidase T, a polyubiquitin-binding DUB, was inhibited by the addition of differently linked dimer analogs and kinetic inhibition constants were determined. The K_i values of dimer analogs were all much lower than the K_i for mono-ubiquitin. Further, there was considerable selectivity, as inhibition constants varied depending on the linkage present in the dimer [62]. This demonstrated that the analogs act as faithful mimics of native polyubiquitin. The other way these chain analogs are used is to synthesize them on affinity supports and analyze cell lysates for DUBs that bind a specific type of polyubiquitin chain. These affinity resins have proven useful in identifying a number of binding proteins, including DUBs, from yeast cell lysates [117]. Analogs with different linkages bind a different subset of proteins, perhaps suggesting a way to identify DUBs acting upon polyubiquitin with linkage specificity. Thus, these analogs are excellent tools for characterizing the substrate preferences of known DUBs and discovering novel ones.

8.9
Conclusion

DUBs are clearly an essential cellular component needed for a variety of pathways including protein degradation, DNA repair, apoptosis, membrane trafficking, stress response, and transcriptional regulation. Not only do they act on various ubiquitin substrates, but they are also needed to process ubiquitin-like substrates. Over 100 DUBs from five major families have been identified and the number is likely to increase. Factors that enhance DUB specificity are the presence of a binding pocket that only accommodates the physiological substrate, the requirement for a substrate-induced conformational change that prevents undesired catalysis, and the recognition of the ubiquitous C-terminal gly–gly motif of all DUB substrates. Subcellular localization to a specific organelle or protein complex and tissue-specific as well as temporal expression are also important components of DUB specificity and function.

In spite of all that has been learned about DUBs and their function, much remains to be discovered. Future studies are likely to focus on identifying *in vivo* DUB substrates, novel DUBs, DUB-binding partners, and phenotypes of DUB deletions. Further development of new reagents, such as the non-hydrolyzable polyubiquitin analogs and active-site-directed inhibitors or substrates will help greatly. Directed proteomics studies should assist in identifying DUBs, loss-of-function phenotypes, and potential binding partners. Despite the major gaps that remain in our understanding of DUBs, our knowledge of their roles and importance has progressed amazingly rapidly. Novel DUB gene families have been identified, new ubiquitin-like DUB substrates have been uncovered, and structural data has been analyzed to elucidate how DUBs perform catalysis and specifically recognize their substrates. *In vivo* substrates of DUBs are beginning to be identified and the tools and techniques needed to search for novel DUBs and analyze known ones for their specificity are rapidly being created. With so much discovered, and yet so much remaining to be found, deubiquitinating enzymes are a vibrant field of study.

References

1 SCHLESINGER, D. H., GOLDSTEIN, G., and NIALL, H. D. The complete amino acid sequence of ubiquitin, an adenylate cyclase stimulating polypeptide probably universal in living cells, *Biochemistry*, **1975**, *14*, 2214–8.

2 HOCHSTRASSER, M. Ubiquitin-dependent protein degradation, *Annu Rev Genet*, **1996**, *30*, 405–39.

3 THROWER, J. S., HOFFMAN, L., RECHSTEINER, M., and PICKART, C. M. *Recognition of the polyubiquitin proteolytic signal*, Embo J, **2000**, *19*, 94–102.

4 WILKINSON, K. D. Ubiquitination and deubiquitination: targeting of proteins for degradation by the proteasome, *Semin Cell Dev Biol*, **2000**, *11*, 141–8.

5 WANG, Q., GOH, A. M., HOWLEY,

P. M., and WALTERS, K. J. *Ubiquitin recognition by the DNA repair protein hHR23a*, Biochemistry, **2003**, *42*, 13529–35.

6 GEWIES, A. and GRIMM, S. UBP41 is a proapoptotic ubiquitin-specific protease, *Cancer Res*, **2003**, *63*, 682–8.

7 KOVALENKO, A., CHABLE-BESSIA, C., CANTARELLA, G., ISRAEL, A., WALLACH, D., and COURTOIS, G. *The tumour suppressor CYLD negatively regulates NF-kappaB signalling by deubiquitination*, Nature, **2003**, *424*, 801–5.

8 HICKE, L. A new ticket for entry into budding vesicles-ubiquitin, *Cell*, **2001**, *106*, 527–30.

9 MASTRANDREA, L. D., YOU, J., NILES, E. G., and PICKART, C. M. *E2/E3-*

mediated assembly of lysine 29-linked polyubiquitin chains, *J Biol Chem*, **1999**, *274*, 27299–306.

10 You, J. and Pickart, C. M. A HECT domain E3 enzyme assembles novel polyubiquitin chains, *J Biol Chem*, **2001**, *276*, 19871–8.

11 Baboshina, O. V. and Haas, A. L. Novel multiubiquitin chain linkages catalyzed by the conjugating enzymes E2EPF and RAD6 are recognized by 26 S proteasome subunit 5, *J Biol Chem*, **1996**, *271*, 2823–31.

12 Mahajan, R., Delphin, C., Guan, T., Gerace, L., and Melchior, F. A small ubiquitin-related polypeptide involved in targeting RanGAP1 to nuclear pore complex protein RanBP2, *Cell*, **1997**, *88*, 97–107.

13 Kamitani, T., Kito, K., Nguyen, H. P., and Yeh, E. T. Characterization of NEDD8, a developmentally down-regulated ubiquitin-like protein, *J Biol Chem*, **1997**, *272*, 28557–62.

14 Narasimhan, J., Potter, J. L., and Haas, A. L. Conjugation of the 15-kDa interferon-induced ubiquitin homolog is distinct from that of ubiquitin, *J Biol Chem*, **1996**, *271*, 324–30.

15 Walden, H., Podgorski, M. S., Huang, D. T., Miller, D. W., Howard, R. J., Minor, D. L., Jr., Holton, J. M., and Schulman, B. A. The structure of the APPBP1-UBA3-NEDD8-ATP complex reveals the basis for selective ubiquitin-like protein activation by an E1, *Mol Cell*, **2003**, *12*, 1427–37.

16 Johnson, E. S., Schwienhorst, I., Dohmen, R. J., and Blobel, G. The ubiquitin-like protein Smt3p is activated for conjugation to other proteins by an Aos1p/Uba2p hetero-dimer, *Embo J*, **1997**, *16*, 5509–19.

17 Hemelaar, J., Borodovsky, A., Kessler, B. M., Reverter, D., Cook, J., Kolli, N., Gan-Erdene, T., Wilkinson, K. D., Gill, G., Lima, C. D., Ploegh, H. L., and Ovaa, H. Specific and covalent targeting of conjugating and deconjugating enzymes of ubiquitin-like proteins, *Mol Cell Biol*, **2004**, *24*, 84–95.

18 Andersen, M. W., Goldknopf, I. L., and Busch, H. Protein A24 lyase is an isopeptidase, *FEBS Lett*, **1981**, *132*, 210–4.

19 Pickart, C. M. and Rose, I. A. Ubiquitin carboxyl-terminal hydrolase acts on ubiquitin carboxyl-terminal amides, *J Biol Chem*, **1985**, *260*, 7903–10.

20 Jentsch, S. and Pyrowolakis, G. Ubiquitin and its kin: how close are the family ties?, *Trends Cell Biol*, **2000**, *10*, 335–42.

21 Li, T., Duan, W., Yang, H., Lee, M. K., Bte Mustafa, F., Lee, B. H., and Teo, T. S. Identification of two proteins, S14 and UIP1, that interact with UCH37, *FEBS Lett*, **2001**, *488*, 201–5.

22 Jensen, D. E., Proctor, M., Marquis, S. T., Gardner, H. P., Ha, S. I., Chodosh, L. A., Ishov, A. M., Tommerup, N., Vissing, H., Sekido, Y., Minna, J., Borodovsky, A., Schultz, D. C., Wilkinson, K. D., Maul, G. G., Barlev, N., Berger, S. L., Prendergast, G. C., and Rauscher, F. J., 3rd. BAP1: a novel ubiquitin hydrolase which binds to the BRCA1 RING finger and enhances BRCA1-mediated cell growth suppression, *Oncogene*, **1998**, *16*, 1097–112.

23 Wilkinson, K. D. Regulation of ubiquitin-dependent processes by deubiquitinating enzymes, *Faseb J*, **1997**, *11*, 1245–56.

24 Larsen, C. N., Krantz, B. A., and Wilkinson, K. D. Substrate specificity of deubiquitinating enzymes: ubiquitin C-terminal hydrolases, *Biochemistry*, **1998**, *37*, 3358–68.

25 Dang, L. C., Melandri, F. D., and Stein, R. L. Kinetic and mechanistic studies on the hydrolysis of ubiquitin C-terminal 7-amido-4-methylcoumarin by deubiquitinating enzymes, *Biochemistry*, **1998**, *37*, 1868–79.

26 Gan-Erdene, T., Kolli, N., Yin, L., Wu, K., Pan, Z. Q., and Wilkinson, K. D. Identification and characterization of DEN1, a deneddylase of the ULP family, *J Biol Chem*, **2003**, *31*, 28892–900.

27 Johnston, S. C., Riddle, S. M., Cohen, R. E., and Hill, C. P.

Structural basis for the specificity of ubiquitin C-terminal hydrolases, Embo J, **1999**, *18*, 3877–87.

28 JOHNSTON, S. C., LARSEN, C. N., COOK, W. J., WILKINSON, K. D., and HILL, C. P. Crystal structure of a deubiquitinating enzyme (human UCH-L3) at 1.8 A resolution, *Embo J,* **1997**, *16*, 3787–96.

29 WILKINSON, K. D., LALELI-SAHIN, E., URBAUER, J., LARSEN, C. N., SHIH, G. H., HAAS, A. L., WALSH, S. T., and WAND, A. J. The binding site for UCH-L3 on ubiquitin: mutagenesis and NMR studies on the complex between ubiquitin and UCH-L3, *J Mol Biol,* **1999**, *291*, 1067–77.

30 WING, S. S. Deubiquitinating enzymes – the importance of driving in reverse along the ubiquitin-proteasome pathway, *Int J Biochem Cell Biol,* **2003**, *35*, 590–605.

31 MALAKHOV, M. P., MALAKHOVA, O. A., KIM, K. I., RITCHIE, K. J., and ZHANG, D. E. UBP43 (USP18) specifically removes ISG15 from conjugated proteins, *J Biol Chem,* **2002**, *277*, 9976–81.

32 GONG, L., KAMITANI, T., MILLAS, S., and YEH, E. T. Identification of a novel isopeptidase with dual specificity for ubiquitin- and NEDD8-conjugated proteins, *J Biol Chem,* **2000**, *275*, 14212–6.

33 WILKINSON, K. D., TASHAYEV, V. L., O'CONNOR, L. B., LARSEN, C. N., KASPEREK, E., and PICKART, C. M. *Metabolism of the polyubiquitin degradation signal: structure, mechanism, and role of isopeptidase T,* Biochemistry, **1995**, *34*, 14535–46.

34 MELANDRI, F., GRENIER, L., PLAMONDON, L., HUSKEY, W. P., and STEIN, R. L. Kinetic studies on the inhibition of isopeptidase T by ubiquitin aldehyde, *Biochemistry,* **1996**, *35*, 12893–900.

35 HU, M., LI, P., LI, M., LI, W., YAO, T., WU, J. W., GU, W., COHEN, R. E., and SHI, Y. Crystal structure of a UBP-family deubiquitinating enzyme in isolation and in complex with ubiquitin aldehyde, *Cell,* **2002**, *111*, 1041–54.

36 LI, M., CHEN, D., SHILOH, A., LUO, J., NIKOLAEV, A. Y., QIN, J., and GU, W. Deubiquitination of p53 by HAUSP is an important pathway for p53 stabilization, *Nature,* **2002**, *416*, 648–53.

37 WU, K., YAMOAH, K., DOLIOS, G., GAN-ERDENE, T., TAN, P., CHEN, A., LEE, C. G., WEI, N., WILKINSON, K. D., WANG, R., and PAN, Z. Q. DEN1 is a dual function protease capable of processing the C-terminus of Nedd8 deconjugating hyper-neddylated CUL1, *J Biol Chem,* **2003**, *31*, 28882–91.

38 MENDOZA, H. M., SHEN, L. N., BOTTING, C., LEWIS, A., CHEN, J., INK, B., and HAY, R. T. NEDP1, a highly conserved cysteine protease that deNEDDylates Cullins, *J Biol Chem,* **2003**, *278*, 25637–43.

39 LI, S. J. and HOCHSTRASSER, M. A new protease required for cell-cycle progression in yeast, *Nature,* **1999**, *398*, 246–51.

40 MOSSESSOVA, E. and LIMA, C. D. Ulp1-SUMO crystal structure and genetic analysis reveal conserved interactions and a regulatory element essential for cell growth in yeast, *Mol Cell,* **2000**, *5*, 865–76.

41 BORODOVSKY, A., OVAA, H., KOLLI, N., GAN-ERDENE, T., WILKINSON, K. D., PLOEGH, H. L., and KESSLER, B. M. *Chemistry-based functional proteomics reveals novel members of the deubiquitinating enzyme family,* Chem Biol, **2002**, *9*, 1149–59.

42 BALAKIREV, M. Y., TCHERNIUK, S. O., JAQUINOD, M., and CHROBOCZEK, J. *Otubains: a new family of cysteine proteases in the ubiquitin pathway,* EMBO Rep, **2003**, *4*, 517–22.

43 EVANS, P. C., SMITH, T. S., LAI, M. J., WILLIAMS, M. G., BURKE, D. F., HEYNINCK, K., KREIKE, M. M., BEYAERT, R., BLUNDELL, T. L., and KILSHAW, P. J. *A novel type of deubiquitinating enzyme,* J Biol Chem, **2003**, *278*, 23180–6.

44 EVANS, P. C., OVAA, H., HAMON, M., KILSHAW, P. J., HAMM, S., BAUER, S., PLOEGH, H. L., and SMITH, T. S. A20, a regulator of inflammation and cell survival, has deubiquitinating activity, *Biochem J,* **2004**, *Pt 3*, 727–34.

45 Cope, G. A., Suh, G. S., Aravind, L., Schwarz, S. E., Zipursky, S. L., Koonin, E. V., and Deshaies, R. J. Role of predicted metalloprotease motif of Jab1/Csn5 in cleavage of Nedd8 from Cul1, *Science*, **2002**, *298*, 608–11.

46 Verma, R., Aravind, L., Oania, R., McDonald, W. H., Yates, J. R., 3rd, Koonin, E. V., and Deshaies, R. J. Role of Rpn11 metalloprotease in deubiquitination and degradation by the 26S proteasome, *Science*, **2002**, *298*, 611–5.

47 Yao, T. and Cohen, R. E. A cryptic protease couples deubiquitination and degradation by the proteasome, *Nature*, **2002**, *419*, 403–7.

48 Berndt, C., Bech-Otschir, D., Dubiel, W., and Seeger, M. Ubiquitin system: JAMMing in the name of the lid, *Curr Biol*, **2002**, *12*, R815–7.

49 Tran, H. J., Allen, M. D., Lowe, J., and Bycroft, M. Structure of the Jab1/MPN domain and its implications for proteasome function, *Biochemistry*, **2003**, *42*, 11460–5.

50 Ambroggio, X. I., Rees, D. C., and Deshaies, R. J. JAMM: A Metalloprotease-Like Zinc Site in the Proteasome and Signalosome, *PLoS Biol*, **2004**, *2*, E2.

51 Mueller, T. D., Kamionka, M., and Feigon, J. Specificity of the interaction between UBA domains and ubiquitin, *J Biol Chem*, **2004**, *12*, 11926–36.

52 Kang, R. S., Daniels, C. M., Francis, S. A., Shih, S. C., Salerno, W. J., Hicke, L., and Radhakrishnan, I. Solution structure of a CUE-ubiquitin complex reveals a conserved mode of ubiquitin binding, *Cell*, **2003**, *113*, 621–30.

53 Prag, G., Misra, S., Jones, E. A., Ghirlando, R., Davies, B. A., Horazdovsky, B. F., and Hurley, J. H. Mechanism of ubiquitin recognition by the CUE domain of Vps9p, *Cell*, **2003**, *113*, 609–20.

54 Fisher, R. D., Wang, B., Alam, S. L., Higginson, D. S., Robinson, H., Sundquist, W. I., and Hill, C. P. Structure and ubiquitin binding of the

ubiquitin interacting motif, *J Biol Chem*, **2003**, *31*, 28976–84.

55 Beal, R. E., Toscano-Cantaffa, D., Young, P., Rechsteiner, M., and Pickart, C. M. The hydrophobic effect contributes to polyubiquitin chain recognition, *Biochemistry*, **1998**, *37*, 2925–34.

56 Lam, Y. A., DeMartino, G. N., Pickart, C. M., and Cohen, R. E. Specificity of the ubiquitin isopeptid-ase in the PA700 regulatory complex of 26 S proteasomes, *J Biol Chem*, **1997**, *272*, 28438–46.

57 Peng, J., Schwartz, D., Elias, J. E., Thoreen, C. C., Cheng, D., Marsischky, G., Roelofs, J., Finley, D., and Gygi, S. P. A proteomics approach to understanding protein ubiquitination, *Nat Biotechnol*, **2003**, *21*, 921–6.

58 Spence, J., Sadis, S., Haas, A. L., and Finley, D. A ubiquitin mutant with specific defects in DNA repair and multiubiquitination, *Mol Cell Biol*, **1995**, *15*, 1265–73.

59 Hofmann, R. M. and Pickart, C. M. In vitro assembly and recognition of Lys-63 polyubiquitin chains, *J Biol Chem*, **2001**, *276*, 27936–43.

60 Varadan, R., Walker, O., Pickart, C., and Fushman, D. Structural prop-erties of polyubiquitin chains in solu-tion, *J Mol Biol*, **2002**, *324*, 637–47.

61 Varadan, R., Assfalg, M., Haririnia, A., Raasi, S., Pickart, C., and Fushman, D. Solution conformation of Lys63-linked di-ubiquitin chain provides clues to functional diversity of polyubiquitin signaling, *J Biol Chem*, **2003**, *8*, 7055–63.

62 Yin, L., Krantz, B., Russell, N. S., Deshpande, S., and Wilkinson, K. D. Nonhydrolyzable diubiquitin analogues are inhibitors of ubiquitin conjugation and deconjugation, *Biochemistry*, **2000**, *39*, 10001–10.

63 Johnson, E. S., Ma, P. C., Ota, I. M., and Varshavsky, A. A proteolytic pathway that recognizes ubiquitin as a degradation signal, *J Biol Chem*, **1995**, *270*, 17442–56.

64 Li, Z., Na, X., Wang, D., Schoen, S. R., Messing, E. M., and Wu, G.

Ubiquitination of a novel deubiquitinating enzyme requires direct binding to von Hippel-Lindau tumor suppressor protein, *J Biol Chem*, **2002**, *277*, 4656–62.

65 CHEN, X., ZHANG, B., and FISCHER, J. A. A specific protein substrate for a deubiquitinating enzyme: Liquid facets is the substrate of Fat facets, *Genes Dev*, **2002**, *16*, 289–94.

66 COHEN, M., STUTZ, F., and DARGEMONT, C. Deubiquitination, a new player in Golgi to endoplasmic reticulum retrograde transport, *J Biol Chem*, **2003**, *278*, 51989–92.

67 IDEGUCHI, H., UEDA, A., TANAKA, M., YANG, J., TSUJI, T., OHNO, S., HAGIWARA, E., AOKI, A., and ISHIGATSUBO, Y. Structural and functional characterization of the USP11 deubiquitinating enzyme, which interacts with the RanGTP-associated protein RanBPM, *Biochem J*, **2002**, *367*, 87–95.

68 KATO, M., MIYAZAWA, K., and KITAMURA, N. A deubiquitinating enzyme UBPY interacts with the Src homology 3 domain of Hrs-binding protein via a novel binding motif PX(V/I)(D/N)RXXKP, *J Biol Chem*, **2000**, *275*, 37481–7.

69 LI, S. J. and HOCHSTRASSER, M. The Ulp1 SUMO isopeptidase: distinct domains required for viability, nuclear envelope localization, and substrate specificity, *J Cell Biol*, **2003**, *160*, 1069–81.

70 KINNER, A. and KOLLING, R. The yeast deubiquitinating enzyme Ubp16 is anchored to the outer mitochondrial membrane, *FEBS Lett*, **2003**, *549*, 135–40.

71 DUPRE, S. and HAGUENAUER-TSAPIS, R. Deubiquitination step in the endocytic pathway of yeast plasma membrane proteins: crucial role of Doa4p ubiquitin isopeptidase, *Mol Cell Biol*, **2001**, *21*, 4482–94.

72 WANG, Y., SATOH, A., WARREN, G., and MEYER, H. H. VCIP135 acts as a deubiquitinating enzyme during p97-p47-mediated reassembly of mitotic Golgi fragments, *J Cell Biol*, **2004**, *164*, 973–8.

73 TIAN, Q. B., OKANO, A., NAKAYAMA, K., MIYAZAWA, S., ENDO, S., and SUZUKI, T. A novel ubiquitin-specific protease, synUSP, is localized at the post-synaptic density and post-synaptic lipid raft, *J Neurochem*, **2003**, *87*, 665–75.

74 FISCHER-VIZE, J. A., RUBIN, G. M., and LEHMANN, R. The fat facets gene is required for Drosophila eye and embryo development, *Development*, **1992**, *116*, 985–1000.

75 LIU, N., DANSEREAU, D. A., and LASKO, P. Fat facets interacts with vasa in the Drosophila pole plasm and protects it from degradation, *Curr Biol*, **2003**, *13*, 1905–9.

76 ZHU, Y., PLESS, M., INHORN, R., MATHEY-PREVOT, B., and D'ANDREA, A. D. The murine DUB-1 gene is specifically induced by the betac subunit of interleukin-3 receptor, *Mol Cell Biol*, **1996**, *16*, 4808–17.

77 ZHU, Y., LAMBERT, K., CORLESS, C., COPELAND, N. G., GILBERT, D. J., JENKINS, N. A., and D'ANDREA, A. D. DUB-2 is a member of a novel family of cytokine-inducible deubiquitinating enzymes, *J Biol Chem*, **1997**, *272*, 51–7.

78 LOEB, K. R. and HAAS, A. L. The interferon-inducible 15-kDa ubiquitin homolog conjugates to intracellular proteins, *J Biol Chem*, **1992**, *267*, 7806–13.

79 FINLEY, D., OZKAYNAK, E., and VARSHAVSKY, A. The yeast polyubiquitin gene is essential for resistance to high temperatures, starvation, and other stresses, *Cell*, **1987**, *48*, 1035–46.

80 FINLEY, D., BARTEL, B., and VARSHAVSKY, A. The tails of ubiquitin precursors are ribosomal proteins whose fusion to ubiquitin facilitates ribosome biogenesis, *Nature*, **1989**, *338*, 394–401.

81 AMERIK, A. Y., LI, S. J., and HOCHSTRASSER, M. Analysis of the deubiquitinating enzymes of the yeast Saccharomyces cerevisiae, *Biol Chem*, **2000**, *381*, 981–92.

82 LAMMER, D., MATHIAS, N., LAPLAZA, J. M., JIANG, W., LIU, Y., CALLIS, J., GOEBL, M., and ESTELLE, M. Modifica-

tion of yeast Cdc53p by the ubiquitin-related protein rub1p affects function of the SCFCdc4 complex, *Genes Dev*, **1998**, *12*, 914–26.

83 LINGHU, B., CALLIS, J., and GOEBL, M. G. Rub1p processing by Yuh1p is required for wild-type levels of Rub1p conjugation to Cdc53p, *Eukaryot Cell*, **2002**, *1*, 491–4.

84 SWAMINATHAN, S., AMERIK, A. Y., and HOCHSTRASSER, M. The Doa4 deubiquitinating enzyme is required for ubiquitin homeostasis in yeast, *Mol Biol Cell*, **1999**, *10*, 2583–94.

85 AMERIK, A. Y., NOWAK, J., SWAMINATHAN, S., and HOCHSTRASSER, M. The Doa4 deubiquitinating enzyme is functionally linked to the vacuolar protein-sorting and endocytic pathways, *Mol Biol Cell*, **2000**, *11*, 3365–80.

86 AMERIK, A., SWAMINATHAN, S., KRANTZ, B. A., WILKINSON, K. D., and HOCHSTRASSER, M. In vivo disassembly of free polyubiquitin chains by yeast Ubp14 modulates rates of protein degradation by the proteasome, *Embo J*, **1997**, *16*, 4826–38.

87 KAWAKAMI, T., CHIBA, T., SUZUKI, T., IWAI, K., YAMANAKA, K., MINATO, N., SUZUKI, H., SHIMBARA, N., HIDAKA, Y., OSAKA, F., OMATA, M., and TANAKA, K. NEDD8 recruits E2-ubiquitin to SCF E3 ligase, *Embo J*, **2001**, *20*, 4003–12.

88 DANIEL, J. A., TOROK, M. S., SUN, Z. W., SCHIELTZ, D., ALLIS, C. D., YATES, J. R., 3rd, and GRANT, P. A. Deubiquitination of histone H2B by a yeast acetyltransferase complex regulates transcription, *J Biol Chem*, **2004**, *279*, 1867–71.

89 HOEGE, C., PFANDER, B., MOLDOVAN, G. L., PYROWOLAKIS, G., and JENTSCH, S. RAD6-dependent DNA repair is linked to modification of PCNA by ubiquitin and SUMO, *Nature*, **2002**, *419*, 135–41.

90 GROLL, M., DITZEL, L., LOWE, J., STOCK, D., BOCHTLER, M., BARTUNIK, H. D., and HUBER, R. Structure of 20S proteasome from yeast at 2.4 A resolution, *Nature*, **1997**, *386*, 463–71.

91 LAM, Y. A., XU, W., DEMARTINO, G. N., and COHEN, R. E. Editing of ubiquitin conjugates by an isopeptidase in the 26S proteasome, *Nature*, **1997**, *385*, 737–40.

92 GUTERMAN, A. and GLICKMAN, M. H. Complementary roles for Rpn11 and Ubp6 in deubiquitination and proteolysis by the proteasome, *J Biol Chem*, **2004**, *279*, 1729–38.

93 EYTAN, E., ARMON, T., HELLER, H., BECK, S., and HERSHKO, A. Ubiquitin C-terminal hydrolase activity associated with the 26 S protease complex, *J Biol Chem*, **1993**, *268*, 4668–74.

94 LEGGETT, D. S., HANNA, J., BORODOVSKY, A., CROSAS, B., SCHMIDT, M., BAKER, R. T., WALZ, T., PLOEGH, H., and FINLEY, D. Multiple associated proteins regulate proteasome structure and function, *Mol Cell*, **2002**, *10*, 495–507.

95 CHERNOVA, T. A., ALLEN, K. D., WESOLOSKI, L. M., SHANKS, J. R., CHERNOFF, Y. O., and WILKINSON, K. D. Pleiotropic effects of Ubp6 loss on drug sensitivities and yeast prion are due to depletion of the free ubiquitin pool, *J Biol Chem*, **2003**, *278*, 52102–15.

96 BAKER, R. T., TOBIAS, J. W., and VARSHAVSKY, A. Ubiquitin-specific proteases of Saccharomyces cerevisiae. Cloning of UBP2 and UBP3, and functional analysis of the UBP gene family, *J Biol Chem*, **1992**, *267*, 23364–75.

97 TOBIAS, J. W. and VARSHAVSKY, A. Cloning and functional analysis of the ubiquitin-specific protease gene UBP1 of Saccharomyces cerevisiae, *J Biol Chem*, **1991**, *266*, 12021–8.

98 KARIN, M. and BEN-NERIAH, Y. Phosphorylation meets ubiquitination: the control of NF-[kappa]B activity, *Annu Rev Immunol*, **2000**, *18*, 621–63.

99 ZHOU, H., WERTZ, I., O'ROURKE, K., ULTSCH, M., SESHAGIRI, S., EBY, M., XIAO, W., and DIXIT, V. M. Bcl10 activates the NF-kappaB pathway through ubiquitination of NEMO, *Nature*, **2004**, *427*, 167–71.

100 YARON, A., GONEN, H., ALKALAY, I.,

HATZUBAI, A., JUNG, S., BEYTH, S., MERCURIO, F., MANNING, A. M., CIECHANOVER, A., and BEN-NERIAH, Y. Inhibition of NF-kappa-B cellular function via specific targeting of the I-kappa-B-ubiquitin ligase, *Embo J*, 1997, *16*, 6486–94.

101 WANG, C., DENG, L., HONG, M., AKKARAJU, G. R., INOUE, J., and CHEN, Z. J. TAK1 is a ubiquitin-dependent kinase of MKK and IKK, *Nature*, 2001, *412*, 346–51.

102 TANG, E. D., WANG, C. Y., XIONG, Y., and GUAN, K. L. A role for NF-kappaB essential modifier/IkappaB kinase-gamma (NEMO/IKKgamma) ubiquitination in the activation of the IkappaB kinase complex by tumor necrosis factor-alpha, *J Biol Chem*, 2003, *278*, 37297–305.

103 TROMPOUKI, E., HATZIVASSILIOU, E., TSICHRITZIS, T., FARMER, H., ASHWORTH, A., and MOSIALOS, G. CYLD is a deubiquitinating enzyme that negatively regulates NF-kappaB activation by TNFR family members, *Nature*, 2003, *424*, 793–6.

104 BRUMMELKAMP, T. R., NIJMAN, S. M., DIRAC, A. M., and BERNARDS, R. Loss of the cylindromatosis tumour suppressor inhibits apoptosis by activating NF-kappaB, *Nature*, 2003, *424*, 797–801.

105 CIECHANOVER, A. and BRUNDIN, P. The ubiquitin proteasome system in neurodegenerative diseases: sometimes the chicken, sometimes the egg, *Neuron*, 2003, *40*, 427–46.

106 LIU, Y., FALLON, L., LASHUEL, H. A., LIU, Z., and LANSBURY, P. T., JR. The UCH-L1 gene encodes two opposing enzymatic activities that affect alpha-synuclein degradation and Parkinson's disease susceptibility, *Cell*, 2002, *111*, 209–18.

107 LEROY, E., BOYER, R., AUBURGER, G., LEUBE, B., ULM, G., MEZEY, E., HARTA, G., BROWNSTEIN, M. J., JONNALAGADA, S., CHERNOVA, T., DEHEJIA, A., LAVEDAN, C., GASSER, T., STEINBACH, P. J., WILKINSON, K. D., and POLYMEROPOULOS, M. H. The ubiquitin pathway in Parkinson's disease, *Nature*, 1998, *395*, 451–2.

108 LOWE, J., McDERMOTT, H., LANDON, M., MAYER, R. J., and WILKINSON, K. D. Ubiquitin carboxyl-terminal hydrolase (PGP 9.5) is selectively present in ubiquitinated inclusion bodies characteristic of human neurodegenerative diseases, *J Pathol*, 1990, *161*, 153–60.

109 WILSON, S. M., BHATTACHARYYA, B., RACHEL, R. A., COPPOLA, V., TESSAROLLO, L., HOUSEHOLDER, D. B., FLETCHER, C. F., MILLER, R. J., COPELAND, N. G., and JENKINS, N. A. Synaptic defects in ataxia mice result from a mutation in Usp14, encoding a ubiquitin-specific protease, *Nat Genet* 2002, *32*, 420–5.

110 GARCIA-MATA, R., GAO, Y. S., and SZTUL, E. Hassles with taking out the garbage: aggravating aggresomes, *Traffic*, 2002, *3*, 388–96.

111 KAWAGUCHI, Y., KOVACS, J. J., McLAURIN, A., VANCE, J. M., ITO, A., and YAO, T. P. The deacetylase HDAC6 regulates aggresome formation and cell viability in response to misfolded protein stress, *Cell*, 2003, *115*, 727–38.

112 SAIGOH, K., WANG, Y. L., SUH, J. G., YAMANISHI, T., SAKAI, Y., KIYOSAWA, H., HARADA, T., ICHIHARA, N., WAKANA, S., KIKUCHI, T., and WADA, K. Intragenic deletion in the gene encoding ubiquitin carboxy-terminal hydrolase in gad mice, *Nat Genet* 1999, *23*, 47–51.

113 KURIHARA, L. J., SEMENOVA, E., LEVORSE, J. M., and TILGHMAN, S. M. Expression and functional analysis of Uch-L3 during mouse development, *Mol Cell Biol*, 2000, *20*, 2498–504.

114 KURIHARA, L. J., KIKUCHI, T., WADA, K., and TILGHMAN, S. M. Loss of Uch-L1 and Uch-L3 leads to neuro-degeneration, posterior paralysis and dysphagia, *Hum Mol Genet*, 2001, *10*, 1963–70.

115 BORODOVSKY, A., KESSLER, B. M., CASAGRANDE, R., OVERKLEEFT, H. S., WILKINSON, K. D., and PLOEGH, H. L. A novel active site-directed probe specific for deubiquitylating enzymes reveals proteasome association of USP14, *Embo J*, 2001, *20*, 5187–96.

116 HEMELAAR, J., LELYVELD, V. S., KESSLER, B. M., and PLOEGH, H. L. A single protease, Apg4B, is specific for the autophagy-related ubiquitin-like proteins GATE-16, MAP1-LC3, GABARAP, and Apg8L, *J Biol Chem*, **2003**, *278*, 51841–50.

117 RUSSELL, N. S. and WILKINSON, K. D. Identification of a novel 29-linked polyubiquitin binding protein, ufd3, using polyubiquitin chain analogues, *Biochemistry*, **2004**, *43*, 4844–54.

118 MOAZED, D. and JOHNSON, D. A deubiquitinating enzyme interacts with SIR4 and regulates silencing in S. cerevisiae, *Cell*, **1996**, *86*, 667–77.

119 NAVIGLIO, S., MATTECUCCI, C., MATOSKOVA, B., NAGASE, T., NOMURA, N., DI FIORE, P. P., and DRAETTA, G. F. UBPY: a growth-regulated human ubiquitin isopeptidase, *Embo J*, **1998**, *17*, 3241–50.

120 VALERO, R., BAYES, M., FRANCISCA SANCHEZ-FONT, M., GONZALEZ-ANGULO, O., GONZALEZ-DUARTE, R., and MARFANY, G. Characterization of alternatively spliced products and tissue-specific isoforms of USP28 and USP25, *Genome Biol*, **2001**, *2*, RESEARCH0043.

121 LEE, E. G., BOONE, D. L., CHAI, S., LIBBY, S. L., CHIEN, M., LODOLCE, J. P., and MA, A. Failure to regulate TNF-induced NF-kappaB and cell death responses in A20-deficient mice, *Science*, **2000**, *289*, 2350–4.

122 SOARES, L., SEROOGY, C., SKRENTA, H., ANANDASABAPATHY, N., LOVELACE, P., CHUNG, C. D., ENGLEMAN, E., and FATHMAN, C. G. Two isoforms of otubain 1 regulate T cell anergy via GRAIL, *Nat Immunol* **2004**, *5*, 45–54.

123 LI, S. J. and HOCHSTRASSER, M. The yeast ULP2 (SMT4) gene encodes a novel protease specific for the ubiquitin-like Smt3 protein, *Mol Cell Biol*, **2000**, *20*, 2367–77.

124 KIM, K. I., BAEK, S. H., JEON, Y. J., NISHIMORI, S., SUZUKI, T., UCHIDA, S., SHIMBARA, N., SAITOH, H., TANAKA, K., and CHUNG, C. H. A new SUMO-1-specific protease, SUSP1, that is highly expressed in reproduc-

tive organs, *J Biol Chem*, **2000**, *275*, 14102–6.

125 BURNETT, B., LI, F., and PITTMAN, R. N. The polyglutamine neuro-degenerative protein ataxin-3 binds polyubiquitylated proteins and has ubiquitin protease activity, *Hum Mol Genet*, **2003**, *12*, 3195–205.

126 WADA, H., KITO, K., CASKEY, L. S., YEH, E. T., and KAMITANI, T. Cleavage of the C-terminus of NEDD8 by UCH-L3, *Biochem Biophys Res Commun*, **1998**, *251*, 688–92.

127 CABALLERO, O. L., RESTO, V., PATTURAJAN, M., MEERZAMAN, D., GUO, M. Z., ENGLES, J., YOCHEM, R., RATOVITSKI, E., SIDRANSKY, D., and JEN, J. Interaction and colocalization of PGP9.5 with JAB1 and p27(Kip1), *Oncogene*, **2002**, *21*, 3003–10.

128 FIORANI, P., REID, R. J., SCHEPIS, A., JACQUIAU, H. R., GUO, H., THIMMAIAH, P., BENEDETTI, P., and BJORNSTI, M. A. The deubiquitinating enzyme Doa4p protects cells from DNA topoisomerase I poisons, *J Biol Chem*, **2004**, *20*, 21271–81.

129 BREW, C. T. and HUFFAKER, T. C. The yeast ubiquitin protease, Ubp3p, promotes protein stability, *Genetics*, **2002**, *162*, 1079–89.

130 VERMA, R., CHEN, S., FELDMAN, R., SCHIELTZ, D., YATES, J., DOHMEN, J., and DESHAIES, R. J. Proteasomal proteomics: identification of nucleotide-sensitive proteasome-interacting proteins by mass spectrometric analysis of affinity-purified proteasomes, *Mol Biol Cell*, **2000**, *11*, 3425–39.

131 HADARI, T., WARMS, J. V., ROSE, I. A., and HERSHKO, A. A ubiquitin C-terminal isopeptidase that acts on polyubiquitin chains. Role in protein degradation, *J Biol Chem*, **1992**, *267*, 719–27.

132 HONG, S., KIM, S. J., KA, S., CHOI, I., and KANG, S. USP7, a ubiquitin-specific protease, interacts with ataxin-1, the SCA1 gene product, *Mol Cell Neurosci*, **2002**, *20*, 298–306.

133 GNESUTTA, N., CERIANI, M., INNOCENTI, M., MAURI, I., ZIPPEL, R., STURANI, E., BORGONOVO, B.,

Berruti, G., and Martegani, E.
Cloning and characterization of
mouse UBPy, a deubiquitinating
enzyme that interacts with the ras
guanine nucleotide exchange factor
CDC25(Mm)/Ras-GRF1, *J Biol Chem*,
2001, *276*, 39448–54.

134 Wang, G., Sawai, N., Kotliarova, S.,
Kanazawa, I., and Nukina, N. Ataxin-
3, the MJD1 gene product, interacts
with the two human homologs of
yeast DNA repair protein RAD23,
HHR23A and HHR23B, *Hum Mol
Genet*, **2000**, *9*, 1795–803.

9

The 26S Proteasome

Martin Rechsteiner

Abstract

The 26S proteasome is a large ATP-dependent protease composed of more than 30 different polypeptide chains. Like the ribosome, the 26S proteasome is assembled from two "subunits", the 19S regulatory complex and the 20S proteasome. The 19S regulatory complex confers the ability to recognize and unfold protein substrates, and the 20S proteasome provides the proteolytic activities needed to degrade the substrates. The 26S proteasome is the only enzyme known to degrade ubiquitylated proteins, and it also degrades intracellular proteins that have not been marked by ubiquitin. The 26S proteasome is located in the nucleus and cytosol of eukaryotic cells, where the enzyme is responsible for the selective degradation of a vast number of important cellular proteins. Because rapid proteolysis is a pervasive regulatory mechanism, the 26S proteasome is essential for the proper functioning of many physiological processes.

9.1
Introduction

It has become apparent since the mid-1990s that the *u*biquitin–*p*roteasome *s*ystem (UPS) plays a major regulatory role in eukaryotic cells. The UPS helps to control such important physiological processes as the cell cycle [1, 2], circadian rhythms [3], axon guidance [4], synapse formation [5] and transcription [6–8], to name just a few. In view of the growing family of ubiquitin-like proteins [9, 10], it is possible that covalent attachment of ubiquitin and its relatives will even surpass phosphorylation as a regulatory mechanism. Although ubiquitin serves non-proteolytic roles, such as histone modification [11, 12] the activation of cell signaling components [13], endocytosis [14], or viral budding [15], the protein's principal function appears to be targeting proteins for destruction [16]. To do this, the C-terminus of ubiquitin is activated by an ATP-consuming enzyme (E1) and transferred to one of several dozen or more small carrier proteins (E2s) in the form of a reactive thiol ester. The E2s collaborate with members of several large families of ubiquitin li-

Protein Degradation. Edited by J. Mayer, A. Ciechanover, M. Rechsteiner
Copyright © 2005 WILEY-VCH Verlag GmbH & Co. KGaA, Weinheim
ISBN: 3-527-30837-7

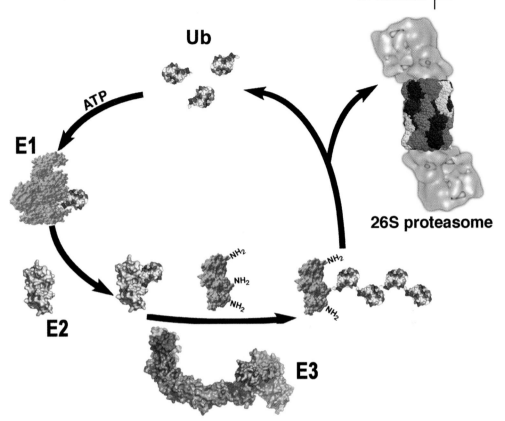

Fig. 9.1. Schematic representation of the ubiquitin–proteasome pathway. Ubiquitin molecules are activated by an E1 enzyme (shown green at 1/3 scale) in an ATP-dependent reaction, transferred to a cysteine residue (yellow) on an E2 or Ub carrier protein and subsequently attached to amino groups (NH₂) on a substrate protein (lysozyme shown in purple) by an E3 or ubiquitin ligase, (the multicolored SCF complex). Note that chains of Ub are generated on the substrate, and these are recognized by the 26S proteasome depicted in the upper right at 1/20 scale.

gases or E3s, and ubiquitin is transferred once again to lysine amino groups on the proteolytic substrates (S) and to itself, thereby generating chains of ubiquitin. The substrate bearing the ubiquitin chains is subsequently recognized and degraded by the 26S proteasome, a large, complex ATP-dependent protease (see Figure 9.1).

Eukaryotic genomes contain information for more than 20 E2s and hundreds of E3s. In contrast to the wealth of components devoted to marking protein substrates for destruction, only one enzyme, the 26S proteasome, has been found to degrade ubiquitylated proteins. However, there is complexity here as well, since the 26S proteasome is an assemblage of at least 30 different subunits. Moreover, there is a growing list of proteins that act as proteasome activators, adapters, or accessory factors. In this chapter I focus on basic biochemical and physiological properties of the 26S proteasome, drawing occasionally from findings on structurally similar prokaryotic, ATP-dependent proteases [17]. Other chapters will provide greater depth to several aspects of the 26S proteasome.

9.2
The 20S Proteasome

9.2.1
Structure

We know the molecular anatomy of archaebacterial, yeast and bovine proteasomes in great detail since high-resolution crystal structures have been determined for all three enzymes [18–20]. The archaebacterial proteasome is composed of two kinds of subunits, called α and β. Each subunit forms heptameric rings that assemble into the 20S proteasome by stacking four deep on top of one another to form a "hollow" cylinder. Catalytically inactive α rings form the ends of the cylinder while proteolytic β subunits occupy the two central rings. The quaternary structure of the 20S proteasome can therefore be described as $\alpha7\beta7\beta7\alpha7$. The active sites of the β-subunits face a large central chamber about the size of serum albumin. The α-rings seal off the central proteolytic chamber and two smaller antechambers from the external solvent. Archaebacterial proteasomes, with their fourteen identical β subunits, preferentially hydrolyze peptide bonds following hydrophobic amino acids and are therefore said to have chymotrypsin-like activity [21]. Eukaryotic proteasomes maintain the overall structure of the archaebacterial enzyme, but they exhibit a more complicated subunit composition. There are seven different α-subunits and at least seven distinct β-subunits arranged in a precise order within their respective rings. Although current evidence indicates that only three of its seven β-subunits are catalytically active, the eukaryotic proteasome cleaves a wider range of peptide bonds, containing, as it does, two copies each of trypsin-like, chymotrypsin-like and post-glutamyl-hydrolyzing subunits. For this reason it is capable of cleaving almost any peptide bond, having difficulty only with proline–X, glycine–X and to a lesser extent with glutamine–X bonds [22].

9.2.2
Enzyme Mechanism and Proteasome Inhibitors

Whereas standard proteases use serine, cysteine, aspartate, or metals to cleave peptide bonds, the proteasome employs an unusual catalytic mechanism. N-terminal threonine residues are generated by self-removal of short peptide extensions from the active β-subunits and act as nucleophiles during peptide-bond hydrolysis [23]. Given its unusual catalytic mechanism, it is not surprising that there are highly specific inhibitors of the proteasome. The fungal metabolite lactacystin and the bacterial product epoxomicin covalently modify the active-site threonines and inhibit the enzyme [24, 25]. Other inhibitors include vinylsulfones [26] and various peptide aldehydes, which are generally less specific. A peptide boronate inhibitor of the proteasome, Velcade, has been approved for the treatment of multiple mye-loma [27].

9.2.3
Immunoproteasomes

Interferon γ (IFNγ) is an immune cytokine that increases expression of a num-ber of cellular components involved in Class I antigen presentation [28]. Among the IFNγ-inducible components are three catalytically active β-subunits of the proteasome, called LMP2, LMP7 and MECL1 [29]. Each replaces its correspond-ing constitutive subunit resulting in altered peptide-bond cleavage preferences of 20S immunoproteasomes. For example, immunoproteasomes exhibit much re-duced cleavage after acidic residues and enhanced hydrolysis of peptide bonds fol-lowing branch-chain amino acids such as isoleucine or valine. Class I molecules preferentially bind peptides with hydrophobic or positive C-termini, and protea-somes generate the vast majority of Class I peptides [28, 30]. Hence, the observed β-subunit exchanges are well suited for producing peptides able to bind Class I molecules.

9.3
The 26S Proteasome

9.3.1
The Ubiquitin–Proteasome System

Bacteria express as many as five ATP-dependent proteases, all of which contain nucleotide-binding domains that belong to the AAA+ family of ATPases [31]. By contrast, the 26S proteasome is the only ATP-dependent protease discovered so far in the nuclear and cytosolic compartments of eukaryotic cells. Because the 20S proteasome's internal cavities are inaccessible to intact proteins, openings must be generated in the enzyme's outer surface for proteolysis to occur. A number of pro-tein complexes have been found to bind the proteasome and stimulate peptide hydrolysis (see Figure 9.2). The most important of the proteasome-associated com-ponents is the 19S regulatory complex (RC) for it is a major part of the 26S ATP-dependent enzyme that degrades ubiquitin-tagged proteins in eukaryotic cells. In Figure 9.2 the 20S proteasome is shown binding only one 19S RC although doubly-capped 26S proteasomes also exist (see Figure 9.3 below). The 20S protea-some also binds activators such as PA28 or PA200. Each of these activators can be present in 26S proteasome complexes forming what are called hybrid proteasomes. Finally a protein called Ecm29p has been found associated with the 26S protea-some. Ecm29p is thought to act as an adapter coupling the 26S enzyme to secre-tory vesicles.

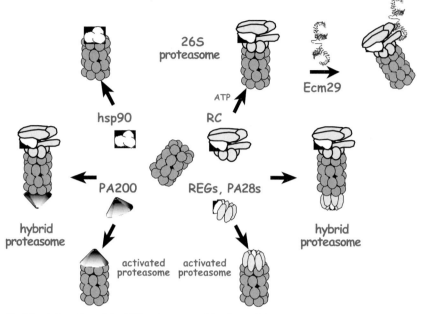

Fig. 9.2. Interaction of the 20S proteasome with other cellular components.

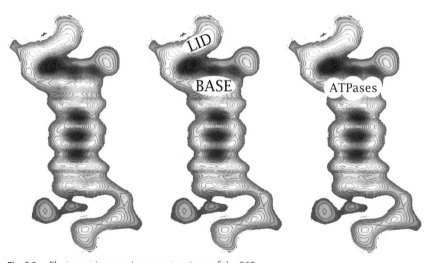

Fig. 9.3. Electron-microscopic reconstructions of the 26S proteasome. Three images of a doubly-capped 26S proteasome are presented to illustrate the positions of the lid and base sub-complexes of the 19S RC and to identify the most probable location of the RC ATPases.

9.3.2
Ultrastructure of the 26S Proteasome and Regulatory Complex

Electron micrographs of purified 26S proteasomes by Baumeister and colleagues [32] reveal a dumbbell-shaped particle approximately 40 nm in length in which the central 20S proteasome cylinder is capped at one or both ends by asymmetric RCs looking much like Chinese dragonheads (Figure 9.3). In doubly capped 26S proteasomes the regulatory complexes face in opposite directions, indicating that contact between the proteasome's α-rings and the RC is highly specific. However, the contacts may not be especially tight since image analysis of Drosophila 26S proteasomes suggests movement of the RCs relative to the 20S proteasome [33]. Electron microscopy (EM) images of the 26S proteasome from several organisms appear similar, indicating that the overall architecture of the enzyme has been conserved in evolution. This conclusion is also supported by sequence conservation among RC subunits (see below).

A yeast mutant lacking the RC subunit Rpn10 contains a salt-labile RC that dissociates into two sub-complexes called the lid and the base [34]. The base contains nine RC subunits, which include six ATPases described below, the two largest RC subunits S1 and S2, and S5a; the lid contains the remaining RC subunits. Thus the RC is composed of two sub-complexes separated on one side by a cavity, i.e. the dragon's mouth (see Figure 9.3). Ultrastructural studies have also been performed on the lid and on a related protein complex called the COP9 signalosome [35]. Both particles lack obvious symmetry. Some particles exhibit a negative stain-filled, central groove; other classes of particles exhibit seven or eight lobes in a disc-like arrangement. Since both the lid and the COP9 signalosome are composed of eight subunits, the lobes may represent individual subunits.

9.3.3
The 19S Regulatory Complex

The regulatory complex is also called the 19S cap, PA700, and the μ particle. As its most commonly used name suggests, the 19S RC is roughly the same size as the 20S proteasome. In fact it is a more complicated protein assembly containing 17 different subunits ranging in size from 25 kDa to about 110 kDa. In animal cells the subunits are designated S1 through S15. Homologs for each of these subunits are present in budding yeast where an alternate nomenclature has been adopted (see Table 9.1). Sequences for the 17 RC subunits permit their classification into a group of 6 ATPases and another group containing the 11 non-ATPases.

9.3.4
ATPases of the RC

The six ATPases belong to the rather large family of AAA ATPases (for *A*TPases *A*ssociated with a variety of cellular *A*ctivities) whose eukaryotic members include the motor protein dynein, the membrane fusion factor NSF, and the chaperone

Tab. 9.1. Subunits of the 19S regulatory complex.

Mammalian nomenclature	Yeast nomenclature	Function	Motifs
S1	Rpn2	Ub/UbL binding	Leu-rich repeats, KEKE
S2	Rpn1	Ub/UbL binding	Leu-rich repeats, KEKE
S3	Rpn3	?	PCI
p55	Rpn5	?	PCI
S4	Rpt2	ATPase	AAA nucleotidase
S5a	Rpn10	polyubiquitin binding	UIM, KEKE
S5b	none	?	
S6	Rpt3	ATPase	AAA nucleotidase
S6'	Rpt5	ATPase	AAA nucleotidase
S7	Rpt1	ATPase	AAA nucleotidase
S8	Rpt6	ATPase	AAA nucleotidase
S9	Rpn6	?	PCI
S10a	Rpn7	?	PCI, KEKE
S10b	Rpt4	ATPase	AAA nucleotidase
S11	Rpn9	?	PCI
S12	Rpn8	?	MPN, KEKE
S13	Rpn11	Isopeptidase	MPN
S14	Rpn12	?	PCI

VCP/Cdc48 and whose prokaryotic members include five ATP-dependent proteases [31]. The six ATPases, denoted S4, S6, S6', S7, S8, and S10b in mammals, are about 400 amino acids in length and homologous to one another. Based on their sequences, one can distinguish three major regions: (1) A central nucleotide-binding domain of about 200 amino acids, which is roughly 60% identical among members of the RC subfamily; (2) the C-terminal region, approximately 100 amino acids in length and with a lesser, though significant, degree of conservation ($\sim$40%); and (3) a highly divergent N-terminal region ($<$20% identity) around 120 amino acids in length; this region contains heptad repeats characteristic of coiled-coil proteins. Despite sequence differences among RC ATPases within an organism, each ATPase has been conserved during evolution with specific subunits being almost 75% identical between yeast and humans. The high degree of conservation encompasses the entire sequence, making it likely that even the divergent N-terminal regions play an important role in RC function. Conceivably they are used to select substrates for degradation by the 26S proteasome [36]. Alternatively the variable N-terminal regions in the RC ATPases may promote assembly of the RC by the specific placement of the ATPase subunits within the complex. In this regard, the six ATPases associate with one another in highly specific pairs: S4 binds S7, S6 binds S8, and S6' binds to S10b. Moreover, the N-terminal regions of RC ATPases are required for partner-specific binding [37].

Staining patterns of two-dimensional gels show the six RC ATPases to be present at comparable levels, and affinity capture of yeast 26S proteasomes indicate the presence of one copy of each in the regulatory complex. Mutational analysis in yeast demonstrates that the ATPases are not functionally redundant since mutation of yeast S4 has a particularly profound effect on peptide hydrolysis [38]. It is probable that the ATPases form a hexameric ring like other members of the AAA family of ATPases such as NSF or VCP/Cdc48. But this assumption has not been experimentally verified, and pentameric or heptameric arrangements of AAA ATPase complexes have been reported [39, 40]. Finally it is quite likely that the ATPases directly bind the α-ring of the 20S proteasome (see Figure 9.3). Evidence favoring this arrangement comes from chemical cross-linking experiments [41] and the presence of the ATPases in the base subcomplex of the yeast 26S proteasome [34].

9.3.5
The non-ATPase Subunits

Whereas the six RC ATPases are homologous and relatively uniform in size, the non-ATPases are heterogeneous in size and sequence. Nonetheless, they can be grouped on the basis of their location, on the presence or absence of certain sequence motifs, and on their affinity for ubiquitin chains or ubiquitin-like proteins. Eight RC subunits are found in the lid subcomplex, each being homologous to a subunit in a separate protein complex called the COP9 Signalosome [42–44]. One of the lid subunits, S13, is a metalloisopeptidase that removes ubiquitin chains from the tagged substrate prior to its translocation into the proteasome for degradation. Six of the eight lid subunits contain PCI domains, stretches of about 200 residues so named from their occurrence in *P*roteasome, *C*op9 signalosome, and the eukaryotic *I*nitiation factor 3 subunits. The PCI domains are thought to mediate subunit–subunit interactions. Two lid subunits contain 140 amino acid-long MPN domains (*M*pr1p and *P*ad1p *N*-terminal regions) with one of these subunits being the S13 isopeptidase. Although several models have been proposed [45–48], the arrangement of subunits within the lid subcomplex is not known. Functionally, only S13 stands out because of its isopeptidase activity. The presence of S13 in the lid explains why the lid is necessary for degradation of ubiquitylated proteins even though the RC base complex supports the ATP-dependent degradation of some small non-ubiquitylated proteins [34]. Interestingly the COP9 signalosome also exhibits isopeptidase activity that removes the ubiquitin-like protein, NEDD8, from certain ubiquitin ligases [49]. Biochemical activity has not been assigned to any of the remaining seven lid subunits, although two lid subunits, S3 and S9, are critical for the degradation of specific substrates [50, 51].

In addition to the six ATPases the base sub-complex contains the two largest RC subunits (S1, S2) and a smaller subunit called S5a [52]. Besides their common location, these three subunits share the property of binding polyubiquitin chains or *ub*iquitin-*l*ike (UbL) domains. S5a binds polyubiquitin chains even after it has been transferred from SDS-PAGE gels and displays many features that match polyubiquitin recognition by the 26S proteasome [52]. However, S5a cannot be the only ubiquitin-recognition component in the 26S proteasome because deletion of the gene encoding yeast S5a (Rpn10) has only a modest impact on proteolysis

[53]. This strongly suggests that there are other ubiquitin-recognition components in the RC, with S1 and S2 being prime candidates. S1 and S2 display significant homology to each other, and both can be modeled as α-helical toroids [54]. They have been shown to bind the UbL domains of RAD23 and Dsk2, adapter proteins that target ubiquitylated proteins to the 26S proteasome [55, 56]. It has also been found that the S6' ATPase can be cross-linked to ubiquitin [57]. Currently then it appears that the RC contains three or possibly four subunits able to recognize ubiquitin or UbL proteins. As discussed below there are other ways in which the RC can select proteins for destruction.

9.3.6
Biochemical Properties of the Regulatory Complex

9.3.6.1 Nucleotide Hydrolysis

Both the 26S proteasome and the RC hydrolyze all four nucleotide triphosphates, with ATP and CTP preferred over GTP and UTP [58]. Although ATP hydrolysis is required for conjugate degradation, the two processes are not strictly coupled. Complete inhibition of the peptidase activity of the 26S proteasome by calpain inhibitor I has little effect on the ATPase activity of the enzyme. The nucleotidase activities of the RC and the 26S proteasome closely resemble those of *E. coli* Lon protease, which is composed of identical subunits that possess both proteolytic and nucleotidase activities in the same polypeptide chain. Like the regulatory complex and 26S proteasome, Lon hydrolyzes all four ribonucleotide triphosphates, but not ADP or AMP [18].

9.3.6.2 Chaperone-like Activity

AAA nucleotidases share the common property of altering the conformation or association state of proteins, so it is not surprising that the RC has been shown to prevent aggregation of several denatured proteins including citrate synthase and ribonuclease A [59–61]. The chaperone activity of the RC may explain why the RC plays a role in transcription apparently in the absence of an attached 20S proteasome [62].

9.3.6.3 Proteasome Activation

The 20S proteasome is a latent protease owing to the barrier imposed by the α-subunit rings on peptide entry. Consequently, a readily measured activity of the RC is activation of fluorogenic peptide hydrolysis by the 20S proteasome. The extent of activation is generally found to be in the range 3- to 20-fold [63]. Activation is relatively uniform for all three proteasome catalytic subunits and presumably reflects opening by the attached RC of a channel leading to the proteasome's central chamber.

9.3.6.4 Ubiquitin Isopeptide Hydrolysis

The channel through the proteasome α-ring into the central chamber measures 1.3 nm in diameter, a size too small to permit passage of a folded protein, even one as small as ubiquitin. This consideration, coupled with the fact that ubiquitin is recycled intact upon substrate degadation, requires an enzyme to remove the polyubiquitin chain prior to or concomitant with proteolysis. Several isopeptidases that remove ubiquitin from substrates have been found associated with the 26S proteasome [64–66]. Of these the ATP-stimulated metalloisopeptidase S13 is an integral component of the enzyme.

9.3.6.5 Substrate Recognition
It is clear that the RC plays a predominant role in selecting proteins for degradation. This important topic is covered below in the context of substrate recognition by both 20S and 26S proteasomes.

9.4
Substrate Recognition by Proteasomes

9.4.1
Degradation Signals (Degrons)

The discovery that proteins possess built-in signals targeting them to specific locations within cells was a major success of twentieth-century cell biology [67]. Selective proteolysis can be considered targeting out of existence, and a number of short peptide motifs have been discovered to confer rapid destruction on proteins that bear them. These motifs include PEST sequences [68], the N-terminal amino acid [69], and destruction and KEN boxes [70]. These motifs are recognized by one or more ubiquitin ligases that mark the substrate protein by addition of a polyubiquitin chain. However some proteins are degraded by the 26S proteasome without prior marking by ubiquitin [71]. Denatured proteins are also selectively degraded by both 26S and 20S proteasomes [72]. It is not clear what features of denatured proteins are recognized by proteasomes or by components of the ubiquitin proteolytic system.

9.4.2
Ubiquitin-dependent Recognition of Substrates

Most well characterized substrates of the 26S proteasome are ubiquitylated proteins so our discussion starts with them. Ubiquitin contains seven lysine residues, and proteomic studies in yeast indicate that chains (or dimers) can be formed using any of them [73]. In higher eukaryotes polyubiquitin chains formed via Lys6, Lys27, Lys29, Lys48, and Lys63 have been observed. Lys6 chains are formed by BRCA/BARD heterodimers where they presumably play a role in DNA repair [74]. Ubiquitin monomers linked to each other through Lys63 are involved in endocytosis and DNA repair, but not apparently in targeting proteins to the 26S proteasome [75, 76]. Lys27 chains have been found on the co-chaperone BAG1, and they target degradation of misfolded proteins bound by the Hsp70 chaperone to the 26S proteasome [77]. Lys29 and Lys48 chains form directly on proteolytic substrates and target them for destruction [78, 79].

Efficient proteolysis of ubiquitylated proteins by the 26S proteasome requires a chain containing at least four ubiquitin monomers [80]. This matches well the ubiquitin-binding characteristics of the RC subunit S5a. It too selectively binds ubiquitin polymers composed of four or more ubiquitin moieties and exhibits increased affinity for longer chains [52]. S5a molecules from a number of higher eukaryotes contain two independent polyubiquitin-binding sites; each is approximately 30 residues long and characterized by five hydrophobic residues that consist of alternating large and small hydrophobic residues, e.g. Leu–Ala–Leu–Ala–Leu [81]. Similar motifs have been found in other proteins of the ubiquitin system and are now called UIMs, (*U*biquitin *I*nteracting *M*otifs) [82]. Two recent NMR studies have demonstrated direct interaction between UIMs and the hydrophobic patch on ubiquitin [83–85]. Whereas S5a provides for direct recognition of polyubiquitylated substrates, a second mechanism involves adapter proteins possessing both a UbL domain and one or more UbA (*ub*iquitin *a*ssociated) domains [86]. UbA domains are polyubiquitin-binding domains found in several presumed adapter proteins of the ubiquitin system. The adapter proteins include RAD23 and Dsk2 in yeast and recruit substrates to the 26S proteasome. The UbL domains of these proteins bind to 26S proteasome subunits while their UbA domains bind substrate-tethered Ub chains. In yeast the RC subunits S1 and S2 serve as UbL-binding components [55, 87]; in mammals S5a serves this purpose [88–90].

The co-chaperone BAG1 illustrates a third way in which polyubiquitin can target substrate proteins to the 26S proteasome. In this case the substrate is not polyubiquitylated; rather it is bound to the chaperone Hsp70. A polyubiquitin chain, linked through Lys27, is attached to the Hsp70-associated co-chaperone BAG1 [77]. Apparently the Lys27 chain promotes association of the chaperone–substrate complex with the 26S proteasome, after which the substrate is degraded while BAG1, Hsp70, and ubiquitin are recycled. Direct interaction between E3 ubiquitin ligases and RC subunits can also deliver ubiquitylated substrates to the protease. The yeast E3 ligase called UFD4 binds RC ATPases [91] and UFD4-mediated delivery of substrates bypasses the requirement for S5a. In what appears to be a similar delivery system, the mammalian E3 Parkin uses a UbL domain to bind the 26S proteasome [89], and the E3 component pVHL binds a 26S proteasome ATPase [92]. Mutational analyses in yeast have shown that whereas deletion of either S5a or Rad23 has a mild impact on proteolysis, loss of both proteins produces a severe phenotype [93]; furthermore, yeasts lacking S5a, RAD23 and Dsk2 are not viable, indicating that direct delivery by E3 ligases cannot compensate for the absence of all three proteins.

9.4.3
Substrate Selection Independent of Ubiquitin

The 26S proteasome also degrades non-ubiquitylated proteins [71]. The short-lived enzyme ornithine decarboxylase (ODC) and the cell-cycle regulator p21Cip provide well documented examples of ubiquitin-independent proteolysis by the 26S enzyme [94, 95]. ODC degradation is stimulated by antizyme, a polyamine-induced protein that binds both ODC and the 26S proteasome. Antizyme functions as an adapter much like RAD23 and Dsk2 except that polyubiquitin chains are not involved. However, free ubiquitin chains do compete with antizyme–ODC for degradation [96]. Other potential adapters, such as gankyrin, may target proteins to the 26S proteasome in the absence of ubiquitin marking [97].

The CDK inhibitor p21Cip is degraded in a nonubiquitin-dependent reaction, as clearly demonstrated by substitution of arginines for all the lysine residues in p21Cip. These modifications prevented ubiquitylation of p21Cip, but the lysine-less protein was still degraded in human fibroblasts by the proteasome [95]. The C-terminal region of p21Cip binds to the proteasome α-subunit C8, and *in vitro* p21Cip is degraded by the 20S proteasome alone [98, 99]. Direct binding of p21Cip to the 20S proteasome may open a channel through the α-ring allowing the loosely structured protein to enter the central proteolytic chamber. c-Jun, "aged" calmodulin, troponin C, and p53 are other proteins that can be degraded by the 26S proteasome absent marking by ubiquitin [71]. Thus, other 20S proteasome substrates, *in vitro* at least, include oxidized proteins, small, denatured proteins and loosely folded proteins such as casein. Whether the 20S proteasome degrades proteins within cells is an unresolved problem.

9.5
Proteolysis by the 26S Proteasome

9.5.1
Presumed Mechanism

Proteolysis of ubiquitylated proteins by the 26S proteasome can be thought to consist of seven steps: (1) chaperone-mediated substrate presentation; (2) substrate association with RC subunits; (3) substrate unfolding; (4) detachment of polyubiquitin from the substrate; (5) translocation of the substrate into the 20S proteasome central chamber; (6) peptide bond cleavage; and (7) release of peptide products as well as polyubiquitin (see Figure 9.4). Step 1 is optional depending on the substrate, and in principle steps 3 and 4 could occur in either order. Step 4 is unnecessary for substrates like ODC that are not ubiquitylated. The other steps almost have to occur as presented. Although it is easy to conceptualize the reaction sequence, few experimental findings bear directly on any of the proposed subreactions, and virtually nothing is known about molecular movements within the 26S proteasome. However several studies on prokaryotic ATP-dependent proteases permit some informed speculation, and it has been shown that step 6 is not required for sequestration of ODC by the 26S proteasome [100].

9.5.2
Contribution of Chaperones to Proteasome-mediated Degradation

Chaperones are connected to proteasomes in at least four ways. First, chaperones can deliver substrates to the proteasome as described above for the co-chaperone

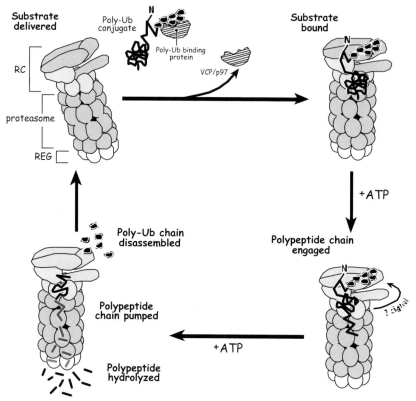

Fig. 9.4. Hypothetical reaction cycle for the 26S proteasome. A polyubiquitylated substrate is delivered to a 26S hybrid proteasome in some cases by chaperones such as VCP/p97 (step 1). Substrate is bound by polyubiquitin recognition components within the regulatory complex (RC) until the polypeptide chain is engaged by the ATPases (step 2). As the polypeptide chain is unfolded and presumably pumped down the central pore of the proteasome, a signal is conveyed to the S13 metallo-isopeptidase to remove the polyubiquitin chain (step 3). The unfolded polypeptide is eventually degraded within the inner chamber of the proteasome (step 4) and peptide fragments exit the enzyme.

BAG1 [77]. In a similar fashion the chaperone VCP/Cdc48 is required for the degradation of several ubiquitin-pathway substrates. VCP, a member of the AAA family of ATPases, is a large hexameric ATPase that appears to function as a protein separase able to remove ubiquitylated monomers from multisubunit complexes [101–104]. In some cases the liberated proteins are degraded by the 26S proteasome; in other cases the separated proteins may change their intracellular location. The proteasome also degrades proteins embedded in the endoplasmic reticulum (ER) membrane. If these ER membrane proteins possess a large cytoplasmic domain, their proteasomal degradation can require Hsps 40, 70, and 90 as well as VCP [105, 106]. Hsp90 is required to assemble and stabilize the yeast 26S proteasome providing a third connection between chaperones and proteasomes; Hsp90 is also able to bind and suppress peptide hydrolysis by the 20S proteasome (see below). Finally both chaperones and proteasomes are induced by the accumulation of denatured proteins within eukaryotic cells.

9.5.2.1 Substrate Binding to the 26S Proteasome

Although chaperones may provide for the recognition of some 26S proteasome substrates [107], there is little doubt that the 26S proteasome recognizes substrates directly. As mentioned, ODC is the best-characterized substrate recognized by the 26S proteasome in the absence of a polyubiquitin chain [96]. Which RC subunits actually recognize ODC–antizyme complexes has not been discovered. Presumably one or more subunits in the RC recognize both the C-terminal degron in ODC and some feature of antizyme. The apparent dual recognition of elements in ODC and antizyme may reflect a general mechanism by which ATP-dependent proteases process substrates. For example a number of studies on substrate recognition by *E. coli* ATP-dependent proteases indicate that substrate adapters provide one recognition site while the substrate provides another [108–110]. Similarly, for the 26S proteasome one recognition element, the attached polyubiquitin chains, may be seen by RC subunits S5a or S2, while the substrate's N- or C-termini are recognized by RC ATPases. Interestingly, the location of the polyubiquitin chain on the substrate can affect rates of degradation as much as five-fold [111] perhaps by altering the rates at which the RC ATPases engage substrate termini.

9.5.2.2 Translocation of the Polypeptide Substrate to the Central Proteolytic Chamber

Translocation is thought to proceed by the six ATPases threading the polypeptide through a channel in the 20S proteasome's α-ring. It is also thought that the RC ATPases processively unravel substrates from degrons within the polypeptide chain and are able to "pump" the polypeptide chain in either the N-terminal to C-terminal direction or the opposite [112–114]. Several studies have estimated that hundreds of ATP molecules are needed to degrade small to medium-sized proteins [115, 116]. Rates of translocation range between 10 and 30 amino acids per second [116, 117]. These values are similar to DNA helicases where rates of 50 bases per second and 1ATP per base have been reported [118]. Current models would suggest that the RC ATPases hydrolyze ATP in a sequential rotary fashion essentially screwing the polypeptide chain into the 20S proteasome [119, 120], but the possibility that convulsive movements transfer the substrate has not been ruled out. The ATPases may even be capable of transferring loops into the 20S enzyme. However, a proteomic screen for ClpXP substrates revealed that degrons were either C-terminal or N-terminal [121], so it is likely that the RC ATPases usually engage polypeptide termini.

The peptide fragments generated in the central chamber are generally 5 to 10 residues in length, but fragments as long as 35 amino acids can be present [64]. How these fragments exit the central chamber is not known. They could diffuse back through the RC, through small side panels in the 20S proteasome or in the case of hybrid proteasomes (see below) through the end capped by a proteasome activator.

9.5.3
Processing by the 26S Proteasome

In some cases the 26S proteasome partially degrades the substrate protein, releasing processed functional domains. The best-studied example of processing involves the transcriptional activator NFκB. The C-terminal half of a 105-kDa precursor is degraded by the 26S proteasome to yield a 50-kDa N-terminal domain that is the active transcription component [122]. A glycine-rich stretch of amino acids at the C-terminal boundary of p50 is an important factor in limiting proteolysis [123]. It is possible that polypeptide translocation by the RC starts at the Gly-rich region and proceeds in only one direction owing to the presence of the tightly folded N-terminal domain. Or the RC may start translocation at the C-terminus and stop when the ATPases encounter the Gly-rich region. Insertion of a Gly–Ala stretch as small as seven residues into the C-terminal degron of ODC is sufficient to prevent complete destruction of ODC leading instead to partial processing of the enzyme. This has led Coffino and colleagues to suggest that the Gly–Ala stretch impairs substrate transfer by the RC ATPases [124]. Another example of partial processing involves SPT23, a yeast protein embedded in the endoplasmic reticulum membrane [125]. SPT23 controls unsaturated fatty acid levels, and membrane fluidity regulates 26S proteasomal generation of a freely diffusible transcription factor from the SPT23 precursor. Partial processing may be a more widespread regulatory mechanism than is currently thought.

9.6
Proteasome Biogenesis

9.6.1
Subunit Synthesis

The synthesis of proteasome subunits is markedly affected by proteasome function. For example, inhibition of proteasome activity by lactacystin induces coordinate expression of both RC and 20S proteasome subunits [66]. Likewise, impaired synthesis of a given RC subunit results in over-expression of all RC subunits [126, 127]. Proteasome subunit synthesis in yeast is controlled by Rpn4p, a short-lived positive transcription factor that binds PACE elements upstream of proteasome genes [128]. Rpn4p is a substrate of the 26S proteasome suggesting that the transcription factor functions in a feedback loop in which proteasome activity limits its concentration thereby regulating proteasome levels [129]. To date an Rpn4-like factor has not been identified in higher eukaryotes, but such a factor is likely to exist.

9.6.2
Biogenesis of the 20S Proteasome

Proteasome β-subunits are synthesized with N-terminal extensions and are inactive because a free N-terminal threonine is required for peptide-bond hydrolysis [130]. The precursor β-subunits assemble with α-subunits to form half proteasomes composed of one α- and one β-ring, which then dimerize to form the 20S particle [131]. The N-terminal extensions are removed thereby generating a new unblocked N-terminal threonine in the catalytically active β-subunits. A small accessory protein called Ump1 in yeast or proteassemblin in mammalian cells assists in the final assembly of the 20S proteasome [132]. Interestingly Ump1/POMP is apparently trapped in the proteasome's central chamber and degraded upon maturation of the enzyme [133].

9.6.3
Biogenesis of the RC

Assembly pathways for the RC are virtually unknown. As mentioned above, the ATP-ases interact with one another and complexes containing all six S4 subfamily members have been observed following *in vitro* synthesis. Impaired synthesis of the yeast lid subunit Rpn6 results in the absence of the entire lid [134], so presumably lid and base subcomplexes assemble independently and associate in the final stages of RC formation cells. In mammalian cells, 26S proteasomes assemble from preformed regulatory complexes and 20S proteasomes [135].

9.6.4
Post-translational Modification of Proteasome Subunits

Proteasome and RC subunits are subjected to a variety of post-translational modifications including phosphorylation, acetylation, myristoylation, and even O-glycosylation [136–140]. In yeast all seven α-subunits are acetylated as well as two β-subunits. Since acetylation of the N-terminal threonine in an active β-subunit would poison catalysis, it has been suggested that the propeptide extensions function to prevent acetylation [130]. Three members of the S4 ATPase subfamily (S4, S6, and S10b) and two 20S α-subunits (C8 and C9) are known to be phosphorylated. Phosphorylation appears to be particularly important for 26S proteasome assembly and stability. The kinase inhibitor staurosporine reduces 26S proteasome levels in mouse lymphoma cells [135] and interferon γ results in reduced phosphorylation of 20S proteasome α-subunits and decreased 26S proteasome levels [141].

9.6.5
Assembly of the 26S Proteasome

The RC and 20S proteasome associate to form the 26S proteasome in the presence of ATP [63]. Comparison of the cross-linking patterns of RC and assembled 26S proteasomes indicates that this association is accompanied by subunit rearrangement [142]. In yeast two proteins play a special role in 26S proteasome assembly or stability. Nob1p is a nuclear protein required for biogenesis of the 26S proteasome and is degraded following assembly of the 26S enzyme. Thus Nob1p suffers the same fate as Ump1 does following 20S maturation [143]. In fission yeast the protein Yin6 regulates both the nuclear localization and the stability of the 26S proteasome [144]. In budding yeast the chaperone Hsp90 also plays a role in the assembly and maintenance of yeast 26S proteasomes, since functional loss of Hsp90 results in 26S proteasome dissociation, indeed even dissociation of the lid subcomplex [145]. The 26S proteasome also dissociates into RC and 20S proteasomes when budding yeast is subjected to long periods of starvation [146].

9.7
Proteasome Activators

In addition to the RC there are two protein complexes, REGαβ and REGγ, and a single polypeptide chain, PA200, that bind the 20S proteasome and stimulate peptide hydrolysis but not protein degradation. Like the RC, proteasome activators bind the ends of the 20S proteasome and, importantly, they can form mixed or hybrid 26S proteasomes in which one end of the 20S proteasome is associated with a 19S RC and the other is bound to a proteasome activator [147–150]. This latter property raises the possibility that proteasome activators serve to localize the 26S proteasome within eukaryotic cells.

9.7.1
REGs or PA28s

9.7.1.1 REGs
There are three distinct REG subunits called αβγ [151, 152]. REGsα and β form donut-shaped heteroheptamers found principally in the cytoplasm, whereas REGγ forms a homoheptamer located in the nucleus. REGαβ is abundantly expressed in immune tissues, while REGγ expression is highest in brain. The REGs also differ in their activation properties. REGαβ activates all three proteasome active sites; REGγ only activates the trypsin-like subunit. There is reasonably solid evidence that REGαβ plays a role in Class I antigen presentation [153], but we have little knowledge concerning REGγ function since REGγ knockout mice have almost no phenotype [154]. The crystal structure of REGα reveals that the seven subunits form a donut-shaped structure with a central aqueous channel, and the structure of a REG–proteasome complex provides important insight into the mechanism by which REGα activates the proteasome. The carboxyl tail on each REG subunit fits into a corresponding cavity on the α-ring of the proteasome. Loops on the REG subunits displace N-terminal strands on several proteasome α-subunits reorienting them upward into the aqueous channel of the REG heptamer thereby opening a continuous channel from the exterior solvent to the proteasome central chamber [155].

9.7.1.2 PA200
A new proteasome activator called PA200 was recently purified from bovine testis [156]. Human PA200 is a nuclear protein of 1843 amino acids that activates all three catalytic subunits with some preference for the PGPH active site. Homologs

of PA200 are present in budding yeast, worms, and plants. A single chain of PA200 can bind each end of the proteasome and, when bound, PA200 molecules look like volcanos in negatively stained EM images. PA200 is thought to play a role in DNA repair, perhaps by recruiting proteasomes to DNA double-strand breaks.

9.7.1.3 Hybrid Proteasomes

As the α-rings at each end of the 20S proteasome are equivalent, the 20S proteasome is capable of binding two RCs, two PA28s, two PA200s, or combinations of these components. In fact 20S proteasomes simultaneously bound to RC and PA28 or PA200 have been observed, and are called hybrid proteasomes [147, 148, 150]. In HeLa cells the levels of hybrid proteasomes containing PA28 at one end and an RC at the other are two-fold higher than 26S proteasomes capped at both ends by 19S RCs [149]. Hybrid 26S proteasomes containing PA200 appear to be much less abundant in HeLa cells [156].

9.7.2
ECM29

Another proteasome-associated protein, called Ecm29, has been identified in several proteomic screens [157, 158]. Ecm29p clearly associates with 26S proteasomes; whether it activates proteasomal peptide hydrolysis is currently unknown. Ecm29p is reported to stabilize the yeast 26S proteasome by clamping the RC to the 20S cylinder [159]. However, in mammalian cells Ecm29p is found mainly associated to secretory and endocytic organelles, a location suggesting a role in secretion rather than 26S proteasome stability. Moreover, the levels of Ecm29p vary markedly among mouse organs, so if mammalian Ecm29p serves as a clamp, some tissues either do not require a clamped 26S proteasome or other proteins function as RC-20S clamps.

9.8
Protein Inhibitors of the Proteasome

A number of proteins have been found to suppress proteolysis by the proteasome. One of these is PI31 [160, 161]; another is the abundant cytosolic chaperone Hsp90 [162], and a third is a proline/arginine-rich 39-residue peptide called PR39 [163]. Both PI31 and Hsp90 may affect how the proteasome functions in Class I antigen presentation. PI31 is a 30-kDa proline-rich protein that inhibits peptide hydrolysis by the 20S proteasome and can block activation by both RC and REGαβ. Although surveys of various cell lines show PI31 to be considerably less abundant than RC or REGαβ, when over-expressed, PI31 is reported to inhibit Class I antigen presentation by interfering with the assembly of immuno-proteasomes [161]. A number of studies have shown that Hsp90 can bind the 20S proteasome and inhibit its chymotrypsin-like and PGPH activities. Interestingly inhibition by Hsp90 is observed with constitutive but not with immuno-proteasomes, a finding consistent with proposals that Hsp90 shuttles immuno-proteasome-generated peptides to the endoplasmic reticulum for Class I presentation [164]. PR39 was originally isolated from bone marrow as a factor able to induce angiogenesis and inhibit inflammation. Two hybrid screens showed that PR39 binds the 20S proteasome. Apparently PR39 affects angiogenesis and inflammation by inhibiting respectively the degradation of HIF1 or IκBα, the latter being an inhibitor of NFκB. Finally HIV's Tat protein inhibits the 20S proteasome's peptidase activity [165]. Tat also competes with REG$\alpha\beta$ for proteasome binding and, by doing so, Tat can inhibit Class I presentation of certain epitopes [166].

9.9
Physiological Aspects

9.9.1
Tissue and Subcellular Distribution of Proteasomes

Proteasomes are found in all organs of higher eukaryotes, but the degree to which the composition of proteasomes and its activators varies among tissues is largely unexplored territory. Proteasomes are very abundant in testis which contains almost five-fold more 20S subunits than skeletal muscle [167]. At the cellular level there are about 800 000 proteasomes in a HeLa cell and roughly 20 000 proteasomes in a yeast cell. At the subcellular level, 26S proteasomes are present in cytosol and nucleus where they appear to be freely diffusible [168, 169]. They are not usually found in the nucleolus [170] and have not been reported in membrane-bound organelles other than the nucleus. When large amounts of misfolded proteins are synthesized by a cell, the aberrant polypeptides often accumulate around the centrosome in what are called "aggresomes" [171]. Under these conditions 26S proteasomes, chaperones, and proteasome activators also redistribute to the aggresomes, presumably to refold and/or degrade the misfolded polypeptides [172].

9.9.2
Physiological Importance

Deletion of yeast genes encoding 20S proteasome and 19S RC subunits is usually lethal, indicating that the 26S proteasome is required for eukaryotic cell viability. Known substrates of the 26S proteasome include transcription factors, cell-cycle regulators, protein kinases, etc., essentially most of the cell's important regulatory proteins. Even proteins secreted into the endoplasmic reticulum are returned to the cytosol for degradation by the 26S proteasome [173, 174]. Given the scope of its substrates it is hardly surprising that in higher eukaryotes the ubiquitin–proteasome system contributes to the regulation of a vast array of physiological processes ranging from cell-cycle traverse to circadian rhythms to learning. Discussion of these fascinating regulatory mechanisms is covered in other chapters of this volume.

Summary

The 20S proteasome was discovered in 1980 and the 26S proteasome six years later. Research since the mid-1980s has made it abundantly clear that the ubiquitin–proteasome system is of central importance in eukaryotic cell physiology. Yet there is much more to discover. A crystal structure of the 19S RC, or better still the 26S proteasome, would surely provide insight into the mechanism by which the 26S proteasome degrades its substrates. How the 26S proteasome is itself regulated and the extent to which proteasomal components vary among tissues in higher eukaryotes are other important unresolved problems. Hopefully, these and other unanswered questions will spark further interest in the proteasome among readers of this book on Intracellular Proteolysis.

References

1 REED, S. I. Ratchets and clocks: the cell cycle, ubiquitylation and protein turnover. *Nat Rev Mol Cell Biol* **2003**, *4*, 855–64.

2 BASHIR, T. and PAGANO, M. Aberrant ubiquitin-mediated proteolysis of cell cycle regulatory proteins and oncogenesis. *Adv Cancer Res* **2003**, *88*, 101–44.

3 VIERSTRA, R. D. The ubiquitin/26S proteasome pathway, the complex last chapter in the life of many plant proteins. *Trends Plant Sci* **2003**, *8*, 135–42.

4 CAMPBELL, D. S. and HOLT, C. E. Chemotropic responses of retinal growth cones mediated by rapid local protein synthesis and degradation. *Neuron* **2001**, *32*, 1013–1026.

5 HEGDE, A. N. and DiANTONIO, A. Ubiquitin and the synapse. *Nat Rev Neurosci* **2002**, *3*, 854–61.

6 MURATANI, M. and TANSEY, W. P. How the ubiquitin-proteasome system controls transcription. *Nat Rev Mol Cell Biol* **2003**, *4*, 192–201.

7 CONAWAY, R. C., BROWER, C. S., and CONAWAY, J. W. Emerging roles of ubiquitin in transcription regulation. *Science* **2002**, *296*, 1254–8.

8 LIPFORD, J. R. and DESHAIES, R. J. Diverse roles for ubiquitin-dependent proteolysis in transcriptional activation. *Nat Cell Biol* **2003**, *5*, 845–50.

9 SEELER, J. S. and DEJEAN, A. Nuclear and unclear functions of SUMO. *Nat Rev Mol Cell Biol* **2003**, *4*, 690–9.

10 SCHWARTZ, D. C. and HOCHSTRASSER, M. A superfamily of protein tags: ubiquitin, SUMO and related modifiers. *Trends Biochem Sci* **2003**, *28*, 321–8.

11 JASON, L. J., MOORE, S. C., LEWIS, J. D., LINDSEY, G., and AUSIO, J. Histone ubiquitination: a tagging tail unfolds? *Bioessays* **2002**, *24*, 166–74.

12 ZHANG, Y. Transcriptional regulation by histone ubiquitination and deubiquitination. *Genes Dev* **2003**, *17*, 2733–40.

13 DENG, L. et al. Activation of the IkappaB kinase complex by TRAF6 requires a dimeric ubiquitin-conjugating enzyme complex and a unique polyubiquitin chain. *Cell* **2000**, *103*, 351–361.

14 HICKE, L. and DUNN, R. Regulation of membrane protein transport by ubiquitin and ubiquitin-binding proteins. *Annu Rev Cell Dev Biol* **2003**, *19*, 141–72.

15 PORNILLOS, O., GARRUS, J. E., and SUNDQUIST, W. I. Mechanisms of enveloped RNA virus budding. *Trends Cell Biol* **2002**, *12*, 569–79.

16 HERSHKO, A., CIECHANOVER, A., and VARSHAVSKY, A. Basic Medical Research Award. The ubiquitin system. *Nat Med* **2000**, *6*, 1073–81.

17 Kessel, M. et al. Homology in structural organization between E. coli ClpAP protease the eukaryotic 26S proteasome. *J Mol Biol* **1995**, *250*, 587–594.

18 Lowe, J. et al. Crystal structure of the 20S proteasome from the archaeon T. acidophilum at 3.4 A resolution. *Science* **1995**, *268*, 533–539.

19 Groll, M. et al. Structure of 20S proteasome from yeast at 2.4 A resolution. *Nature* **1997**, *386*, 463–471.

20 Unno, M. et al. The structure of the mammalian 20S proteasome at 2.75 A resolution. *Structure (Camb)* **2002**, *10*, 609–18.

21 Dahlmann, B. et al. The multicatalytic proteinase (prosome, proteasome): comparison of the eukaryotic and archaebacterial enzyme. *Biomed Biochim Acta* **1991**, *50*, 465–9.

22 Harris, J. L., Alper, P. B., Li, J., Rechsteiner, M., and Backes, B. J. Substrate specificity of the human proteasome. *Chem Biol* **2001**, *8*, 1131–41.

23 Seemuller, E. et al. Proteasome from Thermoplasma acidophilum: a threonine protease. *Science* **1995**, *268*, 579–582.

24 Fenteany, G. et al. Inhibition of proteasome activities and subunit-specific amino-terminal threonine modification by lactacystin. *Science* **1995**, *268*, 726–731.

25 Meng, L. et al. Epoxomicin, a potent and selective proteasome inhibitor, exhibits *in vivo* anti-inflammatory activity. *Proc. Natl. Acad. Sci. USA* **1999**, *96*, 10403–10408.

26 Bogyo, M. et al. Covalent modification of the active site threonine of proteasomal *β* subunits and the *Escherichia coli* homolg HsIV by a new class of inhibitors. *Proc. Natl. Acad. Sci. USA* **1997**, *94*, 6629–6634.

27 Kane, R. C., Bross, P. F., Farrell, A. T., and Pazdur, R. Velcade: U.S. FDA approval for the treatment of multiple myeloma progressing on prior therapy. *Oncologist* **2003**, *8*, 508–13.

28 Yewdell, J. W. and Hill, A. B. Viral interference with antigen presentation. *Nat Immunol* **2002**, *3*, 1019–25.

29 Goldberg, A. L., Cascio, P., Saric, T., and Rock, K. L. The importance of the proteasome and subsequent proteolytic steps in the generation of antigenic peptides. *Mol Immunol* **2002**, *39*, 147–64.

30 Rammensee, H.-G., Falk, K., and Rötzschke, O. Peptides naturally presented by MHC class I molecules. *Annu. Rev. Immunol.* **1993**, *11*, 213–244.

31 Dougan, D. A., Mogk, A., Zeth, K., Turgay, K., and Bukau, B. AAA+ proteins and substrate recognition, it all depends on their partner in crime. *FEBS Lett* **2002**, *529*, 6–10.

32 Peters, J. M., Cejka, Z., Harris, J. R., Kleinschmidt, J. A., and Baumeister, W. Structural features of the 26S proteasome complex. *J Mol Biol* **1993**, *234*, 932–937.

33 Walz, J. et al. 26S proteasome structure revealed by three-dimensional electron microscopy. *J Struct Biol* **1998**, *121*, 19–29.

34 Glickman, M. H. et al. A subcomplex of the proteasome regulatory particle required for ubiquitin-conjugate degradation and related to the COP9-signalosome and elF3. *Cell* **1998**, *94*, 615–623.

35 Kapelari, B. et al. Electron microscopy and subunit-subunit interaction studies reveal a first architecture of COP9 signalosome. *J Mol Biol* **2000**, *300*, 1169–78.

36 Rechsteiner, M., Hoffman, L., and Dubiel, W. The multicatalytic and 26S proteases. *J. Biol. Chem.* **1993**, *268*, 6065–6068.

37 Richmond, C., Gorbea, C., and Rechsteiner, M. Specific interactions between ATPase subunits of the 26S protease. *J. Biol. Chem.* **1997**, *272*, 13403–13411.

38 Rubin, D. M., Glickman, M. H., Larsen, C. N., Dhruvakumar, S., and Finley, D. Active site mutants in the six regulatory particle ATPases reveal multiple roles for ATP in the proteasome. *EMBO J.* **1998**, *17*, 4909–4919.

39 Davey, M. J., Jeruzalmi, D., Kuriyan, J., and O'Donnell, M. Motors and switches: AAA+ machines within the replisome. *Nat Rev Mol Cell Biol* **2002**, *3*, 826–35.

40 LEE, S. Y. et al. Regulation of the transcriptional activator NtrC1: structural studies of the regulatory and AAA+ ATPase domains. *Genes Dev* **2003**, *17*, 2552–63.

41 HARTMANN-PETERSEN, R., TANAKA, K., and HENDIL, K. B. Quaternary structure of the ATPase complex of human 26S proteasomes determined by chemical cross-linking. *Arch Biochem Biophys* **2001**, *386*, 89–94.

42 BERNDT, C., BECH-OTSCHIR, D., DUBIEL, W., and SEEGER, M. Ubiquitin system: JAMMing in the name of the lid. *Curr Biol* **2002**, *12*, R815–7.

43 COPE, G. A. and DESHAIES, R. J. COP9 signalosome: a multifunctional regulator of SCF and other cullin-based ubiquitin ligases. *Cell* **2003**, *114*, 663–71.

44 LI, L. and DENG, X. W. The COP9 signalosome: an alternative lid for the 26S proteasome? *Trends Cell Biol* **2003**, *13*, 507–9.

45 FERRELL, K., WILKINSON, C. R., DUBIEL, W., and GORDON, C. Regulatory subunit interactions of the 26S proteasome, a complex problem. *Trends Biochem Sci* **2000**, *25*, 83–8.

46 FU, H., REIS, N., LEE, Y., GLICKMAN, M. H., and VIERSTRA, R. D. Subunit interaction maps for the regulatory particle of the 26S proteasome and the COP9 signalosome. *Embo J* **2001**, *20*, 7096–107.

47 CAGNEY, G., UETZ, P., and FIELDS, S. Two-hybrid analysis of the Saccharomyces cerevisiae 26S proteasome. *Physiol Genomics* **2001**, *7*, 27–34.

48 DAVY, A. et al. A protein-protein interaction map of the Caenorhabditis elegans 26S proteasome. *EMBO Rep* **2001**, *2*, 821–8.

49 COPE, G. A. et al. Role of predicted metalloprotease motif of Jab1/Csn5 in cleavage of Nedd8 from Cul1. *Science* **2002**, *298*, 608–11.

50 BAILLY, E. and REED, S. I. Functional characterization of rpn3 uncovers a distinct 19S proteasomal subunit requirement for ubiquitin-dependent proteolysis of cell cycle regulatory proteins in budding yeast. *Mol Cell Biol* **1999**, *19*, 6872–90.

51 FONG, A., ZHANG, M., NEELY, J., and SUN, S. C. S9, a 19 S proteasome subunit interacting with ubiquitinated NF-kappaB2/p100. *J Biol Chem* **2002**, *277*, 40697–702.

52 DEVERAUX, Q., USTRELL, V., PICKART, C., and RECHSTEINER, M. A 26S protease subunit that binds ubiquitin conjugates. *J. Biol. Chem.* **1994**, *269*, 7059–7061.

53 VAN NOCKER, S. et al. The Multiubiquitin-Chain-Binding Protein Mcb1 Is a Component of the 26S Proteasome in *Saccharomyces cerevisiae* and Plays a Nonessentail, Substrate-Specific Role in Protein Turnover. *Mol. Cell Biol.* **1996**, *16*, 6020–6028.

54 KAJAVA, A. V. What curves alpha-solenoids? Evidence for an alpha-helical toroid structure of Rpn1 and Rpn2 proteins of the 26S proteasome. *J Biol Chem* **2002**, *277*, 49791–8.

55 ELSASSER, S. et al. Proteasome subunit Rpn1 binds ubiquitin-like protein domains. *Nat Cell Biol* **2002**, *4*, 725–30.

56 SAEKI, Y., SONE, T., TOH-E, A., and YOKOSAWA, H. Identification of ubiquitin-like protein-binding subunits of the 26S proteasome. *Biochem Biophys Res Commun* **2002**, *296*, 813–9.

57 LAM, Y. A., LAWSON, T. G., VELAYUTHAM, M., ZWEIER, J. L., and PICKART, C. M. A proteasomal ATPase subunit recognizes the polyubiquitin degradation signal. *Nature* **2002**, *416*, 763–7.

58 HOFFMAN, L. and RECHSTEINER, M. Activation of the multicatalytic protease. The 11 S regulator and 20S ATPase complexes contain distinct 30-kilodalton subunits. *J. Biol. Chem.* **1994**, *269*, 16890–16895.

59 BRAUN, B. C. et al. The base of the proteasome regulatory particle exhibits chaperone-like activity. *Nat Cell Biol* **1999**, *1*, 221–6.

60 STRICKLAND, E., HAKALA, K., THOMAS, P. J., and DEMARTINO, G. N. Recognition of misfolding proteins by

PA700, the regulatory subcomplex of the 26S proteasome. *J Biol Chem* **2000**, *275*, 5565–72.

61 LIU, C. W. et al. Conformational remodeling of proteasomal substrates by PA700, the 19 S regulatory complex of the 26S proteasome. *J Biol Chem* **2002**, *277*, 26815–20.

62 GONZALEZ, F., DELAHODDE, A., KODADEK, T., and JOHNSTON, S. A. Recruitment of a 19S proteasome subcomplex to an activated promoter. *Science* **2002**, *296*, 548–50.

63 HOFFMAN, L., PRATT, G., and RECHSTEINER, M. Multiple forms of the 20 S multicatalytic and the 26S ubiquitin/ATP-dependent proteases from rabbit reticulocyte lysate. *J. Biol. Chem.* **1992**, *267*, 22362–22368.

64 VERMA, R. et al. Proteasomal proteomics: identification of nucleotide-sensitive proteasome-interacting proteins by mass spectrometric analysis of affinity-purified proteasomes. *Mol Biol Cell* **2000**, *11*, 3425–39.

65 GUTERMAN, A. and GLICKMAN, M. H. Complementary roles for rpn11 and ubp6 in deubiquitination and proteolysis by the proteasome. *J Biol Chem* **2004**, *279*, 1729–38.

66 MEINERS, S. et al. Inhibition of proteasome activity induces concerted expression of proteasome genes and de novo formation of Mammalian proteasomes. *J Biol Chem* **2003**, *278*, 21517–25.

67 BLOBEL, G. Protein targeting (Nobel lecture). *Chembiochem* **2000**, *1*, 86–102.

68 RECHSTEINER, M. and ROGERS, S. W. PEST sequences and regulation by proteolysis. *Trends Biochem. Sci.* **1996**, *21*, 267–271.

69 VARSHAVSKY, A. The N-end rule: Functions, mysteries, uses. *Proc. Natl. Acad. Sci. USA* **1996**, *93*, 12142–12149.

70 ZUR, A. and BRANDEIS, M. Timing of APC/C substrate degradation is determined by fzy/fzr specificity of destruction boxes. *Embo J* **2002**, *21*, 4500–10.

71 ORLOWSKI, M. and WILK, S. Ubiquitin-independent proteolytic functions of the proteasome. *Arch Biochem Biophys* **2003**, *415*, 1–5.

72 GRUNE, T., MERKER, K., SANDIG, G., and DAVIES, K. J. Selective degradation of oxidatively modified protein substrates by the proteasome. *Biochem Biophys Res Commun* **2003**, *305*, 709–18.

73 PENG, J. et al. A proteomics approach to understanding protein ubiquitination. *Nat Biotechnol* **2003**, *21*, 921–6.

74 WU-BAER, F., LAGRAZON, K., YUAN, W., and BAER, R. The BRCA1/BARD1 heterodimer assembles polyubiquitin chains through an unconventional linkage involving lysine residue K6 of ubiquitin. *J Biol Chem* **2003**, *278*, 34743–6.

75 SPENCE, J., SADIS, S., HAAS, A. L., and FINLEY, D. A ubiquitin mutant with specific defects in DNA repair and multiubiquitination. *Mol Cell Biol* **1995**, *15*, 1265–1273.

76 SPRINGAEL, J. Y., GALAN, J. M., HAGUENAUER-TSAPIS, R., and ANDRE, B. NH4+-induced down-regulation of the Saccharomyces cerevisiae Gap1p permease involves its ubiquitination with lysine-63-linked chains. *J Cell Sci* **1999**, *112* (Pt 9), 1375–83.

77 ALBERTI, S. et al. Ubiquitylation of BAG-1 suggests a novel regulatory mechanism during the sorting of chaperone substrates to the proteasome. *J Biol Chem* **2002**, *277*, 45920–7.

78 CHAU, V. et al. A multiubiquitin chain is confined to specific lysine in a targeted short-lived protein. *Science* **1989**, *243*, 1576–1583.

79 JOHNSON, E. S., MA, P. C., OTA, I. M., and VARSHAVSKY, A. A proteolytic pathway that recognizes ubiquitin as a degradation signal. *J. Biol. Chem.* **1995**, *270*, 17442–17456.

80 THROWER, J. S., HOFFMAN, L., RECHSTEINER, M., and PICKART, C. M. Recognition of the polyubiquitin proteolytic signal. *EMBO J.* **2000**, *19*, 94–102.

81 YOUNG, P., DEVERAUX, Q., BEAL, R., PICKART, C., and RECHSTEINER, M. Characterization of two polyubiquitin binding sites in the 26S protease

subunit 5a. *J. Biol. Chem.* **1998**, *273*, 5461–5467.

82 HOFMANN, K. and FALQUET, L. A ubiquitin-interacting motif conserved in components of the proteasomal and lysosomal protein degradation systems. *Trends Biochem Sci* **2001**, *26*, 347–50.

83 SWANSON, K. A., KANG, R. S., STAMENOVA, S. D., HICKE, L., and RADHAKRISHNAN, I. Solution structure of Vps27 UIM-ubiquitin complex important for endosomal sorting and receptor downregulation. *Embo J* **2003**, *22*, 4597–606.

84 MUELLER, T. D. and FEIGON, J. Structural determinants for the binding of ubiquitin-like domains to the proteasome. *Embo J* **2003**, *22*, 4634–45.

85 RYU, K. S. et al. Binding surface mapping of intra- and interdomain interactions among hHR23B, ubiquitin, and polyubiquitin binding site 2 of S5a. *J Biol Chem* **2003**, *278*, 36621–7.

86 HARTMANN-PETERSEN, R., SEEGER, M., and GORDON, C. Transferring substrates to the 26S proteasome. *Trends Biochem Sci* **2003**, *28*, 26–31.

87 SAEKI, Y., SAITOH, A., TOH-E, A., and YOKOSAWA, H. Ubiquitin-like proteins and Rpn10 play cooperative roles in ubiquitin-dependent proteolysis. *Biochem Biophys Res Commun* **2002**, *293*, 986–92.

88 HIYAMA, H. et al. Interaction of hHR23 with S5a. The ubiquitin-like domain of hHR23 mediates interaction with S5a subunit of 26S proteasome. *J Biol Chem* **1999**, *274*, 28019–25.

89 SAKATA, E. et al. Parkin binds the Rpn10 subunit of 26S proteasomes through its ubiquitin-like domain. *EMBO Rep* **2003**, *4*, 301–6.

90 KLEIJNEN, M. F., ALARCON, R. M., and HOWLEY, P. M. The ubiquitin-associated domain of hPLIC-2 interacts with the proteasome. *Mol Biol Cell* **2003**, *14*, 3868–75.

91 XIE, Y. and VARSHAVSKY, A. UFD4 lacking the proteasome-binding region catalyses ubiquitination but is impaired in proteolysis. *Nat Cell Biol* **2002**, *4*, 1003–7.

92 CORN, P. G., McDONALD, E. R., 3rd, HERMAN, J. G., and EL-DEIRY, W. S. Tat-binding protein-1, a component of the 26S proteasome, contributes to the E3 ubiquitin ligase function of the von Hippel-Lindau protein. *Nat Genet* **2003**, *35*, 229–37.

93 LAMBERTSON, D., CHEN, L., and MADURA, K. Pleiotropic defects caused by loss of the proteasome-interacting factors Rad23 and Rpn10 of Saccharomyces cerevisiae. *Genetics* **1999**, *153*, 69–79.

94 MURAKAMI, Y. et al. Ornithine decarboxylase is degraded by the 26S proteasome without ubiquitination. *Nature* **1992**, *360*, 597–9.

95 SHEAFF, R. J. et al. Proteasome turnover of p21^{Cip1} does not require p21^{Cip1} ubiquitination. *Molec. Cell.* **2000**, *5*, 403–410.

96 ZHANG, M., PICKART, C. M., and COFFINO, P. Determinants of proteasome recognition of ornithine decarboxylase, a ubiquitin-independent substrate. *Embo J* **2003**, *22*, 1488–96.

97 REZVANI, K. et al. Proteasomal interactors control activities as diverse as the cell cycle and glutaminergic neurotransmission. *Biochem Soc Trans* **2003**, *31*, 470–3.

98 TOUITOU, R. et al. A degradation signal located in the C-terminus of p21WAF1/CIP1 is a binding site for the C8 alpha-subunit of the 20S proteasome. *Embo J* **2001**, *20*, 2367–75.

99 COLEMAN, M. L., MARSHALL, C. J., and OLSON, M. F. Ras promotes p21(Waf1/Cip1) protein stability via a cyclin D1-imposed block in proteasome-mediated degradation. *Embo J* **2003**, *22*, 2036–46.

100 MURAKAMI, Y., MATSUFUJI, S., HAYASHI, S. I., TANAHASHI, N., and TANAKA, K. ATP-Dependent inactivation and sequestration of ornithine decarboxylase by the 26S proteasome are prerequisites for degradation. *Mol Cell Biol* **1999**, *19*, 7216–27.

101 BAYS, N. W. and HAMPTON, R. Y.

Cdc48-Ufd1-Npl4: stuck in the middle with Ub. *Curr Biol* **2002**, *12*, R366–71.

102 BRUNGER, A. T. and DeLaBARRE, B. NSF and p97/VCP: similar at first, different at last. *FEBS Lett* **2003**, *555*, 126–33.

103 RABINOVICH, E., KEREM, A., FROHLICH, K. U., DIAMANT, N., and BAR-NUN, S. AAA-ATPase p97/Cdc48p, a cytosolic chaperone required for endoplasmic reticulum-associated protein degradation. *Mol Cell Biol* **2002**, *22*, 626–34.

104 WOJCIK, C., YANO, M., and DeMARTINO, G. N. RNA interference of valosin-containing protein (VCP/p97) reveals multiple cellular roles linked to ubiquitin/proteasome-dependent proteolysis. *J Cell Sci* **2004**, *117*, 281–92.

105 IMAMURA, T. et al. Involvement of heat shock protein 90 in the degradation of mutant insulin receptors by the proteasome. *J Biol Chem* **1998**, *273*, 11183–8.

106 TAXIS, C. et al. Use of modular substrates demonstrates mechanistic diversity and reveals differences in chaperone requirement of ERAD. *J Biol Chem* **2003**, *278*, 35903–13.

107 TOKUMOTO, T. et al. Regulated interaction between polypeptide chain elongation factor-1 complex with the 26S proteasome during Xenopus oocyte maturation. *BMC Biochem* **2003**, *4*, 6.

108 GONCIARZ-SWIATEK, M. et al. Recognition, targeting, and hydrolysis of the lambda O replication protein by the ClpP/ClpX protease. *J Biol Chem* **1999**, *274*, 13999–4005.

109 STUDEMANN, A. et al. Sequential recognition of two distinct sites in sigma(S) by the proteolytic targeting factor RssB and ClpX. *Embo J* **2003**, *22*, 4111–20.

110 NEHER, S. B., SAUER, R. T., and BAKER, T. A. Distinct peptide signals in the UmuD and UmuD' subunits of UmuD/D' mediate tethering and substrate processing by the ClpXP protease. *Proc Natl Acad Sci USA* **2003**, *100*, 13219–24.

111 PETROSKI, M. D. and DESHAIES, R. J. Context of multiubiquitin chain attachment influences the rate of Sic1 degradation. *Mol Cell* **2003**, *11*, 1435–44.

112 LEE, C., SCHWARTZ, M. P., PRAKASH, S., IWAKURA, M., and MATOUSCHEK, A. ATP-dependent proteases degrade their substrates by processively unraveling them from the degradation signal. *Mol Cell* **2001**, *7*, 627–37.

113 HOSKINS, J. R., YANAGIHARA, K., MIZUUCHI, K., and WICKNER, S. ClpAP and ClpXP degrade proteins with tags located in the interior of the primary sequence. *Proc Natl Acad Sci USA* **2002**, *99*, 11037–42.

114 LEE, C., PRAKASH, S., and MATOUSCHEK, A. Concurrent translocation of multiple polypeptide chains through the proteasomal degradation channel. *J Biol Chem* **2002**, *277*, 34760–5.

115 BENAROUDJ, N., ZWICKL, P., SEEMULLER, E., BAUMEISTER, W., and GOLDBERG, A. L. ATP hydrolysis by the proteasome regulatory complex PAN serves multiple functions in protein degradation. *Mol Cell* **2003**, *11*, 69–78.

116 KENNISTON, J. A., BAKER, T. A., FERNANDEZ, J. M., and SAUER, R. T. Linkage between ATP consumption and mechanical unfolding during the protein processing reactions of an AAA+ degradation machine. *Cell* **2003**, *114*, 511–20.

117 REID, B. G., FENTON, W. A., HORWICH, A. L., and WEBER-BAN, E. U. ClpA mediates directional translocation of substrate proteins into the ClpP protease. *Proc Natl Acad Sci USA* **2001**, *98*, 3768–72.

118 DILLINGHAM, M. S., WIGLEY, D. B., and WEBB, M. R. Demonstration of unidirectional single-stranded DNA translocation by PcrA helicase: measurement of step size and translocation speed. *Biochemistry* **2000**, *39*, 205–12.

119 LLORCA, O. et al. The 'sequential allosteric ring' mechanism in the eukaryotic chaperonin-assisted folding of actin and tubulin. *Embo J* **2001**, *20*, 4065–75.

120 Laskey, R. A. and Madine, M. A. A rotary pumping model for helicase function of MCM proteins at a distance from replication forks. *EMBO Rep* **2003**, *4*, 26–30.

121 Flynn, J. M., Neher, S. B., Kim, Y. I., Sauer, R. T., and Baker, T. A. Proteomic discovery of cellular substrates of the ClpXP protease reveals five classes of ClpX-recognition signals. *Mol Cell* **2003**, *11*, 671–83.

122 Palombella, V. J., Rando, O. J., Goldberg, A. L., and Maniatis, T. The ubiquitin-proteasome pathway is required for processing the NF-kappa B1 precursor protein and the activation of NF-kappa B. *Cell* **1994**, *78*, 773–785.

123 Ciechanover, A. et al. Mechanisms of ubiquitin-mediated, limited processing of the NF-kappaB1 precursor protein p105. *Biochimie* **2001**, *83*, 341–9.

124 Zhang, M. and Coffino, P. Repeat sequence of Epstein-Barr virus EBNA1 protein interrupts proteasome substrate processing. *J Biol Chem* **2004**, *279*, 8635–8641.

125 Hoppe, T. et al. Activation of a membrane-bound transcription factor by regulated ubiquitin/proteasome-dependent processing. *Cell* **2000**, *102*, 577–86.

126 Wojcik, C. and DeMartino, G. N. Analysis of Drosophila 26S proteasome using RNA interference. *J Biol Chem* **2002**, *277*, 6188–97.

127 Szlanka, T. et al. Deletion of proteasomal subunit S5a/Rpn10/p54 causes lethality, multiple mitotic defects and overexpression of proteasomal genes in Drosophila melanogaster. *J Cell Sci* **2003**, *116*, 1023–33.

128 Mannhaupt, G., Schnall, R., Karpov, V., Vetter, I., and Feldmann, H. Rpn4p acts as a transcription factor by binding to PACE, a nonamer box found upstream of 26S proteasomal and other genes in yeast. *FEBS Lett* **1999**, *450*, 27–34.

129 Xie, Y. and Varshavsky, A. RPN4 is a ligand, substrate, and transcriptional regulator of the 26S proteasome: a negative feedback circuit. *Proc Natl Acad Sci USA* **2001**, *98*, 3056–61.

130 Arendt, C. S. and Hochstrasser, M. Eukaryotic 20S proteasome catalytic subunit propeptides prevent active site inactivation by N-terminal acetylation and promote particle assembly. *Embo J* **1999**, *18*, 3575–85.

131 Nandi, D., Woodward, E., Ginsburg, D. B., and Monaco, J. J. Intermediates in the formation of mouse 20S proteasomes: implications for the assembly of precursor beta subunits. *Embo J* **1997**, *16*, 5363–75.

132 Griffin, T. A., Slack, J. P., McCluskey, T. S., Monaco, J. J., and Colbert, R. A. Identification of proteassemblin, a mammalian homologue of the yeast protein, Ump1p, that is required for normal proteasome assembly. *Mol Cell Biol Res Commun* **2000**, *3*, 212–7.

133 Ramos, P. C., Hockendorff, J., Johnson, E. S., Varshavsky, A., and Dohmen, R. J. Ump1p is required for proper maturation of the 20S proteasome and becomes its substrate upon completion of the assembly. *Cell* **1998**, *92*, 489–99.

134 Santamaria, P. G., Finley, D., Ballesta, J. P., and Remacha, M. Rpn6p, a proteasome subunit from Saccharomyces cerevisiae, is essential for the assembly and activity of the 26S proteasome. *J Biol Chem* **2003**, *278*, 6687–95.

135 Yang, Y., Früh, K., Ahn, K., and Peterson, P. A. *In vivo* assembly of the proteasomal complexes, implications for antigen processing. *J. Biol. Chem.* **1995**, *270*, 27687–27694.

136 Mason, G. G., Hendil, K. B., and Rivett, A. J. Phosphorylation of proteasomes in mammalian cells. Identification two phosphorylated subunits and the effect of phosphorylation on activity. *Eur. J. Biochem.* **1996**, *238*, 453–462.

137 Mason, G. G., Murray, R. Z., Pappin, D., and Rivett, A. J. Phosphorylation of ATPase subunits of the 26S proteasome. *FEBS Lett* **1998**, *430*, 269–74.

138 KIMURA, Y. et al. N(alpha)-acetylation and proteolytic activity of the yeast 20 S proteasome. *J Biol Chem* 2000, *275*, 4635–9.

139 KIMURA, Y. et al. N-Terminal modifications of the 19S regulatory particle subunits of the yeast proteasome. *Arch Biochem Biophys* 2003, *409*, 341–8.

140 SUMEGI, M., HUNYADI-GULYAS, E., MEDZIHRADSZKY, K. F., and UDVARDY, A. 26S proteasome subunits are O-linked N-acetylglucosamine-modified in Drosophila melanogaster. *Biochem Biophys Res Commun* 2003, *312*, 1284–9.

141 BOSE, S., STRATFORD, F. L., BROADFOOT, K. I., MASON, G. G., and RIVETT, A. J. Phosphorylation of 20S proteasome alpha subunit C8 (alpha7) stabilizes the 26S proteasome and plays a role in the regulation of proteasome complexes by gamma-interferon. *Biochem J* Pt 2004, *378*, 177–184.

142 KURUCZ, E. et al. Assembly of the Drosophila 26S proteasome is accompanied by extensive subunit rearrangements. *Biochem J* 2002, *365*, 527–36.

143 TONE, Y. and TOH, E. A. Nob1p is required for biogenesis of the 26S proteasome and degraded upon its maturation in Saccharomyces cerevisiae. *Genes Dev* 2002, *16*, 3142–57.

144 YEN, H. C., GORDON, C., and CHANG, E. C. Schizosaccharomyces pombe Int6 and Ras homologs regulate cell division and mitotic fidelity via the proteasome. *Cell* 2003, *112*, 207–17.

145 IMAI, J., MARUYA, M., YASHIRODA, H., YAHARA, I., and TANAKA, K. The molecular chaperone Hsp90 plays a role in the assembly and maintenance of the 26S proteasome. *Embo J* 2003, *22*, 3557–67.

146 BAJOREK, M., FINLEY, D., and GLICKMAN, M. H. Proteasome disassembly and downregulation is correlated with viability during stationary phase. *Curr Biol* 2003, *13*, 1140–4.

147 HENDIL, K. B., KHAN, S. and TANAKA, K. Simultaneous binding of PA28 and

PA700 activators to 20 S proteasomes. *Biochem. J.* 1998, *332*, 749–754.

148 KOPP, F., DAHLMANN, B., and KUEHN, L. Reconstitution of hybrid proteasomes from purified PA700–20 S complexes and PA28alphabeta activator: ultrastructure and peptidase activities. *J Mol Biol* 2001, *313*, 465–71.

149 TANAHASHI, N. et al. Hybrid proteasomes. Induction by interferon-gamma and contribution to ATP-dependent proteolysis. *J. Biol. Chem.* 2000, *275*, 14336–14345.

150 CASCIO, P., CALL, M., PETRE, B., T., W., and GOLDBERG, A. L. Properties of the hybrid form of the 26S proteasome containing both 19S and PA28 complexes. *EMBO J.* 2002, *21*, 2636–2645.

151 RECHSTEINER, M., REALINI, C., and USTRELL, V. The proteasome activator 11S REG (PA28) and class I antigen presentation. *Biochem. J.* 2000, *345*, 1–15.

152 HILL, C. P., MASTERS, E. I., and WHITBY, F. G. The 11S regulators of 20S proteasome activity. *Curr Top Microbiol Immunol* 2002, *268*, 73–89.

153 MURATA, S. et al. Immunoproteasome assembly and antigen presentation in mice lacking both PA28alpha and PA28beta. *Embo J* 2001, *20*, 5898–907.

154 MURATA, S. et al. Growth retardation in mice lacking the proteasome activator PA28gamma. *J Biol Chem* 1999, *274*, 38211–5.

155 WHITBY, F. G. et al. Structural basis for the activation of 20 S proteasomes by 11 S regulators. *Nature* 2000, *408*, 115–120.

156 USTRELL, V., HOFFMAN, L., PRATT, G., and RECHSTEINER, M. PA200, a nuclear proteasome activator involved in DNA repair. *EMBO J.* 2002, *21*, 3403–3412.

157 HO, Y., GRUHLER, A., HEILBUT, A., BADER, G. D., and MOORE, L. Systematic identification of protein complexes in *Saccharomyces cerevisiae* by mass spectrometry. *Nature* 2002, *415*, 180–183.

158 GAVIN, A. C. et al. Functional organization of the yeast proteome by

systematic analysis of protein complexes. *Nature* **2002**, *415*, 141–7.

159 LEGGETT, D. S. et al. Multiple associated proteins regulate proteasome structure and function. *Mol Cell* **2002**, *10*, 495–507.

160 McCUTCHEN-MALONEY, S. L. et al. cDNA cloning, expression, and functional characterization of PI31, a proline-rich inhibitor of the proteasome. *J Biol Chem* **2000**, *275*, 18557–65.

161 ZAISS, D. M., STANDERA, S., KLOETZEL, P. M., and SIJTS, A. J. PI31 is a modulator of proteasome formation and antigen processing. *Proc Natl Acad Sci USA* **2002**, *99*, 14344–9.

162 LU, D. C. et al. A second cytotoxic proteolytic peptide derived from amyloid beta-protein precursor. *Nat. Medicine* **2000**, *6*, 397–404.

163 GACZYNSKA, M., OSMULSKI, P. A., GAO, Y., POST, M. J., and SIMONS, M. Proline- and arginine-rich peptides constitute a novel class of allosteric inhibitors of proteasome activity. *Biochemistry* **2003**, *42*, 8663–70.

164 YAMANO, T. et al. Two distinct pathways mediated by PA28 and hsp90 in major histocompatibility complex class I antigen processing. *J Exp Med* **2002**, *196*, 185–96.

165 APCHER, G. S. et al. Human immunodeficiency virus-1 Tat protein interacts with distinct proteasomal alpha and beta subunits. *FEBS Lett* **2003**, *553*, 200–4.

166 HUANG, X. et al. The RTP site shared by the HIV-1 Tat protein and the 11S regulator subunit alpha is crucial for their effects on proteasome function including antigen processing. *J Mol Biol* **2002**, *323*, 771–82.

167 FAROUT, L. et al. Distribution of proteasomes and of the five proteolytic activities in rat tissues. *Arch Biochem Biophys* **2000**, *374*, 207–12.

168 BROOKS, P. et al. Subcellular localization of proteasomes and their regulatory complexes in mammalian cells. *Biochem. J.* **2000**, *346*, 155–161.

169 REITS, E. A., BENHAM, A. M., PLOUGASTE, B., NEEFJES, J., and TROWSDALE, J. Dynamics of proteasome distribution in living cells. *EMBO J.* **1997**, *16*, 6087–94.

170 ARABI, A., RUSTUM, C., HALLBERG, E., and WRIGHT, A. P. Accumulation of c-Myc and proteasomes at the nucleoli of cells containing elevated c-Myc protein levels. *J Cell Sci* **2003**, *116*, 1707–17.

171 KOPITO, R. R. Aggresomes, inclusion bodies and protein aggregation. *Trends Cell Biol* **2000**, *10*, 524–30.

172 WOJCIK, C. and DeMARTINO, G. N. Intracellular localization of proteasomes. *Int J Biochem Cell Biol* **2003**, *35*, 579–89.

173 HAMPTON, R. Y. ER-associated degradation in protein quality control and cellular regulation. *Curr Opin Cell Biol* **2002**, *14*, 476–82.

174 McCRACKEN, A. A. and BRODSKY, J. L. Evolving questions and paradigm shifts in endoplasmic-reticulum-associated degradation (ERAD). *Bioessays* **2003**, *25*, 868–77.

10
Molecular Machines for Protein Degradation

Matthias Bochtler, Michael Groll, Hans Brandstetter, Tim Clausen,
and Robert Huber

10.1
Introduction

The action of intracellular proteolytic enzymes is tightly controlled to avoid destruction of properly folded and functional proteins essential for cell viability and to restrict their activity towards sick molecules or/and those marked for destruction. The four (five) proteases discussed in this chapter display different regulatory mechanisms but show sequestration of their active sites inside molecular cages as a common structural principle, albeit being assembled from different building blocks in different shapes and with varying symmetries. The chapter focuses on structural, functional, and mutational studies from our laboratory. We are aware of other cage-forming proteases and their regulatory components that have been structurally characterized and which are mentioned in brief later in the context of our studies. The chapter is arranged in four main sections focused on the proteases HslVU, proteasome, tricorn (DPPIV), and DegP.

10.2
The ATP-dependent Protease HslVU

ATP-dependent proteases are complex proteolytic machines. They are present in eubacteria, archaebacteria, in eukaryotic organelles and, as the 20S or 26S proteasome, in the eukaryotic cytosol and nucleoplasm. The activators of all known ATP-dependent proteases are related. They all contain an AAA(+) ATPase domain as a module (Neuwald et al. 1999) and are thought to assemble into hexameric particles or, in the case of 26S proteasomes, are present in six variants in the 19S activators (Glickman et al. 1999). Like the ATPases, the proteolytic components of the ATP-dependent proteases form higher order complexes, but unlike for the ATPases, the symmetry of the protease assemblies varies, and the folds of the subunits need not be related. ClpP is a serine protease, FtsH a metalloprotease, and HslV and the proteasomes from archaebacteria and eubacteria are threonine proteases.

Protein Degradation. Edited by J. Mayer, A. Ciechanover, M. Rechsteiner
Copyright © 2005 WILEY-VCH Verlag GmbH & Co. KGaA, Weinheim
ISBN: 3-527-30837-7

Although extensive biochemical data on both the bacterial and eukaryotic ATP-dependent proteases are available, the characterization of these proteolytic machines at atomic resolution has proven difficult, because of both the large size of these complexes and their lability to proteolysis and dissociation. No structural data at all are currently available for Lon and the mitochondrial ATP-dependent proteases. In the case of the cytosolic, membrane-integrated bacterial protease FtsH, atomic resolution data are available only for the ATPase domain (Krzywda et al. 2002; Niwa et al. 2002). In contrast, the ATP-dependent activators of the ClpAP and ClpXP proteolytic machines have so far resisted crystallization. Atomic resolution data are available only for the proteolytic component ClpP (Wang et al. 1997), and separately for a ClpX monomer (Kim and Kim 2003) and a ClpA monomer (Guo et al. 2002b).

The bacterial protease HslVU is unique in two respects: at present, it is the only ATP-dependent protease to have atomic coordinates of the full complex determined; secondly, and in contrast to all other bacterial ATP-dependent proteases, it contains a proteolytic core that is related to the 20S proteolytic core of archaebacterial and eukaryotic proteasomes. The following sections summarize our understanding of HslVU biochemistry, crystallography, and enzymology and end with some speculation on the implications of these results for other ATP-dependent proteases.

10.2.1
HslVU Physiology and Biochemistry

In *E. coli* and most, but not all, other bacteria, the HslU (ATPase) and HslV (protease) genes are found in one operon under the control of a heat-shock promoter. The operon was first found (Chuang and Blattner 1993) and sequenced (Chuang et al. 1993) in the course of a search for new heat-shock genes. It was later independently isolated again in screens for proteins that could down-regulate the heat-shock response (Missiakis et al. 1996) and for suppressors of the SOS-mediated inhibition of cell division in *E. coli* (Khattar 1997). The observed biological responses in HslVU over-expression or deletion strains result from a decrease or increase in the steady levels of HslVU substrates (Kanemori et al. 1997). HslVU, itself a heat-shock protein, affects the heat-shock response by degradation of the heat-shock factor σ32 (Missiakis et al. 1996; Kanemori et al. 1997) and the SOS response via the degradation of the cell-division inhibitor SulA (Kanemori et al. 1999; Seong et al. 1999).

The physiological role of HslVU seems to be limited, probably because of overlapping substrate specificity with other ATP-dependent proteases in bacteria. The *E. coli* HslVU deletion strain has no phenotype at standard growth temperature, and it appears that HslVU is required for normal growth only at very high temperatures (Kanemori et al. 1997). According to the protease database MEROPS, some bacterial species appear to lack an HslVU-type peptidase altogether (Rawlings et al. 2002). Therefore, it came as a surprise that some HslV and HslU homologs were recently found in primordial eukaryotes where they appear to be simultaneously present with genuine 20S proteasomes (Couvreur et al. 2002).

Low expression levels and the lability of the HslVU complex make work with proteins from wild-type strains difficult. Gratifyingly, the active protease can be reconstituted *in vitro* from over-expressed and purified components (Rohrwild et al. 1996). It requires ATP for the degradation of folded substrates and ATP or some of its analogs for the purification of small chromogenic peptides. As expected, ATP-hydrolysis and proteolysis activities are mutually dependent (Seol et al. 1997). In addition, the peptidase activity was found to depend in complex ways on the presence of various cations, especially K^+ in the buffers (Huang and Goldberg 1997).

10.2.2
HslV Peptidase

On the sequence level, HslV shows sequence similarity with the β-subunits of archaebacterial and eukaryotic proteasomes, a fact that was immediately noticed when the *E. coli* gene was sequenced (Chuang et al. 1993) and was later shown to extend to other related eubacterial sequences (Lupas et al. 1994). Electron microscopy (EM) of recombinant HslV subsequently suggested that the particle formed a dimer of hexamers that appeared to enclose only one central cavity without antechambers as in proteasomes (Rohrwild et al. 1997). The unexpected six-fold symmetry of HslV and the similarity in subunit fold with eukaryotic proteasomes were subsequently confirmed by X-ray crystallography (Bochtler et al. 1997). The crystallographic data also showed that the contracted ring compared to proteasomes resulted from small changes to the subunit–subunit interface, not from an entirely new mode of oligomerization (Bochtler et al. 1997) (see Figure 10.1).

All 12 active sites of HslV are located on the inner walls of the hollow particle. In the *E. coli* particle, each active site has neighboring active sites 28 Å away on the same ring and 22 Å and 26 Å away on the opposite ring. The environment of the nucleophilic Thr1 looks similar to that in proteasomes, and the presence of a (putatively protonated) lysine residue near the active site probably helps to lower the pK_a of the N-terminal α-amino group so that it is present in the unprotonated form, which can act as the general base to accept a proton from Thr1.

Since the determination of the HslV crystal structure, two additional crystal structures from other species have been determined. The highest resolution structure available to date is the crystal structure of the *Haemophilus influenzae* enzyme (Sousa and McKay 2001). Intriguingly, this structure showed the presence of cation-binding sites near the active centers (Sousa and McKay 2001), a finding that could subsequently be confirmed for the *Thermotoga maritima* enzyme (Song et al. 2003) and that explains, at least in qualitative terms, the dependence of HslVU activity on various cations in solution. Overall, the crystal structures of the enzymes from *H. influenzae* (Sousa and McKay 2001) and *T. maritima* (Song et al. 2003) are very similar to the original structure of the *E. coli* enzyme. Therefore, it came as a surprise that the HslV homolog known as CodW from *Bacillus subtilis* behaves rather differently. Although the enzyme contains a threonine residue that aligns with the active site threonine of the *E. coli*, *H. influenzae*, and *T. maritima* enzymes, it does not contain the glycine at the C-terminus of the profragment

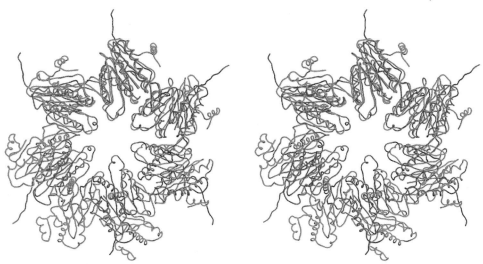

Fig. 10.1. *E. coli* HslV *vs. T. acidophilum* proteasome. Superposition of one hexameric ring of *E. coli* HslV (red) and of one heptameric ring of *T. acidophilum* proteosome β-subunits (green) in stereo representation. The subunits at the "top" of the ring have been overlayed optimally. Note that the "tails" of the HslV subunits that point radially outwards are histidine tags and thus cloning artefacts.

that is believed to be required for efficient autocatalytic processing, and indeed the polypeptide chain is processed five residues upstream of the conserved threonine that is the active-site nucleophile in other species to expose an N-terminal serine residue (Kang et al. 2001). Whether this implies that the serine is in the spatial position normally filled by threonine, implying a discrepancy between the sequence-based and structure-based alignments, or whether it means that the accessory catalytic residues are either dispensable or anchored elsewhere on the sequence is currently not clear.

10.2.3
HslU ATPase

Based on the sequence, HslU can be easily classified as an ATPase by the presence of conventional Walker A (phosphate binding loop or P-loop) and Walker B (magnesium binding) motifs. Beyond this simple classification, two competing models for HslU were proposed, classifying the enzyme either as a PDZ-domain containing ATPase (Levchenko et al. 1997) or alternatively as a AAA(+)-type ATPase (Neuwald et al. 1999). The crystal structure settled the issue in favour of the AAA(+) model (Bochtler et al. 2000). AAA(+)-ATPases consist essentially of two structural domains that are connected through a short linker. A nucleotide binds at the interface of the two domains. As first observed with HslU, the presence or absence of a nucleotide induces different relative orientations between the two domains (Bochtler et al. 2000). With the availability of many different nucleotide states of HslU, the model was later refined to include a dependence on the state of hydrolysis of the nucleotide (Wang et al. 2001b).

The nucleotide is in a strategic position both at the interface of the N- and C-domains of one subunit and at the interface of adjacent subunits. A combination of mostly conserved residues from the two subunits around the nucleotide creates a highly polar environment (Bochtler et al. 2000). Two arginine residues have attracted particular attention: R393 of *E. coli* HslU is thought to act as the "sensor" that transmits information on the presence or absence of nucleotide, and possibly on its identity, to the C-domain and thus controls the relative orientation of N- and C-domains in HslU. Another conserved arginine residue, R325, is anchored on the subunit that makes fewer contacts with the nucleotide and is the homolog of the proposed "arginine finger" in FtsH (Karata et al. 1999). Although the term "arginine finger" (taken from small GTP-binding proteins Ras and Rho) implies a direct catalytic role for this residue, its distance from the nucleotide phosphates argues more for an indirect role. A similar conclusion has since been reached for ClpA (Guo et al. 2002b). Experimentally, mutation of either of the two arginine residues abolishes all ATP-dependent proteolysis activity (Song et al. 2000). Loss of subunit interactions plays a major role in the loss of function: The "arginine sensor" mutant R325E is fully and the "arginine finger" mutant R393 is partially dissociated in gel-filtration experiments in the presence of salt (Song et al. 2000).

A very complex picture has emerged from biochemical and crystallographic studies designed to characterize the substrate-binding sites in HslU. An essential role for the C-terminus of HslU was first suggested on the basis of experimental studies that were designed based on the prediction of PDZ-like domains at the C-terminus of HslU (Levchenko et al. 1997). Although the prediction of PDZ-like domains later turned out to be in error, the conclusion about the role of the C-terminus of HslU in substrate recognition was later corroborated with the definition of a biochemically defined *s*ensor and *s*ubstrate *d*iscrimination domain (SSD) (Smith et al. 1999) that is also present in other AAA(+) ATPases and was suggested to act as the "hook" for substrates (Wickner and Maurizi 1999). When the crystal structure of HslU became available, the SSD domain turned out to coincide with its C-domain. This finding is remarkable and not fully understood, since in all crystal structures of HslU the C-domain primarily mediates oligomerization contacts between HslU subunits. Its solvent-accessible regions are found far on the periphery of the HslU ring, far outside the cavity that is formed by the protruding I-domains (see Figure 10.2). From the crystal structure (Bochtler et al. 2000), it would appear likely that the protruding I-domains rather than the SSD domains act as the "hook", although the ill-defined tertiary structure of the I-domains makes specific interactions unlikely (see Figure 10.2). Consistent with this model, it was found experimentally that the I-domains are essential for the degradation of the folded substrate MBP-sulA (Song et al. 2000). A recent two-hybrid study is consistent with both points of view. It confirms the essential role of the C-domain (SSD-domain) in oligomerization, but also supports a role for the I-domain and the SSD-domain in substrate degradation (Lee et al. 2003). Presumably, if sub-

A)

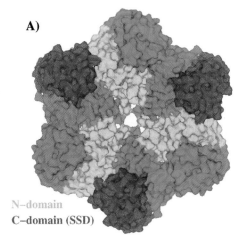

N–domain
C–domain (SSD)

B)

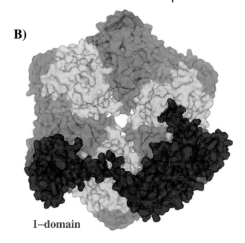

I–domain

Fig. 10.2. HslU surface colored according to domain. (A) View along the six-fold axis, seen from the side opposite to the I-domains. (B) View along the six-fold axis, seen from the side of the I-domains. Every second subunit of the ring is colored according to domain (N-domain yellow, I-domain blue, C-domain or SSD-domain red), the other subunits are colored in green. The diagram is based on the trigonal crystals of *E. coli* HslU that contain nucleotide in every other subunit. This asymmetry and crystallographic packing effects are responsible for the broken six-fold symmetry of the I-domains. Note that the I-domains of three subunits at the top of the figure have been cut away in (B) to allow a view on the globular N- and C-domains.

strates are translocated through the central pore in HslU as EM data suggest for ClpXP (Ortega et al. 2000), both the I-domains and the globular part of the ring would come into contact with substrate during substrate translocation, although the location of the C-domains (SSD-domains) on the periphery of the HslU ring then still needs to be reconciled with this model.

The precise mode of recognition of substrates is even less clear for CodX, the HslU homolog from *B. subtilis*. In the absence of detergent, the I-domains of two hexameric CodX rings contact each other, leading to a head-to-head stacking of CodX rings and presumably the formation of a central cavity loosely surrounded by I-domains. As the dimer of CodX rings can associate with CodW protease on either side, repetitive, chain-type structures with alternating double rings of the peptidase CodW and the ATPase CodX can be formed (Kang et al. 2003). The physiological significance of these high molecular weight assemblies is currently not clear.

10.2.4
The HslVU–Protease Complex

Over the years, a key theme in ATP-dependent proteolysis has been the issue of "symmetry-matched" *vs.* "symmetry-mismatched" complexes. In the light of the clearly established symmetry mismatch of the ClpAP (Kessel et al. 1995) and ClpXP (Grimaud et al. 1998) complexes, the very clear EM data on the six-fold symmetry of HslV (Kessel et al. 1996; Rohrwild et al. 1997) and reports about a predominant species of HslU with six-fold symmetry (Kessel et al. 1996; Rohrwild et al. 1997) came as a surprise because they implied that a "ratcheting mechanism" of ATP-dependent proteolysis, if it existed at all, could not be operating in the HslVU system. This conclusion has since been confirmed by all HslU and HslVU crystal structures (Bochtler et al. 2000; Sousa et al. 2000; Wang et al. 2001a; Kwon et al. 2003). In all cases, HslU is hexameric and matches the oligomerization state of HslV. For the first crystal structure of an HslU–HslV co-crystal, a controversial I-domain-mediated contact between HslU and HslV was reported (Bochtler et al. 2001; Wang 2001; 2003). The contact was suspicious from the very beginning because of poor contact area, but seemed compatible with the known low affinity between HslV and HslU and appeared to explain how the symmetry mismatched ClpXP and ClpAP complexes could be formed. Although a crystallographic reinterpretation of our original data that attributed this docking mode to overlooked twinning (Wang 2001; 2003) turned out to be itself in error (Bochtler et al. 2001), it is now clear from the combined results of cryoelectron microscopy (Ishikawa et al. 2000), small-angle scattering (Sousa et al. 2000) and several additional crystal structures of the complex (Sousa et al. 2000; Wang et al. 2001a) that the physiological mode of interaction between HslU and HslV is with HslU I-domains distal to HslV (see Figure 10.3).

10.2.4.1 Allosteric Activation

In the absence of a nucleotide, HslVU has residual activity at best, but the presence of several non-hydrolyzable ATP-analogs is sufficient to stimulate HslVU-driven proteolysis activity against substrates that do not require unfolding, suggesting an allosteric effect of nucleotide on HslU and via HslU on HslV. This was further corroborated by the observation that a peptide vinyl sulfone formed a covalent complex with HslV only in the presence of HslU and a nucleotide (Bogyo et al. 1997). The details of this allosteric mechanism emerged from the crystal structure of HslVU from *H. influenzae*. In this case, but not in other crystal structures of the HslVU complex, the normally buried C-termini of HslU distend and insert into active-site clefts in HslV to reach out almost to the HslV active centers (Sousa et al. 2000) (see Figure 10.4). The crystal structure of HslVU in complex with a peptide vinyl sulfone inhibitor (Sousa et al. 2002) and two independent biochemical studies (Ramachandran et al. 2002; Seong et al. 2002) that demonstrated the activatory properties of the C-terminal tails of HslU further corroborated this mechanism. In the light of these data, it is remarkable that wild-type HslU in the presence of ADP does not act as an activator for HslV, not even against unfolded or chromogenic substrates. A possible, but experimentally untested explanation could be that the C-termini of HslU are available for HslV binding only in the presence of activatory nucleotides.

Whatever the details of the allosteric activation mechanism, it is already clear

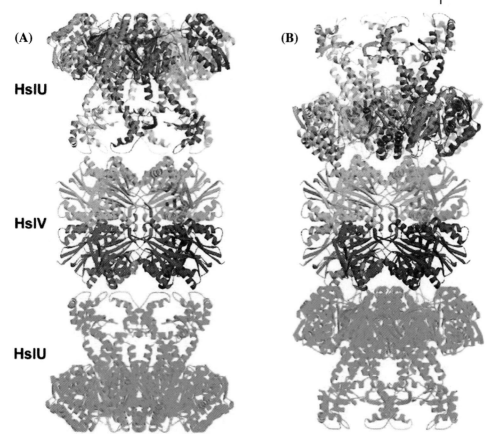

(A)

HsIU

HsIV

HsIU

(B)

Fig. 10.3. HslVU complex. Originally reported (A) and physiologically relevant (B) docking mode between HslV and HslU. In the physiologically relevant docking mode, the I-domains of HslU point away from the HslV.

that HslU affects primarily the conformation of HslV active sites and not the accessibility of the HslV proteolytic chamber. Two independent lines of *in vitro* evidence support this conclusion. Firstly, an HslV mutant with a widened entrance channel does not show increased proteolytic activity in the absence of HslU, although it can still be activated like wild-type HslV by the presence of HslU (Ramachandran et al. 2002). Secondly, the crystal structure of the *H. influenzae* asymmetric HslVU protease in complex with an inhibitory peptide vinyl sulfone has the inhibitor bound only in HslV subunits that are in contact with HslU (Kwon et al. 2003), strongly arguing against accessibility of the proteolytic chamber as the rate-limiting factor at least under the experimental conditions.

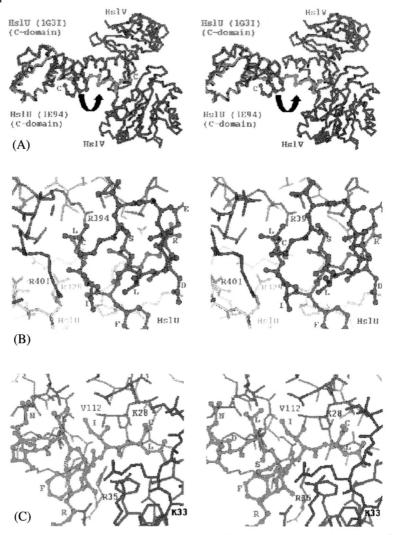

Fig. 10.4. HslVU activation mechanism. (A) Stereo view (C^z-trace) of the superposition of the C-domain of an HslU subunit (red) from the original *E. coli* HslVU complex onto that of the *H. influenzae* HslU subunit (green) from its complex Two HslV subunits (pink and blue) from the *H. influenzae* complex are also shown to illustrate the binding of the C-terminal segment of *H. influenzae* HslU to the pocket between the HslV subunits (indicated also by a black curved arrow). (B) Stereo view of the close-up of the C-terminal residues of an *E. coli* HslU subunit (red) from the complex. The carboxylate of the terminal leucine residue forms salt bridges with R394 of the same subunit and with R329 of an adjacent (yellow) HslU subunit that is not illustrated in (A) for clarity. (C) Stereo view of the close-up of the C-terminal residues of an HslU subunit from the *H. influenzae* HslVU complex.

10.2.5
A Comparison of HslVU with ClpXP and ClpAP

The protease core particles HslV and ClpP are assembled from subunits of entirely different fold and catalytic mechanism. ClpP is a serine protease that belongs to the crotonase superfamily of enzymes, a large class of enzymes that catalyze a variety of chemical reactions that all require the stabilization of an intermediate by an oxyanion hole (Babbitt and Gerlt 1997). HslV belongs to the family of Ntn-hydrolases that share the fold and use the N-terminal residue as the nucleophile (Brannigan et al. 1995). Both ClpP (Wang et al. 1997) and HslV subunits assemble into large oligomers that enclose a central proteolytic chamber, although the symmetry is different, since HslV is a dimer of hexamers and ClpP is a dimer of heptamers.

In contrast to the protease components, which have varying symmetry, all known Clp ATPases are assembled from six identical subunits. These subunits contain either one (HslU and ClpX) or two (ClpA) copies of the AAA(+) module. In addition, the Clp ATPases contain additional domains that are unique for each ATPase. In HslU, a mostly helical I-domain is inserted into the AAA(+) module. ClpX contains an N-terminal domain that was shown to bind zinc (Banecki et al. 2001) and act as a dimerization module (Donaldson et al. 2003). Based on the latter result, a model of ClpX as a trimer of dimers was proposed, with N- and C-domains of ClpX forming a regular hexamer and the zinc-binding modules pairing into dimers (Donaldson et al. 2003). In this context, it is remarkable that freshly isolated HslU behaves as a hexamer, but migrates with the apparent molecular weight of a dimer or trimer in gel filtration after a freeze–thaw cycle (Bochtler 1999). Like ClpX, ClpA contains a non-AAA(+) domain at its N-terminus, and like the N-terminal domain in ClpX, this domain also has the capability to bind zinc (Guo et al. 2002a). However, unlike the N-domain of ClpX, the N-domain of ClpA is almost entirely helical and consists of two four-helix tandem motifs.

In both ClpA and ClpX, the N-terminal non-AAA(+) motifs serve as "docking modules" for accessory proteolysis factors that modulate or change the activity of the proteolytic complex itself (Dougan et al. 2002a). The N-domain of ClpX interacts specifically with the adapter protein SspB that stimulates the degradation of SsrA-tagged proteins (Dougan et al. 2003). The tag is jointly recognized by the SspB-ClpX complex, where SspB interacts with the N-terminal and central region of the SsrA tag and leaves the C-terminal region for interaction with ClpX (Levchenko et al. 2003; Song and Eck 2003). The C-terminal region of SspB shares considerable homology with the corresponding region in RssB, another ClpX adapter protein (Dougan et al. 2003). It appears that RssB promotes the degradation of a specific substrate, namely a subunit of RNA polymerase known as σ^S (Zhou et al. 2001). It has been shown biochemically that in this case again the adapter protein and the ATPase recognize distinct sites in the substrate (Studemann et al. 2003). ClpA has its own adapter protein, ClpS. At least *in vitro*, ClpS switches ClpAP activity away from SsrA tagged towards heat-aggregated proteins (Dougan et al. 2002b). The independent crystal structures of ClpS in complex with the N-domain of ClpA are available (Guo et al. 2002a; Zeth et al. 2002). They explain the specificity of ClpS for ClpA over other related Clp proteins, especially ClpB (Zeth et al. 2002). HslU lacks a domain upstream of the AAA(+) module. Consistent with this, no adapter proteins for HslU have been found so far, to the best of our knowledge.

Mechanistically, there are important differences between the protease activation mechanisms of ClpXP/AP and HslVU. Most importantly, HslVU is a symmetry-matched proteolytic machine. In contrast, the definitive seven-fold symmetry of the ClpP protease and the well-established six-fold symmetry of ClpA and ClpX imply that ClpXP is a symmetry-mismatched system. It is hard to imagine how such an arrangement would be compatible with an HslVU-style activation mechanism with insertion of the C-termini of all activator subunits into clefts in protease. Moreover, the C-termini of ClpX particles from various species are very poorly conserved, arguing against their involvement in any allosteric activation mechanism (Ramachandran et al. 2002). Experimental evidence implicates an internal loop of ClpX in ClpP binding (Kim et al. 2001). This loop is required for ClpXP proteolytic activity and may well be the functional equivalent of the C-terminus of HslU. If so, then the symmetry mismatch in ClpXP would suggest that only a subset of ClpX loops could insert into ClpP clefts at any given time.

10.2.6
HslVU Peptidase as a Model for the Eukaryotic 26S Proteasome?

On the sequence level, HslV shows sequence similarity with the β-subunits of archaebacterial and eukaryotic proteasomes. The crystal structure of *E. coli* HslV confirmed that individual subunits share the Ntn-hydrolase fold with Thr1 at the N-terminus as the nucleophile, just as in proteasomes. Despite these similarities, there are substantial differences between bacterial HslVU and archaebacterial and eukaryotic 20S proteasomes. In contrast to HslVU, 20S proteasomes are assembled from four rings of seven subunits each, that build up a central proteolytic chamber and two flanking antechambers.

The essential role of the C-terminus of HslU has its direct counterpart in the essential role of the C-terminus of the ATP-dependent proteasome activator PA28 (Wilk and Chen 1997). A complex of the yeast 20S proteasome with PA26, the *Trypanosoma brucei* homolog of PA28 has been crystallized and shows that the C-termini of PA28 insert into clefts in the 20S core particle (Whitby et al. 2000), leading to an opening of the gates in the antechambers. So far, there is no evidence for allosteric activation in the 20S proteasome–PA26 complex or the 26S proteasome, where channel "gating" appears to be important as discussed below in Section 10.3 on the yeast 20S proteaseome.

Currently, high-resolution EM image reconstructions for the 26S proteasome (Walz et al. 1998), but no atomic-resolution crystallographic data are available for any complex of 20S proteasomes with ATP-dependent activators. The expected assembly of PAN, the archaebacterial AAA(+) activator of proteasomes (Zwickl et al. 1999) into hexamers suggests a symmetry-mismatched complex in archaebacteria.

The ATP-dependent proteasome activators of eukaryotic proteasomes known as PA700 and the 19S cap also contain six AAA(+) ATPases in a subcomplex of the 19S complex known as the "base" (Glickman et al. 1999). *A priori*, one would expect the six ATPases to form a ring with pseudo six-fold symmetry similar to the six-fold ring seen in bacterial AAA(+) activators, but two-hybrid experiments have suggested alternative models (Richmond et al. 1997). It is currently not clear whether the C-termini of the AAA(+) ATPases in the 19S cap play a similar role as in the ATP-independent PA28–20S proteasome complex. There is no consensus in the C-terminal sequences of different proteasomal AAA(+) ATPases of any particular species, but consensus sequences for the C-termini of any particular subunit from different species can be defined. Unfortunately, the overall sequence similarity of homologous sequences from different species is too high to infer a functional role of the AAA(+) C-termini from sequence similarity.

As discussed above, the eubacterial HslVU is distantly related in structure to the proteasome found in archaea and eukaryotes. Surprisingly, however, the structural relationship is not reflected in the regulatory properties as will be described in Section 10.3, which focuses on structural studies of the yeast 20S proteasome and its activation, activity, and inhibition.

10.3
The Yeast 20S Proteasome

The most elaborate version of the proteasome core particle (CP) is found in eukaryotes as shown by the crystal structure of the yeast 20S proteasome (Groll et al. 1997) (see Figure 10.5B). Here, the α- and β-subunits have diverged into seven different subunits each as compared to the archaeal enzyme which mostly consists of two components (Löwe et al. 1995). The subunits are present in two copies and occupy precisely defined positions within the 20S complex. As in the archaeal proteasome, the α-subunits are inactive and contribute to the antechambers of the particle, whereas the β-subunits set up the inner hydrolytic chamber. Remarkably, four of the seven different eukaryotic β-subunits lack residues that are essential for propeptide autolysis and are therefore proteolytically inactive. As in the archaeal CP, the remaining three subunits mature autoproteolytically to active threonine proteases but with a caspase-like (β1), trypsin-like (β2), and chymotrypsin-like (β5) activity, they exhibit different cleavage potentials (Groll et al. 1997). The crystal structure of the bovine 20S proteasome (Unno et al. 2002) demonstrated that yeast and mammalian CPs are highly homologous in their structural architecture, quaternary assembly, and active-site geometry. However, in mammalian cells, three additional non-essential subunits, β1i, β2i and β5i, respectively, can replace their constitutive counterparts upon induction by the cytokine γ-interferon. The interchange of active subunits modifies the CPs peptidase specificity and is important for the function of the immunoproteasome. The specificity of the S1 pockets of the induced subunits increases the yield of peptides favored for binding to MHC class I molecules and antigen presentation. The bovine proteasome structure provides

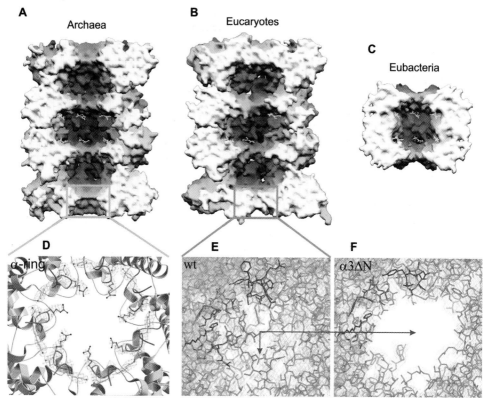

Fig. 10.5. Molecular surface of the archaeal (A), the eukaryotic 20S (B) and the HslV proteasome (C). The accessible surface is colored in blue, the clipped surface (along the cylinder axis) in white. To mark the position of the active sites, the complexes are shown with the bound inhibitor calpain (yellow). (A) The disorder of the first N-terminal residues in the archaeal α-subunits generates a channel in the structure of the CP, (B) whereas the asymmetric but well-defined arrangement of the α N-terminal tails seals the chamber in eukaryotic CPs. (C) The eubacterial "miniproteasome" has an open channel through which unfolded proteins and small peptides can access the proteolytic sites. (D) Ribbon plot of the free

α-ring from *A. fulgidus* focusing on the defined N-termini (red). Tyr8 of each N-terminal part makes hydrogen bonds to Asp9 of the adjacent α-subunit (yellow), Arg10 (red) points toward the channel, generating a 13-Å entrance. The final $2F_O - F_C$ electron density map, contoured at 1σ is shown for the YDR-motif. (E, F) Electron density maps of the yeast core particle from wild type and α3ΔN mutant, respectively. The individual N-terminal tails of the α-subunits are drawn in different colours. Asp9 of subunit α3 plays a key role in stabilizing the closed state of the channel and is marked with a black arrow. In the α3ΔN mutant, an open axial channel is visible, whose dimensions are comparable to those of the archaeal CP channel.

an explanation of how the constitutive and inducible β-subunits can be mutually interchanged. In the yeast proteasome, these subunits are held in place by several specific interactions, which are absent in the mammalian homolog (Unno et al. 2002).

10.3.1
The Proteasome, a Threonine Protease

As mentioned in Section 10.3, proteasomes are threonine proteases. Accordingly, a proton acceptor is required to activate the hydroxylic group of Thr1. Although there are several potential acid–base catalysts around Thr1, activation seems to proceed by its own terminal amino group (Arendt and Hochstrasser 1999; Groll et al. 1999). A positively charged side chain of Lys33 lowers the pK_a of the Thr1 amino group (see Figure 10.6). The assignment of the N-terminus as the catalytic base in proteolysis is further supported by the fact that all Ntn-hydrolases share a common fold, but generally do not display any kind of active-site consensus. A second essential factor for proteolysis is a catalytic water molecule that has been observed in the high-resolution structures of CPs (Groll et al. 1997; Sousa and McKay 2001; Unno et al. 2002; Groll et al. 2003). The solvent molecule is ideally positioned to shuttle protons between the Thr1O$^\gamma$ and the N-terminus. Furthermore it could act as the base for the cleavage of the acyl ester intermediate, thereby releasing Thr1O$^\gamma$ for the next catalytic cycle (Ditzel et al. 1998).

In general, the cleavage products of the proteasome vary in length between 3 and 25 amino acids with an average length of 7 to 8 amino acids. Several models have

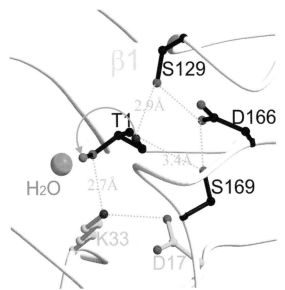

Fig. 10.6. Proteolytic site of the yeast β1 subunit. The protein backbone is drawn as a white coil with the active-site residues Thr1, Asp17, Lys33, Ser129, Asp166 and Ser169 shown in ball-and-stick mode. Owing to the salt-bridge with Asp17, the amino group of Lys33 should be positively charged and thus be able to lower the pK_a of Thr1O$^\gamma$ electrostatically. The other active-site residues Ser129, Asp166 and Ser169 define the orientation of Thr1. A water molecule (green sphere) is located properly to shuttle protons between the terminal amino group and Thr1O$^\gamma$ during proteolysis.

been suggested for the "molecular ruler" that determines fragment length. On the basis of structural studies of mutants unable to autolyse, we suggest that it is determined by the substrate-binding clefts designed for peptides 7–9 amino acids long (Groll et al. 1999). The likelihood of substrate cleavage depends on the mean residence time at the proteolytic sites, which is maximal if all binding sites are filled (Dick et al. 1998; Nussbaum et al. 1998). The active subunits in eukaryotic CPs differ mainly in their binding pockets, yielding different cleavage specificities. However, it must be emphasized that the proteasome complex does not represent a simple collection of chymotrypsin-like, trypsin-like, and caspase-like enzymes. In fact it is the structural architecture of the proteolytic chamber that determines specificity. The local structure around each active site imposes a physical constraint on the peptide substrates, whereas the selectivity of the S1 pockets is less relevant. The mechanistic importance of the inner chamber can also be seen from the fact that substrate residues other than P1 influence degradation (Cardozo et al. 1994; Bogyo et al. 1998; Groll et al. 2002) and that neighboring subunits interfere with the functions of the catalytic subunits (Heinemeyer et al. 1997; Groll et al. 1999; Jäger et al. 1999). However, inhibitor binding and mutational studies indicate that allosteric interactions between individual subunits are insignificant (Wenzel et al. 1994; Heinemeyer et al. 1997; Groll et al. 1999; Jäger et al. 1999; Groll et al. 2002) contrasting the allosteric activation of HslV by HslU described in the previous chapter.

10.3.2
Inhibiting the Proteasome

Proteasome inhibitors have been instrumental in identifying numerous protein substrates and in elucidating the importance of the proteasome/ubiquitin pathway in many biological processes. Initially, non-specific cell-penetrating peptide aldehydes were used for this purpose. More recently, it became possible to synthesize compounds with increased potency and selectivity (Adams et al. 1998; Elofsson et al. 1999). Furthermore, based on the crystal structure of the yeast and bovine liver CP (Groll et al. 1997; Unno et al. 2002), molecular modeling can now be used to engineer improved inhibitors.

Besides the synthetic inhibitors, a variety of natural compounds is known to inhibit the CP. One of these natural inhibitors, lactacystin, was discovered by its ability to induce neurite outgrowth in a murine neuroblastoma cell line. Incubation of cells in the presence of radioactive lactacystin leads to the labelling of the β5 subunit (Fenteany et al. 1995) and to irreversible inhibition of the CP. As shown by X-ray analysis, the inhibitor is covalently attached to subunit β5 by an ester bond with the N-terminal Thr1O$^\gamma$ (Groll et al. 1997) (see Figure 10.7A). The subunit selectivity of lactacystin can be attributed to its dimethyl group, which mimics a valine or a leucine side chain and closely interacts with Met45 in the hydrophobic S1 pocket of subunit β5.

Epoxomicin, an α',β'-epoxyketone peptide, is a natural compound that potently and irreversibly inhibits the catalytic activity of the CP (Meng et al. 1999). Unlike most other proteasome inhibitors, epoxomicin is highly specific for the proteasome and does not inhibit any other protease. The crystal structure of epoxomicin bound to the yeast CP explained the unique selectivity of the inhibitor (see Figure 10.7B). Adduct formation yields an unexpected morpholino ring, which is formed between the Thr1O$^\gamma$, the N-terminus, and the epoxy group of the inhibitor (Groll et al. 2000b). However it should be noted that proteasome inhibitors that covalently bind to the active β-subunits, usually cause apoptosis and cell death (Kloetzel 2001). They are therefore cytotoxic and thus may not be pharmaceutically relevant.

Recently, it was shown that certain natural products from *Apiospora montagnei*, TMC-95s, block the proteolytic activity of the CP selectively and reversibly in the low nanomolar range (Koguchi et al. 2000; Kohno et al. 2000). The TMC-95s represent a novel class of proteasome inhibitors consisting of modified amino acids, which form a heterocyclic ring system. The crystal structure of the yeast CP in complex with TMC-95A shows the inhibitor non-covalently bound to all active sites (Groll et al. 2001) (see Figure 10.7C). TMC-95A was anchored by several specific hydrogen bonds, which are formed with main-chain atoms and strictly conserved residues of the β-subunits. The structures of TMC-95s contain a crosslink between a tyrosine and an oxoindol side chain, resulting in a strained conformation that fits ideally to the CP active site. Thus the entropic penalty of binding is lower than for more-flexible ligands, which in turn explains the specificity and high affinity of the TMC-95s. Modeling studies indicate that it is possible to generate a TMC95 scaffold with a variety of functional groups attached to target the proteasomal S1 and S3 pockets and generate subunit specificity (Kaiser et al. 2002; Lin and Danishefsky 2002).

All agents that specifically inhibit the proteasome are potentially of great pharmacological interest (Loidl et al. 1999). As the CP plays a dominant role in generating antigenic peptides, which are subsequently bound by MHC I molecules, compounds that block this activity might serve as a basis for the development of immunosuppressive drugs. Attempts are being made to design synthetic proteasome inhibitors using the discussed natural inhibitors as lead structures. In addition, proteasomal inhibitors represent powerful tools in molecular biology and can be utilized to identify novel cellular roles of the proteasome.

10.3.3
Access to the Proteolytic Chamber

In the *Thermoplasma acidophilum* and *Archaeoglobus fulgidus* CP, two narrow entry ports of ∼13-Å diameter exist at both ends of the cylinder, which prevent folded proteins entering (Löwe et al. 1995; Groll et al. 2003) (see Figure 10.5A). Many archaebacteria, such as *Methanococcus jannaschii*, contain a gene named PAN (*p*roteasome-*a*ctivating *n*ucleotidase), which is highly homologous to the six ATPases in the 19S-component of the eukaryotic 26S proteasome (Zwickl et al. 1999). It was shown that PAN selectively stimulates the degradation of unfolded proteins, whereas the digestion of small peptides is not enhanced. The threading of specific protein substrates into the lumen of the CP requires the action of an

(A)

(B)

(C)

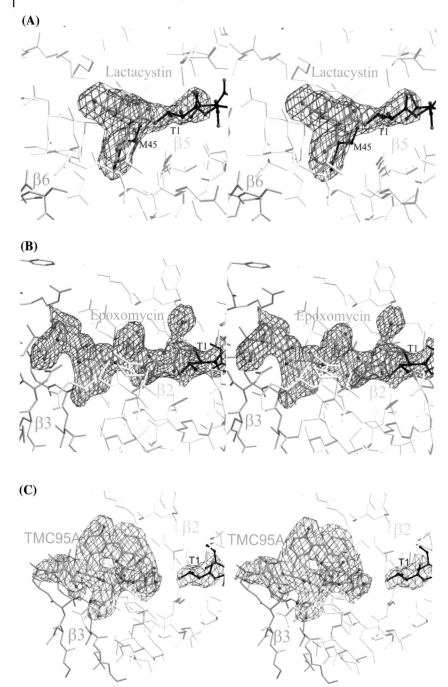

ATPase. The corresponding translocation process catalyzed by PAN follows ATP-dependent unfolding (Navon and Goldberg 2001; Benaroudj et al. 2003). However, it still remains to be clarified whether complex formation is a prerequisite for the cooperation between PAN and CP or whether the two systems work independently. In contrast to the archaeal CPs, the hydrolytic chamber of the eukaryotic 20S proteasome is tightly sealed (see Figure 10.5B). The N-termini of the α-subunits project down and across the axial pore and block the entrances by several layers of interdigitating side chains, which form a lattice-like structure (Groll et al. 1997) (see Figure 10.5E). Thus activation of the eukaryotic CPs requires substantial structural rearrangements of the N-terminal tails to open the molecular gate. This regulatory principle has been confirmed by a yeast CP mutant, in which the first nine amino acids of subunit α3 were deleted (α3ΔN-mutant) (Groll et al. 2000a). The α3-N-terminal tail was chosen for deletion because it traverses the pore of the CP and contacts all other N-termini that are involved in the structural organization of the plug (Groll et al. 1997; Unno et al. 2002). In the crystal structure of the mutant, open axial pores were observed that were equivalent in size to those seen in the archaebacterial CP (see Figure 10.5F). Several points of evidence indicate that opening of the gate is indeed essential for catalytic activation, as all proteolytically active sites are simultaneously activated and no significant structural changes can be seen between mutant and wild-type CP excluding allosteric effects. Furthermore, addition of the synthetic α3-N-terminal peptide to the α3ΔN mutant restores wild-type behavior. An alanine scan of this peptide revealed that α3-Asp9 is essential for stabilizing the closed state of the channel. This aspartate residue closely interacts with Tyr8 and Arg10 of the neighboring subunit α4 (see Figure 10.5E). The strict conservation of this YDR motif and of other α-N-terminal residues suggests a universal mechanism for opening gates in eukaryotic CPs that has been conserved during evolution. We suggest that binding of regulatory proteins to the CP triggers the rearrangement of the α3-tail and thus opens the gate.

This notion was further confirmed by the crystal structure of a 20S/11S heterologous complex between the yeast 20S-proteasome and the Trypanosoma 11S-regulator (Whitby et al. 2000). This approach was justified by the ability of 11S reg-

---

Fig. 10.7. Inhibitor binding to individual active sites of the yeast 20S proteasome. The inhibitors lactacystin (A), epoxomicin (B) and TMC95A (C) are colored green and are shown in stereo mode together with their unbiased electron densities. The active-site Thr1 is highlighted in black. (A) Covalent binding of the Streptomyces metabolite lactacystin to the active site of β5. The S1 pockets of the active subunits β1 and β2 differ from that of β5 and are not suitably constructed to bind the inhibitor. As discussed in the text, Met45 (black), which is located at the bottom of the β5-S1 pocket, makes the difference for inhibitor binding. (B) Covalent binding of the proteasome inhibitor epoxomicin to β2. The electron density reveals the presence of a unique six-membered ring. The morpholino derivative results from adduct formation between epoxomycin and the proteasomal Thr1O$^\gamma$ and amino terminus (pink sticks) and explains the specificity of the inhibitor towards Ntn-hydrolases. (C) Noncovalent binding of the specific proteasome inhibitor TMC-95A from *Apiospora montagnei* to β2. TMC-95A binds near the proteolytic centre in all active subunits in the extended substrate binding site.

ulators to activate 20S proteasomes from widely divergent species. The structure of the chimeric complex showed that one cylindrical regulator was bound at each end of the 20S barrel structure. Unlike the uncomplexed proteasome, all of the seven α-subunit N-terminal tails extend away from the CP in the complex and project towards the pore of the regulator. This rearrangement provides access to the proteolytic chamber and is basically achieved by two features: Firstly, the 11S-"activation loops" impose a more stringent seven-fold symmetry on the CP thereby straightening out the asymmetrically oriented α-tails and removing them from the entrance/exit gates. Secondly, the high-affinity binding between CP and 11S is accomplished by the C-terminal sequences of the regulator, which insert into pockets formed between the 20S α-subunits. The major contact is observed between the C-terminal main-chain carboxylate of the 11S regulator and the entry of an internal helix of the CP α-subunit. The strength of this interaction is amplified by the heptameric assembly of the 20S/11S complex (Whitby et al. 2000).

Activation of the CP by the 19S-complex is also regulated by controlling access to the proteolytic chamber, but the gating mechanism differs from that seen in the 11S/20S-complex. The 19S RP consists of two subcomplexes termed lid and base (Glickman et al. 1998). The base appears to form a ring like structure, including six conserved ATPase subunits (Rpt1–6), and Rpn1 and Rpn2, which are located proximal to the CP's α-ring. Mutation studies indicated that the ATPase domain of Rpt2 plays a major role in regulating peptidase activity and that the 19S RP opens the gate to the protease in an ATP-dependent manner (Köhler et al. 2001). No detailed structural information is as yet available that could provide further insight into how the 19S RP controls proteasomal activity.

In some but not all archaea the proteasome is accompanied by another large cage-forming protease, the tricorn protease. Tricorn functionally interacts with the proteasome by cleaving the proteasomal peptide products into smaller peptides, which are further degraded into single amino acids by associated factors. Structural and functional aspects of tricorn are described in Section 10.4. The unexpected relationship between tricorn and the eukaryotic dipeptidyl peptidase IV revealed by these structural studies is also discussed in brief.

10.4
The Tricorn Protease and its Structural and Functional Relationship with Dipeptidyl Peptidase IV

Each living cell is a complex system and needs to continuously clear unnecessary or defective components. Within this context, the importance of the proteasome is well established (see Section 10.2). It predominantly carries out the degradation of cytosolic proteins and generates peptides varying in length between 3 and 25 amino acids. In order to be useful resources to the cell, these products need to be further degraded to eventually yield single amino acids. In the model organism *Thermoplasma acidophilum* a proteolytic system has been identified that does indeed perform this processing (Tamura et al. 1996a). Based on the crystal structure of the tricorn protease (Brandstetter et al. 2001), we provide evidence of how the tricorn protease accomplishes efficient turnover of the proteasome-generated peptides. The structure of tricorn reveals a complex mosaic protein whereby five domains combine to form one of six subunits, which further assemble to form the *D*3 symmetric core protein. The structure shows how the individual domains coordinate the specific steps of substrate processing, including channeling of both the substrate to and the product from the catalytic site. Moreover, the structure shows how accessory protein components might additionally contribute to an even more complex protein machinery that efficiently collects the tricorn-released products.

10.4.1
Architecture of the Tricorn Protease

The hexameric *D*3-symmetric tricorn protein is assembled by two perfectly staggered and interdigitating trimeric rings with every subunit of one ring forming contacts almost exclusively with the two subunits of the other ring related by the molecular diads. The toroid structure has the shape of a distorted hexagon formed by a trimer of dimers (see Figure 10.8). The overall dimensions of the molecule are 160 Å within the plane normal to the three-fold axis and 88 Å parallel to it. The conically shaped central pore connects with additional cavities formed by the indi-

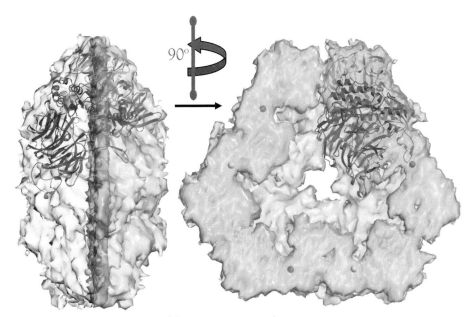

Fig. 10.8. Surface representation of the tricorn protease with the ribbon model of one subunit superimposed. The two orthogonal views are along the molecular two-fold and three-fold axis, respectively. The six solid spheres indicate the active-site positions.

vidual subunits like spokes of a wheel (see Figure 10.8). A single subunit is further divided into five sequential sub-domains, namely the N-terminal six-bladed β-propeller (β6) followed sequentially by a seven-bladed β-propeller (β7). Both β6 and β7 are topologically unclosed, an extremely rare feature observed only in the prolyl oligopeptidase (POP) (Fülöp et al. 1998) and DPIV protease (Engel et al. 2003; Rasmussen et al. 2003). A PDZ-like domain (R761-D855) is interspersed between the two C-terminal mixed α-β domains. These C-terminal domains harbor the catalytic residues and exhibit the α-β hydrolase fold again underlining the relationship of tricorn with DPIV and POP.

10.4.2
Catalytic Residues and Mechanism

To elucidate the amino acids crucial for its catalytic activity, we have co-crystallized tricorn with a series of chloromethyl ketone-based inhibitors, including TLCK and TPCK for which we have confirmed inhibitory efficacy. For all of these inhibitors, we have observed continuous electron density connecting to the side chain of S965 which was unambiguously fitted by the respective inhibitor. S965 is positioned at the entrance to helix H3 within sub-domain C2. The uncapped amino group of D966 forms, together with that of G918, the oxyanion hole, which is occupied by a water molecule in the uninhibited structure. H746 is ideally positioned to activate the catalytic S965 at a hydrogen-bonding distance of 2.7 Å. However, in none of the inhibitor complexes could we observe a covalent linkage between H746 and the inhibitor, as observed in the trypsin-like serine proteases (Bode et al. 1989). Tricorn is related to the cysteine proteinases in this respect (Eichinger et al. 1999). We confirmed that both residues are crucial for catalysis by constructing the single-site mutants S965A and H746A, both of which are amidolytically inactive. The H746 is correctly oriented by the O^γ of S745 which in turn is polarized by E1023.

The arrangement of S965, H746, and the oxyanion hole suggests that the classical steps of peptide-bond hydrolysis follow the sequence of the trypsin-like serine proteases, namely the formation of the tetrahedral adduct, the acyl–enzyme complex, and hydrolysis. Tricorn has been shown to exhibit both tryptic and chymotryptic specificities (Tamura et al. 1996a). The X-ray structure reveals that specificity for basic P1 residues is conferred by D936 which is provided by the diad-related subunit (see Figures 10.9 and 10.10).

In this way, the previously described structural linkage (trimer of dimers) is translated into functional cooperativity within the dimers. Intriguingly, in the uninhibited high-resolution crystal structure, the acidic S1 specificity-determinant residue D936 was mobile. Consistent with this, the side chain of D936 in the TPCK complex structure adopts an alternative rotamer to allow the TPCK phenyl ring to freely access the hydrophobic niche formed by Y946, I969, V991, and F1013. D936 thus serves as a substrate-specificity switch accommodating both hydrophobic and basic P1 residues. The SO$_2$ group of TPCK and TLCK interacts with the NH moiety of I994, thereby already suggesting the strand E993–P996 as the unprimed-substrate docking site. These substrate-recognition sites are rather un-

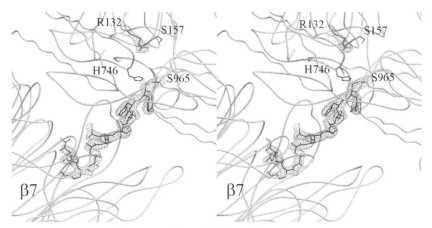

Fig. 10.9. Stereo view of a 13-mer chloromethyl ketone bound to the active site. The electron density of the peptide directs to the β7 propeller.

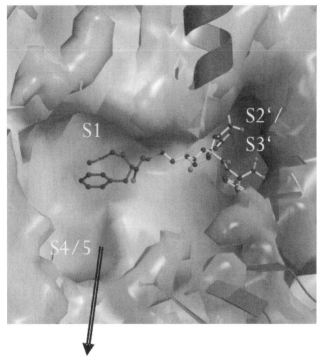

b7 propeller

Fig. 10.10. Detailed active-site view and substrate recognition as deduced from experimental complex structures. The substrate C-terminus is anchored by R131 and R132.

restricted in accord with tricorn's broad substrate specificity (Tamura et al. 1996a; Tamura et al. 1996b).

This situation is contrasted by the length restriction of the primed-substrate recognition site. A prominent cluster of basic residues (R131, R132) delineates the binding site of the substrate C-terminus. These basic residues, positioned on a flexible loop as discussed in detail below, together with the primed-site topology, clearly mark tricorn as a carboxypeptidase. The geometric dimensions explain tricorn's preferential di- and tri-carboxypeptidase activity, while the cleavage of longer peptides will require some conformational rearrangement and is energetically less favorable. By contrast, single amino acids cannot be cleaved off a substrate, because the P1' residue is unable to anchor its carboxylate-group on the basic backstop residues (see Figures 10.9 and 10.10).

A negative charge was not tolerated at positions P3, P4, and P5 of a synthetic fluorogenic AMC-substrate (Tamura et al. 1996b). The crystal structure did not indicate any steric or electrostatic conflicts, if a canonical binding mode of these substrates was assumed. Owing to their lack of a free C-terminus, the charge polarity of N-terminally succinylated fluorogenic substrates is inverted with respect to an unmodified peptide substrate and may lead to unproductive binding with inverted strand polarity.

Each of the three C-terminal domains (C1, PDZ, C2) is remarkably similar to the respective domains (A, B, and C) found in the D1-processing protease (D1P) of photosystem II. The rms deviations between the Cα positions of these domains are 2.2, 2.3, and 2.7 Å with 84, 86, and 135 matching amino acids, respectively. A weak homology between these domains is recognizable in the primary sequences (11, 19, and 20% identities). The relative arrangement of these domains, however, differs very much between tricorn and D1P. With the C2 domain aligned to the C domain of D1P, the orientation of the C1 domain differs from that of the D1P A-domain by 35°. Analogously, the required transformation to align the PDZ-like domains includes a 96° rotation. The rotation axes of these transformations are unrelated to each other. In addition, proper alignment of the PDZ domain requires a 30-Å translation. The catalytic serine residues (S965 and S372, respectively) are positioned on topologically equivalent positions at the helix entrance in the C2 (C) domain (D1P). Further, the amides forming the oxyanion hole (G918, D966, and G318, A373 in tricorn and D1P, respectively) superimpose to within 1 Å. As in other Tsp-like proteases, the residue serving as general base in D1P is a lysine (K372) residing within the C domain of D1P, while it is a histidine in tricorn (H746) which resides on tricorn's C1 domain. The relative arrangement of the C1 and C2 domains in tricorn must, for that reason, remain very restricted to allow for proper catalysis.

One role of the PDZ domain in substrate recognition has been shown for Tsp (Beebe et al. 2000) and was analogously suggested for the tricorn protease (Ponting and Pallen 1999). While the GLGF substrate recognition element is structurally conserved ($R^{764}IAC^{767}$ in tricorn), as pointed out earlier, it appears for a number of reasons unlikely that the tricorn PDZ will participate in substrate recognition in the same way as suggested for D1P (Liao et al. 2000): (1) The putative substrate-binding site as defined by the crystal structures of the C-terminal peptides complexed with PDZ domains (Cabral et al. 1996; Doyle et al. 1996) is partly occupied by outer strands of blade 3 of β6 within the same subunit; (2) the generally conserved arginine (R247) involved in recognition of the carboxylate of the peptide C-terminus corresponds to a hydrophobic residue in tricorn (I851); (3) the orientation and position of tricorn PDZ differs so strongly from that seen in D1P that any analogy based on the sequential domain arrangement is invalidated on the basis of their respective three-dimensional domain arrangement. Instead, the PDZ domain mainly serves to scaffold the sub-domains as described earlier and, in addition, might be involved in recognition of associating component proteins.

10.4.3
Substrate Access and Product Egress Through β-propellers

The comparison with POP, including the open Velcro-topology (Fülöp et al. 1998), suggests an important role of the β-propellers for substrate access to and product exit from the active site (Engel et al. 2003)(see Figure 10.11). Both the β6 and β7 propeller axes are directed towards the active site of the protein, almost intersecting near S965. The arginine anchor (R131, R132) obstructs the otherwise direct connection from the active-site chamber to the exterior through the β6 propeller. Given these observations, we propose that the β6-propeller channel represents one, if not the major, rear exit from the catalytic chamber. This is consistent with the point mutation L184C, positioned within the β6 propeller. The introduced thiol group was modified with maleimide, partially blocking the β6 propeller. The activity of this mutant enzyme towards fluorogenic substrates is significantly reduced (< 50%) compared with the wild-type protein (Brandstetter et al. 2001; Kim et al. 2002). The substrate entrance and product exit paths are indicated in Figure 10.11.

The chloromethyl ketone-based inhibitor-complex crystal structures suggested the strand E993–P996 as a recognition strand for the unprimed-substrate residues.

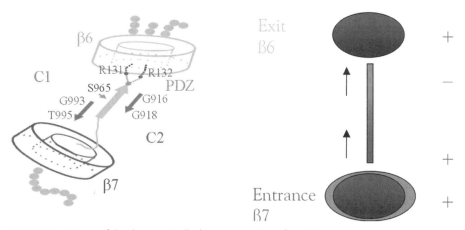

Fig. 10.11. Cartoon of the electrostatically driven processive substrate turnover.

This strand extrapolates towards the β7 channel (see Figure 10.9). The channel through the β7 propeller provides a significantly shorter route from the catalytic chamber to the outside of the protein (60 Å) as compared to the alternative route through the central pore (83 Å). The latter path to the active site has multiple branchings and dead ends. Therefore, the β7 channel might be utilized by the enzyme for the preferred substrate passage to the active site. It is wide open but capped on its outside by four basic residues (R369, R414, R645, K646) which are only partially charge-compensated by one acidic residue (D456). This locally positive lid to the β7 propeller channel is encircled by acidic residues (D333, D335, D372, D456, D506, D508, E592, and E663). Except for E663, which is located on the hairpin connecting strand 3 and 4 of blade 7, all these charged amino acids are positioned between strands 1 and 2 of the respective β7 blades. The resulting charge distribution mimics an electrostatic lens, whereby peptides are pre-oriented with their C-termini towards the central basic propeller lid. Once the entrance to the β7 channel is opened by a concerted side-chain movement of R369, R414, R645, and K646, and possibly assisted by main-chain movements of A643–K646 (blade 7), a peptide is able to enter the channel in an extended conformation where it will find multiple docking sites at the unsaturated inner strands of the β7 propeller blades. A similar substrate-gating filter mechanism through a seven-bladed β propeller has been suggested for the prolyl oligopeptidase (Fülöp et al. 1998; Fülöp et al. 2000), and there is precedent for a β-hairpin binding into a seven-bladed propeller (Ito et al. 1991). The preferred substrate entry through the β7 propeller channel is in line with the point mutation R414C, located in the β7 channel. Derivatization of this introduced thiol group with maleimide markedly reduced the fluorogenic activity of this mutant to about 50% of the wild-type activity (Brandstetter et al. 2001).

Tricorn cleaves substrates in a processive mode (Kim et al. 2002), indicating that only completely digested products will leave the inner protein chambers while larger products will be retained and processed as preferred substrates. The structure suggests several mechanisms to maintain "one way" processing. Basic lids (R414, R645, K646 and R131, R132) are present at the entrances to the β6 and β7 channels. The topology and size of the inner cavities favor an extended conformation of the substrate and the C-terminus of the substrate will be attracted to the basic β6 lid, thereby presenting the substrate's scissile bond at the active site S965 for proteolyis. In one possible scenario, the primed product residues are released by the enzyme through the "rear exit" to the active site formed by the β6 propeller, which is gated by R131–R132. The arginine gate is located on a helical loop containing three glycines (G126, G130, G139) and not restrained to its position via any protein contacts. These glycines might function as hinge residues allowing the gate to move into a sufficiently voluminous cavity of mixed polarity (see Figure 10.10).

The unprimed side of the substrate is held in place by a series of interactions with the protein. In addition to the observed ionic (D936) or hydrophobic S1 interaction site (Y946, I969, V991, F1013), the P1 main chain is held by its interaction with the oxyanion hole (G918, D966). P2–P4 residues will presumably utilize unsaturated main-chain hydrogen bonds at the strand I994–P996 (see Figure 10.10) and further interactions might occur in the β7 propeller channel as described in galactose oxidase (Ito et al. 1991). The modeling studies and suggested substrate binding at the primed and unprimed sides are fully experimentally confirmed by crystal-structural studies using C- and N-terminally extended covalently bound inhibitors (Kim et al. 2002).

Tricorn reportedly cooperates with three additional proteins, termed interacting factors F1, F2, and F3, to degrade oligopeptides sequentially to yield free amino acids (Goettig et al. 2002) . F1 is a prolyl iminopeptidase with 14% sequence identity to the catalytic domain of prolyl oligopeptidase POP, which has an additional propeller domain (Fülöp et al. 1998; Goettig et al. 2002). Guided by this structural scaffold of the latter structure, we speculate that F1 docks onto the six-bladed β-propeller of the tricorn core protein. As in POP, substrate would enter F1 through the propeller channel in this model. While a physical interaction of F1 with tricorn has been suggested (Tamura et al. 1998), the exact mode of interaction of tricorn with F1, F2, and F3 has not been detailed so far

Similarly, there is evidence for functional but not physical interaction of tricorn with the proteasome (Tamura et al. 1998) A physical interaction between these molecules by aligning their respective central pores would imply a symmetry mismatch. While such a physical interaction would be consistent with the geometric dimensions of both molecules, its existence needs to be experimentally confirmed and characterized.

10.4.4
Structural and Functional Relationship of Tricorn and DPIV

The situation in the tricorn protease is closely resembled by *d*ipeptidyl *p*eptidase IV (DPIV) where an eight-bladed topologically open β propeller and a side opening provide entrance to and exit from the active site (see Figure 10.12). Similar to tricorn, DPIV is a serine protease with low but significant structural homology to the family of α/β-hydrolases. We superimposed the catalytic core elements, including the active-site serine and histidine, the strictly conserved helix following the active-site serine (Ser630–Ala642 and Ser965–Leu977, respectively), and tricorn's five-stranded parallel β-sheet onto the equivalent strands of the eight-stranded DPIV-sheet. Both sheets have identical polarity. Significantly, both tricorn propellers come to superimpose onto the two DPIV-openings, the tricorn β7 propeller onto the DPIV β8 propeller, and the tricorn β6 propeller onto the side exit, as schematically indicated in Figure 10.13. This similarity suggests that the β8 propeller provides substrate access to, and the side opening product release from, the DPIV active site. This tricorn-derived model is able to explain the high substrate selectivity critical for DPIV-function to activate or inactivate regulatory peptides. Passage through the β propeller tunnel requires the substrates to unfold thereby providing their "fingerprint" to DPIV. Once the amino terminus of the peptide approaches the active site, it is still held in place by its C-terminus interacting with the β propeller which may contribute to conformationally activate the substrate for cleav-

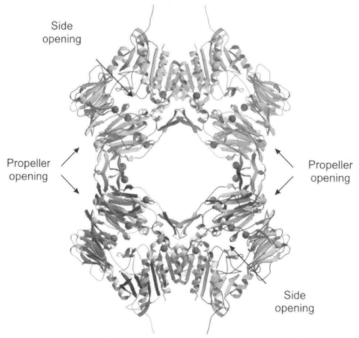

Fig. 10.12. Ribbon representation of the tetrameric DPIV.

age. After the nucleophilic attack the acyl–enzyme intermediate forms, while the primed product is directly released through the side exit. This explains why degradation of glucagon by DPIV is not processive, but occurs sequentially in two independent steps (glucagon 3–29, glucagon 5–29) (Pospisilik et al. 2001). Clearly, the

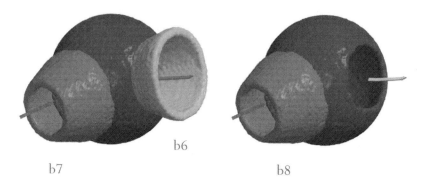

Tricorn   DP IV

Fig. 10.13. Schematic representation of the active-site access and product egress in tricorn and DP IV.

final determination of the functional roles of the DPIV openings awaits further experiments.

10.5
The DegP Protease Chaperone: A Molecular Cage with Bouncers

In this section we describe DegP, a bacterial cage-forming protease, which has homologs in all kingdoms of life. It is distinguished in essential ways from the previous systems by exhibiting extreme flexibility and potential to change its overall shape and its internal structure. Structural flexibility is translated into function and the unique property of DegP to act predominantly as a chaperone or as a protease dependent on temperature. DegP is a Janus-faced molecule appearing as helper or killer as cells need it.

Cells have developed a sophisticated system of molecular chaperones and proteases to reduce the amount of unfolded or aggregated proteins (Wickner et al. 1999). Chaperones recognize hydrophobic stretches of polypeptides that become surface exposed as a consequence of misfolding or unfolding. If refolding attempts fail, irreversibly damaged polypeptides are removed by proteases.

E. coli contains several intracellular proteases that recognize and degrade abnormally folded proteins. The biochemical and structural features of these ATP-dependent proteases have been studied extensively (see Section 10.2). However, relatively little is known about proteases that are responsible for the degradation of non-native proteins in the periplasmic compartment of gram-negative bacteria. Such function has been attributed to the heat-shock protein DegP, also commonly referred to as HtrA or Protease Do. While most factors involved in protein quality control are ATP-dependent heat-shock proteins (Gottesman et al. 1997), DegP fulfills this role without consuming chemical energy (Lipinska et al. 1990). DegP homologs are found in bacteria, fungi, plants, and mammals. Some, but not all, are classical heat-shock proteins. They are localized in extracytoplasmic compartments and have a modular architecture composed of an N-terminal segment believed to have regulatory functions, a conserved trypsin-like protease domain and one or two PDZ domains at the C-terminus (Clausen et al. 2002). PDZ domains are protein modules that mediate specific protein–protein interactions and bind preferentially to the C-terminal 3–4 residues of the target protein (Sheng and Sala 2001). Prokaryotic DegPs have been attributed to the tolerance against thermal, osmotic, oxidative, and pH stress as well as to pathogenicity (Pallen and Wren 1997). A number of DegP substrates are known. These are either largely unstructured proteins such as casein, small proteins that tend to denature, hybrid proteins, or proteins that entered a non-productive folding pathway (Lipinska et al. 1990; Kolmar et al. 1996; Spiess et al. 1999). Stably folded proteins are normally not degraded. In addition to its protease activity, DegP has a general chaperone function. The dual functions switch in a temperature-dependent manner, the protease activity being most apparent at elevated temperatures (Spiess et al. 1999). The ability to switch between refolding and degradation activity and the large variety of known substrates make DegP a key factor controlling protein stability and turnover.

10.5.1

The DegP Protomer, a PDZ Protease

DegP from *E. coli* was crystallized at low temperatures in its chaperone conformation and analyzed (Krojer et al. 2002). The protomer can be divided into three functionally distinct domains, namely a protease and two PDZ domains, PDZ1 and PDZ2 (see Figure 10.14). Like other members of the trypsin family, the protease domain of DegP has two perpendicular β-barrel lobes with a C-terminal helix. The catalytic triad is located in the crevice between the two lobes. While the core of the protease domain is highly conserved, there are striking differences in the surface loops L1, L2, and L3 (for nomenclature see Perona and Craik 1995), which are important for the adjustment of the catalytic triad (Asp105, His135, Ser210) and the

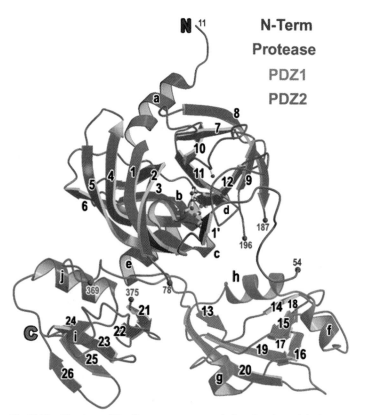

Fig. 10.14. Structure of DegP protomer. Ribbon presentation of the monomer, in which the individual domains are colored differently. Residues of the catalytic triad are shown in a ball-and-stick model. The nomenclature of secondary structure elements, the termini of the protein and regions that were not defined by electron density are indicated.

specificity pocket S1. The enlarged loop LA protrudes into the active site of a molecular neighbor, where it intimately interacts with loops L1 and L2. The resulting loop triad LA*–L1–L2 completely blocks the substrate-binding cleft and results in a severe deformation of the proteolytic site abolishing formation of the catalytic triad, the oxyanion hole, and the S1 specificity pocket. Thus the protease domain of the DegP chaperone is present in an inactive state, in which substrate binding as well as catalysis is prevented. (Krojer et al. 2002).

The structure of the PDZ domains of DegP is similar to PDZ domains of bacterial origin (Liao et al. 2000). Compared to the canonical 4+2 PDZ β-sandwich (Cabral et al. 1996), the DegP PDZ domains show a circularly permuted secondary structure, in which the N- and C-terminal strands are exchanged. Furthermore, they contain a 20-residue insertion following the first β-strand (including helix f) that is important for inter- and intramolecular contacts within the oligomer. In analogy to other PDZ domains, PDZ1 and/or PDZ2 should be involved in substrate binding. PDZ1 contains a deep binding cleft for substrate, which is mainly constructed by strand 14, its N-terminal loop (the so-called carboxylate-binding loop) and helix h. The carboxylate-binding loop is located in a highly positively charged region and is formed by an E–L–G–I motif, which is similar to the frequently observed G–L–G–F motif (Cabral et al. 1996). Binding specificity is mainly conferred by the specific configuration of the 0, −2, and −3 binding pockets (Songyang et al. 1997), where pocket 0 anchors the side chain of the C-terminal residue. In PDZ1, all pockets are built by mainly hydrophobic residues. The thermal motion factors point to the flexibility of strand 14 and its associated carboxylate-binding loop, indicating the plasticity of the binding site. Thus PDZ1 seems to be well adapted to bind various stretches of hydrophobic peptide ligands. Different from PDZ1, the occluded binding site of PDZ2 is unlikely to be involved in substrate recognition.

10.5.2
The Two Forms of the DegP Hexamer

In the crystallographic asymmetric unit, two DegP molecules (A and B) were observed, which build up two distinct hexamers (see Figure 10.15). Both hexamers are formed by staggered association of two trimeric rings. Hexamer A is a largely open structure with a wide lateral passage penetrating the entire complex (see Figure 10.15A), whereas hexamer B corresponds to the closed form, in which a cylindrical 45-Å cavity containing the proteolytic sites is completely shielded from solvent (see Figure 10.15B). In both cases, the top and bottom of the DegP cage are constructed by the six protease domains, whereas the twelve PDZ domains generate the mobile sidewalls. The height of the cavity is determined by three molecular pillars, which are formed by the enlarged LA loops of the protease domain. The PDZ domains are able to adopt different conformations and represent side doors that may open. This en-bloc mobility enables the PDZ domains to function as tentacular arms capturing substrates and delivering them to the inner cavity. This structural organization is strikingly different from the other cage-forming

(A)

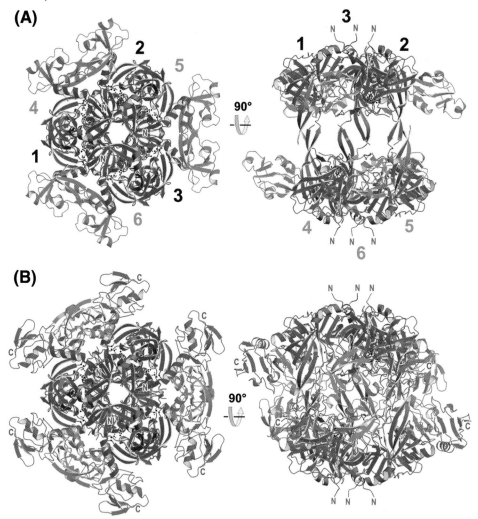

(B)

Fig. 10.15. Structure of the DegP hexamer. Ribbon presentation of the monomer, in which the individual domains are colored differently. Residues of the catalytic triad are shown in a ball-and-stick model. The nomenclature of secondary structure elements, the termini of the protein and regions that were not defined by electron density are indicated.

proteases, where substrates enter the central cavity through narrow axial or lateral pores as described in Sections 10.2 to 10.4.

10.5.3
DegP, a Chaperone

E. coli DegP has the ability to stabilize and support the refolding of several non-native proteins *in vivo* and *in vitro* (Spiess et al. 1999; Misra et al. 2000). Possible

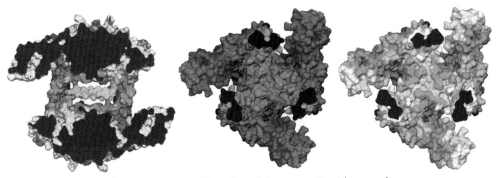

Fig. 10.16. Properties of the inner cavity. Half cut presentations of molecule A (left: side view, center and right: top views) with cut regions shown in dark gray. (Left) Surface representation of the internal tunnel illustrating its molecular-sieve character. Access is restricted to single secondary structure elements as shown by the modeled polyalanine helix, which is colored yellow. (Center) Top view on the ceiling of the inner cavity with mapped thermal motion factors to show its plasticity. Flexible regions are colored red, rigid regions are blue. (Right) Formation of the hydrophobic binding patches within the cavity. Hydrophobic residues of the protease domain are shown in cyan, and the non-polar peptide-binding groove of PDZ1 in green.

binding sites for misfolded proteins are located within the inner cavity (see Figure 10.16). The solvent-accessible height of this chamber is 15 Å at its center and increases to 18 Å near the outer entrance. Owing to these geometric constrictions, substrates must be partially unfolded to reach the active site (see Figure 10.15). As in other chaperones of known structure, the DegP cavity is lined by hydrophobic residues. Two major hydrophobic grooves can be distinguished, which are mainly constructed by residues of loop LA and L2. Notably, the hydrophobic binding sites of the PDZ1 domains are properly oriented to augment the number of potential binding patches. The alternating arrangement of polar and hydrophobic surfaces, both within one trimeric ring and between trimeric rings, should allow the binding of exposed hydrophobic side chains as well as of the peptide backbone of substrates. Taken together, the ceilings of the DegP cavity represent docking platforms for partially misfolded proteins. Both platforms are structurally flexible and should thus allow binding of diverse polypeptides.

10.5.4
The Protease Form

The protease conformation of DegP is still elusive as crystallization of a substrate-like inhibitor complex has failed and maintenance of a stably folded protein precludes long-term experimentation at elevated temperatures where it displays protease activity. We propose a profound rearrangement of the LA*–L1–L2 loop triad into the canonical conformation of active serine proteases competent for substrate binding. This may be initiated by a collapse of the hydrophobic LA platforms and an enlargement of the hydrophobic contacts caused at high temperature.

10.5.5
Working Model for an ATP-independent Heat-shock Protein

Cage-forming proteases and chaperones can be ATP-dependent or -independent. In the former group, ATPase activity is important for recognition of target proteins, their dissociation and unfolding, their translocation within the complex, and for various gating mechanisms. The present crystal structure indicates why these functions are not relevant for DegP. DegP preferably degrades substrates, which are *per se* partially unfolded and which might accumulate under extreme conditions (Swamy et al. 1983; Strauch et al. 1989; Lipinska et al. 1990). Alternatively, threading of substrate through the inner chamber could promote unfolding into an extended conformation. Removal of higher-order structural elements may reinitiate substrate folding after exit from DegP. By binding to the C-terminus or a *β*-hairpin loop of a protein, the PDZ domains could properly position the substrate for threading it into the central cavity. After accessing this chamber, the fate of the unfolded protein depends on the interplay and structural organization of loops LA, L1, and L2. Recruitment of PDZ domains for the gating mechanism should permit a direct coupling of substrate binding and translocation within the DegP particle. This two-step binding process is similar to that of other cage-forming proteins such as the proteasome or the Clp proteases. Here, two binding sites (chambers) exist, the first of which primarily determines substrate specificity.

Acknowledgment

M.B. thanks Dr. Ravishankar Ramachandran, Molecular and Structural Biology division, Central Drug Research Institute, Lucknow, India for a critical reading of his contribution to the manuscript.

References

ADAMS, J., BEHNKE, M., CHEN, S., CRUICKSHANK, A. A., DICK, L. R., GRENIER, L., KLUNDER, J. M., MA, Y.-T., PLAMONDON, L., and STEIN, R. L. Potent and selective inhibitors of the proteasome: dipeptidyl boronic acids. *Bioorg. Med. Chem. Lett.* **1998**, *8*, 333–338.

ARENDT, C. S. and HOCHSTRASSER, M. Eukaryotic 20S proteasome catalytic subunit propeptides prevent active site inactivation by N-terminal acetylation and promote particle assembly. *EMBO J.* **1999**, *18*, 3575–3585.

BABBITT, P. C. and GERLT, J. A. Understanding enzyme superfamilies. Chemistry as the fundamental determinant in the evolution of new catalytic activities. *J. Biol. Chem.* **1997**, *272*, 30591–30594.

BANECKI, B., WAWRZYNOW, A., PUZEWICZ, J., GEORGOPOULOS, C., and ZYLICZ, M. Structure-function analysis of the zinc-binding region of the Clpx molecular chaperone. *J. Biol. Chem.* **2001**, *276*, 18843–18848.

BEEBE, K. D., SHIN, J., PENG, J., CHAUDHURY, C., KHERA, J., and PEI, D. Substrate Recognition through a PDZ Domain in Tail-Specific Protease. *Biochemistry* **2000**, *39*, 3149–3155.

BENAROUDJ, N., ZWICKL, P., SEEMULLER, E., BAUMEISTER, W., and GOLDBERG, A. L. ATP hydrolysis by the proteasome regulatory complex PAN serves multiple functions in protein degradation. *Mol Cell* **2003**, *11*, 69–78.

BOCHTLER, M., PH. D. THESIS, TU München **1999**.

BOCHTLER, M., DITZEL, L., GROLL, M., and HUBER, R. Crystal structure of heat shock locus V (HslV) from Escherichia coli. *Proc. Natl. Acad. Sci. USA* **1997**, *94*, 6070–6074.

BOCHTLER, M., HARTMANN, C., SONG, H. K., BOURENKOV, G. P., BARTUNIK, H. D., and HUBER, R. The structures of HslU and the ATP-dependent protease HslU-HslV. *Nature* **2000**, *403*, 800–805.

BOCHTLER, M., SONG, H. K., HARTMANN, C., RAMACHANDRAN, R., and HUBER, R. The quaternary arrangement of HslU and HslV in a cocrystal: a response to Wang, Yale. *J Struct Biol* **2001**, *135*, 281–293.

BODE, W., MAYR, I., BAUMANN, U., HUBER, R., STONE, S. R., and HOFSTEENGE, J. The refined 1.9 A crystal structure of human alpha-thrombin: interaction with D-Phe-Pro-Arg chloromethylketone and significance of the Tyr-Pro-Pro-Trp insertion segment. *EMBO J.* **1989**, *8*, 3467–3475.

BOGYO, M., McMASTER, J. S., GACZINSKA, M., TORTORELLA, D., GOLDBERG, A. L., and PLOEGH, H. Covalent modification of the active site threonine of proteasomal β-subunits and the Escherichia coli homolog HslV by a new class of inhibitors. *Proc. Nat. Acad. Sci. USA* **1997**, *94*, 6629–6634.

BOGYO, M., SHIN, S., McMASTER, J. S., and PLOEGH, H. L. Substrate binding and sequence preference of the proteasome revealed by active-site-directed affinity probes. *Chem Biol* **1998**, *5*, 307–320.

BRANDSTETTER, H., KIM, J.-S., GROLL, M., and HUBER, R. The crystal structure of the tricorn protease reveals a protein disassembly line. *Nature* **2001**, *414*, 466–469.

BRANNIGAN, J. A., DODSON, G., DUGGLEBY, H. J., MOODY, P. C. E., SMITH, J. L., TOMCHICK, D. R., and MURZIN, A. G. A protein catalytic framework with an N-terminal nucleophile is capable of self-activation. *Nature* **1995**, *378*, 416–419.

CABRAL, J. H. M., PETOSA, C., SUTCLIFFE, M. J., RAZA, S., BYRON, O., POY, F., MARFATIA, S. M., CHISHTI, A. H., and LIDDINGTON, R. C. Crystal structure of a PDZ domain. *Nature* **1996**, *382*, 649–652.

CARDOZO, C., VINITSKY, A., MICHAUD, C., and ORLOWSKI, M. Evidence that the nature of amino acid residues in the P3 position directs substrates to distinct catalytic sites of the pituitary multicatalytic proteinase complex (proteasome). *Biochemistry* **1994**, *33*, 6483–6489.

CHUANG, S. E., and BLATTNER, F. R. Characterization of twenty-six new heat shock genes of Escherichia coli. *J Bacteriol* **1993**, *175*, 5242–5252.

CHUANG, S. E., BURLAND, V., PLUNKETT, G., DANIELS, D. L., and BLATTNER, F. R. Sequence analysis of four new heat shock genes constituting the hslu and hslv operons in Escherichia coli. *Gene* **1993**, *134*, 1–6.

CLAUSEN, T., SOUTHAN, C., and EHRMANN, M. The HtrA family of proteases: implications for protein composition and cell fate. *Mol Cell* **2002**, *10*, 443–455.

COUVREUR, B., WATTIEZ, R., BOLLEN, A., FALMAGNE, P., LE RAY, D., and DUJARDIN, J. C. Eubacterial HslV and HslU subunits homologs in primordial eukaryotes. *Mol Biol Evol* **2002**, *19*, 2110–2117.

DICK, T. P., NUSSBAUM, A. K., DEEG, M., HEINEMEYER, W., GROLL, M., SCHIRLE, M., KEILHOLZ, W., STEVANOVIC, S., WOLF, D. H., HUBER, R., et al. Contribution of proteasomal beta-subunits to the cleavage of peptide substrates analyzed with yeast mutants. *J. Biol. Chem.* **1998**, *273*, 25637–25646.

DITZEL, L., HUBER, R., MANN, K., HEINEMEYER, W., WOLF, D. H., and GROLL, M. Conformational constraints for protein self-cleavage in the proteasome. *J. Mol. Biol.* **1998**, *279*, 1187–1191.

DONALDSON, L. W., WOJTYRA, U., and HOURY, W. A. Solution structure of the dimeric zinc binding domain of the chaperone ClpX. *J. Biol. Chem.* **2003**, *278*, 48991–48996.

DOUGAN, D. A., MOGK, A., ZETH, K., TURGAY, K., and BUKAU, B. AAA+ proteins and substrate recognition, it all depends on their partner in crime. *FEBS Lett.* **2002a**, *529*, 6–10.

DOUGAN, D. A., REID, B. G., HORWICH, A. L., and BUKAU, B. ClpS, a substrate modulator of the ClpAP machine. *Mol Cell* **2002b**, *9*, 673–683.

DOUGAN, D. A., WEBER-BAN, E., and BUKAU, B. Targeted delivery of an ssrA-tagged substrate by the adapter protein SspB to its cognate AAA+ protein ClpX. *Mol Cell* **2003**, *12*, 373–380.

DOYLE, D. A., LEE, A., LEWIS, J., KIM, E., SHENG, M., and MACKINNON, R. Crystal structures of a complexed and peptide-free membrane protein-binding domain: Molecular basis of peptide recognition by PDZ. *Cell* **1996**, *85*, 1067–1076.

EICHINGER, A., BEISEL, H. G., JACOB, U., HUBER, R., MEDRANO, F. J., BANBULA, A., POTEMPA, J., TRAVIS, J., and BODE, W. Crystal structure of gingipain R: an Arg-specific bacterial cysteine proteinase with a caspase-like fold. *EMBO J.* **1999**, *18*, 5453–5462.

ELOFSSON, M., SPLITTGERBER, U., MYUNG, J., MOHAN, R., and CREWS, C. M. Towards subunit-specific proteasome inhibitors: synthesis and evaluation of peptide alpha′,beta′-epoxyketones. *Chem Biol* **1999**, *6*, 811–822.

ENGEL, M., HOFFMANN, T., WAGNER, L., WERMANN, M., HEISER, U., KIEFERSAUER, R., HUBER, R., BODE, W., DEMUTH, H.-U., and BRANDSTETTER, H. The crystal structure of dipeptidyl peptidase IV (CD26) reveals its functional regulation and enzymatic mechanism. *Proc. Natl. Acad. Sci. USA* **2003**, *100*, 5063–5068.

FENTEANY, G., STANDAERT, R. F., LANE, W. S., CHOI, S., COREY, E. J., and SCHREIBER, S. L. Inhibition of proteasome activities and subunit-specific amino-terminal threonine modification by lactacystin. *Science* **1995**, *268*, 726–731.

FÜLÖP, V., BÖCSKEI, Z., and POLGÁR, L. Prolyl Oligopeptidase: An Unusual *β*-Propeller Domain Regulates Proteolysis. *Cell* **1998**, *94*, 161–170.

FÜLÖP, V., SZELTNER, Z., and POLGAR, L. Catalysis of serine oligopeptidases is controlled by a gating filter mechanism. *EMBO Reports* **2000**, *1*, 277–281.

GLICKMAN, M. H., RUBIN, D. M., COUX, O., WEFES, I., PFEIFER, G., CJEKA, Z., BAUMEISTER, W., FRIED, V. A., and FINLEY, D. A subcomplex of the proteasome regulatory particle required for ubiquitin-conjugate degradation and related to the COP9-signalosome and eIF3. *Cell* **1998**, *94*, 615–623.

GLICKMAN, M. H., RUBIN, D. M., FU, H., LARSEN, C. N., COUX, O., WEFES, I.,

PFEIFER, G., CJEKA, Z., VIERSTRA, R., BAUMEISTER, W., et al. Functional analysis of the proteasome regulatory particle. *Mol. Biol. Rep.* **1999**, *26*, 21–28.

GOETTIG, P., GROLL, M., KIM, J.-S., HUBER, R., and BRANDSTETTER, H. Crystal structures of the tricorn interacting aminopeptidase F1 with various ligands reveal its catalytic mechanism. *EMBO J.* **2002**, 21.

GOTTESMAN, S., WICKNER, S., and MAURIZI, M. R. Protein quality control: Triage by chaperones and proteases. *Genes & Development* **1997**, *11*, 815–823.

GRIMAUD, R., KESSEL, M., BEURON, F., STEVEN, A. C., and MAURIZI, M. R. Enzymatic and structural similarities between the Escherichia coli ATP-dependent proteases, ClpXP and ClpAP. *J. Biol. Chem.* **1998**, *273*, 12476–12481.

GROLL, M., DITZEL, L., LÖWE, J., STOCK, D., BOCHTLER, M., BARTUNIK, H. D., and HUBER, R. Structure of 20S proteasome from yeast at 2.4 Å resolution. *Nature* **1997**, *386*, 463–471.

GROLL, M., HEINEMEYER, W., JAGER, S., ULLRICH, T., BOCHTLER, M., WOLF, D. H., and HUBER, R. The catalytic sites of 20S proteasomes and their role in subunit maturation: a mutational and crystallographic study. *Proc. Natl. Acad. Sci. USA* **1999**, *96*, 10976–10983.

GROLL, M., BAJOREK, M., KÖHLER, A., MORODER, L., RUBIN, D. M., HUBER, R., GLICKMAN, M. H., and FINLEY, D. A gated channel into the proteasome core particle. *Nat. Struct. Biol.* **2000a**, *7*, 1062–1067.

GROLL, M., KIM, K. B., KAIRIES, N., HUBER, R., and CREWS, C. M. Crystal structure of epoxomicin: 20S proteasome reveals a molecular basis for selectivity of alpha′,beta′-epoxyketone proteasome inhibitors. *J. Am. Chem. Soc.* **2000b**, *122*, 1237–1238.

GROLL, M., KOGUCHI, Y., HUBER, R., and KOHNO, J. Crystal structure of the 20 S proteasome:TMC-95A complex: a non-covalent proteasome inhibitor. *J. Mol. Biol.* **2001**, *311*, 543–548.

GROLL, M., NAZIF, T., HUBER, R., and BOGYO, M. Probing structural determinants distal to the site of hydrolysis that control substrate specificity of the 20S proteasome. *Chem. Biol.* **2002**, *9*, 1–20.

GROLL, M., BRANDSTETTER, H., BARTUNIK, H., BOURENKOW, G., and HUBER, R. Investigations on the maturation and regulation of archaebacterial proteasomes. *J Mol Biol.* **2003**, *327*, 75–83.

GUO, F., ESSER, L., SINGH, S. K., MAURIZI, M. R., and XIA, D. Crystal structure of the heterodimeric complex of the adapter, ClpS, with the N-domain of the AAA+ chaperone, ClpA. *J. Biol. Chem.* **2002a**, *277*, 46753–46762.

GUO, F., MAURIZI, M. R., ESSER, L., and XIA, D. Crystal structure of ClpA, an Hsp100 chaperone and regulator of ClpAP protease. *J. Biol. Chem.* **2002b**, *277*, 46743–46752.

HEINEMEYER, W., FISCHER, M., KRIMMER, T., STACHON, U., and WOLF, D. H. The active sites of the eukaryotic 20S proteasome and their involvement in subunit precursor processing. *J. Biol. Chem.* **1997**, *272*, 25200–25209.

HUANG, H.-C., and GOLDBERG, A. L. Proteolytic activity of the ATP-dependent protease HslVU can be uncoupled from ATP-hydrolysis. *J. Biol. Chem.* **1997**, *272*, 21364–21372.

ISHIKAWA, T., MAURIZI, M. R., BELNAP, D., and STEVEN, A. C. Docking of components in a bacterial complex. *Nature* **2000**, *408*, 667–668.

ITO, N., PHILLIPS, S. E. V., STEVENS, C., OGEL, Z. B., McPHERSON, M. J., KEEN, J. N., YADAV, K. D. S., and KNOWLES, P. F. Novel thioether bond revealed by a 1.7 A crystal structure of galactose oxidase. *Nature* **1991**, *35*, 87–90.

JÄGER, S., GROLL, M., HUBER, R., WOLF, D. H., and HEINEMEYER, W. Proteasome beta-type subunits: unequal roles of propeptides in core particle maturation and a hierarchy of active site function. *J. Mol. Biol.* **1999**, *291*, 997–1013.

KAISER, M., GROLL, M., RENNER, C., HUBER, R., and MORODER, L. The core structure of TMC-95A is a promising lead for reversible proteasome inhibition. *Angew. Chem. Int. Ed.* **2002**, *41*, 780–783.

KANEMORI, M., NISHIHARA, K., YANAGI, H., and YURA, T. Synergistic roles of HslVU and other ATP-dependent proteases in controlling in vivo turnover of σ32 and abnormal proteins in Escherichia coli. *J. Bact.* **1997**, *179*, 7219–7225.

KANEMORI, M., YANAGI, H., and YURA, T. The ATP-dependent HslVU/ClpQY protease participates in turnover of cell division

inhibitor SulA in Escherichia coli. *J Bacteriol* **1999**, *181*, 3674–3680.

KANG, M. S., KIM, S. R., KWACK, P., LIM, B. K., AHN, S. W., RHO, Y. M., SEONG, I. S., PARK, S. C., EOM, S. H., CHEONG, G. W., et al. Molecular architecture of the ATP-dependent CodWX protease having an N-terminal serine active site. *EMBO J.* **2003**, *22*, 2893–2902.

KANG, M. S., LIM, B. K., SEONG, I. S., SEOL, J. H., TANAHASHI, N., TANAKA, K., and CHUNG, C. H. The ATP-dependent CodWX (HslVU) protease in Bacillus subtilis is an N-terminal serine protease. *EMBO J.* **2001**, *20*, 734–742.

KARATA, K., INAGAWA, T., WILKINSON, A. J., TATSUTA, T., and OGURA, T. Dissecting the role of a conserved motif (the second region of homology) in the AAA family of ATPases. Site-directed mutagenesis of the ATP-dependent protease FtsH. *J. Biol. Chem.* **1999**, *274*, 26225–26232.

KESSEL, M., MAURIZI, M. R., KIM, B., KOCSIS, E., TRUS, B. L., SINGH, S. K., and STEVEN, A. C. Homology in structural organization between E. coli ClpAP protease and the eukaryotic 26S proteasome. *J. Mol. Biol* **1995**, *250*, 587–594.

KESSEL, M., WHU, W. F., GOTTESMAN, S., KOCSIS, E., STEVEN, A. C., and MAURIZI, M. R. Six-fold rotational symmetry of ClpQ, the E. coli homolog of the 20S proteasome, and its ATP-dependent activator, ClpY. *FEBS Lett.* **1996**, *398*, 274–278.

KHATTAR, M. M. Overexpression of the hslVU operon suppresses SOS-mediated inhibition of cell division in Escherichia coli. *FEBS Lett.* **1997**, *414*, 402–404.

KIM, D. Y. and KIM, K. K. Crystal structure of ClpX molecular chaperone from Helicobacter pylori. *J. Biol. Chem.* **2003**, *278*, 50664–50670.

KIM, J.-S., GROLL, M., MUSIOL, H. J., BEHRENDT, R., KAISER, M., MORODER, L., HUBER, R., and BRANDSTETTER, H. Navigation inside a protease: substrate selection and product exit in the tricorn protease from Thermoplasma acidophilum. *J. Mol. Biol.* **2002**, *324*, 1041–1050.

KIM, Y. I., LEVCHENKO, I., FRACZKOWSKA, K., WOODRUFF, R. V., SAUER, R. T., and BAKER, T. A. Molecular determinants of complex formation between Clp/Hsp100 ATPases and the ClpP peptidase. *Nat. Struct. Biol.* **2001**, *8*, 230–233.

KLOETZEL, P. Antigen processing by the proteasome. *Nat Rev Mol Cell Biol.* **2001**, *2*, 179–187.

KOGUCHI, Y., KOHNO, J., NISHIO, M., TAKAHASHI, K., OKUDA, T., OHNUKI, T., and KOMATSUBARA, S. TMC-95A, B, C, and D, novel proteasome inhibitors produced by Apiospora montagnei Sacc. TC 1093. Taxonomy, production, isolation, and biological activities. *J Antibiot (Tokyo)* **2000**, *53*, 105–109.

KÖHLER, A., CASCIO, P., LEGGETT, D. S., WOO, K. M., GOLDBERG, A. L., and FINLEY, D. The axial channel of the proteasome core particle is gated by the Rpt2 ATPase and controls both substrate entry and product release. *Molecular Cell* **2001**, *7*, 1143–1152.

KOHNO, J., KOGUCHI, Y., NISHIO, M., NAKAO, K., KURODA, M., SHIMIZU, R., OHNUKI, T., and KOMATSUBARA, S. Structures of TMC-95A-D: novel proteasome inhibitors from Apiospora montagnei sacc. TC 1093. *J Org Chem* **2000**, *65*, 990–995.

KOLMAR, H., WALLER, P. R. H., and SAUER, R. T. The DegP and DegQ periplasmic endoproteases of Escherichia coli: Specificity for cleavage sites and substrate conformation. *J. Bacteriol.* **1996**, *178*, 5925–5929.

KROJER, T., GARRIDO-FRANCO, M., HUBER, R., EHRMANN, M., and CLAUSEN, T. Crystal structure of DegP (HtrA) reveals a new protease-chaperone machine. *Nature* **2002**, *416*, 455–459.

KRZYWDA, S., BRZOZOWSKI, A. M., VERMA, C., KARATA, K., OGURA, T., and WILKINSON, A. J. The crystal structure of the AAA domain of the ATP-dependent protease FtsH of Escherichia coli at 1.5 A resolution. *Structure (Camb)* **2002**, *10*, 1073–1083.

KWON, A. R., KESSLER, B. M., OVERKLEEFT, H. S., and McKAY, D. B. Structure and reactivity of an asymmetric complex between HslV and I-domain deleted HslU, a prokaryotic homolog of the eukaryotic proteasome. *J. Mol. Biol.* **2003**, *330*, 185–195.

LEE, Y. Y., CHANG, C. F., KUO, C. L., CHEN, M. C., YU, C. H., LIN, P. I., and WU, W. F. Subunit oligomerization and substrate recognition of the Escherichia coli ClpYQ (HslUV) protease implicated by in vivo protein-protein interactions in the yeast two-hybrid system. *J. Bacteriol.* **2003**, *185*, 2393–2401.

LEVCHENKO, I., GRANT, R. A., WAH, D. A., SAUER, R. T., and BAKER, T. A. Structure of a delivery protein for an AAA+ protease in complex with a peptide degradation tag. *Mol Cell* **2003**, *12*, 365–372.

LEVCHENKO, I., SMITH, C. K., WALSH, N. P., SAUER, R. T., and BAKER, T. A. PDZ-like domains mediate binding specificity in the Clp/Hsp100 family of chaperones and protease regulatory subunits. *Cell* **1997**, *91*, 939–947.

LIAO, D. I., QIAN, J., CHISHOLM, D. A., JORDAN, D. B., and DINER, B. A. Crystal structures of the photosystem II D1 C-terminal processing protease. *Nat. Struct. Biol.* **2000**, *7*, 749–753.

LIN, S. and DANISHEFSKY, S. The total synthesis of proteasome inhibitors TMC-95A and TMC-95B: discovery of a new method to generate cis-propenyl amides. *Angew Chem Int Ed Engl.* **2002**, *41*, 512–515.

LIPINSKA, B., ZYLICZ, M., and GEORGOPOULOS, C. The Htra (Degp) Protein, Essential For Escherichia-Coli Survival At High-Temperatures, Is an Endopeptidase. *J. Bacteriol.* **1990**, *172*, 1791–1797.

LOIDL, G., GROLL, M., MUSIOL, H. J., HUBER, R., and MORODER, L. Bivalency as a principle for proteasome inhibition. *Proc. Natl. Acad. Sci. USA* **1999**, *96*, 5418–5422.

LÖWE, J., STOCK, D., JAP, B., ZWICKL, P., BAUMEISTER, W., and HUBER, R. Crystal structure of the 20S proteasome from the archaeon T. acidophilum at 3.4 Å resolution. *Science* **1995**, *268*, 533–539.

LUPAS, A., ZWICKL, P., and BAUMEISTER, W. Proteasome sequences in eubacteria. *Trends Biochem. Sci.* **1994**, *19*, 533–534.

MENG, L., MOHAN, R., KWOK, B. H., ELOFSSON, M., SIN, N., and CREWS, C. M. Epoxomicin, a potent and selective proteasome inhibitor, exhibits in vivo antiinflammatory activity. *Proc. Natl. Acad. Sci. USA* **1999**, *96*, 10403–10408.

MISRA, R., CASTILLOKELLER, M., and DENG, M. Overexpression of protease-deficient DegP(S210A) rescues the lethal phenotype of Escherichia coli OmpF assembly mutants in a degP background. *J. Bacteriol.* **2000**, *182*, 4882–4888.

MISSIAKIS, D., SCHWAGER, F., BETTON, J.-M., GEORGOPOULOS, C., and RAINA, S. Identification and characterizaton of HslV HslU (ClpQ ClpY) proteins involved in overall proteolysis of misfolded proteins in Escherichia coli. *EMBO J.* **1996**, *15*, 6899–6909.

NAVON, A., and GOLDBERG, A. L. Proteins are unfolded on the surface of the ATPase ring before transport into the proteasome. *Mol Cell* **2001**, *8(6)*, 1339–1349.

NEUWALD, A. F., ARAVIND, L., SPOUGE, J. L., and KOONIN, E. V. AAA+: A class of chaperone-like ATPases associated with the assembly, operation, and disassembly of protein complexes. *Genome Res* **1999**, *9*, 27–43.

NIWA, H., TSUCHIYA, D., MAKYIO, H., YOSHIDA, M., and MORIKAWA, K. Hexameric ring structure of the ATPase domain of the membrane-integrated metalloprotease FtsH from Thermus thermophilus HB8. *Structure (Camb)* **2002**, *10*, 1415–1423.

NUSSBAUM, A. K., DICK, T. P., KEILHOLZ, W., SCHIRLE, M., STEVANOVIC, S., DIETZ, K., HEINEMEYER, W., GROLL, M., WOLF, D. H., HUBER, R., et al. Cleavage motifs of the yeast 20S proteasome beta subunits deduced from digests of enolase 1. *Proc. Natl. Acad. Sci. USA* **1998**, *95*, 12504–12509.

ORTEGA, J., SINGH, S. K., ISHIKAWA, T., MAURIZI, M. R., and STEVEN, A. C. Visualization of substrate binding and translocation by the ATP-dependent protease, ClpXP. *Mol Cell* **2000**, *6*, 1515–1521.

PALLEN, M. J. and WREN, B. W. The HtrA family of serine proteases. *Molec. Microbiol.* **1997**, *26*, 209–221.

PERONA, J. J. and CRAIK, C. S. Structural basis of substrate specificity in the serine proteases. *Protein Sci* **1995**, *4*, 337–360.

PONTING, C. P. and PALLEN, M. J. β-Propeller repeats and a PDZ domain in the tricorn protease: predicted self-compartmentalisation and C-terminal polypeptide-binding strategies of substrate selection. *FEMS Microbiol. Lett.* **1999**, *179*, 447–451.

POSPISILIK, J. A., HINKE, S. A., PEDERSON, R. A., HOFFMANN, T., ROSCHE, F., SCHLENZIG, D., GLUND, K., HEISER, U., McINTOSH, C. H., and DEMUTH, H. Metabolism of glucagon by dipeptidyl peptidase IV (CD26). *Regul Pept* **2001**, *96*, 133–141.

RAMACHANDRAN, R., HARTMANN, C., SONG, H. K., HUBER, R., and BOCHTLER, M. Functional interactions of HslV (ClpQ) with the ATPase HslU (ClpY). *Proc. Natl. Acad. Sci. USA* **2002**, *99*, 7396–7401.

RASMUSSEN, H. B., BRANNER, S., WIBERG, F. C., and WAGTMANN, N. Crystal structure of human dipeptidyl peptidase IV/CD26 in complex with a substrate analog. *Nat. Struct. Biol.* **2003**, *9*.

RAWLINGS, N. D., O'BRIEN, E., and BARRETT, A. J. MEROPS: the protease database. *Nucleic Acids Res* **2002**, *30*, 343–346.

RICHMOND, C., GORBEA, C., and RECHSTEINER, M. Specific interactions between the ATPase subunits of the 26S protease. *J. Biol. Chem.* **1997**, *272*, 13403–13411.

ROHRWILD, M., COUX, O., HUANG, H. C., MOERSCHELL, R. P., SOON, J. Y., SEOL, J. H., CHUNG, C. H., and GOLDBERG, A. L. HslV-HslU: A novel ATP-dependent protease complex in Escherichia coli related to the eukaryotic proteasome. *Proc. Natl. Acad. Sci. USA* **1996**, *93*, 5808–5813.

ROHRWILD, M., PFEIFER, G., SANTARIUS, U., MÜLLER, S. A., HUANG, H. C., ENGEL, A., BAUMEISTER, W., and GOLDBERG, A. L. The ATP-dependent HslVU protease from Escherichia coli is a four-ring structure resembling the proteasome. *Nat. Struct. Biol* **1997**, *4*, 133–139.

SEOL, J. H., YOO, S. J., SHIN, D. H., SHIM, Y. K., KANG, M.-S., GOLDBERG, A. L., and CHUNG, C. H. The heat shock protein HslVU from Escherichia coli is a protein activated ATPase as well as an ATP-dependent proteinase. *Eur. J. Biochem.* **1997**, *247*, 1143–1150.

SEONG, I. S., KANG, M. S., CHOI, M. K., LEE, J. W., KOH, O. J., WANG, J., EOM, S. H., and CHUNG, C. H. The C-terminal tails of HslU ATPase act as a molecular switch for activation of HslV peptidase. *J. Biol. Chem.* **2002**, *277*, 25976–25982.

SEONG, I. S., OH, J. Y., YOO, S. J., SEOL, J. H., and CHUNG, C. H. ATP-dependent degradation of SulA, a cell division inhibitor, by the HslVU protease in Escherichia coli. *FEBS Lett.* **1999**, *456*, 211–214.

SHENG, M. and SALA, C. PDZ domains and the organization of supramolecular complexes. *Annu. Rev. Neurosci.* **2001**, *24*, 1–29.

SMITH, C. K., BAKER, T. A., and SAUER, R. T. Lon and Clp family proteases and chaperones share homologous substrate-recognition domains. *Proc. Natl. Acad. Sci. USA* **1999**, *96*, 6678–6682.

Song, H., Bochtler, M., Azim, M., Hartmann, C., Huber, R., and Ramachandran, R. Isolation and characterization of the prokaryotic proteasome homolog HslVU (ClpQY) from Thermotoga maritima and the crystal structure of HslV. *Biophys Chem.* **2003**, *100*, 437–452.

Song, H. K. and Eck, M. J. Structural basis of degradation signal recognition by SspB, a specificity-enhancing factor for the ClpXP proteolytic machine. *Mol Cell* **2003**, *12*, 75–86.

Song, H. K., Hartmann, C., Ramachandran, R., Bochtler, M., Behrendt, R., Moroder, L., and Huber, R. Mutational studies on HslU and its docking mode with HslV. *Proc. Natl. Acad. Sci. USA* **2000**, *97*, 14103–14108.

Songyang, Z., Fanning, A. S., Fu, C., Xu, J., Marfatia, S. M., Chishti, A. H., Crompton, A., Chan, A. C., Anderson, J. M., and Cantley, L. C. Recognition of unique carboxyl-terminal motifs by distinct PDZ domains. *Science* **1997**, *275*, 73–77.

Sousa, M. C., Kessler, B. M., Overkleeft, H. S., and McKay, D. B. Crystal structure of HslUV complexed with a vinyl sulfone inhibitor: corroboration of a proposed mechanism of allosteric activation of HslV by HslU. *J. Mol. Biol.* **2002**, *318*, 779–785.

Sousa, M. C. and McKay, D. B. Structure of Haemophilus influenzae HslV protein at 1.9 A resolution, revealing a cation-binding site near the catalytic site. *Acta Crystallogr D Biol Crystallogr* **2001**, *57*, 1950–1954.

Sousa, M. C., Trame, C. B., Tsuruta, H., Wilbanks, S. M., Reddy, V. S., and McKay, D. B. Crystal and solution structures of an HslUV protease-chaperone complex. *Cell* **2000**, *103*, 633–643.

Spiess, C., Beil, A., and Ehrmann, M. A temperature-dependent switch from chaperone to protease in a widely conserved heat shock protein. *Cell* **1999**, *97*, 339–347.

Strauch, K. L., Johnson, K., and Beckwith, J. Characterization of Degp, a Gene Required For Proteolysis in the Cell-Envelope and Essential For Growth of Escherichia-Coli At High-Temperature. *J. Bacteriol.* **1989**, *171*, 2689–2696.

Studemann, A., Noirclerc-Savoye, M., Klauck, E., Becker, G., Schneider, D., and Hengge, R. Sequential recognition of two distinct sites in sigma(S) by the proteolytic targeting factor RssB and ClpX. *EMBO J.* **2003**, *22*, 4111–4120.

Swamy, K. H. S., Chin, H. C., and Goldberg, A. L. Isolation and Characterization of Protease Do From Escherichia-Coli, a Large Serine Protease Containing Multiple Subunits. *Arch. Biochem. Biophys.* **1983**, *224*, 543–554.

Tamura, N., Lottspeich, F., Baumeister, W., and Tamura, T. The role of tricorn protease and its aminopeptidase-interacting factors in cellular protein degradation. *Cell* **1998**, *95*, 637–648.

Tamura, T., Tamura, N., Cejka, Z., Hegerl, R., Lottspeich, F., and Baumeister, W. Tricorn Protease – The Core of a Modular Proteolytic System. *Science* **1996a**, *274*, 1385–1389.

Tamura, T., Tamura, N., Lottspeich, F., and Baumeister, W. Tricorn protease (TRI) interacting factor 1 from *Thermoplasma acidophilum* is a proline iminopeptidase. *FEBS Lett.* **1996b**, *398*, 101–105.

Unno, M., Mizushima, T., Morimoto, Y., Tomisugi, Y., Tanaka, K., Yasuoka, N., and Tsukihara, T. The Structure of the Mammalian 20S Proteasome at 2.75 A Resolution. *Structure* **2002**, *10(5)*, 609–618.

Walz, J., Erdmann, A., Kania, M., Typke, D., Koster, A. J., and Baumeister, W. 26S proteasome structure revealed by three-dimensional electron microscopy. *J Struct Biol* **1998**, *121*, 19–29.

Wang, J. A corrected quaternary arrangement of the peptidase HslV and atpase HslU in a cocrystal structure. *J Struct Biol* **2001**, *134*, 15–24.

Wang, J. A second response in correcting the HslV-HslU quaternary structure. *J Struct Biol* **2003**, *141*, 7–8.

Wang, J., Hartling, J. A., and Flanagan, J. M. The structure of ClpP at 2.3 A resolution suggests a model for ATP-dependent proteolysis. *Cell* **1997**, *91*, 447–456.

Wang, J., Song, J. J., Franklin, M. C., Kamtekar, S., Im, Y. J., Rho, S. H., Seong, I. S., Lee, C. S., Chung, C. H., and Eom, S. H. Crystal structures of the HslVU peptidase-ATPase complex reveal an ATP-dependent proteolysis mechanism. *Structure (Camb)* **2001a**, *9*, 177–184.

WANG, J., SONG, J. J., SEONG, I. S., FRANKLIN, M. C., KAMTEKAR, S., EOM, S. H., and CHUNG, C. H. Nucleotide-dependent conformational changes in a protease-associated ATPase HsIU. *Structure (Camb)* **2001b**, *9*, 1107–1116.

WENZEL, T., ECKERSKORN, C., LOTTSPEICH, F., and BAUMEISTER, W. Existence of a molecular ruler in proteasomes suggested by analysis of degradation products. *FEBS Lett.* **1994**, *349*, 205–209.

WHITBY, F. G., MASTERS, E. I., KRAMER, L., KNOWLTON, J. R., YAO, Y., WANG, C. C., and HILL, C. P. Structural basis for the activation of 20S proteasomes by 11S regulators. *Nature* **2000**, *408*, 115–120.

WICKNER, S. and MAURIZI, M. R. Here's the hook: similar substrate binding sites in the chaperone domains of Clp and Lon. *Proc. Natl. Acad. Sci. USA* **1999**, *96*, 8318–8320.

WICKNER, S., MAURIZI, R., and GOTTESMAN, S. Posttranslational Quality Control: Folding, Refolding, and Degrading Proteins. *Science* **1999**, *286*, 1888–1893.

WILK, S. and CHEN, W.-E. Synthetic peptide based activators of the proteasome. *Mol. Biol. Rep.* **1997**, *24*, 119–124.

ZETH, K., RAVELLI, R. B., PAAL, K., CUSACK, S., BUKAU, B., and DOUGAN, D. A. Structural analysis of the adapter protein ClpS in complex with the N-terminal domain of ClpA. *Nat. Struct. Biol.* **2002**, *9*, 906–911.

ZHOU, Y., GOTTESMAN, S., HOSKINS, J. R., MAURIZI, M. R., and WICKNER, S. The RssB response regulator directly targets sigma(S) for degradation by ClpXP. *Genes Dev* **2001**, *15*, 627–637.

ZWICKL, P., NG, D., WOO, K. M., KLENK, H. P., and GOLDBERG, A. L. An archaebacterial ATPase, homologous to ATPases in the eukaryotic 26 S proteasome, activates protein breakdown by 20 S proteasomes. *J. Biol. Chem.* **1999**, *274*, 26008–26014.

After completion of this manuscript, several crystallographic studies on fragments of protease Lon have appeared (*J. Struct. Biol.* **2004**, *146*, 113–122 and *J. Biol. Chem.* **2004**, *279*, 8140–8148).

11
Proteasome Regulator, PA700 (19S Regulatory Particle)

George N. DeMartino and Cezary Wojcik

11.1
Overview

The proteasome is responsible for the degradation of most intracellular proteins in eukaryotic cells [1, 2]. It functions as part of a modular system whereby a protease module, the 20S proteasome, forms larger complexes with one or more of a group of regulatory protein modules [3, 4]. The general function of these regulatory proteins is to impart specific catalytic and regulatory features to the resulting proteasome complexes. This chapter describes PA700 (*P*roteasome *A*ctivator of *700* kDa), also known as the 19S RP (*19S R*egulatory *P*article), a multisubunit ATPase regulatory complex, that binds to one or both ends of the cylinder-shaped 20S proteasome [4, 5–10]. The resulting "singly capped" or "doubly capped" complexes are both referred to as 26S proteasomes (see Figure 11.1), although it is unclear whether these forms differ in function and/or abundance in cells [3, 11]. The 26S proteasome is responsible for the selective degradation of polyubiquitin-modified

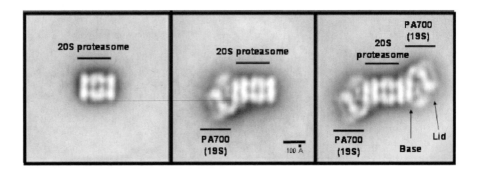

20S proteasome   **26S proteasomes**

Fig. 11.1. 20S and 26S proteasomes. Image-averaged electron micrographs of 20S and 26S proteasomes from bovine red blood cells. 20S proteasome capped on one or both ends by PA700/19S RP [11].

Protein Degradation. Edited by J. Mayer, A. Ciechanover, M. Rechsteiner
Copyright © 2005 WILEY-VCH Verlag GmbH & Co. KGaA, Weinheim
ISBN: 3-527-30837-7

proteins, the most extensively studied and, in our current understanding, the most physiologically important pathway for proteasome-dependent proteolysis [12]. Our presentation will focus on the structure and function of PA700/19S RP and the mechanisms by which this regulatory complex mediates selective degradation of ubiquitinated proteins. We also will present emerging data about "non-canonical" functions of PA700, including its role in ubiquitin-independent proteolysis by the proteasome, and its participation in non-proteolytic processes. Space limitations prohibit a complete or inclusive presentation of many interesting and important topics related to PA700/19S RP biochemistry and physiology. For detailed information about the 20S proteasome and for a specific discussion of the 26S proteasome, readers are referred to Chapters 10 and 9, respectively.

11.2
Structure

11.2.1
Component Subunits of PA700/19S RP

PA700/19S RP is a 700-kDa complex composed of approximately 18 distinct gene products. The best evidence indicates that each gene product is present in a single copy per complex [5, 6]. The overall structure and function of PA700/19S RP are highly conserved in eukaryotes, and only minor differences in subunit composition appear to exist among species (see below). Numerous proteins have been shown to associate physically with PA700/19S RP. Although many of these interactions are of unknown or questionable biological significance, it is likely that many others are physiologically meaningful. As described in detail in later sections, the functions of some of these interacting proteins raise questions about whether they should be classified as authentic PA700 subunits. In this chapter, we distinguish arbitrarily between PA700 subunits and PA700-associated proteins, but recognize that this distinction may be artificial and could be altered as a more complete understanding is obtained about the function and regulation of PA700/19S RP.

PA700/19S RP has been studied in many different species, and many individual subunits were identified prior to realization that they were components of PA700. These factors have resulted in a diverse and confusing subunit nomenclature (Table 11.1). The rational and increasingly accepted "Rpt/Rpn" and "S" nomenclatures (see below) will be used in the current presentation. An introductory description of PA700/19S PR subunits follows immediately below. Additional details of subunit functions and regulatory features in proteolysis by the 26S proteasome are presented in later sections.

The component subunits of PA700 range in size from 112 to 28 kDa. The "S" nomenclature identifies subunits on the basis of their relative mobility during SDS polyacrylamide-gel electrophoresis, whereas the Rpt/Rpn nomenclature distinguishes between the AAA ATPase subunits (**R**egulatory **p**article **t**riple-A protein) and the non-AAA ATPase subunits (**R**egulatory **p**article **n**on-ATPase) (see below).

Tab. 11.1. Component Subunits of PA700/19S RP.

Rpn	S	Other common names	Subcomplex	Approximate MW (Daltons)	Reported Post-translation modifications	Structural Features	Function
Rpt1	S7	CIM5	Base	48,500	O-Glc-NAc	AAA domain	ATPase
Rpt2	S4	YTA5, p56	Base	49,000	Phosphorylation, O-Glc-NAc	AAA domain	ATPase; gate regulation
Rpt3	S6b	TBP7, p48	Base	47,000	Phosphorylation	AAA domain	ATPase
Rpt4	S10b	SUG2, p42	Base	44,000	Phosphorylation	AAA domain	ATPase
Rpt5	S6a	TBP1, p50	Base	49,000		AAA domain	ATPase; polybubiquitn binding
Rpt6	S8	SUG1, CIM3, p45	Base	45,500	Phosphorylation, O-Glc-NAc	AAA domain	ATPase
Rpn1	S2	HRD2, NAS1, p97	Base	100,000		LRR, KEKE motifs	UBL binding
Rpn2	S1	SEN3, p112	Base	106,000		LRR, KEKE motifs	UBL binding
Rpn3	S3	SUN2, p58	Lid	61,000		PCI domain	
Rpn4		SON1		60,000			Tanscriptional regulation
	S5b	p50.5	Lid	55,000			
		p42E	Lid	42,000			
Rpn5	–	p55	Lid	53,000		PCI domain	
Rnp6	S9	p44.5	Lid	47,500		PCI domain	
Rpn7	S10a	p44	Lid	45,500		PCI domain	
Rpn8	S12	Mov34, p40	Lid	37,000	Phosphorylation		
Rpn9	S11	Nas7, p40.5	Lid	43,000		PCI domain	
Rpn10	S5a	Mcb1, Mbp1, p54	Base-lid interface	41,000		UIM motif	Polybubiquitin binding
Rpn11	S13	Poh1, Pad1	Lid	35,000		JAMM/MPN domain	Deubiquitinating metalloprotease
Rpn12	S14	NIN1, p31	Lid	31,000			
Rpn13		Daq1					
		UCH37, p37	Lid	37,000			Deubiquitining
		NAS6, p28		28,000		Ankyrin repeats	
	S15					PDZ domain	

The primary structures of all subunits have been determined, but most reveal little detailed insight about their roles. The notable exceptions are six homologous subunits of the AAA (*A*TPases *A*ssociated with various cellular *A*ctivities, [13–16] ATPase family (Rpt1/S7, Rpt2/S4, Rpt3/S6, Rpt4/S10b, Rpt5/S6a, and Rpt6/S8). These proteins have similar molecular weights (49–44 kDa) and contain a 200-amino acid domain characteristic of the AAA protein family [14]. These homologous domains contain Walker A and Walker B nucleotide-binding motifs of P-loop ATPases [16]. The remaining portions of the AAA PA700 subunits are divergent. Intact PA700 is an ATPase, but the relative contributions of individual ATPase subunits to this overall activity and to ATP-dependent functions of the 26S proteasome remain incompletely defined [17, 18]. Nevertheless, current evidence indicates that they play distinct and non-redundant roles (see below).

The non-ATPase subunits of PA700 (Rpn1–Rpn13) represent a diverse group of proteins. Rpn1/S2 and Rpn2/S1 share a low degree of sequence similarity and are probably evolved from a common protein; they contain leucine-rich repeats likely involved in protein–protein interactions [17, 19–21]. The remaining subunits have little similarity to one another, and their primary structures generally provide little specific information about their functions. Rpn10/S5a binds polyubiquitin chains, and features of this property helped to establish a motif for this function [22, 23]. Rpn11/S13 contains a JAMM domain characteristic of metalloproteases, and this subunit has been shown to display deubiquitinating activity [24, 25]. Uch37, a non-universal subunit, also functions as a deubuiqutinating enzyme and contains a conserved cysteine residue characteristic of the active-site-family enzymes [26–28].

11.2.2
Non-universal Subunits of PA700/19S RP

Several subunits of PA700 have not been identified universally (see Table 11.1). It is unclear if these discrepancies reflect authentic distinctions among species or whether they result from differences in experimental procedures and/or analysis. For example, Rpn4 was identified as a subunit of yeast 26S proteasome, but an ortholog of this protein has not been identified in other species [29]. Subsequent work has shown the special nature of this component, and calls into question its identity as an authentic PA700 subunit. Rpn4 is constitutively short-lived and is a proteasome substrate. Inhibition of the proteasome activity results in accumulation of Rpn4, which functions as a positive transcriptional factor for global expression of proteasome subunits [30, 31]. An unidentified functional counterpart of Rpn4 may exist in higher eukaryotes because inhibition of proteasome function can up-regulate expression of proteasome subunits under certain conditions [32, 33]. As described above, Uch37, a subunit with deubiquitinating activity, has been identified in PA700 from all examined sources except *Saccharomyces cerevisiae*. A Drosophila subunit termed p42E has not been identified in yeast or mammals [34], and a mammalian subunit, S5b has not been identified in yeast [6].

11.2.3
General Architecture of PA700/19S RP: The Base and the Lid

Unlike the 20S proteasome, a crystal structure has not been solved for either the 26S proteasome or isolated PA700/19S RP. Nevertheless, the general architecture of PA700/19S RP has been established, including most subunit–subunit interactions [35]. A major advance in understanding the general architecture of PA700/19S RP has been the identification and characterization of two component sub-complexes, termed the "base" and the "lid" [36]. The base sub-complex contains eight subunits, including the six AAA ATPases (Rpt1–Rpt6), and two non-ATPases, Rpn1/S2 and Rpn2/S1. The ATPases form a heterologous six-membered ring that directly abuts the terminal α-ring of the 20S proteasome. It seems likely that the center of the ATPase ring is coaxial with the annulus of the proteasome α-ring and that substrates must pass through it to enter the proteasome (see below). The exact orientation of Rpn1/S2 and Rpn2/S1 relative to the ATPase ring is uncertain. One modeling study has suggested that these Rpn subunits form an α-helical toroid with a central pore that extends the axial channel of the proteasome and ATPase ring [37]. We are unaware of any direct experimental evidence to support this attractive model, and interpret other available data to argue against it; additional structural studies should resolve this issue. As described below, the base probably serves multiple roles in degradation of polyubiquitinated proteins and mediates the overall ATP dependence of 26S proteasome function.

The lid sub-complex contains the remaining Rpn subunits, and is linked to the base via Rpn10/S5a. Yeast with a disrupted Rpn10 gene contain proteasomes from which the lid readily dissociates [38]. The precise function of the lid is poorly understood. Rpn10/S5a is a polyubiquitin-chain-binding protein, but curiously, this property is dispensable for most normal proteasome functions [39]. The only established enzymatic activity of the lid is that of deubiquitination, as expressed by Rpn11/S13 and Uch37 subunits. Rpn11 is a Zn^{2+} metalloprotease that cleaves polyubiquitin chains from their attachment points on proteins. Uch37 is a cysteine protease that cleaves ubiquitin monomers from the distal ends of polyubiquitin chains, thereby progressively decreasing the length of the chain [26]. The relative roles of these subunits are described below. The remaining subunits of the lid have uncharacterized functions. Remarkably, however, most cells also contain an eight-membered complex termed the COP9 signalosome (CSN) with subunit-for-subunit homology to the proteasome lid [36]. CSN may play multiple roles in cellular function, and has been proposed to be physically interchangeable with the lid of the proteasome. In our judgment this exciting possibility has not been established conclusively. A critical review of CSN structure and function with respect to PA700 is beyond the scope of this chapter, but future work will clarify the relationship of CSN to proteasome function.

11.3
Post-translational Modifications of PA700

11.3.1
Overview

Given the structural and functional complexity of PA700, it is not surprising that component subunits are subject to various types of post-translational modifications. Several types of post-translational modifications of PA700 subunits have been described, but in general, the functional significance of these modifications is in an early stage of investigation.

11.3.2
Phosphorylation of PA700/19S RP

Several subunits of PA700 including Rpt2/S4, Rpt3/S6', Rpt4/S10b, Rpt6/S8, and Rpn8/S12 have been shown to be phosphorylated [40]. The physiological significance of these modifications is poorly understood. One study has shown that treatment of cells with interferon-γ decreased PA700 phosphorylation and increased proteasome binding to PA28 (another proteasome regulator) in favor of PA700. These results suggest that phosphorylation of PA700 might alter its interaction with the 20S proteasome [41].

11.3.3
Glycosylation of PA700/19S RP

Multiple subunits of PA700 including, Rpt1/S7, Rpt2/S4, Rpt6/S8, and Rpn3/S3 are modified by O-linked N-acetylglucosamine [34, 42]. Modification of at least one PA700 ATPase subunit, Rtp2/S4, inhibits both ATPase activity and certain peptidase activities of the 26S proteasome *in vitro*, and certain proteasome functions in intact cells [42]. These results suggest that glycosylation of PA700 may be a mechanism to regulate its cellular function. Additional work will be required to understand the physiological significance of these early results.

11.4
Function of PA700/19S RP

11.4.1
A Model for PA700 Regulation of Proteasome Function

To focus our discussion of the function of PA700/19S RP in control of proteasome activity, we outline a "canonical" model for the ATP-dependent degradation of ubiquitinated proteins by the 26S proteasome, the best studied and most clearly established role of the 26S proteasome (Figure 11.2). PA700/19S RP serves multiple roles in mediating proteasomal degradation of ubiquitinated proteins. First, PA700 relieves the structurally imposed inhibition of proteolytic activity by opening the blocked gates at the α-terminal rings of the 20S proteasome. Second, PA700/19S RP serves as the recognition and binding element for the polyubiquitin degradation tag. Third, PA700/19S RP prepares the protein substrate for degradation by destabilizing its tertiary and/or quaternary structure, and translocating the unfolded polypeptide chain through the proteasome's open ends to the central chamber containing catalytic sites responsible for peptide-bond hydrolysis. Fourth, PA700 removes the polyubiquitin chain from the protein substrate, an essential function for complete proteolysis. PA700-catalyzed ATP hydrolysis is obligatory for overall proteolysis and each of the PA700-mediated processes listed above may

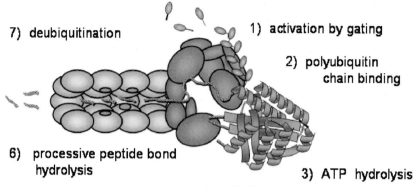

7) **deubiquitination**

1) **activation by gating**

2) **polyubiquitin chain binding**

6) **processive peptide bond hydrolysis**

3) **ATP hydrolysis**

4) **substrate unfolding**

5) **processive translocation of unfolded polypeptide chain**

Fig. 11.2. Canonical model for the degradation of polyubiquitinated proteins by the 26S proteasome. The canonical model for the degradation of polyubiquitinated proteins by the 26S proteasome involves multiple functions of PA700 including: 1) PA700 activates the 20S proteasome by opening the gate on the terminal rings of the proteasome; 2) PA700 specifies selectivity degradation of polyubiquitinated proteins by direct recognition and binding of polyubiquitin chains; 3, 4 and 5) PA700 is an ATPase that utilizes ATP hydrolysis to unfold and translocate substrates to the proteasome; 6) PA700 initiates processive proteolysis by translocating the N- or C-terminus of the substrates to the proteasome; and 7) PA700 removes the polyubiquitin chain from the substrate; for the purpose of this illustration, the deubiquitination is shown as a processive process starting at the distal end of the chain, but the deubiquitinating activity most closely linked to substrate degradation probably occurs at the isopeptide bond proximal to the substrate. See text for details on each of these processes.

be coupled to ATPase activity. In the following sections, we present detailed features for each of these PA700 functions. For convenience and clarity, we present these functions separately and in arbitrary order. We recognize, however, that they probably are temporally coordinated and mechanistically linked. Following a description of this canonical model for PA700/19S RP function, we present emerging data about non-canonical functions for PA700/19S RP in the degradation of non-ubiquitinated proteins, and for non-proteolytic roles of PA700/19S RP.

11.4.2
Roles of ATPase Activity in PA700 Function

The degradation of ubiquitinated proteins by the 26S proteasome requires PA700-catalyzed ATP hydrolysis. How is proteolysis mechanistically linked to ATPase activity? Despite the fundamental nature of this question, a detailed answer is unknown. As described below, ATP hydrolysis probably mediates multiple elements of 26S proteasome function, including assembly of the complex from 20S proteasome and PA700 sub-complexes, proteasome activation, polyubiquitin-chain binding, and substrate unfolding, translocation, and deubiquitination. Although ATPase activity of PA700/19S RP is catalyzed by the AAA ATPase subunits of the base, the relative roles and contributions of the six different ATPase subunits to various functions remain unclear. Available biochemical and genetic evidence suggests that these subunits play distinct non-redundant roles [43], but there are insufficient existing data to be certain that this represents a complete division of labor for these ATPases among different ATP-dependent processes. In any case, the hexameric ring-shaped structure of the ATPases and the position of this ring on the α-ring of the proteasome are almost certainly critical to their functions. Interestingly, the same ring-to-ring orientation of protease and AAA ATPase is found in non-proteasomal ATP-dependent proteases in bacteria, suggesting that this topology is intimately linked to regulation of proteolysis [44].

11.4.3
Proteasome Activation by PA700

The 20S proteasome displays inherently low catalytic activity because of its structure, which excludes substrates from interacting with the catalytic sites. Substrates reach the sequestered catalytic sites only after passing through a narrow 13-Å annulus formed by the terminal α-rings of the proteasome [45–47]. This structural feature is sufficient to prevent entry of substrates with appreciable tertiary structure, but even short or unfolded polypeptides must overcome a second structural impediment posed by the proteasome. Specifically, the annulus is physically occluded by N-terminal peptides of four α-subunits that project across it [46]. Activation of the proteasome by regulatory proteins such as PA700 involves clearing the occlusion by conformational rearrangement of the α-ring peptides. Thus, the terminal rings of the proteasome act as a regulated gate and a critical role of PA700/19S RP is to activate the proteasome by opening the gate. A detailed molecular explanation for PA700-induced proteasome activation is lacking in the absence of the crystal structure of the 26S proteasome, but it seems highly likely that activation is generally analogous to proteasome activation by a distinct proteasome activator, PA28. PA28 is a heptameric ring-shaped protein that activates the proteasome's hydrolysis of short peptides. A co-crystal structure of yeast 20S proteasome and PA26, an ortholog of PA28 from *Trypanosome brucei*, has been determined [48]. Binding of C-termini of PA28 subunits to the α-ring of 20S proteasome promotes an interaction between an "activation" domain within individual PA28 subunits and the annulus-occluding peptides of proteasome subunits. This interaction conformationally rearranges the occluding polypeptides from a position roughly perpendicular to the central proteasome channel to one roughly parallel to the channel and projecting into PA28, thereby opening a pore through which substrates may pass. Unlike PA700, PA28 does not promote the proteasome's degradation of ubiquitinated proteins, presumably because PA28 lacks other essential features for processing such substrates (see below). This example highlights the multiple functions that PA700 must conduct for degradation of polyubiquitinated proteins and distinguishes it from simpler regulators such as PA28.

Several lines of biochemical evidence strongly indicate that PA700 also activates the proteasome by relieving occlusion of the proteasome pore, and involves the physical interaction between the heterohexameric AAA ATPase ring of the base and the heteroheptameric α-ring of the proteasome. First, binding of PA700 to the proteasome greatly activates the hydrolysis of short peptide substrates, suggesting that PA700 binding increases access of these substrates to the catalytic sites [8]. This effect can be accomplished entirely by the base sub-complex, indicating that the interaction of the ATPase ring is sufficient for activation [36]. Second, deletion of the pore-occluding peptide of the α3 proteasome subunit results in a constitutively active 20S proteasome, whose activity is not stimulated further by binding to PA700 [49]. The amino terminus of the α3 subunit contacts all other α-subunit peptides that block the pore and seems to be particularly important for the general organization of the structure that obstructs the annulus of the proteasome. Addition of the deleted α3 peptide to the mutant 20S proteasome in *trans* restores the "latent" inhibited state to the enzyme, further demonstrating the critical role of this peptide. Third, mutations in the ATP-binding domain of the Rpt2 ATPase subunit of PA700 have no effect on binding of PA700 to either wild-type or mutant proteasomes with constitutively high activity due to N-terminal deletions of α3 and α7 subunits [50]. Nevertheless, the 26S proteasome composed of mutant Rpt2 and wild-type (i.e. "latent") 20S had low protease activity, whereas 26S proteasome composed of both mutant PA700 and mutant 20S proteasome had high proteasome activity. Thus, a mutant Rpt2 cannot activate wild-type proteasome, whereas a constitutively active mutant 20S proteasome can suppress the inhibitory effect of PA700/19S RP with mutant Rpt2. These results indicate that the Rpt2 subunit of PA700 activates the proteasome by a mechanism involving gating of the annulus [50].

Despite some generally similar features between PA700- and PA28-mediated proteasome activation, detailed mechanisms of these processes are likely to have significant differences. In the case of PA28, binding to the proteasome is necessary and sufficient for activation and neither binding nor activation requires other cofactors such as ATP. In contrast, PA700 binds to the proteasome by an ATP-dependent process whose molecular mechanism is undefined. Moreover, PA700 may utilize ATP for an additional role in proteasome activation *per se*, as suggested by the inhibited states of 26S proteasomes containing Rpt2 mutants [50]. Further work will be required to define the exact role of ATP in proteasome activation by PA700. It is also interesting to note that unlike the proteasome–PA28 interaction, binding and activation of the proteasome by PA700 likely involves an initial symmetry mismatch between the seven-membered proteasome ring and the six-membered ATPase ring of PA700. Such a mismatch also occurs between the homohexamer ATPase rings and the homoheptameric protease rings of bacterial ClpAP protease, and may play a significant role in the mechanism of proteasome activation.

11.4.4
Polyubiquitin-chain Binding

A major physiological role of PA700 is the recognition of polyubiquitinated proteins. This function serves as the principal determinant for the selectivity of ubiquitin-modified proteins for degradation by the 26S proteasome. PA700 binds K48–G76 linked polyubiquitin chains composed of four or more ubiquitin moieties with high affinity, but the exact molecular basis for this property remains poorly understood [51]. Two subunits of PA700, Rpn10/S5a and Rpt5/S6a, have been identified as polyubiquitin-chain-binding proteins, and the features of these properties are described immediately below. Moreover, cells contain other polyubiquitin-binding proteins that interact with the 26S proteasome and may function to deliver substrates to it for degradation (described in later sections).

Rpn10/S5a was the first PA700 subunit to be identified as a polyubiquitin-chain-binding protein [22, 52]. It selectively binds K48–G76 polyubiquitin chains, as demonstrated by "far western" methodology [22, 39, 53–55] or by affinity chromatography with tagged recombinant Rpn10/S5a. Such isolated versions of Rpn10/ S5a bind polyubiuquitin chains composed of four or more ubiquitin moieties, thereby mimicking the features of the chain requirements for overall degradation of polyubiquitin-modified proteins. Soluble recombinant Rpn10/S5a inhibits the degradation of ubiquitinated proteins in cell-free extracts, presumably by competing with polyubiquitinated proteins for the 26S proteasome [56, 57]. Structure–function analysis of Rpn10/S5a from various sources has identified a short hydrophobic sequence responsible for polyubiquitin binding. The motif, termed the *u*biquitin-*i*nteracting *m*otif (UIM), is found in many proteins, including others involved in various aspects of ubiquitin metabolism [23]. Rpn10/S5a from *Saccharomyces cerevisiae* and *Arabidopsis* contain a C-terminal UIM [58], whereas human and *Drosophila* Rpn10/S5a contain two UIMs [57]. Each isolated site can bind polyubiquitin independently, albeit with very different affinities. The two UIM sites may bind polyubiquitin with some degree of cooperation in intact Rpn10/S5a. In any case, UIMs appear to bind to hydrophobic patches composed of side chains from ubiquitin residues L8, I44, and V70. Structure–function analysis of ubiquitin has established the importance of these residues for ubiquitin's role in targeting proteins for degradation [53]. Therefore, a reasonable model suggests that complementary hydrophobic patches on ubiquitin and the Rpn10/S5a form the interaction site for these proteins [53, 58, 59]. The topology of Rpn10/S5a as a structural link between the base and lid sub-complexes of PA700 further suggests that this subunit would be well positioned to deliver bound polyubiquitinated substrates to the base for ATP-dependent unfolding and translocation into the proteasome.

Despite the initial indication that Rpn10/S5a was the polyubiquitin-chain-binding subunit of the 26S proteasome, subsequent studies showed that it cannot be the principal recognition element for this process. First, disruption of the Rpn10/S5a gene in yeast is not lethal and inhibits the degradation of only a sub-class of ubiquitinated proteins [39]. Likewise, RNA interference-inhibited expression of Rpn10/S5a in Drosophila cells does not inhibit growth or overall ubiquitin-dependent protein degradation [32]. These features are not expected for a protein with an essential role in ubiquitin-dependent proteolysis. Second, deletion of the single conserved UIM in yeast has no effect on the degradation of Ub–Pro–β-galactosidase, a substrate whose cellular degradation otherwise requires expression of Rpn10/S5a [58]. Remarkably, the degradation of this model substrate of the ubiquitin-fusion-dependent pathway of the ubiquitin–proteasome system does require the N-terminal domain of Rpn10/S5a, demonstrating that this region of the Rpn10/S5a, but not the UIM, is responsible for a critical function in the degradation of at least one cognate substrate. Notably, the N-terminus is highly conserved in all Rpn10/S5a sequences, even though this domain is not required for assembly of Rpn10/S5a into PA700. In sum, these findings indicate that Rpn10/S5a does not serve an obligatory role as an exclusive polyubiquitin-chain-binding component of the 26S proteasome, and the exact significance of the polyubiquitin-chain-binding properties of Rpn10/S5a remains unclear. As described below, Rpn10/S5a can also interact with proteins containing **ub**iquitin-like domains (UBLs), and could mediate interactions between PA700 and such proteins. Moreover, Rpn10/S5a appears to exist in a non-proteasome-associated form in some cells. This finding could indicate alternatively that Rpn10/S5a has non-proteolytic roles or that it could function as a polyubiquitin-transfer factor that carries substrates to the proteasome (see below for a discussion of other polyubiquitin-chain-binding proteins that may also serve such a function). In either case, we note that studies demonstrating the binding of polyubiquitin to Rpn10/S5a have been conducted with the isolated protein, a limitation imposed by certain technical constraints of the experiments. In contrast, some evidence indicates that the proteasome-associated Rpn10/S5a may not be competent to bind polyubiquitin [60] (although other data appear to contradict this conclusion [38]); the different results may reflect differences in experimental details. In summary, although most available evidence argues against an obligatory role for Rpn10/S5a in polyubiquitin binding by 26S proteasome, we believe that it is premature to dismiss its role in this process. Additional work will be required to delineate the nature and contribution of Rpn10/S5a to polyubiquitin-chain binding of PA700.

The "far-western" methodology that originally identified Rpn10/S5a as a polyubiquitin-chain-binding protein failed to identify other components of the 26S proteasome with this property. This result, however, may reflect unique features of Rpn10/S5a that permit it to retain this function under the harsh conditions required for the far-western binding analysis. In fact, the short UIM responsible for binding is unlikely to have significant tertiary structure [59]. Nevertheless, these results do not exclude the possibility that other PA700 subunits bind polyubiquitin chains under native conditions. Recently, chemical cross-linking has been used to identify Rpt5/S6a as a second polyubiquitin-chain-binding protein of PA700 [60]. Rpt5/S6a is an AAA ATPase of the base and was identified as the cross-linked product of a photoactivatable variant of K48–G76-linked tetra-ubiquitin and purified 26S proteasome. This study produced several important and surprising findings. First, of all 26S proteasome subunits only Rpt5/S6a was labeled specifically by the tetra-ubiquitin. Purified recombinant Rpn10 was also efficiently labeled by this method, further supporting the view that it is a polyubiquitin-chain-binding protein, but loses this property when it is a constituent of the intact PA700 structure. Second, the interaction between Rpt5/S6' and polyubiquitin required ATP hydrolysis. This effect could not be attributed to the role of ATP in maintaining the structural integrity of the 26S proteasome. Electron paramagnetic resonance, a direct indicator of binding, confirmed the ATP-dependent physical interaction between tetra-ubiquitin and Rpt5/S6a. Importantly, $(Ub)_5$-dihydrofolate reductase, an established proteolytic substrate of the 26S proteasome, competitively inhibited both the cross-linking and the EPR-monitored binding of tetra-ubiquitin to 26S proteasome. These results establish Rpt5/S6a as a polyubiquitin-chain-binding protein. The molecular basis for the interaction between Rpt5/S6a and polyubiquitin is unknown, and we are unaware of similarities in the structure of this protein with motifs known to interact with polyubiquitin. Unlike Rpt10/S5a, Rpt5/S6a is an essential protein in yeast, and RNAi of Rpt5/S6a significantly reduced growth of Drosophila S2 cells [32, 43]. The selective cross-linking of Rpt5/S6a provides another example of the non-redundant properties of the six ATPases that compose the base structure of PA700. Although these results suggest that polyubiquitin binding at the base of PA700 positions the substrate near to the axial channel of the proteasome, "lidless" proteasomes from yeast lacking Rpn10, have defective degradation of model ubiquitinated proteins, suggesting that other features of the lid are important for manifestation of normal degradation of ubiquitinated proteins [36]. These various results highlight that the molecular basis of polyubiquitin-chain binding to PA700 remains poorly understood and is likely to be highly complex.

11.4.5
Unfolding/Modification of Substrates

Many proteins degraded by the proteasome, including those of high regulatory significance, retain most or all of their native tertiary structure after they are ubiquitinated. Because of structural features of the proteasome described above, it is clear that the tertiary structures of these proteins must be destabilized prior to their proteolysis. Moreover, some proteins are ubiquitinated and selectively degraded while they are components of multi-subunit complexes. In such cases, the quaternary structure of the complex must be destabilized to allow the ubiquitinated subunit to be selectively dislodged and degraded. The canonical model of 26S proteasome function posits that PA700/19S RP directly destabilizes substrates under each of these conditions. In other words, once a protein is targeted to PA700/19S RP by its polyubiquitin chain, PA700/19S RP unfolds the protein to allow its transit through the opened annulus of the proteasome for degradation. An excellent example of the ability of PA700/19S RP to accomplish this function directly is the selective degradation of ubiquitinated Sic1 from a Sic1/Cdk/cyclin complex by purified 26S proteasome [61]. An analogous example is the selective degradation of IκB, an inhibitor of the heterodimeric NFκB transcription factor complex, although the direct action of PA700 is this case has not been demonstrated. The exact mechanisms by which PA700/19S RP carries out protein unfolding of monomeric or multimeric proteins are unclear. Several studies have demonstrated that PA700/ 19S RP has chaperone-like properties that would likely participate in substrate unfolding. For example, PA700/19S RP inhibits the aggregation of misfolding proteins and catalyzes the refolding of certain heat- and chemically denatured proteins [62–64]. These properties are manifested by the base sub-complex, indicating a role for ATPase subunits in these functions; at least one study has shown that these features are indeed regulated by ATP [62]. Such results indicate that in addition to polyubiquitin chains, PA700 also recognizes and interacts with certain structural features of non-native proteins. Such features are likely to occur transiently or in specific limited regions of proteins with otherwise high global stability. Thus, proteins initially targeted to PA700/19S RP by a polyubiquitin chain would be subject to this type of secondary interaction. Once such features are recognized by PA700, further destabilization could occur, perhaps linked to cycles of ATP hydrolysis and/or processive proteolysis. Support for this model has been obtained by examining ubiquitin/proteasome-dependent degradation of stable model proteins in reticulocyte extracts [65]. These elegant experiments indicate that 26S proteasome unfolds and degrades proteins processively starting at a point near the polyubiquitination site. Replication of these results in a defined system of purified proteins will confirm PA700/19S RP as the agent that directly unfolds the substrate. Thus, despite considerable overall progress, a detailed molecular mechanism for protein unfolding by PA700 remains very poorly understood. Some of the outstanding issues yet to be resolved include: the exact nature of the interaction between PA700 and the substrate, the role of ATP hydrolysis in substrate unfolding, and details of the likely mechanistic linkage between substrate unfolding, translocation, and proteolysis.

11.4.6
Translocation of Substrates from PA700/19S RP to the Proteasome

With few known exceptions, substrates of the 26S proteasome are degraded completely to short peptides and amino acids once they are engaged by the proteasome. This feature suggests that proteolysis is processive, and examples of processive proteolysis by the proteasome have been demonstrated for several model substrates [65–67]. To satisfy the requirement for unstructured proteins as suitable proteasome substrates, processive proteolysis may often be initiated at a free N- or C-terminus, which could pass easily through the opened annulus of the proteasome. Processive proteolysis could then proceed by a mechanism linked to successive unfolding and translocation of the rest of the substrate. This model is also compatible with the few known examples of limited proteasomal proteolysis, in that degradation could start at one terminus of a protein and proceed processively until reaching a "stop translocation/degradation" site dictated by a structural feature of the substrate. Stalled substrates could be released from the proteasome, thereby generating the mature processed protein. The p105 subunit of NFκB is the best example of a protein processed by this mechanism. In this instance, the C-terminal half of p105 is degraded to yield the mature p50 subunit of NFκB [68]. Although it seems clear that processive proteolysis initiates from a free terminus of many proteins, the proteasome can also catalyze endoproteolysis [69, 70]. This activity requires an unstructured region, presumably to form a loop that can be accommodated by the opened proteasome annulus. Such a structure could be assumed by certain disordered proteins or could be present in certain regions of otherwise well-folded proteins. In any case, once endoproteolysis is achieved, processive proteolysis could proceed from newly generated N- or C-termini, leading either to complete or to limited proteolysis. We assume that PA700/19S RP plays an important role in driving substrate translocation for all processive proteolysis, regardless of its mode of initiation. However, as with unfolding, considerably more work will be required to understand the detailed mechanisms of PA700-mediated substrate translocation.

11.4.7
Deubiquitination of Substrates by PA700/19S RP

Degradation of polyubiquitinated proteins by the 26S proteasome results in the proteolysis of the substrate, but not of ubiquitin moieties that compose the polyubiquitin chain. Instead, the polyubiquitin chain is removed from the substrate during proteolysis and this process appears to be obligatory for, and coupled to, substrate degradation. Two PA700 lid subunits, Rpn11/S13 and Uch37, as well as several PA700-associated proteins, (see below), catalyze deubiquitinating activity. Rpn11/S13 contains a JAMM domain characteristic of a Zn^{2+} metalloprotease catalytic site [24, 25]. Rpn11/S13 occurs in PA700 from all sources, and is an essential protein [24, 25, 71]. Rpn11/S13 cleaves the isopeptide bond linking the polyubiquitin chain to the substrate. This reaction is catalyzed by both free PA700 and 26S proteasome, but curiously depends on ATP hydrolysis only with the 26S proteasome. This feature highlights another likely role of ATPase activity in proteolysis by the 26S proteasome. Because there is no reason to believe that the isopeptide bond catalysis *per se* requires energy, the energy dependence of Rpn11-catalyzed deubiquitination may be linked to translocation or unfolding of the substrate whereby the isopeptide bond is made available for cleavage. Removal of the polyubiquitin chain is probably important for overall substrate degradation on steric considerations because the bulky chain would impede translocation of the attached polypeptide substrate through the opened pore of the proteasome. In fact, inhibition of Rpn11 severely reduces rates of proteolysis by the 26S proteasome [24, 25]. Uch37 is the second deubiquitinating subunit of PA700 [25, 26]. This protein does not exist in budding yeast, but is found in *Schizosaccharomyces pombe*, *Drosophila*, and all mammals [72]. Like Rpn11 it is found in the lid, and immunoelectron microscopy has localized it to a peripheral site on PA700 [73]. Uch37 cleaves ubiquitin from the distant end of polyubiquitin chains [27]. The exact significance of this type of activity is unclear. It is conceivable that it provides an "editing" function whereby tagged proteins that do not become engaged in degradation within a reasonable time are deubiqutinated and released from the proteasome. Unlike Rpn11, decreased expression of Uch37 has little effect on cell viability, proteasome function, or global ubiquitin-dependent protein degradation [32, 72].

11.5
Interaction of PA700 with Non-proteasomal Proteins

11.5.1
Overview

Numerous studies have identified interactions between individual subunits of PA700 and non-proteasomal proteins. In fact, many subunits of PA700 were first described as interacting proteins of non-proteasomal bait proteins in yeast two-hybrid screens whose purpose was unrelated to direct investigation of the proteasome or protein degradation. Other interactions have been found by various approaches between 26S proteasome, PA700, or individual PA700 subunits and proteins both with and without obvious relationship to the ubiquitin–proteasome system [74, 75]. It is difficult to judge the physiological significance of many of these interactions, which could reflect authentic, but currently unrecognized regulatory interactions between PA700 and proteins of the ubiquitin system or other cellular process, interactions between the 26S proteasome and proteolytic substrates, or spurious interactions with no physiological significance. It is beyond the scope of this chapter to review and evaluate all of these reports. Instead, we will focus on selected examples from several classes of PA700-interacting proteins whose identification seems firmly established and/or that have a rationale for or a promise of physiological significance. Many of these PA700-interacting proteins contain a UBL domain, which appears to serve as a common structural element for the interaction of these proteins with the proteasome (see below). UBL domains have low primary sequence similarity to ubiquitin, but assume a general three-dimensional structure remarkably like that of ubiquitin. UBL domains occur in many proteins, most of which do appear to interact with PA700. As noted above, we arbitrarily classify these proteins as "non-proteasomal" but recognize that additional work may alter this classification. (For an overview see Figure 11.3.)

11.5.2
26S Proteasome Assembly/Stability Proteins

The exact cellular process by which the 26S proteasome is assembled remains unknown, but the best evidence suggests that it results from binding of independently assembled 20S proteasome and PA700/19S RP [76]. There is considerable, but incomplete, information about the assembly of the 20S proteasome [77, 78], whereas very little is known about the assembly of PA700. Formation of 26S proteasome from purified 20S proteasome and PA700 can be achieved *in vitro* by an

A

Non-proteasomal
polyubiquitin chain binding protein

B

Substrate unfolding
by non-proteasomal protein

C

Ubiquitin-independent
Substrate targeting

D

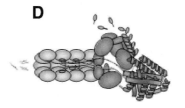

Endoproteolysis

Fig. 11.3. Non-canonical functions of PA700 in the regulation of proteasome activity. A) Certain polyubiquitinated proteins may be recognized by and bound to non-proteasomal proteins that subsequently transfer them to the proteasome after binding to PA700. B) Certain substrates may be unfolded by non-PA700 proteins and then transferred to PA700. C) PA700 may promote degradation of some non- ubiquitinated proteins by recognizing and binding unstructured regions of these proteins. D) PA700 may promote endoproteolysis of certain substrates through recognition of unstructured portions of those substrates. Permutations of these and other functions, not shown explicitly here, likely provide multiple variations of PA700-mediated processes. See text for details.

ATP-dependent process. This suggests that PA700 and 20S proteasome are sufficient for assembly of 26S proteasome [8]. This process, however, is inefficient *in vitro*, and one study has indicated that the Hsp90 mediates 26S proteasome assembly [79]. Recently, a protein termed ecm29 was identified as a stoichiometric component of the 26S proteasome from *Saccharomyces cerevisiae* purified by affinity chromatography without exposure to high salt concentrations [80]. Ecm29 is a 200-kDa protein that binds to both the 20S proteasome and PA700 and has been proposed to tether the two sub-complexes. Electron microscopy reveals a V-shaped protein that may act as "clip" between the α-rings of the 20S proteasome and the base of PA700, but little is known about the molecular basis of such binding. Ecm29 stabilizes 26S proteasome in the absence of ATP, further supporting a role for it in physically linking 20S proteasome to PA700. Orthologs of ecm29 are widely distributed among species, but further study of the protein in yeast and other organism will be required to establish its precise function in 26S proteasome structure and function.

11.5.3
Deubiquitinating Enzymes

Deubiquitinating enzymes constitute a large class of proteins in the ubiquitin system, and play a largely unexplored role in ubiquitin biology (29). Ubp6/Usp14 is a widely distributed deubiquitinating enzyme that appears to be a stoichiometric component of yeast 26S proteasomes isolated without exposure to high salt [80]. This finding suggests that Ubp6 may be an authentic, but easily dissociated, PA700 subunit. Ubp6 contains an N-terminal UBL domain that is responsible for binding to PA700 via the Rpn1/S2 subunit of the base [80]; Rpn1/S2 interacts with several other proteins containing UBL domains (see below). Ubp6 has low catalytic activity as an isolated protein. However, the deubiquitinating activity of Ubp6 is enhanced over 100-fold when the protein is associated with yeast PA700, suggesting that catalytic function is restricted to PA700. Ubiquitin vinyl sulfone, an agent that covalently modifies the active sites of many deubiquitinating enzymes, was used to identify Ubp6/Usp14 as a component of mammalian 26S proteasome [81]. This result also supports the conclusion that Ubp6/Usp14 is active only when associated with the proteasome. Yeast from which Ubp6 has been deleted are viable, but are defective for degradation of some, although not all, model substrates of the ubiquitin system. Interestingly, ubiquitin itself is destabilized in this mutant, suggesting that failure of Ubp6 to deubiquitinate modified substrates at the proteasome results in proteolysis of ubiquitin itself. Inhibition of deubiquitinating activity of PA700 *in vitro* also leads to degradation of ubiquitin attached to protein substrates [25].

In addition to Ubp6, Doa4 is a deubiquitinating protein that associates with the 26S proteasome [82]. Unlike Ubp6, Doa4 is present as a sub-stoichiometric component [83]. The N-terminus of this protein is required for proteasome binding and is sufficient to direct association of otherwise non-interacting proteins to the complex. Yeasts from which Doa4 are deleted are viable, but are defective in degradation of model proteins in several pathways of the ubiquitin–proteasome system. Moreover, over-expression of Doa4 increases rates of degradation of certain substrates, suggesting that it can enhance the function of the proteasome. In contrast, Doa4 also functions in the vacuolar protein-sorting and endocytic pathways, perhaps acting on ubiquitin-modified plasma membrane proteins targeted to the vacuole [84, 85]. Although these later data imply a non-proteasomal function for Doa4, it is possible that Doa4 plays multiple distinct roles, or that the proteasome is involved in aspects of ubiquitin-dependent membrane/vacuolar processes.

11.5.4
Ubiquitin-conjugating Machinery

Numerous components of the ubiquitin-conjugating system, including various E2-conjugating enzymes and E3 ligases, have been identified as PA700-interacting proteins [86]. Such findings raise the intriguing possibility that substrate ubiquitination and degradation are spatially linked. Although available data do not provide the precise physiological significance or mechanistic details of such linkage, it is easy to imagine that spatial coupling of ubiquitination and degradation could improve the efficiency of substrate targeting to and processing by the proteasome. We briefly list some components of the ubiquitin-conjugating machinery reported as proteasome interacting proteins.

Hul5 is a stoichiometric component of affinity-purified yeast 26S proteasomes purified under low-salt conditions [80]. Hul5 is a HECT-domain E3 ligase known as KIAA10 in mammals. It assembles both K48- and K29-linked polyubiquitin chains and binds to PA700 and to isolated Rpn1/S2 via an N-terminal domain [87].

Ubr1, the E3 ligase of the N-end rule, and Ufd4 a ligase of the Ufd pathway, associate with PA700 [88]. Ubr1 interacts with Rpn2/S1, whereas Ufd4 interacts with the Rpt6/S8 ATPase of the base.

CHIP is a U-box E3 ligase that ubiquitylates misfolded proteins [89]. CHIP forms a complex with Hsp70 and BAG1, a UBL-domain protein. BAG1, like some other UBL-domain proteins, associates with PA700 [90]. These findings suggest an attractive model in which a complex of CHIP, Hsp70, and BAG1 binds to the 26S proteasome to partition misfolded Hsp70 substrates to ubiquitylation and degradation instead of refolding. The physical association of these components might improve the efficiency of degradation. Moreover, because these substrates are misfolded prior to reaching PA700, they might be particularly susceptible to aggregation, and therefore harmful to the cell if left unescorted at each step of degradative process. Formation of a complex to achieve all of the functions of the degradative pathway would likely be of considerable benefit to the cell.

E6-AP, a HECT-domain E3 ligase, and *β*TrcCP, the F-box component of an SCF-type E3 ligase, associate with hPLIC proteins, the mammalian versions of yeast Dsk2 [91] (see below). PLIC proteins contain both UBL and UBA (*ub*iquitin pathway *a*ssociated) domains that mediate PLIC interactions with PA700 [92]. This finding raises the possibility that other examples of the many cellular F-box proteins of SCF complexes associate with PA700.

11.5.5
Polyubiquitin-chain-binding Proteins

The canonical model of 26S proteasome function described above invokes the direct recognition of polyubiquitin-modified substrates by polyubiquitin-chain-binding subunits of PA700, such as Rpt5/S6′ and/or Rpn10/S5a. However, cells contain a variety of non-proteasomal proteins that also bind polyubiquitin chains. Many of these proteins have established roles in ubiquitin conjugation and deconjugation, whereas others have unknown functions. Although some of the latter proteins likely mediate non-proteolytic roles of ubiquitin, others also bind to the 26S proteasome via PA700. Such a property suggests a model in which certain polyubiquitinated proteins are initially recognized by and bound to non-proteasomal proteins, and subsequently delivered to the proteasome for degradation [38, 93]. This general mechanism could provide additional regulation and versatility for substrate selection. Evidence in support of this model has been obtained for several polyubiquitin-chain-binding proteins including, Rad23, Dsk2, and VCP[Ufd1/Npl4], as described below.

Rad23 and Dsk2 are widely distributed eukaryotic proteins, identified originally in yeast as mediators of DNA repair and spindle-pole duplication, respectively. Rad23 and Dsk2 have similar domain structures; each contains an N-terminal UBL domain, which is responsible for their respective binding to PA700 [38, 93–95] and UBA domains near their C-termini (Rad23 contains two UBA domains, whereas Dsk2 contains only one). UBA domains are found in numerous proteins, and like the UIM of Rpn10/S5a, function as polyubiquitin-chain-binding elements [96, 97]. Thus Rad23 and Dsk2 can bind to both PA700 and polyubiquitin. Several early reports indicated that the UBL of Rad23 binds to the second UIM of human Rpn10/S5a, suggesting that Rpn10/S5a was the Rad23-receptor of PA700 [98]. Although this interaction has been verified [59], 26S proteasomes lacking Rpn10/S5a still bind Rad23, indicating that another subunit also serves this function. Moreover, yeast lacks the second UIM of mammalian Rpn10/S5, making it unlikely that this interaction is of general significance. More recently, Rad23 and Dsk2 have been shown to bind to PA700 via leucine-rich repeats of Rpn1/S2 and Rpn2/S1, the non-ATPase components of the base [38, 94]. This finding has interesting mechanistic implications, because substrates bound to Rad23 or Dsk2 at the base of the 26S proteasome would be positioned for unfolding and translocation into the proteasome by the AAA ATPase subunits. Several lines of evidence involving expression of wild-type and mutant variants of Rad23 and Dsk2 in yeast support the general model of these proteins as carriers of polyubiquitinated substrates to the 26S proteasome [38, 93, 99, 100]. Despite these data, the mechanisms by which substrates would be transferred from carrier proteins to the proteasome remain unclear. Moreover, other results are inconsistent with a carrier model. For example, Rad23 inhibits 26S proteasome-dependent degradation of an otherwise susceptible polyubiquitinated protein *in vitro* [101]. Rad23 also inhibits polyubiquitin-chain formation. Finally, several studies indicate that the interaction between Rad23 and PA700 mediates DNA repair by a mechanism independent of proteolysis [102, 103] (and see below). These various results indicate that Rad23 biology is likely to be complex and may affect PA700 function by multiple mechanisms. Detailed mechanistic information about this relationship will be required to clearly interpret the results of cellular experiments.

The VCP/Npl4/Ufd1 complex represents another prominent example of a polyubiquitin-chain-binding protein that may aid targeting of certain substrates to the 26S proteasome. VCP, known as Cdc48 in yeast, is a ring-shaped homohexamer of 90-kDa AAA ATPase subunits [104]. Thus, VCP assumes an architecture similar to that of the AAA ATPases of the PA700 base and some other AAA ATPases [105]. VCP binds polyubiquitin chains [106, 107], and appears to play diverse cellular roles, determined in part by its differential association with various proteins that modulate its function. For example, VCP forms a complex with two additional polyubiquitin-chain-binding proteins, Npl4 and Ufd1, to mediate ERAD (**e**ndoplasmic **r**eticulum-**a**ssociated **d**egradation) [107, 108]. The role of the ubiquitin–proteasome system in ERAD are covered in detail elsewhere. In brief, ERAD is the process by which endogenous ER proteins, or proteins that transit through the ER, are constitutively or conditionally degraded. ERAD plays a critical role in protein quality control; mutant proteins that fail to fold properly in the ER are retrotranslocated to the cytoplasm, ubiquitylated, and degraded by the proteasome. VCP$^{\text{Npl4/Ufd1}}$ is required for degradation of both normal and mutant ER proteins, and may couple its polyubiquitin-chain-binding properties to ATPase activity to assist in the translocation of substrates across the membrane [105, 108]. Certain VCP mutants as well as RNAi-decreased expression of VCP inhibits ubiquitin-dependent proteolysis, suggesting that VCP may effect degradation of both ERAD and non-ERAD substrates [106, 109]. Despite the strong evidence for a critical role of VCP$^{\text{Npl4/Ufd1}}$ in ERAD, the exact physical relationship between it and the 26S proteasome remains unclear. At least one report has indicated that VCP binds to the proteasome, but it is unclear whether this interaction is direct [110].

In summary, despite emerging evidence that certain polyubiquitinated proteins are targeted to the 26S proteasome by carrier proteins, additional molecular details will be required for verification of this attractive model. These details include the manner in which specific substrates are selected by different polyubiquitin-chain-binding proteins, and how the substrates are transferred from these proteins to PA700.

11.5.6
Roles of PA700 in Ubiquitin-independent Proteolysis

The 26S proteasome is the only identified protease that selectively degrades polyubiquitinated proteins. Although many, and perhaps most, cellular proteins are degraded by the ubiquitin–proteasome system, certain non-ubiquitinated proteins are also substrates for the 26S proteasome. This latter function suggests that PA700 can recognize and interact with features other than polyubiquitin chains for selection of certain substrate proteins. Ornithine decarboxylase (ODC) is the best example of a non-ubiquitinated protein to be degraded by the 26S proteasome [111, 112]. ODC degradation requires ODC binding to antizyme, an endogenous protein inhibitor. Antizyme, however, does not appear to interact directly with PA700. Instead, antizyme probably induces a conformational state of ODC that permits interaction of its C-terminus with PA700 [113]. Polyubiquitin chains competitively inhibit antizyme-induced ODC degradation, indicating that the same element of PA700 recognizes both features [114]. Once the C-terminus of ODC is engaged by PA700, ODC is inactivated and unfolded prior to degradation, and either or both of these processes require PA700-catalyzed ATP hydrolysis in a mechanism that probably is related to the chaperone-like properties of PA700 [115]. Interestingly, the C-terminus of ODC is probably disordered, a structural feature that may dictate its initial interaction with PA700. Other unstructured non-ubiquitinated proteins also interact with PA700 and are degraded by the 26S proteasome *in vitro* [69]. Thus, it is reasonable to believe that the ability of PA700 to interact with features of unstructured proteins (such as might be part of the unfolding or translocation processes) could dictate targeting of certain proteins to the complex for degradation. It is unclear to what extent this type of process occurs in intact cells.

11.6
Roles for PA700 in Non-proteolytic Processes

11.6.1
Overview

The presentation above has focused on the role of PA700 in regulating proteolysis by the proteasome. Although this is undoubtedly a major function of PA700, emerging evidence indicates that PA700, either as part of the intact 26S proteasome or as a separate complex or sub-complex, may mediate non-proteolytic processes. The extent to which PA700 mediates non-proteolytic functions in cells, and the exact mechanisms by which they occur, are unclear at present, but we believe that such functions will prove to be an important part of PA700 biology.

11.6.2
The Role of PA700 in Nucleotide-excision Repair

Rad23 was originally identified as a component of the *n*ucleotide-*e*xcision-*r*epair (NER) process in yeast; deletion of Rad23 increases UV sensitivity of yeast and extracts of these strains are defective in NER. As noted above, Rad23 is a polyubiquitin-chain-binding protein that also binds directly to PA700. Several lines of evidence indicate that the Rad23–PA700 interaction is required for normal NER. Thus, Rad23 might recruit PA700 to the site of NER through a PA700-UBL domain. The precise role of PA700 in NER is controversial [102, 103, 106]. Some studies indicate that NER requires ubiquitin–proteasome-dependent proteolysis [93, 116]; others indicate that PA700 mediates NER by a mechanism that is independent of proteolysis [102, 103]. In the latter instance, PA700 might provide general chaperone-like properties required for remodeling of proteins of the NER complex.

11.6.3
The Role of PA700 in Transcription

The ubiquitin–proteasome system regulates many aspects of transcription via regulated degradation of specific transcription factors and/or their regulatory proteins [117]. By this traditional mechanism, proteolysis regulates a process, in this case transcription, by determining the content of proteins that mediate it. However, components of the ubiquitin–proteasome system, including PA700, also appear to serve as fundamental elements of the transcriptional machinery [118, 119]. The exact roles of ubiquitin, 26S proteasome, and PA700 in the mechanism of transcription are not yet clear, but the available evidence indicates that they are multifaceted and include both proteolytic and non-proteolytic aspects. In retrospect, an essential role for the 26S proteasome, and PA700 in particular, is not surprising because several AAA ATPases of PA700 were identified originally as putative transcription factors by genetic analysis in yeast [120]. It is beyond the scope of this chapter to thoroughly and critically review the mechanism by which ubiquitin and the proteasome regulate transcription; we focus here on proposed roles for PA700 and highlight possible non-proteolytic functions.

Activation of transcription by interactions between activator proteins and the general transcriptional machinery is closely linked to proteolytic destruction of the activator [118, 121]. This process is probably triggered by recruitment of ubiquitin ligases to the promoter, followed by destruction of the ubiquitinated activator by the 26S proteasome. Nevertheless, several studies indicate that a sub-complex of PA700, similar or identical to the base, is recruited initially to the GAL promoter of yeast by the Gal4 transactivator, thus forming a PA700–transcription factor complex competent for transcription [122]. This type of process has been found for other transcription factors in yeast and humans [123, 124]. Surprisingly, the PA700 base, perhaps after recruitment of other components of PA700 and proteasome, converts the polymerase to which is it bound to an elongation-competent form [125]. The exact molecular basis for these various non-proteolytic roles of PA700 is unclear, but could be related to PA700's general chaperone-like function, which might be required for restructuring of transcriptional complexes and/or alteration of the chromatin structure. Obviously, these early studies represent only the beginning of our understanding of the role of PA700 in transcription.

11.7
Summary and Perspective

PA700/19S RP is a multifunctional protein complex that plays essential roles in both the birth and death of cellular proteins. The best understood function of PA700 is its regulation of 26S proteasome function in ubiquitin-dependent protein degradation. However, despite remarkable progress, detailed mechanisms for fundamental features of this function, such as how and where polyubiquitin binds to PA700, the role of ATP hydrolysis in substrate degradation, the roles of individual PA700 subunits, and the extent to which PA700 mediates ubiquitin-independent protein degradation, remain largely unknown. Studies showing that PA700 and PA700 sub-complexes may function in various non-proteolytic processes such as transcription open a rich new area of investigation. A crystal structure of PA700 in the presence and absence of a 20S proteasome will be an important advance for understanding the function and regulation of this complex in various cellular processes.

References

1 CRAIU, A., GACZYNSKA, M., AKAPIAN, T., GRAMM, C. F., FENTEANY, G., GOLDBERG, A. L., and ROCK, K. L. Lactacystin and *clasto*-lactacystin β-lactone modify multiple proteasome β-subunits and inhibit intracellular protein degradation and major histocompatibility complex class I antigen presentation. *J. Biol. Chem.* **1997**, *272*, 13437–13445.

2 ROCK, K. L., GRAMM, C., ROTHSTEIN, L., CLARK, K., STEIN, R., DICK, L., HWANG, D., and GOLDBERG, A. L. Inhibitors of the proteasome block the degradation of most cell proteins and the generation of peptides presented on MHC class I molecules. *Cell* **1994**, *78*, 761–771.

3 COUX, O., TANAKA, K., and GOLDBERG, A. L. Structure and functions of the 20S and 26S proteasomes. *Ann. Rev. Biochem.* **1996**, *65*, 801–847.

4 DEMARTINO, G. N. and SLAUGHTER, C. A. The proteasome, a novel protease regulated by multiple mechanisms. *J. Biol. Chem.* **1999**, *274*, 22123–22126.

5 VOGES, D., ZWICKL, P., and BAUMEISTER, W. The 26S proteasome: a molecular machine designed for controlled proteolysis. *Ann. Rev. Biochem.* **1999**, *68*, 1015–1068.

6 GLICKMAN, M. H., RUBIN, D. M., FRIED, V. A., FISCHER, J. E., and FINLEY, D. The regulatory particle of the *Saccharomyces cerevisiae* proteasome. *Mol. Cell. Biol.* **1998**, *18*, 3149–3162.

7 HOFFMAN, L., PRATT, G., and RECHSTEINER, M. Multiple forms of the 20S multicatalytic and the 26S ubiquitin-ATP-dependent proteases from rabbit reticulocyte lysate. *J. Biol. Chem.* **1992**, *267*, 22362–22368.

8 MA, C.-P., VU, J. H., PROSKE, R. J., SLAUGHTER, C. A., and DEMARTINO, G. N. Identification, purification, and characterization of a high-molecular weight, ATP-dependent activator (PA700) of the 20S proteasome. *J. Biol. Chem.* **1994**, *269*, 3539–3547.

9 UDVARDY, A. Purification and characterization of a multiprotein component of the Drosophila 26S (1500 kDa) proteolytic complex. *J. Biol. Chem.* **1993**, *268*, 9055–9062.

10 PETERS, J.-M., WALSH, M. J., and FRANKE, W. W. An abundant and ubiquitous homo-oligomeric ring-shaped ATPase particle related to the putative vesicle fusion proteins Sec 18p and NSF. *EMBO J.* **1990**, *9*, 1757–1767.

11 ADAMS, G. M., FALKE, S., GOLDBERG, A. L., SLAUGHTER, C. A., DEMARTINO, G. N., and GOGOL, E. P. Structural and functional effects of PA700 and modulator protein on proteasomes. *J. Mol. Biol.* **1997**, *273*, 646–657.

12 GLICKMAN, M. H. and CIECHANOVER, A. The ubiquitin-proteasome system: destruction for the sake of construction. *Physiol. Rev.* **2002**, *82*, 373–428.

13 BEYER, A. Sequence analysis of the AAA protein family. *Protein Sci.* **1997**, *6*, 2043–2058.

14 CONFALONIERI, F. and DUGUET, M. A 200-amino acid ATPase module in search of a basic function. *BioEssays* **1995**, *17*, 639–650.

15 VALE, R. D. AAA proteins: lords of the ring. *J. Cell Biol.* **2000**, *150*, F13–F19.

16 OGURA, T. and WILKINSON, A. J. AAA+ superfamily ATPases: common structure-diverse function. *Genes to Cells* **2001**, *6*, 575–597.

17 DEMARTINO, G. N., MOOMAW, C. R., ZAGNITKO, O. P., PROSKE, R. J., MA, C.-P., AFENDIS, S. J., SWAFFIELD, J. C., and SLAUGHTER, C. A. PA700, an ATP-dependent activator of the 20S proteasome, is an ATPase containing multiple members of a nucleotide-binding protein family. *J. Biol. Chem.* **1994**, *269*, 20878–20884.

18 HOFFMAN, L. and RECHSTEINER, M. Nucleotidase activities of the 26S proteasome and its regulatory complex. *J. Biol. Chem.* **1996**, *271*, 32538–32545.

19 DEMARINI, D. J., PAPA, F. R., SWAMINATHAN, S., URSIC, D.,

RASMUSSEN, T. P., CULBERTSON, M. R., and HOCHSTRASSER, M. The yeast SEN3 gene encodes a regulatory subunit of the 26S proteasome complex required for ubiquitin-dependent protein degradation in vivo. *Mol. Cell. Biol.* 1995, *15*, 6311–6321.

20 HAMPTON, R. Y., GARDNER, R. G., and RINE, J. Role of the 26S proteasome and *HRD* genes in the degradation of 3-hydroxy-3methylglutaryl-CoA reductase, an integral endoplasmic reticulum membrane protein. *Mol. Biol. Cell* 1996, *7*, 2029–2044.

21 YOKOTA, K., KAGAWA, S., SHIMIZU, Y., AKIOKA, H., TSURUMI, C., NODA, C., FUJIMURA, M., YOKOSAWA, H., FUJIWARA, T., TAKAHASHI, E., OHBA, M., YAMASAKI, M., DEMARTINO, G. N., SLAUGHTER, C. A., TOH-E, A., and TANAKA, K. cDNA cloning of p112, the largest regulatory subunit of the human 26S proteasome, and functional analysis of its yeast homologue, Sen3p. *Mol. Biol. Cell* 1996, *7*, 853–870.

22 DEVERAUX, Q., USTRELL, V., PICKART, C., and RECHSTEINER, M. A 26S protease subunit that binds ubiquitin conjugates. *J. Biol. Chem.* 1994, *269*, 7059–7061.

23 HOFMANN, K. and FALQUET, L. A ubiquitin-interacting motif conserved in components of the proteasomal and lysosomal protein degradation systems. *Trends Biochem. Sci.* 2001, *26*, 347–350.

24 VERMA, R., ARAVIND, L., OANIA, R., MCDONALD, W. H., YATES, J. R., KOONIN, E. V., and DESHAIES, R. J. Role of Rpn11 metalloprotease in deubiquitination and degradation by the 26S proteasome. *Science* 2002, *298*, 611–615.

25 YAO, T. and COHEN, R. E. A cryptic protease couples deubiquitination and degradation by the proteasome. *Nature* 2002, *419*, 403–407.

26 LAM, Y. A., XU, W., DEMARTINO, G. N., and COHEN, R. E. Editing of ubiquitin conjugates by an isopeptidase in the 26S proteasome. *Nature* 1997, *385*, 737–740.

27 LAM, Y. A., DEMARTINO, G. N., PICKART, C. M., and COHEN, R. E. Specificity of the ubiquitin isopeptidase in the PA700 regulatory complex of 26S proteasomes. *J. Biol. Chem.* 1997, *272*, 28483–28446.

28 WING, S. S. Deubiquitinating enzymes-the importance of driving in reverse along the ubiquitin-proteasome pathway. *Inter. J. Biochem. Cell Biol.* 2003, *35*, 590–605.

29 FUJIMURO, M., TANAKA, K., YOKOSAWA, H., and TOH-E, A. Son1p is a component of the 26S proteasome of the yeast *Saccharomyces cerevisiae*. *FEBS Lett.* 1998, *423*, 149–154.

30 XIE, Y. and VARSHAVSKY, A. RPN4 is a ligand, substrate, and transcriptional regulator of the 26S proteasome: a negative reedback circuit. *Proc. Natl. Acad. Sci. (USA)* 2001, *98*, 3056–3061.

31 MANNHAUPT, G., SCHNALL, R., KARPOV, V., VETTER, I., and FELDMANN, H. Rpn4 acts as a transcriptional factor by binding to PACE, a nonamer box found upstream of 26S proteasomal and other genes in yeast. *FEBS Lett.* 1999, *450*, 27–34.

32 WOJCIK, C. and DEMARTINO, G. N. Analysis of *Drosophila* 26S proteasome using RNA interference. *J. Biol. Chem.* 2002, *277*, 6188–6197.

33 MEINERS, S., HEYKEN, D., WELLER, A., LUDWIG, A., STANGL, K., and KLOETZEL, P.-M. Inhibition of proteasome activity induces concerted expression of proteasome genes and *de novo* formation of mammalian proteasomes. *J. Biol. Chem.* 2003, *278*, 21517–21525.

34 SÜMEGI, M., HUNYADI-GULYÁS, E., MEDZIHRADSZKY, K. F., and UDVARDY, A. 26S proteasome subunits are O-linked *N*-acetylglucosamine-modified in *Drosophila melanogaster*. *Biochem Biophys. Res Commun.* 2003, *312*, 1284–1289.

35 FERRELL, K., WILKINSON, C. R. M., DUBIEL, W., and GORDON, C. Regulatory subunit interactions of the 26S proteasome, a complex problem. *Trends Biochem. Sci.* 2000, *25*, 83–88.

36 GLICKMAN, M. H., RUBIN, D. M., COUX, O., WEFES, I., PFEIFER, G., CJEKA, Z., BAUMEISTER, W., FRIED, V.,

and FINLEY, D. A sub-complex of the proteasome regulatory particle required for ubiquitin-conjugate degradation and related to the COP9-signalosome and eIF3. *Cell* **1998**, *94*, 615–623.

37 KAJAVA, A. V. What curves alpha-solenoids? Evidence for an alpha-helical toroid structure of Rpn1 and Rpn2 proteins of hte 26S proteasome. *J. Biol. Chem.* **2002**, *277*, 49791–49798.

38 ELSASSER, S., GALI, R. R., SCHWICKART, M., LARSEN, C. N., LEGGETT, D. S., MULLER, B., FENG, M. T., TUBING, F., DITTMAR, G. A., and FINLEY, D. Proteasome subunit Rpn1 binds ubiquitin-like protein domains. *Nat. Cell Biol.* **2002**, *4*, 725–730.

39 VAN NOCKER, S., SADIS, S., RUBIN, D. M., GLICKMAN, M., FU, H., COUX, O., WEFES, I., FINLEY, D., and VIERSTRA, R. D. The multiubiquitin-chain-binding protein Mcb1 is a component of the 26S proteasome in *Saccharomyces cerevisiae* and plays a nonessential, substrate-specific role in protein turnover. *Mol. Cell. Biol.* **1996**, *16*, 6020–6028.

40 MASON, G. G. F., MURRAY, R. Z., PAPPIN, D., and RIVETT, A. J. Phosphorylation of ATPase subunits of the 26S proteasome. *FEBS Lett.* **1998**, *430*, 269–274.

41 BOSE, S., BROOKS, P., MASON, G. G. F., and RIVETT, A. J. γ-Interferon decreases the level of 26S proteasomes and changes the pattern of phosphorylation. *Biochem. J.* **2001**, *353*, 291–297.

42 ZHANG, F., SU, K., YANG, X., BOWE, D. B., PATERSON, A. J., and KUDLOW, J. E. O-GlcNAc modification is an endogenous inhibitor of the proteasome. *Cell* **2003**, *115*, 715–725.

43 RUBIN, D. M., GLICKMAN, M. H., LARSEN, C. N., DRUVAKUMAR, S., and FINLEY, D. Active site mutants in the six regulatory particle ATPases reveal multiple roles for ATP in the proteasome. *EMBO J.* **1998**, *17*, 4909–4919.

44 KESSEL, M., MAURIZI, M. R., KIM, B., KOCSIS, E., TRUS, B. L., SINGH, S. K., and STEVEN, A. C. Homology in structural organization between E. coli ClpAP protease and the eukaryotic 26 S proteasome. *J. Mol. Biol.* **1995**, *250*, 587–594.

45 BOCHTLER, M., DITZEL, L., GROLL, M., HARTMANN, C., and HUBER, R. The proteasome. *Ann. Rev. Biophys. Biomol. Struct.* **1999**, *28*, 295–317.

46 GROLL, M., DITZEL, L., LOWE, J., STOCK, D., BOCHTLER, M., BARTUNIK, H. D., and HUBER, R. Structure of the 20S proteasome from yeast at 2.4A resolution. *Nature* **1997**, *386*, 463–471.

47 LÖWE, J., STOCK, D., JAP, B., ZWICKL, P., BAUMEISTER, W., and HUBER, R. Crystal structure of the 20S proteasome from the archaeon T. acidophilum at 3.4 Å resolution. *Science* **1995**, *268*, 533–539.

48 WHITBY, F. G., MASTERS, E. I., KRAMER, L., KNOWLTON, J. R., YAO, Y., WANG, C. C., and HILL, C. P. Structural basis for the activation of 20S proteasomes by 11S regulators. *Nature* **2001**, *408*, 115–120.

49 GROLL, M., BAJOREK, M., KOHLER, A., MORODER, L., RUBIN, D. M., HUBER, R., GLICKMAN, M. N., and FINLEY, D. A gated channel into the proteasome core particle. *Nat. Struct. Biol.* **2001**, *11*, 1062–1067.

50 KÖHLER, A., CASCIO, P., LEGGETT, D. S., WOO, K. M., GOLDBERG, A. L., and FINLEY, D. The axial channel of the proteasome core particle is gated by the Rpt2 ATPase and controls both substrate entry and product release. *Mol. Cell* **2001**, *7*, 1143–1152.

51 THROWER, J. S., HOFFMAN, L., RECHSTEINER, M., and PICKART, C. M. Recognition of the polyubiquitin proteolytic signal. *EMBO J.* **2001**, *19*, 94–102.

52 FERRELL, K., DEVERAUX, Q., VAN NOCKER, S., and RECHSTEINER, M. Molecular cloning and expression of a multiubiquitin chain binding subunit of the human 26S protease. *FEBS Lett.* **1996**, *381*, 143–148.

53 BEAL, R., DEVERAUX, Q., XIA, G., RECHSTEINER, M., and PICKART, C. Surface hydrophobic residues of multiubiquitin chains essential for proteolytic targeting. *Proc. Natl. Acad. Sci. (USA)* **1996**, *93*, 861–866.

54 van Nocker, S., Deveraux, Q., Rechsteiner, M., and Vierstra, R. D. *Arabidopsis MBP1* gene encodes a conserved ubiquitin recognition component of the 26S proteasome. *Proc. Natl. Acad. Sci. (USA)* **1996**, *93*, 856–860.

55 Kominami, K., Okura, N., Kawamura, M., DeMartino, G. N., Slaughter, C. A., Simbara, N., Chung, C. H., Fujimuro, M., Yokosawa, H., Shimizu, Y., Tanahashi, N., Tanaka, K., and Toh-e, A. Yeast counterparts of sub-units S5a and p58 (S3) of the human 26S proteasome are encoded by two multicopy suppressors of *nin1–1*. *Mol. Biol. Cell* **1997**, *8*, 171–187.

56 Deveraux, Q., van Nocker, S., Mahaffey, D., Vierstra, R., and Rechsteiner, M. Inhibition of ubiquitin-mediated proteolysis by the *Arabidopsis* 26S protease subunit S5a. *J. Biol. Chem.* **1995**, 29660–29663.

57 Young, P., Deveraux, Q., Beal, R. E., Pickart, C. M., and Rechsteiner, M. Characterization of two polyubiquitin binding sites in the 26S protease subunit 5a. *J. Biol. Chem.* **1998**, *273*, 5461–5467.

58 Fu, H., Sadis, S., Rubin, D. M., Glickman, M., van Nocker, S., Finley, D., and Vierstra, R. D. Multiubiquitin chain binding and protein degradation are mediated by distinct domains within the 26S proteasome subunit Mcb1. *J. Biol. Chem.* **1998**, *273*, 1970–1981.

59 Mueller, T. D. and Feigon, J. Structural determinants for the binding of ubiquitin-like domains to the proteasome. *EMBO J.* **2003**, *22*, 4634–4645.

60 Lam, Y. A., Lawson, T. G., Velayultham, M., Zwierm J. L., and Pickart, C. M. A proteasomal ATPase subunit recognizes the polyubiquitin degradation signal. *Nature (London)* **2002**, *416*, 763–767.

61 Verma, R., McDonald, H., Yates, J. R., and Deshaies, R. J. Selective degradation of ubiquitinated Sic1 by purified 26S proteasome yields active S phase cyclin-Cdk. *Mol. Cell* **2001**, *8*, 439–448.

62 Braun, B. C., Glickman, M., Kraft, R., Dahlmann, B., Kloetzel, P.-M., Finley, D., and Schmidt, M. The base of the proteasome regulatory particle exhibits chaperone-like activity. *Nat. Cell Biol.* **1999**, *1*, 221–226.

63 Strickland, E., Hakala, K., Thomas, P. J., and DeMartino, G. N. Recognition of misfolding proteins by PA700, the regulatory sub-complex of the 26S proteasome. *J. Biol. Chem.* **2000**, *275*, 5565–5572.

64 Liu, C., Millen, L., Roman, T. B., Xiong, H., Gilbert, H. F., Novia, R., DeMartino, G. N., and Thomas, P. J. Conformational remodeling of proteasomal substrates by PA700, the 19S regulatory complex of the 26S proteasome. *J. Biol. Chem.* **2002**, *277*, 26815–26820.

65 Lee, C., Schwartz, M. P., Prakash, S., Iwakura, M., and Matouschek, A. ATP-dependent proteases degrade their substrates by processively unraveling them from the degradation signal. *Mol. Cell* **2001**, *7*, 627–737.

66 Akopian, T. N., Kisselev, A. F., and Goldberg, A. L. Processive degradation of proteins and other catatlytic properties of the proteasome from *Thermoplasma acidophilum*. *J. Biol. Chem.* **1997**, *272*, 1791–1798.

67 Dick, L. R., Moomaw, C. R., DeMartino, G. N., and Slaughter, C. A. Degradation of oxidized insulin B chain by the multiproteinase complex macropain (proteasome). *Biochemistry* **1991**, *30*, 2725–2734.

68 Palombella, V. J., Rando, O. J., Goldberg, A. L., and Maniatis, T. The ubiquitin-proteasome pathway is required for processing th NF-kB1 precursor protein and the activation of NF-kB. *Cell* **1994**, *78*, 773–785.

69 Liu, C.-W., Corboy, M. J., DeMartino, G. N., and Thomas, P. J. Endoproteolytic activity of the proteasome. *Science* **2004**, *299*, 408–411.

70 Rape, M. and Jentsch, S. Taking a bite: proteasomal protein processing. *Nat. Cell Biol.* **2002**, *4*, E113–E116.

71 LUNDGREN, J., MASSON, P., REALINI, C., and YOUNG, P. Use of RNA interference and complementation to study the function of the *Drosophila* and human 26S proteasome subunits. *Mol. Cell Biol.* **2004**, *23*, 5320–5330.

72 LI, T., NAQVI, N. I., YANF, H., and TEO, T. S. Identification of a 26S proteasome-associated UCH in fission yeast. *Biochem Biophys. Res Commun.* **2000**, *272*, 270–275.

73 HÖLZL, H., KAPELARI, B., KELLERMANN, J., SEEMÜLLER, E., SÜMEGI, M., UDVARDY, A., MEDALIA, O., SPERLING, J., MÜLLER, S. A., ENGEL, A., and BAUMEISTER, W. The regulatory complex of *Drosophila melanogaster* 26S proteasomes: subunit composition and localization of a deubiquitylating enzyme. *J. Cell Biol.* **2000**, *150*, 119–129.

74 GORDON, C., McGURK, G., DILLON, P., ROSEN, C., and HASTLE, N. D. Defective mitosis due to a mutation in the gene for a fission yeast 26S protease subunit. *Nature* **1993**, *366*, 355–357.

75 VERMA, R., CHEN, S., FELDMAN, R., SCHIELTZ, D., YATES, J., DOHMEN, J., and DESHAIES, R. J. Proteasomal proteomics: identification of nucleotide-sensitive proteasomal-interacting proteins by mass spectrometric analysis of affinity-purified proteasomes. *Mol. Biol. Cell* **2000**, *11*, 3425–3439.

76 YANG, Y., FRÜH, K., AHN, K., and PETERSON, P. A. *In vivo* assembly of the proteasomal complexes, implications for antigen processing. *J. Biol. Chem.* **1995**, *270*, 27687–27694.

77 BAUMEISTER, W., WALZ, J., ZÜHL, F., and SEEMÜLLER, E. The proteasome: paradigm of a self-compartmentalizing protease. *Cell* **1998**, *92*, 367–380.

78 SCHMIDTKE, G., KRAFT, R., KOSTKA, S., HENKLEIN, P., FRÖMMEL, C., LÖWE, J., HUBER, R., KLOETZEL, P.-M., and SCHMIDT, M. Analysis of mammalian 20S proteasome biogenesis: the maturation of β-subunits is an ordered two-step mechanism involving autocatalysis. *EMBO J.* **1996**, *15*, 6887–6898.

79 IMAI, J., MARUYA, M., YASHIRODA, H., YAHARA, I., and TANAKA, K. The molecular chaperone Hsp90 plays a role in the assembly and maintenance of the 26S proteasome. *EMBO J.* **2003**, *22*, 3557–3567.

80 LEGGETT, D. S., HANNA, J., BORODOVSKY, A., CROSAS, B., SCHMIDT, M., BAKER, R. T., WALTZ, T., PLOEUGH, H., and FINLEY, D. Multiple associated proteins regulate proteasome structure and function. *Mol. Cell* **2002**, *10*, 498–507.

81 BORODOVSKY, A., KESSLER, B. M., CASAGRANDE, R., OVERKLEEFT, H. S., WILKINSON, K. D., and PLOEGH, H. L. A novel active site-directed probe specific for deubiquitylating enzymes reveals proteasome association of USP14. *EMBO J.* **2001**, *20*, 5187–5196.

82 PAPA, F. R. and HOCHSTRASSER, M. The yeast DOA4 gene encodes a deubiquitinating enzyme related to a product of the humun tre-2 oncogene. *Nature* **1993**, *366*, 313–319.

83 PAPA, F. R., AMERIK, A., and HOCHSTRASSER, M. Interaction of the Doa4 deubiquitinating enzyme with the yeast 26S proteasome. *Mol. Biol. Cell* **1999**, *10*, 741–756.

84 SWAMINATHAN, S., AMERIK, A., and HOCHSTRASSER, M. The Doa4 deubiquitinating enzyme is required for ubiquitin homeostasis in yeast. *Mol. Biol. Cell* **1999**, *10*, 2583–2594.

85 AMERIK, A., NOWAK, J., SWAMINATHAN, S., and HOCHSTRASSER, M. The Doa4 deubiquitinating enzyme is functionally linked to the vacuolar protein-sorting and endocytic pathways. *Mol. Biol. Cell* **2000**, *11*, 3365–3380.

86 TONGAONKAR, P., CHEN, L., LAMBERTSON, D., KO, B., and MADURA, K. Evidence for an interaction between ubiquitin-conjugating enzymes and the 26S proteasome. *Mol. Cell Biol.* **2000**, *20*, 4691–4698.

87 YOU, J. and PICKART, C. M. A HECT domain E3 enzyme assembles novel polyubiquitin chains. *J. Biol. Chem.* **2001**, *276*, 19871–19878.

88 XIE, Y. and VARSHAVSKY, A. Physical association of ubiquitin ligases and the 26S proteasome. *Proc. Natl. Acad. Sci. (USA)* **2000**, *97*, 2497–2502.

89 MURATA, S., MINAMI, Y., CHIBA, T., and TANAKA, K. CHIP is a chaperone-dependent E3 ligase that ubiquitylates unfolded protein. *EMBO Rep.* **2001**, *2*, 1133–1138.

90 LUDERS, J., DEMAND, J., and HOHFELD, J. The ubiquitin-related BAG-1 provides a link between the molecular chaperones Hsc70/Hsp70 and the proteasome. *J. Biol. Chem.* **2000**, *275*, 4613–4617.

91 KLEIJNEN, M. F., SHIH, A. H., ZHOU, P., KUMAR, S., SOCCIO, R. E., KEDERSHA, N. L., GILL, G., and HOWLEY, P. M. The hPLIC proteins may provide a link between the ubiquitination machinery and the proteasome. *Mol. Cell* **2000**, *6*, 409–419.

92 KLEIJNEN, M. F., ALARCON, R. M., and HOWLEY, P. M. The ubiquitin-associated domain of hPLIC-2 interacts with the proteasome. *Mol. Biol. Cell* **2003**, *14*, 3868–3875.

93 CHEN, L. and MADURA, K. Rad23 promotes the targeting of proteolytic substrates to the proteasome. *Mol. Cell. Biol.* **2002**, *22*, 4902–4913.

94 SAEKI, Y., SONE, T., TOH-E, A., and YOKOSAWA, H. Identification of ubiquitin-like protein-binding subunits of the 26S proteasome. *Biochem Biophys. Res Commun.* **2002**, *296*, 813.

95 FUNAKOSHI, M., SASAKI, T., NISHIMOTO, T., and KOBAYASHI, H. Budding yeast Dsk2p is a polyubiquitin-binding protein that can interact with the proteasome. *Proc. Natl. Acad. Sci. (USA)* **2002**, *99*, 745–750.

96 HARTMANN-PETERSEN, R., SEMPLE, C., PONTING, C. P., HENDIL, K. B., and GORDON, C. UBA domain containing proteins in fission yeast. *Inter. J. Biochem. Cell Biol.* **2003**, *35*, 629–636.

97 WILKINSON, C. R. M., SEEGER, M., HARTMANN-PETERSEN, R., STONE, M., WALLACE, M., SEMPLE, C., and GORDON, C. Proteins containing the UBA domain are able to bind to multi-

ubiquitin chains. *Nat. Cell Biol.* **2001**, *3*, 939–943.

98 HIYAMA, H., YOKOI, M., MASUTANI, C., SUGASAWA, K., MAEKAWA, T., TANAKA, K., HOEIJAMAKERS, J. H. J., and HANAOKA, F. Interaction of hHR23 with S5a: the ubiquitin-like domain of hHR23 mediates interaction with S5a subunit of 26S proteasome. *J. Biol. Chem.* **1999**, *274*, 28019–28025.

99 LAMBERTSON, D., CHEN, L., and MADURA, K. Investigating the importance of proteasome-interaction for Rad23 function. *Curr. Genet.* **2003**, *42*, 199–208.

100 RAO, H. and SASTRY, A. Recognition of specific ubiquitin conjugates is important for the proteolytic functions of the ubiquitin-associated domain proteins Dsk2 and Rad23. *J. Biol. Chem.* **2002**, *277*, 11691–11695.

101 RASSI, S. and PICKART, C. M. Rad23 ubiquitin-associated domains (UBA) inhibit 26S proteasome-catalyzed proteolysis by sequestering lysine 48-linked polyubiquitin chains. *J. Biol. Chem.* **2003**, *278*, 8951–8959.

102 GILLETTE, T. G., HUANG, W., RUSSELL, S. J., REED, S. H., and JOHNSTON, S. A. The 19S complex of the proteasome regulates nucleotide excision repair in yeast. *Genes. Dev.* **2001**, *15*, 1528–1539.

103 RUSSELL, S. J., REED, S. H., HUANG, W., FRIEDBERG, E. C., and JOHNSTON, S. A. The 19S regulatory complex of the proteasome functions independently of proteolysis in nucleotide excision repair. *Mol. Cell* **1999**, *3*, 687–695.

104 HUYTON, T., PYE, V. E., BRIGGS, L. C., FLYNN, T. C., BEURON, F., KONDO, H., MA, J., ZHANG, X., and FREEMONT, P. S. Crystal structure of murine p97/VCP at 3.6 Å. *J. Struct Biol.* **2003**, *144*, 337–348.

105 ROUILLER, I., DELABARRE, B., MAY, A. P., WEIS, W. I., BRUNGER, A. T., MILLIGAN, R. A., and WILSON-KUBALEK, E. M. Conformational changes of the multifunction p97 AAA ATPase during its ATPase cycle. *Nat. Struct. Biol.* **2002**, *9*, 950–957.

106 DAI, R. M. and LI, C.-C. H. Valosin-containing protein is a multi-ubiquitin chain-targeting factor required in ubiquitin-proteasome degradation. *Nat. Cell Biol.* **2001**, *3*, 740–744.

107 YE, Y., MEYER, H. H., and RAPOPORT, T. A. The AAA ATPase Cdc48/p97 and its partners transport proteins from the ER into the cytoplasm. *Nature* **2001**, *414*, 652–656.

108 YE, Y., MEYER, H. H., and RAPOPORT, T. Function of the p97-Ufd1-Npl4 complex in retrotranslocation from the ER to the cytosol: dual recognition of nonubiquitinated polypeptide segments and polyubiquitin chains. *J. Cell Biol.* **2003**, *162*, 71–84.

109 WOJCIK, C., YANO, M., and DEMARTINO, G. N. RNAi interference of valosin-containing protein (VCP/p97) reveals multiple cellular roles linked to ubiquitin/proteasome dependent proteolysis. *J. Cell Sci.* **2004**, *117*, 281–292.

110 DAI, R. M., CHEN, E., LONGO, D. L., GORBEA, C. M., and LI, C.-C. H. Involvement of valsoin-containing protein, an ATPase co-purified with IκBα and 26S proteasome, in ubiquitin-proteasome-mediated degradation of IκBα. *J. Biol. Chem.* **1998**, *273*, 3562–3573.

111 MURAKAMI, Y., MATSUFUJI, S., KAMEJI, T., HAYASHI, S., IGARASHI, K., TAMURA, T., TANAKA, K., and ICHIHARA, A. Ornithine decarboxylase is degraded by the 26S proteasome without ubiquitination. *Nature* **1992**, *360*, 597–599.

112 TOKUNAGA, F., GOTO, T., KOIDE, T., MURAKAMI, Y., HAYASHI, S., TAMURA, T., TANAKA, K., and ICHIHARA, A. ATP- and antizyme-dependent endoproteolysis of ornithine decarboxylase to oligopeptides by the 26S proteasome. *J. Biol. Chem.* **1994**, *269*, 17382–17385.

113 MURAKAMI, Y., MATSUFUJI, S., HAYASHI, S., TANAHASHI, N., and TANAKA, K. Degradation of ornithine decarboxylase by the 26S proteasome. *Biochem. Biophys. Res. Commun.* **2000**, *267*, 1–6.

114 ZHANG, M., PICKART, C. M., and COFFINO, P. Determinants of proteasome recognition of ornithine decarboxylase, a ubiquitin-independent substrate. *EMBO J.* **2003**, *22*, 1488–1496.

115 MURAKAMI, Y., MATSUFUJI, S., HAYASHI, S., TANAHASHI, N., and TANAKA, K. ATP-dependent inactivation and sequestration of ornithine decarboxylase by the 26S proteasome are prerequisites for degradation. *Mol. Cell Biol.* **1999**, *19*, 7216–7227.

116 SCHAUBER, C., CHEN, L., TONGAONKAR, L., VEGA, I., LAMBERTSON, D., POTTS, W., and MADURA, K. Rad23 links DNA repair to the ubiquitin/proteasome pathway. *Nature* **1998**, *391*, 715–718.

117 CIECHANOVER, A., ORIAN, A., and SCHWARTZ, A. L. The ubiquitin-mediated proteolytic pathway: mode of action and clinical implications. *J. Cell Biochem. Suppl.* **2002**, *34*, 40–51.

118 MURATANI, M. and TANSEY, W. P. How the ubiquitin-proteasome system controls transcription. *Nat. Rev. Mol. Cell Biol.* **2003**, *4*, 1–10.

119 CONAWAY, R. C., BROWER, C. S., and CONAWAY, J. W. Emerging roles of ubiquitin in transcription regulation. *Science* **2002**, *296*, 1254–1558.

120 SWAFFIELD, J. C., BROMBER, J. F., and JOHNSTON, S. A. Alterations in a yeast protein resembling HIV Tat-binding protein relieve requirement for an acidic activation domain in GAL4. *Nature* **1992**, *357*, 698–700.

121 SALGHETTI, S. E., MURATANI, M., WIJNEN, H., FUTCHER, B., and TANSEY, W. P. Functional overlap of sequences that activate transcription and signal ubiquitin-mediated proteolysis. *Proc. Natl. Acad. Sci. (USA)* **2000**, *97*, 3118–3123.

122 GONZALEZ, F., DELAHODDE, A., KODADEK, T., and JOHNSTON, S. A. Recruitment of a 19S proteasome subcomplex to an activated promoter. *Science* **2002**, *286*, 548–550.

123 FERDOUS, A., KODAKEK, T., and JOHNSTON, S. A. A nonproteolytic function of the 19S regulatory subunit of the 26S proteasome is required by efficient activated transcription by human RNA polymerase II. *Biochemistry* **2002**, *41*, 12798–12805.

124 SUN, L., JOHNSTON, S. A., and KODADEK, T. Physical association of the APIS complex and general transcription factors. *Biochem. Biophys. Res. Commun.* **2002**, *296*, 991–999.

125 FERDOUS, A., GONZALEZ, F., SUN, L., KODADEK, T., and JOHNSTON, S. A. The 19S regulatory particle of the proteasome is required for efficient transcription elongation by RNA polymerase II. *Mol. Cell* **2001**, *7*, 981–991.

12

Bioinformatics of Ubiquitin Domains and Their Binding Partners

Kay Hofmann

12.1
Introduction

Since its discovery more than 25 years ago, the small protein ubiquitin has been found to be involved in nearly every important aspect of cell biology. Originally, the covalent attachment of ubiquitin to intracellular proteins was thought to invariably label these proteins for degradation by the proteasome. Since then, there has been a dramatic development in our understanding of both the mechanism and the regulation of protein ubiquitination, at least partially due to the increased application of genomics and bioinformatics techniques. We now known that protein ubiquitination regulates not only proteasomal degradation but also gene expression, chromatin structure, DNA repair, protein sorting, endocytosis, and protein degradation by the lysosome and vacuole. Many of these processes involve a multitude of ubiquitination targets, which have to be recognized with high specificity and whose modification is strictly regulated in space and time. Additional complexity comes from the fact that there are multiple ways to modify a protein: Besides the canonical signal for proteasomal degradation, consisting of a chain of at least four ubiquitin molecules linked via Lys-48, other signals use mono-ubiquitination, multiple mono-ubiquitination, or polyubiquitin chains linked by different isopeptide bonds, e.g. involving Lys-63 or Lys-29. Besides ubiquitin, there are a number of other ubiquitin-related modifiers that appear to work in an analogous fashion but convey signals with a very different meaning.

Our current knowledge of the major protein classes acting in the ubiquitin system, together with the availability of genome-wide sequence data, allows us to appreciate the vast complexity of this signalling network. By using methods of bioinformatical analysis, which will be explained in the following paragraphs, it can be estimated that several hundreds of proteins have a role related to the attachment, removal, or recognition of protein modifications by ubiquitin and its relatives. Only for a small fraction of these proteins do we have experimental data confirming their involvement in the ubiquitin system. For a large number of cellular proteins, their role in ubiquitin-mediated processes can be inferred from their molecular architecture. This prediction is typically based on the presence of "functional do-

Protein Degradation. Edited by J. Mayer, A. Ciechanover, M. Rechsteiner
Copyright © 2005 WILEY-VCH Verlag GmbH & Co. KGaA, Weinheim
ISBN: 3-527-30837-7

mains", a concept that will be explained in Section 12.2. Subsequently the section gives a brief introduction to the bioinformatical methods used to identify such functional domains in a given query sequence or in the genome sequence of an organism, and indicates the advantages of this approach.

Every molecular signaling system consists of a number of major components: the signal itself, a mechanism for generating the signal, a mechanism for detecting the signal, and, finally, a mechanism for resetting or destroying the signal. Section 12.3 discusses the signal – ubiquitin and its relatives – and summarizes our current knowledge about the architecture, properties, and evolution of the ubiquitin family. Section 12.4 deals with the various classes of ubiquitin-recognition domains, including UBA, CUE-Ub, UIM, UEV, and GAT domains. Finally, Section 12.5 tries to put the "parts list" of the previous sections into context. Several examples demonstrate how nature has used the "domain-shuffling" mechanism to generate the vast complexity found in the ubiquitin system, but also show how the bioinformatical detection of homology domains can be useful in understanding the function of complex proteins from their modular architecture.

12.2
The Concept of Functional Domains

The whole domain concept originates from the analysis of three-dimensional protein structures. Typical small proteins have a "monolithic" structure that consists of a single fold with several secondary structure elements such as α-helices or β-strands. A structural fold has a hydrophobic core and hydrophilic regions exposed to the solvent. Larger proteins can follow two different architectural principles. Some large proteins just form larger monolithic structures, similar to the situation seen in small proteins. Most large proteins, however, consist of several smaller folding units, the so-called "domains". Each of these domains can fold independently of the rest of the protein and has its own hydrophobic core region. Structural domains can be regarded as self-sufficient mini-proteins that are connected to each other by inter-domain linkers. As a consequence of their autonomous folding capabilities, domains can often be excised from their host protein and pasted into a different protein context, without major changes in fold or function. In the course of evolution, such events have happened several times for many domain types. Evolutionary processes, such as exon shuffling and the duplication, fusion and fission of genes and gene regions, have helped to create the multi-domain "mosaic" structure found in many extant proteins.

In cases where no structural information is available, the presence of domains can frequently be detected just by analyzing the protein sequences. When comparing two otherwise unrelated sequences that have both acquired a particular domain by shuffling events, this domain appears as a region of localized sequence similarity. Such conserved sequence regions are often called "homology domains". A region of localized sequence homology does not always represent a true homology domain. It is also possible that the detected similarity region is just the best-conserved part of two proteins that are distantly related in their entirety. A true ho-mology domain can be assumed when the boundaries of the similarity region are well defined, e.g. if they are delimited by the N- or C-terminus of the protein or by an adjacent well-characterized domain. Evidently, not all local similarities claimed to be "homology domains" in the literature are true domains in the structural sense, and, even if they are, the position of the domain boundaries can deviate. Nevertheless, most homology domains that occur in diverse sets of proteins have been found to correspond nicely to structural domains. Even after multiple rounds of evolutionary shuffling, most domains preserve not only their structure but also the fundamental aspects of their function. Thus, it is frequently possible to attach functional labels to particular domain types. In favorable situations, the property of a novel protein can be predicted from those "functional domains" contained in its sequence.

The term "homology domain" or "functional domain" should be used only for those protein regions that either are known to be domains in the structural sense or that are at least predicted to fulfil that condition. Conserved sequence regions that are too short to fold independently of the rest of the protein should rather be referred to as "motifs". A considerable number of those "functional motifs" have important roles, e.g. by being responsible for specific domain- or protein-recognition events.

12.2.1
Bioinformatical Methods for Domain Detection

As mentioned above, homology domains can be detected by sequence analysis, where they appear as regions of locally confined sequence similarity embedded into an otherwise dissimilar context. Any tool for local sequence alignment, e.g. those using the Smith and Waterman algorithm [1], is suitable for detecting ho-mology domains, at least if they are moderately well conserved. It is generally as-sumed that the structure of a protein is much better conserved than its sequence. A similar observation can be made for a protein's function, whose key features are frequently maintained even at evolutionary distances where sequences no longer look similar. Thus, it can be expected that domains exist – in both the structural and the functional sense – that cannot be spotted easily by sequence comparison alone but which are readily visible in a structural comparison. On the other hand, there is only a limited number of energetically favorable protein folds, especially for very short domains. A similarity of two protein folds does not necessarily imply a common evolutionary origin and similar folds can be found in proteins with to-tally unrelated functions. Evidently, structural comparisons do also have disadvan-tages and are also hampered by the lack of genome-wide structural data. There are even a number of documented examples where sophisticated methods of sequence comparison are more sensitive than structural comparisons [2].

Over the years there have been considerable improvements in the available se-quence analysis techniques. In particular the "sequence profile" method [3] with its more recent extension to "generalized profiles" [4], and various "*H*idden *M*arkov *M*odel" (HMM) methods [5–7] have proved very useful for detecting very weak sequence similarities. In these methods, the increased sensitivity is made possible by accounting for the fact that not all positions in a protein sequence are equally important and thus equally well conserved. Profile and HMM searches do not start with a single-query sequence but rather with a multiple alignment of established members of a protein family. The relative sequence conservation of the alignment positions is an indication of how important these residues are and how much weight is given to them in the sequence-comparison step. Another important feature of profile-based methods is a sophisticated score statistics that allows a reliable assessment of how trustworthy a newly found sequence similarity really is. The aforementioned properties make profile and HMM methods well suited for domain-detection purposes [8]. A large number of homology domains have been identified by each method, including several of the functional domains discussed below.

When a novel homology domain has been discovered, it is possible to store the corresponding domain descriptor (profile or HMM) in a number of dedicated domain databases, which can be used to analyze newly identified sequences for their domain content [9, 10]. Several competing domain- and motif-databases exist, including PROSITE, PFAM, SMART, and Superfam, which contain descriptors for most, if not all, of the known domains involved in the ubiquitin system [11–14]. Recently, a new meta-database named INTERPRO has been established, which tries to combine the descriptors of several domain databases under a single user interface [15]. Pointers to the very useful search engines of the domain databases are provided in Table 12.1.

12.2.2
Advantages of Studying Domains in the Ubiquitin System

A modular architecture consisting of several functional domains appears to be a hallmark of proteins participating in intracellular signal transduction. The classical signaling paradigm involves protein kinases for generating the signal, a number of specialized domains for recognizing the phospho-Tyr or phospho-Ser/Thr signal, and phosphatases for removing the signal. As the phosphorylating and dephosphorylating enzymes are substrate-specific and stringently regulated, their catalytic domains are frequently associated with specialized targeting or scaffolding do-

Tab. 12.1. WWW-servers for detection of homology domains.

Database	URL
PROSITE	http://www.expasy.ch/prosite
PFAM	http://www.sanger.ac.uk/Pfam
SMART	http://smart.embl-heidelberg.de
Superfamily	http://supfam.mrc-lmb.cam.ac.uk
INTERPRO	http://www.ebi.ac.uk/interpro

mains. As we know now, the situation in the ubiquitin system is perfectly analogous, with the tasks of ubiquitination, deubiquitination, and ubiquitin recognition being executed by specialized types of functional domains. Interestingly, this analogy extends to the ubiquitin-related modifiers and their corresponding modification systems.

Most of the domain types used in the ubiquitin system are found exclusively in this pathway. Thus, a newly identified protein containing one of the ubiquitination-specific domains can with high reliability be considered a new component of the ubiquitin system. In some instances, the presence of a particular domain type does not allow prediction of whether the protein is active against ubiquitin or one of its close relatives. Nevertheless, the mining of sequence databases for new proteins containing ubiquitination-specific domains has been a rich source of new components and regulators of the ubiquitin system. Over the recent years, the bioinformatical analysis of those proteins has been instrumental in (i) the discovery of the evolutionary origins of the ubiquitin system, (ii) the transfer of information from better studied model systems to uncharacterized but evolutionary related systems, (iii) the identification of novel components of the ubiquitin system, (iv) the functional elucidation of complex proteins (and protein complexes) by studying their content of functional domains. The following sections will give an overview of the major types of functional domains with a specific role in the ubiquitin system.

12.3
Ubiquitin and Ubiquitin-like Domains

Ubiquitin is a small protein of 76 amino acids that got its name for its ubiquitous distribution in all eukaryotic kingdoms. Owing to its exceptionally high degree of sequence conservation, it is easy to detect even in most remote species – frequently simply by antibody cross-reactivity. Most genomes harbour multiple copies of the ubiquitin gene, which encode identical proteins. Typically, the ubiquitin is not translated in its mature form – most organisms contain genes with multiple ubiquitin copies fused to a single open reading frame. In addition, there are frequently fusion proteins with an N-terminal ubiquitin moiety and a C-terminal "carrier-protein". In both cases, ubiquitin must first be cleaved from the remaining protein before it can be activated and attached to target proteins. Interestingly, the nature of the "carrier proteins" varies from species to species. Typical examples are ribosomal proteins, but other proteins with a high expression level have also been observed. The genomic organization of ubiquitin genes suggests that cells require high amounts of the protein and the need for ubiquitin seems to be coupled to the amount of protein synthesis.

As will be discussed in more detail elsewhere in this book, ubiquitin attachment to proteins gives rise to a number of different signals. Although it was initially thought to be required only for proteasomal degradation, we now know that there are other ubiquitin-based signals, such as mono-ubiquitination, or

Tab. 12.2. Known ubiquitin-like modification systems.

Modifier	Substrate	Process
Ubiquitin	Many	Protein degradation, sorting, regulation
Nedd8 (Rub1)	Cullins	SCF regulation
Sumo (Smt3, Sentrin)	Many	Nuclear transport, localization, regulation
ISG15 (UCRP)	Stat1, others	Immune response (interferon)
Hub1	Hbt1, Sbh1	Polarized morphogenesis (yeast)
Fat10	Unknown	Apoptosis, interferon response
MNSF (FUBI, FAU)	Bcl-G, others?	T-cell activation
Urm1	Ahp1	Stress, invasive growth (yeast)
Apg8 (Atg8, Aut7)	Phosphatidylethanolamine	Autophagy, CVT pathway
Apg12 (Atg12)	Apg5	Autophagy, CVT pathway

multi-ubiquitination with different chain architectures, which all signal different events. In addition to ubiquitin, there are a number of related systems for protein labeling that use ubiquitin-related protein modifiers and similar components of the conjugation and deconjugation pathways.

12.3.1
Ubiquitin and Related Modifiers

A survey of the human genome, or that of model organisms, shows that there are multiple proteins with readily detectable similarity to ubiquitin. Even before the advent of genome-wide bioinformatics, we knew about several other ubiquitin-like proteins that become attached to proteins in a similar fashion, yet do not signal proteasomal degradation, the most prominent example being Smt3/Sumo [16]. Currently, 10 different ubiquitin-like protein modifications systems have been described; a complete list is found in Table 12.2. All of the proteins shown in the first column of this table are related to ubiquitin, although for some of them the similarity is quite subtle and either structural comparisons or sophisticated profile-based bioinformatics methods are required to obtain a decent alignment. Structurally, all ubiquitin-like domains adopt the extremely robust "β-grasp" fold [17], and high-resolution structures for a large number of ubiquitin-like molecules are available.

A detailed discussion of the physiological relevance of the ubiquitin-like modification systems is beyond the scope of this chapter, but there is a large body of literature on the pathways involving Sumo [18, 19], Nedd8 [20], ISG15 [21], Hub1 [22], Apg8 and Atg12 [23, 24], Fat10 [25], Urm1 [26], and MNSF [27]. There are also several excellent reviews providing detailed comparisons of the ubiquitin-like modifiers [28–30]. So far, ubiquitin is the only modifier known to form chains. ISG15 (UCRP) consists of two ubiquitin-like domains that do not appear to be cleaved, "ISGylated" proteins are thus modified with the equivalent of two ubiquitin units linked by a true peptide bond.

Besides the characterized modifiers, the genome contains a large number of other proteins with ubiquitin-like domains. It cannot be excluded that there are additional modification systems hidden among them. Thus, it is an important question if there is a way to discriminate these functions purely by bioinformatical methods. Until recently, the answer would have been probably yes: there seemed to be a strict requirement for a C-terminal Gly–Gly motif, whose terminal carboxyl group is involved in forming the isopeptide bond to the substrate. This diglycine does not necessarily form the C-terminus of the open reading frame, as the cleavage of a C-terminal extension by specialized proteases is commonplace. However, the two glycine residues are invariably located at the C-terminal end of the ubiquitin homology domain. Nowadays, it is not clear if the answer is so easy: the small modifier Hub1, a recent addition to the ranks of ubiquitin-like modifiers, completely lacks this diglycine in all species studied [22]. So far, it is not clear if Hub1 is a singular exception to the rule, or if other modification systems have eluded us because of their non-canonical C-terminus. Hub1 appears to be atypical in the sense that it also modifies proteins in a non-covalent fashion [31].

12.3.2
The Evolutionary Origin of the Ubiquitin System

Until recently, it was assumed that ubiquitin and its conjugation system were restricted to eukaryotes. In fact, ubiquitin has even served as a paradigm for gene lineages completely absent in bacteria [32]. Today, we know that ubiquitin and at least some components of the ubiquitination system have evolutionary ancestors that predate the prokaryote/eukaryote separation. It is probably true that the use of ubiquitin as a protein-tagging system is a eukaryotic invention; however, there are clear structural and functional similarities to two bacterial metabolic systems, the biosynthesis of the thiazole moiety of thiamine pyrophosphate [33] and the biosynthesis of molybdopterin [34]. The two bacterial pathways employ a combination of a ubiquitin-related protein (MoaD and ThiS) and an E1-type activating protein (MoeB and ThiF). Similar to ubiquitin, the MoaD and ThiS proteins both end with a glycine residue, which becomes C-terminally adenylated in a typical E1-like reaction by MoeB and ThiF, respectively. In the case of thiazole synthesis, this activation process is followed by the formation of a covalent conjugate between the C-terminus of ThiS and Cys-184 of ThiF [35], underscoring the analogy to the E1 reaction. So far, a similar conjugate has not been detected in the MoaD/MoeB system. However, the crystal structure of the MoaD–MoeB complex shows a striking analogy to the ubiquitin activation system [36]. After this point, the two biosynthesis pathways diverge from the protein-tagging systems, as there is no further conjugation of ThiS or MoaD to a target protein but rather an incorporation of the acyl-bound sulfur into the biosynthesis product.

While the functional analogy between ThiS, MoaD, and ubiquitin-like modifiers is widely accepted, the two protein classes are frequently described as unrelated; sometimes even a convergent evolution to the energetically favorable ubiquitin fold is discussed. Despite these claims, there is a statistically significant sequence similarity between all of these proteins, which can be detected by standard profile methods described in Section 12.2.1. In particular the Urm1 family serves as a missing link and facilitates the discovery of this distant sequence relationship. Urm1 (*u*biquitin-*r*elated *m*odifier) is an evolutionary conserved protein found in all major eukaryotic lineages [37]. In yeast, Urm1 is activated by Uba4, one of the four E1 enzymes found in that organism. "Urmylation" is clearly a protein-modification system; it has been shown to be responsive to stress, and one of the modified targets is the antioxidant protein Ahp1 [26, 38]. When analyzing the sequences of the Urm1 family, it becomes obvious that this protein is more closely related to the bacterial MoaD and ThiS proteins than to the typical ubiquitin-like modifiers. However, a significant similarity to ubiquitin can also be established.

Apart from ThiS, MoaD, and their activating enzymes, no other bacterial homologs of ubiquitination components have been described. There do not seem to be bacterial E2 or E3 enzymes, and also the UCH and USP types of deubiquitinating enzymes appear to be absent. Eubacteria and archaea do have compartmentalizing proteases that resemble the eukaryotic proteasomes to a varying degree. However, the targeting of substrates to the bacterial "proteasomes" uses other signals, although one interesting parallel has been reported [39].

12.3.3
Ubiquitin Domains in Complex Proteins

A current census of ubiquitin relatives reveals 18 genes encoded by the budding yeast genome and about 75 in the human genome. These numbers do not include Apg8, Apg12, UBX proteins, or any other protein discussed in Section 12.3.4 below. It does, however, include several proteins of much larger size than the ubiquitin-related modifiers, which contain the ubiquitin similarity region as an integral, non-cleavable part of the protein itself. A small selection of those proteins is shown in Figure 12.1. Here, only proteins of a recognizable modular nature are shown, i.e. proteins that contain other domains in addition to the ubiquitin-like (UbL) domain.

When looking at the examples shown in Figure 12.1, or at the large set of proteins not shown here, two general trends are obvious: (i) the ubiquitin domain tends to be localized at the extreme N-terminus, and (ii) the host protein is typically involved in the ubiquitin system. The first observation has been interpreted as an evolutionary remnant of earlier ubiquitin-fusion proteins [40]. As mentioned above, ubiquitin is typically expressed as a precursor protein, wherein the ubiquitin moiety is localized at the N-terminus and has to be liberated by dedicated ubiquitin hydrolases. It is certainly possible that many extant proteins with ubiquitin-like domains used to be alternative ubiquitin precursors but have lost their cleavability. The second observation will be discussed in more detail in Sections 12.5.1 and 12.5.2.

From the viewpoint of bioinformatics, the second observation has turned out to be most useful. The identification of a ubiquitin-like domain in a protein makes it a good candidate for a new component of the ubiquitin regulatory system. In addi-

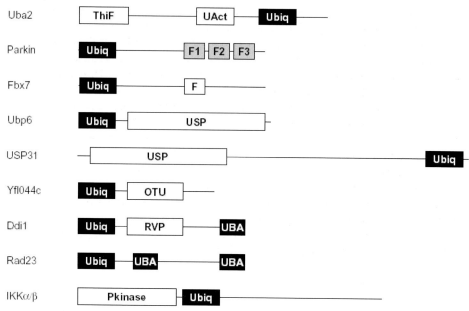

Fig. 12.1. Domain scheme of selected proteins with internal ubiquitin-like domains. Ubiquitin-like domains are indicated by black boxes. Other domains are abbreviated as follows: ThiF, NAD-binding domain in ubiquitin activating enzymes; UAct, 2nd conserved domain in ubiquitin activating enzymes; F1/F2/F3, triad of RING-finger-like domains; F, F-box domain; USP, deubiquitinase catalytic domain; OTU, a particular class of cystein protease domains; RVP, retroviral protease domain; UBA, ubiquitin-associated domain; Pkinase, protein kinase catalytic domain.

tion, other uncharacterized homology domains found in those proteins are good candidates for being ubiquitin-binding proteins or for other functionalities in this pathway. In fact, most of the ubiquitin-recognition domains discussed in Section 12.4 have been discovered by that route.

12.3.4
Other Members of the Ubiquitin Fold

Finally, it should be mentioned that there are a number of protein domains that have some structural resemblance to ubiquitin, although a sequence similarity cannot be established – not even by the most sophisticated methods available today. It cannot be excluded that there are true instances of convergent evolution among these cases. However, it appears more likely that these proteins and domains represent distant members of the ubiquitin superfamily, which have undergone a fundamental change of function and no longer need to conserve sequence positions that are considered hallmarks of ubiquitin-like molecules. In particular three domain classes should be mentioned in this context. The FERM domain (4.1, ezrin, radixin, moesin) is a widespread module found predominantly in actin-organizing proteins. It consists of several sub-domains, the N-terminal of which has a strong resemblance to the ubiquitin fold [41]. There is no indication that FERM domains have a role in the ubiquitin system. The Ras-binding domain of the Raf-kinase is another module assuming the ubiquitin fold. The structure of this domain in complex with a Ras-like GTPase shows that the binding surface for Ras lies on the opposite face from the surface used by ubiquitin for binding to its recognition molecules, discussed in Section 12.4 [42]. There is a second Ras-binding module, the RA-domain, which also has a ubiquitin-like fold but no detectable sequence similarity to ubiquitin or to the Ras-binding domain of Raf [43].

In addition, there are a number of borderline cases, whose sequence relationship to ubiquitin is hard to establish but most probably is real, as these proteins perform a similar function. One well-known example is the UBX domain [44, 45], which seems to replace an internal ubiquitin domain in a certain class of adapter proteins (see Section 12.5.2). Other examples are the autophagy proteins Apg8 and Apg12 [23, 24] which act as ubiquitin-like modifiers.

12.4
Ubiquitin-recognition Domains

Given the widespread role of protein ubiquitination as a signal, there must be a mechanism to detect whether a protein carries a ubiquitin modification. As poly- and mono-ubiquitination appear to signal different conditions, there ought to be specific detection systems that are sensitive to the chain length and probably also to the mode of polyubiquitin linkage. Since the bioinformatical discovery of the UBA domain as the first "professional" ubiquitin-binding domain in 1996, a number of other domains and motifs have been found to bind specifically to ubiquitin and thus to serve as general ubiquitin-recognition modules. Unfortunately, there is still insufficient data to address the important question of whether there really are separate recognition domains for mono- and polyubiquitin, or if ubiquitin-chain recognition requires the cooperation of multiple recognition domains. Several of these domain classes also contain members that bind to ubiquitin-related domains rather than ubiquitin itself. This finding suggests that at least some of the elusive recognition components of ubiquitin-like modifiers might be recruited from the same domain classes. The following sections give a brief overview of the most important ubiquitin recognition domains.

12.4.1
The Ubiquitin-associated (UBA) Domain

The UBA domain was initially identified as a short homology region of about 40 residues, which is found in a multitude of proteins involved in the ubiquitin system [46]. An alignment of some representative UBA domains is shown in Figure 12.2. For a more comprehensive overview, there are a number of excellent reviews

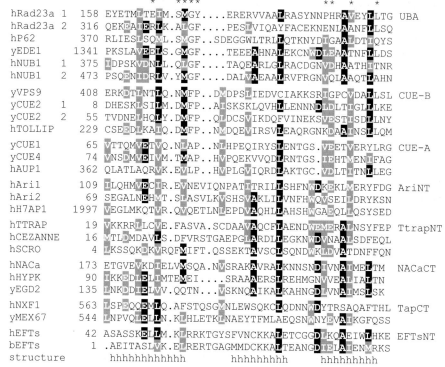

```
                            *    ****                    **    *    *
hRad23a 1   158  EYETML TEIM.SMGY....ERERVVAAIRASYNNPHRAVEYLLTG  UBA
hRad23a 2   316  QEKEAIERLK.AIGF....PESIVIQAYFACEKNENLAANFLLSQ
hP62        370  RLIESLSOVL.SMGF..SDEGGWLTRLLQTKNYDIGAALDTLQYS
yEDE1      1341  PKSLAVEELS.GMGF....TEEEAHNALEKCNMDLEAATNFLLDS
hNUB1   1   375  IDPSKVDNLL.QLGF....TAQEARLGLRACDGNVDHAATHITNR
hNUB1   2   473  PSQENLDRLV.YMGF....DALVAEAALRVFRGNVQLAAQTLAHN

yVPS9       408  ERKDTLNTLQ.NMFP..DMDPSLIEDVCIAKKSRIGPCVDALLSL  CUE-B
yCUE2   1     8  DHESKLSILM.DMFP..AISKSKLQVHLLENNNDLDLTLIGLLLKE
yCUE2   2    55  TVDNELHQLY.DMFP..QLDCSVIKDQFVINEKSVESTLSDLLNY
hTOLLIP     229  CSEEDLKAIQ.DMFP..NLDQEVIRSVLEAQRGNKDAALNSLLQM

yCUE1        65  VTTQMVETVQ.NLAP..NLHPEQIRYSLENTGS.VEETVERYLRG  CUE-A
yCUE4        74  VNSDMVEIVM.TMAP..HVPQEKVVQDLRNTGS.IEHTMENLFAG
hAUP1       362  QLATLAQRVK.EVLP..HVPLGVIQRDLAKTGC.VDLTITNLLEG

hAri1       109  ILQHMVECIR.EVNEVIQNPATITRILLSHFNWDKEKLVERYFDG  AriNT
hAri2        69  SEGALNEHMT.SLASVLKVSHSVAKLIILVNFHWQVSEILDRYKSN
hH7AP1     1997  VEGLMKQTVR.QVQETLNLEPDVAQHLLAHSHWGAEQLLQSYSED

hTTRAP       19  VKKRRLLCVE.FASVA.SCDAAVAQCFLAENDWEMERALNSYFEP  TtrapNT
hCEZANNE     16  MTLDMDAVLS.DFVRSTGAEPGLARDLLEGKNWDVNAALSDFEQL
hSCRO         4  LKSSQKDKVRQFMIFT.QSSEKTAVSCLSQNDWKLDVATDNFFQN

hNACa       173  ETGVEVKDIELVMSQA.NVSRAKAVRALKNNSNDIVNALMEITM  NACaCT
hHYPK        90  IKKEDLELIMTEMEI....SRAAAERSLREHMGNVVEALIALTN
yEGD2       135  LNKDDIELVV.QQTN...VSKNQAIKALKAHNGDIVNALMSLSK

hNXF1       563  LSPEQQEMLQ.AFSTQSGYNLEWSQKCLQDNNWDYTRSAQAFTHL  TapCT
yMEX67      544  LNPVQLELLN.KIHLETKINAEYTFMLAEQSNWNYEVALKGFQSS

hEFTs        42  ASASSKELLM.KIRRKTGYSFVNCKKALETCGGDIKQAEIWLHKE  EFTsNT
bEFTs         1  .AEITASLVK.ELRERTGAGMMDCKKALTEANGDIELALENVRKS
structure        hhhhhhhhhhhh        hhhhhhhhh        hhhhhhhhh
```

Fig. 12.2. Alignment of some representative members of the various classes of UBA-like domains. Positions invariant or conservatively substituted in at least 40% of the sequences are shown on black and gray background, respectively. The UBA-like domain classes are indicated at the right. In the top line, the positions that interact with ubiquitin in the CUE-domain of Cue2 [64] are labeled by asterisks. The bottom line indicates the position of the α-helices found in that structure.

available, which cover both structural and functional aspects of UBA domains and UBA-containing proteins [45, 47, 48]. UBA domains are found, amongst other examples, in selected ubiquitin-conjugating enzymes (E2), in ubiquitin ligases of both HECT and RING type, and also in several ubiquitin-hydrolyzing enzymes; they are particularly widespread in proteins that also contain ubiquitin-like domains. Based on this observation, a general role of the UBA domain in ubiquitin binding was proposed [46] and this prediction was soon confirmed for the UBA domain of the p62 protein [49]. By now, a large number of UBA domains have been shown to bind to ubiquitin and ubiquitin-like domains. Most data is available on the UBA domains of the Rad23 protein, mainly because of this protein's important role in DNA damage repair.

The three-dimensional structure of both UBA domains of the human Rad23a proteins have been solved and reveal a conserved three-helix bundle fold [50, 51]. So far, no structure of a UBA–ubiquitin complex is known and we can only speculate on their mode of interaction. The original UBA structures revealed two hydrophobic surface patches as candidate regions for binding to ubiquitin. Recent experiments for mapping the interaction surface have made use of the NMR chemical shift perturbations seen upon binding of UBA domains to ubiquitin or the ubiquitin-like domain of hRad23B [52]. Apparently, UBA domains bind to the Ile-44-containing surface patch of ubiquitin, and to also to a corresponding region of the Rad23-UbL domain. The interaction surface of UBA domains used for ubiquitin binding is more difficult to judge, as even completely buried residues showed a strong chemical shift perturbation [52]. As UBA domains are very small and probably quite flexible, this might be an indication of subtle structural rearrangements during the binding process. A deeper understanding of UBA–ubiquitin binding will have to await the elucidation of the complex structure. Some general ideas, however, can also be derived from looking at the interaction properties of the related CUE domains, which will be discussed in Section 12.4.2.

Typical UBA domains are thought to bind specifically to Lys-48-linked multi-ubiquitin chains, an idea based on experiments using tetra-ubiquitin and a variety of UBA domains, e.g. those of fission yeast Ucp1 and Mud1 [53], Dsk2 [54] or budding yeast Rad23 [55]. There have also been anecdotal reports of UBA domains binding to free mono-ubiquitin in the case of budding yeast Ddi1 [56] and of a specific binding to Lys-29-linked ubiquitin chains [57, 58]. The canonical mode of UBA binding to multi-ubiquitin chains probably involves the "closed" conformation found in Lys-48 linked di-ubiquitin [59]. The recently published structure of a Lys-63-linked di-ubiquitin would allow the simultaneous and semi-independent binding of multiple UBA domains to the different ubiquitin moieties [60]. As there are several proteins containing multiple UBA domains, it can be envisaged that some of those can specifically interact with ubiquitin chains of a particular linkage.

12.4.2
The CUE Domain

The CUE domain was first described in 2000 by bioinformatical means as a short domain conserved in the yeast protein Cue1 and a number of other proteins [61]. Cue1 has a role in the *ER associated degradation* pathway (ERAD), which is also based on ubiquitin signals. As the apparent role of Cue1 is the recruitment of the ubiquitin-conjugating enzyme UBC7, a general role for CUE domains as a UBC-binding module was proposed. Unpublished work from my group (Hartmut Scheel and K.H) showed that the domains originally classified as CUE-domains should be subdivided into two major groups called Cue-A and Cue-B (Figure 12.3), and that both groups are related in sequence to the UBA-domain family. This finding became significant when in 2002 two independent groups reported that the yeast endocytosis regulator Vps9 binds to mono-ubiquitin by its C-terminal CUE domain [62, 63], and that this binding preference also holds for selected other members of the CUE family. Interestingly, there is a good match between the ubiquitin-binding CUE domains and the bioinformatically defined CUE-B subfamily (Figure 12.3). An alignment of selected CUE domains with some UBA representatives is shown in Figure 12.2.

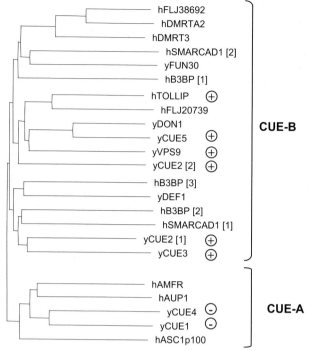

Fig. 12.3. Subclasses of the CUE domain family. All human and budding yeast members of the CUE family have been aligned and subjected to neighbor-joining dendrogram analysis. In proteins having multiple domains, the domain number is indicated in square brackets. The subfamilies CUE-A and CUE-B are indicated at the right border. Domains tested positive or negative for ubiquitin binding [64] are labeled with a circled + or − sign, respectively.

Like the UBA domain, the CUE-A and CUE-B domains occur in a wide variety of seemingly unrelated proteins, although their connection to the ubiquitination pathway is not so obvious, as many CUE-domain proteins still await characterization. Besides the vacuolar sorting protein Vps9 and many uncharacterized ORFs, CUE-B domains are also found in human Tollip, a regulator of interleukin-1 signalling, and in the SWI/SNF helicase SMARCAD1. CUE-A domains are found in the ERAD protein Cue1, in the putative RING-finger ubiquitin ligase AMFR, and in the integrin interactor AUP1.

The CUE domain's propensity to bind ubiquitin was a quite recent discovery, and relatively little is known about its physiological role. Nevertheless, structural work done on this domain type has been instrumental for our understanding of ubiquitin recognition in general. Two independently solved structures of different CUE domains have been reported, both in isolation and in complex with ubiquitin [64, 65]. The NMR structure of the first CUE domain of the uncharacterized budding yeast protein Cue2 shows a three-helix bundle fold resembling that of the

UBA domain [64], nicely confirming the bioinformatical prediction of a common evolutionary history of those two domain classes. Surprisingly, the X-ray structure of the single CUE domain of budding yeast Vps9 is markedly different: here, two CUE domains form a domain-swapped dimer with two bundles of three helices each. One three-helix bundle is formed by α1 and α2 of the first molecule and α3′ of the second one, while the other bundle is formed by α1′, α2′ and α3 [65]. In the complex structure, one molecule of ubiquitin is bound by one CUE dimer. It is still an open question whether this different arrangement reflects a physiological difference between Cue2 and Vps9 CUE domains, or if it is rather caused by the different detection methodologies.

Figure 12.4A shows the interaction of the first CUE domain of Cue2 interacting with ubiquitin, which might serve as a general model for the interaction mode of other UBA-like domains. The CUE domain binds to the Ile-44 patch of ubiquitin, in accordance with the chemical shift perturbation results of the UBA:ubiquitin interaction [52]. On the side of the CUE domain, residues of the first and third helix participate in this interaction surface. These residues include the Phe–Pro and Leu–Leu motifs, which had been predicted to be important for ubiquitin binding, based on comparative sequence analysis of CUE-A and CUE-B domains [62]. Positions in close contact with ubiquitin are also indicated in the alignment of Figure 12.3. The two available structures of the CUE:ubiquitin complexes offer little expla-

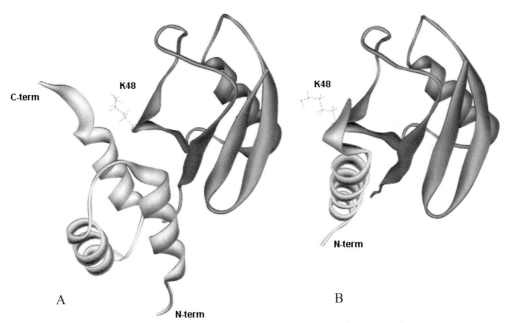

Fig. 12.4. CUE and UIM bind to the same site on ubiquitin. Schematic representation of (A) ubiquitin in complex with the CUE domain of Cue2 [64] and (B) in complex with the N- terminal UIM of Vps27 [79]. In both panels, ubiquitin is rendered in darker colour and the position of Lys-48 is indicated.

nation why CUE domains should prefer mono-ubiquitin. In the Cue2 structure, the C-terminal end of the third helix is in close contact to Lys-48 of ubiquitin, but is unlikely to interfere with the attachment of a further ubiquitin unit to this position. It is, however, conceivable that a CUE domain bound to ubiquitin prevents access of ubiquitin-conjugating enzymes to Lys-48 and thus prevents chain elongation. Other lysine residues used for alternative ubiquitin-chain formation are not within reach of a single monomeric CUE domain. Similar to the situation observed with UBA domains, several proteins contain multiple CUE domains; their relative arrangement might thus be able to specifically recognize different ubiquitin chain topologies.

12.4.3
Other UBA-related Domains

UBA domains are relatively short domains with a high degree of sequence divergence. This combination is quite unfavorable for reliable domain detection with bioinformatical methods, as there is no sharp line between the profile-scores of genuine UBA domains and those of unrelated sequences, particularly if they are of similar amino acid composition. From the bioinformatical point of view, the term "UBA domain" is operationally defined: a UBA domain is every sequence region that gives a significant score with a profile or HMM derived from trustworthy UBA domains. In the biological sense, e.g. when looking for the prediction of protein function or structure, this definition is not totally appropriate. The positive cases, i.e. those sequences giving significant UBA scores, will typically behave as "biological" UBA domains, at least if the statistics has been handled properly. The converse, however, is not always true: there will be a number of sequences that are functionally and structurally related to UBA domains, but which have diverged by such a degree that they are no longer caught by any UBA profile. A good example for such a situation is the CUE domain mentioned above: no CUE domain protein reaches a significant score with canonical UBA profiles, while some established UBA domains reach a quite convincing score with CUE-derived profiles. Together with secondary-structure prediction methods, this behavior has prompted the prediction of CUE domains as ubiquitin-binding modules, which soon turned out to be correct.

In this respect, the CUE domain is not a isolated case. There are a number of other domain families, each of them only defined in the bioinformatical sense, that have significant matches within established UBA or CUE domain regions. Based on this similarity and on secondary-structure predictions, it can be expected that all of those domain types assume the typical UBA-like three-helix bundle fold. However, it is not clear if all of those domains also bind to ubiquitin, or if they have evolved to different binding properties. Many of the UBA-like domain classes are unpublished. Nevertheless, they should be briefly discussed here, as they are a logical extension of the UBA/CUE paradigm.

AriNT: A novel UBA-like domain is found in certain RING-finger type proteins related to the ariadne protein of the fruit fly. Here, the AriNT domain is invariably located upstream of the RING-finger triad, while there is a second conserved domain (AriCT, not UBA related) found C-terminally of the Zn fingers. As all known AriNT proteins (four human proteins and the yeast ORF Ykr017c) are putative ubiquitin ligases, a role of this UBA-like domain in ubiquitin binding appears likely.

TtrapNT: A further UBA-like domain is found at the N-terminus of the TNF- and TRAF-associated protein Ttrap, as well as a number of other sequences including eight other human proteins and the yeast ORF Ylr128w. The scope of proteins harboring the TtrapNT domain resembles that of the UBA proteins. The "Cezanne"-like proteins combine the TtrapNT module with an OUT-type protease domain, while other proteins also contain UIM or UBX domains. Most TtrapNT proteins have an established or predicted role in the ubiquitin pathway, making it likely that TtrapNT serves as a recognition module for ubiquitin or ubiquitin-like domains.

NACaCT: Yet another UBA-like domain is found at the C-terminus of the α-subunit of the nascent polypeptide-associated protein complex (NAC-α). This protein has some properties of a chaperone and regulates the attachment of loaded ribosomes to the ER membrane [66], a process that is not known to involve the ubiquitin system. Further NACaCT domains are found in the huntingtin-interactor HYPK, in the human KIAA0363 protein, and in many NAC-α related proteins including the yeast protein Egd2. It is unlikely that the NACaCT domain has a general role in ubiquitin binding, considering that this domain is also found in archaea, which are devoid of ubiquitin.

EFTsNT: A UBA-like domain with a clear role outside of ubiquitin binding is found at the N-terminus of EF-Ts proteins. The relationship of this region to genuine UBA domains is well established as there is a structure of full-length EF-Ts available [67]. Nevertheless, this domain is widespread in bacteria and archaea, which obviously lack a proper ubiquitin system. The physiological role of the EFTsNT domain is rather in the binding to the elongation factor EF-Tu, which has no resemblance to ubiquitin.

TapCT: The C-terminus of the mammalian nuclear RNA export factor NXF1/2 (also known as Tap) contains a sequence region with significant similarity to UBA-like domains. This region is also found in the yeast RNA export factor Mex67. A three-dimensional structure of this domain is available and confirms its similarity to the UBA domain [68]. This UBA-like domain does not appear to bind to ubiquitin but rather to the Phe–Gly repeat motif found in a number of nucleoporins. The interaction surface of the UBA-like TapCT domain with a Phe–Gly-containing loop was mapped by an NMR/X-Ray combination technique and shown to be different from the ubiquitin-binding mode: the Phe–Gly loop binds on the "backside" of the UBA-like domain and is in contact with helices α2 and α3 [68].

12.4.4
The Ubiquitin-interacting Motif (UIM)

The classical receptor for ubiquitinated proteins is the 26S proteasome, although the true nature of the ubiquitin-sensing subunit – at least the physiologically important one – has always been and is still a matter of discussion. At least three subunits have been suggested to target ubiquitin or UbL domains to the proteasome: S5a/Rpn10 [69], S6′/Rpt5 [70], and Rpn1 [71]. In the case of S5a/Rpn10, the interacting region could be narrowed down to two conserved motifs containing the residues Leu–Ala–Leu–Ala–Leu, termed the "LALAL-motif" [69]. Bioinformatical attempts to identify other ubiquitin interactors using a similar sequence motif were initially unsuccessful. During sequence analysis of the ataxin-3 protein mutated in Machado Joseph disease, a repeat motif was identified which later turned out to also include the proteasomal LALAL-motif [72]. The high prevalence of this motif in proteins known to interact with ubiquitin immediately suggested a role of this motif in ubiquitin binding, hence the name UIM for *u*biquitin-*i*nteracting *m*otif.

The UIM is a short motif spanning only 16 consecutive residues. An alignment of some representative UIMs is shown in Figure 12.5A. The most prominent fea-

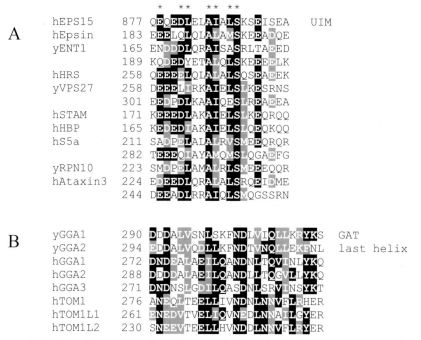

Fig. 12.5. Comparison between the UIM and the ubiquitin-binding part of the GAT domain. Positions invariant or conservatively substituted in at least 50% of the sequences are shown on black and gray background, respectively. (A) The top panel shows some representative members of the UIM family. Positions that in the Vps27 structure [79] have contact with ubiquitin are labelled by asterisks in the top line. (B) The bottom panel shows an alignment of representative GAT-domain members. Only the last helix, which contains the ubiquitin-interaction site, is depicted.

ture of the UIM is the almost invariant Ala–X–X–X–Ser sub-motif, where the conserved Ala corresponds to the second "A" of the original LALAL motif. In addition to this feature, the initial Leu is the only other LALAL residue well conserved in the functional UIMs; the other Leu and Ala residues are more or less specific to the S5a family. This high degree of divergence explains why the UIM was not discovered by motif-searching techniques starting from S5a/Rpn10 but required the more sophisticated profile-HMM techniques.

Probably the most interesting aspect of UIM identification was the presence of four different protein classes working in the receptor endocytosis and protein-sorting pathway. It had been known for a long time that those processes were regulated by mono-ubiquitination of both cargo proteins and signalling components, but the nature of the ubiquitin receptor had been elusive. Immediately after its discovery, a series of studies showed convincingly that the UIM fills this role. A number of reviews cover the field of UIMs in endocytosis, and also highlight the possible mechanisms by which the UIM can contribute to keeping proteins in a mono-ubiquitinated state [73–76]. Besides the proteasomal S5a component and the endocytosis proteins, copies of the UIM can also be found in ubiquitin ligases, UBA-domain-containing adapter proteins, ubiquitin proteases, selected chaperones, and a large collection of uncharacterized proteins. Based on current data, it appears likely that most if not all UIM-containing proteins will turn out to be part of the ubiquitin system. A good example is the identification of the functionally uncharacterized UIM-protein ataxin-3 as a novel ubiquitin protease – another bioinformatical prediction that was followed by experimental confirmation [77, 78].

The sequence properties of the UIM initially suggested the structure of a single helix, which would exclude the UIM from being a "domain" in the strict sense of Section 12.1.1; for that reason the UIM should preferably be called an interaction "motif" instead of an interaction domain. By now, a body of structural data on various UIMs is available, confirming the helical nature of this motif and demonstrating its interaction mode with ubiquitin and UbL domains. Available structures include the UIMs of yeast Vps27 [79, 80] and the complex of the C-terminal UIM of S5a with the UbL of hRad23B [81] and of hRad23A [82]. Also interesting in this respect is the comparison of chemical shift perturbations seen in ubiquitin and UbLs complexed with UIMs and various UBA domains [52]. As shown in Figure 12.4B, the UIM also binds to the Ile-44 patch of ubiquitin, at a similar position to that used by the UBA-type interactors. However, the orientation of the single UIM helix is quite different from the helix-bundle of the UBA-like domains. The UIM:ubiquitin complex structures do not give a clear picture of how UIMs might prevent ubiquitin-chain elongation or how they could discern between different chain topologies. The Lys-48 residue is not part of the interaction surface and a direct interference appears unlikely. It should be noted, however, that UIMs frequently occur in narrowly spaced tandems. Like the multi-UBA and multi-CUE proteins, the UIM tandems might be one way to the specific recognition of certain linkage types.

12.4.5
The UEV Domain

A number of enzyme families are known to contain members that have lost the residues important for catalysis. The maintenance of those catalytically inactive proteins in the translated part of the genome probably means that these proteins have acquired a different function. In some instances this new function is totally unrelated to the original enzymatic activity, e.g. the crystallins of the eye lens contain inactive members of various enzyme families. In other examples, including some inactive kinases and phosphatases, the newly acquired function is related to the original catalysis, e.g. by binding to the substrate or by acting as a heterodimerization partner for a catalytically active version of the enzyme. The latter situation seems to apply for catalytically inactive versions of ubiquitin-conjugating (E2) enzymes, containing a so-called UEV domain (for *u*biquitin-conjugating *e*nzyme *v*ariant). The existence of this homology-domain was first demonstrated bioinformatically in the candidate tumor suppressor TSG101 [83, 84] and a role for this domain as a regulator of ubiquitination was proposed. A second prominent protein with a UEV domain is the DNA-damage-repair protein Mms2, which forms a heterodimer with the active E2 enzyme Ubc13. As this dimer is involved in the creation of the unusual Lys-63-linked multi-ubiquitin chains [85], the presence of a second ubiquitin-binding site in this complex has been proposed, to which the UEV is likely to contribute. Two crystal structures of Mms2/Ubc13 dimers are available [86, 87], but do not give a clear indication where this binding site is located.

More functional and structural information is available on TSG101, whose yeast ortholog, Vps23, is part of the ESCRT-1 complex; both proteins probably play a physiological role in the ubiquitin-dependent sorting of proteins to multivesicular bodies [88]. TSG101 has been shown to bind to ubiquitin *in vitro* and it is also able to bind to certain proline-rich peptides, most importantly the Pro–Thr–Ala–Pro peptide found in the Gag protein of the HIV virus [89]. An NMR structure of TSG101 has been described, which also allowed estimation of the binding sites for the PTAP peptide and for ubiquitin by chemical shift mapping [90]. Somewhat surprisingly, the predicted ubiquitin-binding site does not correspond to the vestigial catalytic site.

12.4.6
The GAT Domain

The GAT domain (GGA and Tom1) has recently joined the ranks of ubiquitin-binding domains [91]. As the name implies, this domain is found in the GGA- and Tom1-like proteins, two regulator classes of clathrin-mediated vesicular traffic. All proteins harboring the GAT domain also contain an N-terminal VHS domain, which is named after the *V*ps27, *H*rs, and *S*TAM proteins. Interestingly, these latter proteins are known to contain a ubiquitin-binding UIM motif, which appears to be replaced by the GAT domain in the GGA and Tom1-like proteins (Figure 12.6).

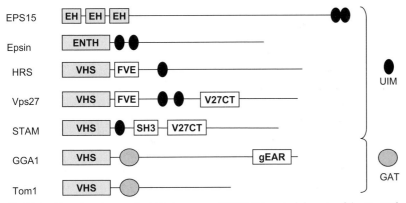

Fig. 12.6. Domain scheme of selected members of the UIM and GAT family. Domain abbreviations are as follows: EH, Eps15-homology domain; ENTH, Epsin N-terminal homology domain; VHS, N-terminal domain of Vps27, HRS, STAM; FVE, FYVE-finger domain; V27CT, C-terminal domain of the Vps27 family; SH3, Src-homology 3 domain; gEAR, γ-adaptin ear domain. The UIM motifs are shown as small black ellipses, the GAT domains as larger gray ones.

The GGA proteins (three in mammals, two in yeast) also contain a so-called "γ-adaptin ear domain" at their C-terminus, which is lacking in the three mammalian Tom1-like proteins.

Initially, the role of the GAT domain in GGA proteins was seen in the binding of small GTPases of the Arf family, a critical step in the recruitment of clathrin to the TGN membrane [92]. However, the GAT domains of the Tom1-like family do not bind to Arf. Recently, GAT domains of both protein classes were found to bind to ubiquitin and it was possible to separate the two binding sites to different subdomains of the GAT domain [91]. A number of X-ray structures of GAT domains are available [93–95], presenting the domain as an elongated three-helix bundle. Unlike the UBA-like structures, the GAT helices are almost parallel and considerably longer. As a prominent feature, the N-terminal helix is much longer than the others; this N-terminal extension contains the Arf interaction site and is not conserved in the Tom1-family [93].

The ubiquitin-interaction site of the GGA3 GAT domain was mapped to the C-terminal helix by deletion and mutation experiments. The Ile-44 patch appears to be the likely interaction site of ubiquitin, as an Ile44Ala mutation abolished the binding [93]. The C-terminal helix bears some resemblance to the UIM: it is an amphipathic helix of similar length that is preceded by a cluster of 3–5 acidic residues (Figure 12.5B). However, the GAT domain lacks the Ala–X–X–X–Ser motif, which is the hallmark of the UIM motif. A conserved Leu–X–X–X–Asp motif that points away from the helix bundle – and thus probably towards the bound ubiquitin – might fill this role in the GAT domain. Taken together, the similarity between GAT and UIM might not be restricted to the scope of the two domains but also include a conserved binding mode.

12.4.7
Other Ubiquitin-binding Domains

In addition to the well-established and widely distributed ubiquitin-interaction domains described above, there are several other domains with a more limited scope or with binding properties that are just beginning to be uncovered. Two interesting candidates are the NZF and ZnF-UBP/PAZ domains.

The NZF domain (*N*pl4 *Z*n *f*inger) is a mononucleate Zn-finger domain with four cysteine ligands, which occurs – amongst others – at the C-terminus of the Cdc48/p97 adapter protein Npl4 and in the vacuolar sorting protein Vps36. In both proteins, the region corresponding to the NZF has been shown to bind to ubiquitin [96, 97]. The structure of the NZF domain does not resemble any of the known ubiquitin-interaction domains but rather looks like a typical C4 zinc finger [96]. Bioinformatically, the NZF domain is identical to the Ran-binding Zn finger found in RanBP proteins [98]. It is currently not clear if all members of this quite large family bind to ubiquitin, or if the ubiquitin-binding Zn fingers are just a small subset of this domain class. It should be noted that the two NZF fingers of Npl4 and Vps36 are not particularly closely related.

A second Zn finger with a putative role in ubiquitin binding is found in various ubiquitin proteins of the USP type and in the histone deacetylase HDAC6. Two recent studies have shown that the corresponding region of HDAC6 binds to polyubiquitin [99, 100]. The Zn finger is referred to as Znf-UBP or as PAZ domain (for *p*olyubiquitin-*a*ssociated *Z*n finger). The latter acronym is somewhat unfortunate, as there is another unrelated domain that goes by this name (*p*iwi, *a*rgonaut, *z*wille). So far, it is not clear if the ubiquitin-binding propensity is specific to HDAC6 or applies to the other family members as well. Since most of the Znf-UBP domains are found in ubiquitin proteases, a general role of this domain in the ubiquitin system is likely.

Obviously, not all proteins known to interact with ubiquitin or ubiquitin-like domains contain one of the "professional" ubiquitin-interaction domains. Rpt5 and Rpn1, two subunits of the proteasome that bind to ubiquitin and UbLs, respectively, do not belong to any of the classes described above. Most probably, a large number of uncharacterized proteins with high affinity and specificity for ubiquitin are still waiting to be discovered. The bioinformatical tools described in the early sections of this chapter will be instrumental for this task.

12.5
Building Complex Systems From Simple Domains

Considering the multitude of ubiquitin domains and their cognate recognition modules, the next question to address is how cells make use of these building blocks to form the highly complex ubiquitination system. Before starting with the discussion of some prominent modular protein architectures, we should bear in mind that there are a number of other functional domains participating in this pathway. So far, little attention has been paid to the fact that the enzymatic functions of ubiquitin activation, ubiquitin conjugation, ubiquitin ligation, and deubiquitination are also encoded by functional domains. Many enzymes of ubiquitin metabolism are architecturally simple and consist only of a single domain. However, several enzymes do have a modular architecture; those will be discussed in the following section.

12.5.1
Enzymes with Additional Ubiquitin-binding Sites

The first hint that the UBA domain might have a role in the ubiquitin system came from its frequent association with known enzymatic components of protein ubiquitination and deubiquitination, particularly with those that confer specificity to the reaction. The same trend is seen with other ubiquitin-binding domains, such as the UIM motif or the NZF domain. From genome-wide sequence-analysis studies, we know that the multiplicity – and also the architectural complexity – of the ubiquitination system increases from E1 via E2 to E3. The human genome appears to encode only one ubiquitin-activating enzyme (E1), plus three paralogs active against ubiquitin-related modifiers. By contrast, there seem to be 34 active members of the ubiquitin-conjugating enzyme family (E2). Since many of these proteins lack any biochemical characterization, it is not clear how many of them actually conjugate ubiquitin, or which ones are active against ubiquitin relatives. Most E2 enzymes appear to be monolithic, but there is at least one (E2-25K) that harbors a UBA domain. The multiplicity of ubiquitin ligases (E3) is even higher: the human genome encodes 27 proteins with a HECT domain, forming the "classical" E3 superfamily [101]. In addition, there are 259 RING-finger proteins, most (if not all) of which can also be assumed to be ubiquitin ligases. Even more variability comes from the existence of composite ubiquitin ligases, such as the SCF complexes [102], which use a common RING-finger component but a multitude of specificity factors (chosen from a set of 58 F-box proteins, 17 SOCS-box proteins, and perhaps 190 BTB proteins). Many of the E3 components contain additional functional domains, frequently including those with ubiquitin-binding properties. The deubiquitination branch consists of at least 61 ubiquitin proteases belonging to different classes. In particular the enzymes of the USP class frequently harbor ubiquitin-binding domains in addition to their enzymatic function.

Obviously all enzymes of ubiquitin metabolism have to recognize ubiquitin. Why do some – but not all – of them contain dedicated ubiquitin-binding domains? I should like to propose the hypothesis that ubiquitin-binding domains occur only in those enzymes that not only transfer (or remove) ubiquitin but also act on ubiquitin as a *substrate*. These would be ubiquitin ligases with a role in chain elongation, or ubiquitin proteases active in chain trimming, but never those that transfer ubiquitin directly onto a non-ubiquitin substrate. There are some data to support this hypothesis. E2-25K, the only UBC enzyme with a UBA domain, is able to catalyze the unusual formation of unattached polyubiquitin chains in solution [103]. Moreover, human isopeptidase T and yeast UBP14, two UBA-containing ubiquitin proteases, hydrolyse only polyubiquitin chains that are not attached to a substrate [104, 105]. Unfortunately, the catalytic properties of other enzymes containing ubiquitin-interaction motifs are not characterized in sufficient detail. Nevertheless, the presence of single or multiple ubiquitin recognition modules should allow a ubiquitin ligase or hydrolase to require a certain minimal chain length or a particular chain topology.

12.5.2
The UbL/UBA Adapter Paradigm

Even more mysterious than the large number of ubiquitin-recognition domains is the equally large number of internal UbL domains found in a diverse set of proteins. A genome-wide survey shows that UbL domains are not randomly distributed throughout the proteome but rather are highly enriched in proteins known or suspected to act in the ubiquitin system. Interestingly, this seems to include several kinases whose activity is required for a subsequent ubiquitination, such as, for example, the IκB-Kinase subunits IKKα and IKKβ (Figure 12.1). For most of the cases, we do not know what the UbL domains are doing, although it is tempting to speculate that they are specific interaction partners for selected members of the ubiquitin-binding domain families.

One class of UbL proteins has been the focus of investigation for their crucial role in the targeting of substrates to the proteaseome, and possibly also to the Cdc48/p97 complex. These proteins, with Rad23 being the most prominent member, have a particular architecture with a UbL domain at one end (typically the N-terminus) and a UBA domain in the C-terminal region. These proteins appear to work as "adapters" by shuttling ubiquitinated substrates to the proteasome without requiring a direct interaction of the proteasome with the ubiquitin signal [106]. In the Rad23 proteins, the N-terminal UbL domain is able to interact with the proteasome [107–109], while the two UBA domains specifically recognize polyubiquitin signals [53, 54]. Recently, the proteasome component Rpn1 has been identified as a receptor for the Rad23 UbL domain [71].

A related adapter family combines the UBA domain with a UBX domain, which is much more distantly related to ubiquitin than the true UbL domains [44, 45]. In these proteins, the UBA domain is frequently found at the N-terminus while the UBX domain forms the C-terminus of the protein; an example is the yeast Shp1 protein and its mammalian homolog p47. So far, there is limited data on the function of this protein family though they appear to shuttle ubiquitinated proteins to the Cdc48/p97 complex instead of to the proteasome [47, 106].

12.5.3
Non-orthologous Domain Replacement

When analyzing proteins of the ubiquitin system from a genomic perspective, there are a number of interesting examples where in the course of evolution one

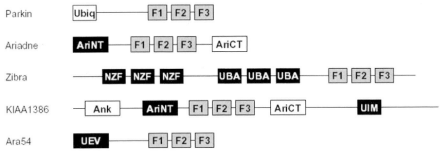

Fig. 12.7. Domain scheme of selected parkin-like ubiquitin ligases. Black boxes represent the ubiquitin-interacting domains discussed in the text. The three gray boxes labeled F1, F2 and F3 represent the triad of RING-finger like domains that define the parkin superfamily. Other domain abbreviations are as follows: Ubiq, ubiquitin-like domain; AriCT, C-terminal domain of the ariadne family; Ank, ankyrin repeats.

domain type has been replaced by a member of a different domain class. A well-known example is the EPS15/Ede1 pair: when analysing the N-terminal region, the human EPS15 and yeast Ede1 proteins appear to be orthologs within the EH-domain family, suggesting a common function. However, the human protein contains two UIM motifs at the C-terminus, while the yeast protein has a UBA domain at an equivalent position [72]. This evolutionary replacement suggests that – at least in this case – the two domain types are functionally equivalent.

A second example, which is quite intriguing although its functional significance is not yet clear, is shown in Figure 12.7. The human genome contains a large number of putative ubiquitin ligases related to the parkin protein. This protein family is characterized by a triad of Zn fingers related to the RING finger. In parkin, the only other domain is a UbL domain at the N-terminus, whose function is unknown. Most other proteins of this family lack the UbL, but several members have a ubiquitin-binding function instead: The Ariadne proteins carry a UBA-like AriNT domain upstream of the Zn-finger triad, the protein KIAA1386 also contains this AriNT domain but has an additional UIM at the C-terminus, the Zibra protein has no AriNT domain but three true UBA domains plus three additional ubiquitin-binding NZF domains. Finally, the Ara54 protein does not have a UBA-like domain but rather a UEV domain filling that position. Apparently, multiple members of the parkin family require the binding of ubiquitin or UbL domains, but evolution has chosen different solutions for that task.

A quite mysterious finding is the occasional positional replacement of ubiquitin-binding domains by ubiquitin-like domains. Parkin in Figure 12.7 is one example, another one is the UBX domain found instead of the UIMs in the ataxin-3 protein from *Plasmodium falciparum* [77]. There is no reason to assume that UbL domains might have a role in ubiquitin binding. A more likely explanation would be the requirement of these proteins to "look like ubiquitin", irrespective of whether ubiquitin is part of the protein itself or rather bound to it non-covalently.

12.6
Outlook

The previous paragraphs were meant to give a brief overview of what we know about ubiquitin-like domains and their recognition by specialized ubiquitin-binding modules. Sequence analysis has proven to be a valuable tool for the discovery of such domains and the identification of new components of the ubiquitination pathway. There is still a big gap between the relatively facile identification of those proteins and the tedious functional characterization of their biochemical and physiological properties. Striving to reach a deeper understanding of the ubiquitin system and the intricate interplay of all its components, we are barely scratching the surface. Nevertheless, there are some general principles that can be derived from the existing data, and I have tried to make a case for bioinformatics as a powerful tool to reach that goal.

References

1 SMITH, T. F. and WATERMAN, M. S., Identification of common molecular subsequences, *J. Mol. Biol.*, **1981**, *147*, 195.

2 TOMIUK, S. and HOFMANN, K., Sequence similarity in structurally dissimilar proteins, *Curr. Biol.*, **2003**, *13*, R124.

3 GRIBSKOV, M., MCLACHLAN, A. D., and EISENBERG, D., Profile analysis: detection of distantly related proteins, *Proc. Natl. Acad. Sci. USA*, **1987**, *84*, 4355.

4 BUCHER, P., KARPLUS, K., MOERI, N., and HOFMANN, K., A flexible motif search technique based on generalized profiles, *Comput. Biol.*, **1996**, *20*, 3.

5 BALDI, P., CHAUVIN, Y., HUNKAPILLER, T., and MCCLURE, M. A., Hidden Markov models of biological primary sequence information, *Proc. Natl. Acad. Sci. USA*, **1994**, *91*, 1059.

6 KROGH, A., BROWN, M., MIAN, I. S., SJOLANDER, K., and HAUSSLER, D., Hidden Markov models in computational biology. Applications to protein modeling, *J. Mol. Biol.*, **1994**, *235*, 1501.

7 EDDY, S. R., Profile hidden Markov models, *Bioinformatics*, **1998**, *14*, 755.

8 HOFMANN, K., Sensitive protein comparisons with profiles and hidden Markov models, *Brief. Bioinform.*, **2000**, *1*, 167.

9 HOFMANN, K., Protein classification and functional assignment, *Trends Guide Bioinformatics*, **1998**, 18.

10 ATTWOOD, T. K., The quest to deduce protein function from sequence: the role of pattern databases, *Int. J. Biochem. Cell Biol.*, **2000**, *32*, 139.

11 BATEMAN, A., BIRNEY, E., DURBIN, R., EDDY, S. R., HOWE, K. L., and SONNHAMMER, E. L., The Pfam Protein Families Database, *Nucleic Acids Res.*, **2000**, *28*, 263.

12 HOFMANN, K., BUCHER, P., FALQUET, L., and BAIROCH, A., The PROSITE database, its status in 1999, *Nucleic Acids Res.*, **1999**, *27*, 215.

13 SCHULTZ, J., MILPETZ, F., BORK, P., and PONTING, C. P., SMART, a simple modular architecture research tool: identification of signaling domains, *Proc. Natl. Acad. Sci. USA*, **1998**, *95*, 5857.

14 GOUGH, J., KARPLUS, K., HUGHEY, R., and CHOTHIA, C., Assignment of homology to genome sequences using a library of hidden Markov models that represent all proteins of known structure, *J. Mol. Biol.*, **2001**, *313*, 903.

15 MULDER, N. J., APWEILER, R., ATTWOOD, T. K., BAIROCH, A.,

Barrell, D., Bateman, A., Binns, D., Biswas, M., Bradley, P., Bork, P., Bucher, P., Copley, R. R., Courcelle, E., Das, U., Durbin, R., Falquet, L., Fleischmann, W., Griffiths-Jones, S., Haft, D., Harte, N., Hulo, N., Kahn, D., Kanapin, A., Krestyaninova, M., Lopez, R., Letunic, I., Lonsdale, D., Silventoinen, V., Orchard, S. E., Pagni, M., Peyruc, D., Ponting, C. P., Selengut, J. D., Servant, F., Sigrist, C. J., Vaughan, R., and Zdobnov, E. M., The InterPro Database, 2003 brings increased coverage and new features, *Nucleic Acids Res.*, **2003**, *31*, 315.

16 Matunis, M. J., Coutavas, E., and Blobel, G., A novel ubiquitin-like modification modulates the partitioning of the Ran-GTPase-activating protein RanGAP1 between the cytosol and the nuclear pore complex, *J. Cell. Biol.*, **1996**, *135*, 1457.

17 Lo Conte, L., Ailey, B., Hubbard, T. J., Brenner, S. E., Murzin, A. G., and Chothia, C., SCOP: a structural classification of proteins database, *Nucleic Acids Res.*, **2000**, *28*, 257.

18 Melchior, F., Schergaut, M., and Pichler, A., SUMO: ligases, isopeptidases and nuclear pores, *Trends Biochem. Sci.*, **2003**, *28*, 612.

19 Muller, S., Hoege, C., Pyrowolakis, G., and Jentsch, S., SUMO, ubiquitin's mysterious cousin, *Nat. Rev. Mol. Cell Biol.*, **2001**, *2*, 202.

20 Ohh, M., Kim, W. Y., Moslehi, J. J., Chen, Y., Chau, V., Read, M. A., and Kaelin, W. G., Jr., An intact NEDD8 pathway is required for Cullin-dependent ubiquitylation in mammalian cells, *EMBO Rep.*, **2002**, *3*, 177.

21 Kim, K. I. and Zhang, D. E., ISG15, not just another ubiquitin-like protein, *Biochem. Biophys. Res. Commun.*, **2003**, *307*, 431.

22 Dittmar, G. A., Wilkinson, C. R., Jedrzejewski, P. T., and Finley, D., Role of a ubiquitin-like modification in polarized morphogenesis, *Science*, **2002**, *295*, 2442.

23 Mizushima, N., Yoshimori, T., and Ohsumi, Y., Role of the Apg12 conjugation system in mammalian autophagy, *Int. J. Biochem. Cell Biol.*, **2003**, *35*, 553.

24 Wang, C. W. and Klionsky, D. J., The molecular mechanism of autophagy, *Mol. Med.*, **2003**, *9*, 65.

25 Raasi, S., Schmidtke, G., and Groettrup, M., The ubiquitin-like protein FAT10 forms covalent conjugates and induces apoptosis, *J. Biol. Biol.*, **2001**, *276*, 35334.

26 Goehring, A. S., Rivers, D. M., and Sprague, G. F., Jr., Urmylation: a ubiquitin-like pathway that functions during invasive growth and budding in yeast, *Mol. Biol. Cell*, **2003**, *14*, 4329.

27 Nakamura, M. and Tanigawa, Y., Characterization of ubiquitin-like polypeptide acceptor protein, a novel pro-apoptotic member of the Bcl2 family, *Eur. J. BioBiol.*, **2003**, *270*, 4052.

28 Jentsch, S. and Pyrowolakis, G., Ubiquitin and its kin: how close are the family ties?, *Trends Cell Biol.*, **2000**, *10*, 335.

29 Schwartz, D. C. and Hochstrasser, M., A superfamily of protein tags: ubiquitin, SUMO and related modifiers, *Trends Biochem. Sci.*, **2003**, *28*, 321.

30 Larsen, C. N. and Wang, H., The ubiquitin superfamily: members, features, and phylogenies, *J. Proteome Res.*, **2002**, *1*, 411.

31 Luders, J., Pyrowolakis, G., and Jentsch, S., The ubiquitin-like protein HUB1 forms SDS-resistant complexes with cellular proteins in the absence of ATP, *EMBO Rep.*, **2003**, *4*, 1169.

32 Doolittle, R. F., The origins and evolution of eukaryotic proteins, *Philos. Trans. R. Soc. Lond. B. Biol. Sci.*, **1995**, *349*, 235.

33 Park, J. H., Dorrestein, P. C., Zhai, H., Kinsland, C., McLafferty, F. W., and Begley, T. P., Biosynthesis of the thiazole moiety of thiamin pyrophosphate (vitamin B1), *Biochemistry*, **2003**, *42*, 12430.

34 Rudolph, M. J., Wuebbens, M. M., Turque, O., Rajagopalan, K. V., and Schindelin, H., Structural studies of molybdopterin synthase provide insights into its catalytic mechanism, *J. Biol. Biol.*, **2003**, *278*, 14514.

35 Xi, J., Ge, Y., Kinsland, C., McLafferty, F. W., and Begley, T. P., Biosynthesis of the thiazole moiety of thiamin in Escherichia coli: identification of an acyldisulfide-linked protein–protein conjugate that is functionally analogous to the ubiquitin/E1 complex, *Proc. Natl. Acad. Sci. USA*, **2001**, *98*, 8513.

36 Lake, M. W., Wuebbens, M. M., Rajagopalan, K. V., and Schindelin, H., Mechanism of ubiquitin activation revealed by the structure of a bacterial MoeB-MoaD complex, *Nature*, **2001**, *414*, 325.

37 Furukawa, K., Mizushima, N., Noda, T., and Ohsumi, Y., A protein conjugation system in yeast with homology to biosynthetic enzyme reaction of prokaryotes, *J. Biol. Biol.*, **2000**, *275*, 7462.

38 Goehring, A. S., Rivers, D. M., and Sprague, G. F., Jr., Attachment of the ubiquitin-related protein Urm1p to the antioxidant protein Ahp1p, *Eukaryot. Cell*, **2003**, *2*, 930.

39 Lupas, A. N. and Koretke, K. K., Bioinformatic analysis of ClpS, a protein module involved in prokaryotic and eukaryotic protein degradation, *J. Struct. Biol.*, **2003**, *141*, 77.

40 Hochstrasser, M., Biochemistry. All in the ubiquitin family, *Science*, **2000**, *289*, 563.

41 Hamada, K., Shimizu, T., Matsui, T., Tsukita, S., and Hakoshima, T., Structural basis of the membrane-targeting and unmasking mechanisms of the radixin FERM domain, *Embo J.*, **2000**, *19*, 4449.

42 Nassar, N., Horn, G., Herrmann, C., Scherer, A., McCormick, F., and Wittinghofer, A., The 2.2 A crystal structure of the Ras-binding domain of the serine/threonine kinase c-Raf1 in complex with Rap1A and a GTP analogue, *Nature*, **1995**, *375*, 554.

43 Huang, L., Hofer, F., Martin, G. S., and Kim, S. H., Structural basis for the interaction of Ras with RalGDS, *Nat. Struct. Biol.*, **1998**, *5*, 422.

44 Buchberger, A., Howard, M. J., Proctor, M., and Bycroft, M., The UBX domain: a widespread ubiquitin-like module, *J. Mol. Biol.*, **2001**, *307*, 17.

45 Buchberger, A., From UBA to UBX: new words in the ubiquitin vocabulary, *Trends Cell Biol.*, **2002**, *12*, 216.

46 Hofmann, K. and Bucher, P., The UBA domain: a sequence motif present in multiple enzyme classes of the ubiquitination pathway, *Trends Biochem. Sci.*, **1996**, *21*, 172.

47 Hartmann-Petersen, R., Semple, C. A., Ponting, C. P., Hendil, K. B., and Gordon, C., UBA domain containing proteins in fission yeast, *Int. J. Biochem. Cell Biol.*, **2003**, *35*, 7629.

48 Madura, K., The ubiquitin-associated (UBA) domain: on the path from prudence to prurience, *Cell Cycle*, **2002**, *1*, 235.

49 Vadlamudi, R. K., Joung, I., Strominger, J. L., and Shin, J., p62, a phosphotyrosine-independent ligand of the SH2 domain of p56lck, belongs to a new class of ubiquitin-binding proteins, *J. Biol. Biol.*, **1996**, *271*, 20235.

50 Dieckmann, T., Withers-Ward, E. S., Jarosinski, M. A., Liu, C. F., Chen, I. S., and Feigon, J., Structure of a human DNA repair protein UBA domain that interacts with HIV-1 Vpr, *Nat. Struct. Biol.*, **1998**, *5*, 1042.

51 Mueller, T. D. and Feigon, J., Solution structures of UBA domains reveal a conserved hydrophobic surface for protein-protein interactions, *J. Mol. Biol.*, **2002**, *319*, 1243.

52 Ryu, K. S., Lee, K. J., Bae, S. H., Kim, B. K., Kim, K. A., and Choi, B. S., Binding surface mapping of intra- and interdomain interactions among hHR23B, ubiquitin, and polyubiquitin binding site 2 of S5a, *J. Biol. Biol.*, **2003**, *278*, 36621.

53 Wilkinson, C. R., Seeger, M., Hartmann-Petersen, R., Stone, M., Wallace, M., Semple, C., and Gordon, C., Proteins containing the UBA domain are able to bind to multiubiquitin chains, *Nat. Cell Biol.*, **2001**, *3*, 939.

54 Funakoshi, M., Sasaki, T., Nishimoto, T., and Kobayashi, H.,

Budding yeast Dsk2p is a polyubiquitin-binding protein that can interact with the proteasome, *Proc. Natl. Acad. Sci. USA*, **2002**, *99*, 745.

55 CHEN, L., SHINDE, U., ORTOLAN, T. G., and MADURA, K., Ubiquitin-associated (UBA) domains in Rad23 bind ubiquitin and promote inhibition of multi-ubiquitin chain assembly, *EMBO Rep.*, **2001**, *2*, 933.

56 BERTOLAET, B. L., CLARKE, D. J., WOLFF, M., WATSON, M. H., HENZE, M., DIVITA, G., and REED, S. I., UBA domains of DNA damage-inducible proteins interact with ubiquitin, *Nat. Struct. Biol.*, **2001**, *8*, 417.

57 ORTOLAN, T. G., TONGAONKAR, P., LAMBERTSON, D., CHEN, L., SCHAUBER, C., and MADURA, K., The DNA repair protein rad23 is a negative regulator of multi-ubiquitin chain assembly, *Nat. Cell Biol.*, **2000**, *2*, 601.

58 RAO, H. and SASTRY, A., Recognition of specific ubiquitin conjugates is important for the proteolytic functions of the ubiquitin-associated domain proteins Dsk2 and Rad23, *J. Biol. Biol.*, **2002**, *277*, 11691.

59 VARADAN, R., WALKER, O., PICKART, C., and FUSHMAN, D., Structural properties of polyubiquitin chains in solution, *J. Mol. Biol.*, **2002**, *324*, 637.

60 VARADAN, R., ASSFALG, M., HARIRINIA, A., RAASI, S., PICKART, C., and FUSHMAN, D., Solution conformation of Lys63-linked di-ubiquitin chain provides clues to functional diversity of polyubiquitin signaling, *J. Biol. Biol.*, **2003**, *25*, 25.

61 PONTING, C. P., Proteins of the endoplasmic-reticulum-associated degradation pathway: domain detection and function prediction, *Biochem. J.*, **2000**, *351 Pt 2*, 527.

62 SHIH, S. C., PRAG, G., FRANCIS, S. A., SUTANTO, M. A., HURLEY, J. H., and HICKE, L., A ubiquitin-binding motif required for intramolecular monoubiquitylation, the CUE domain, *Embo J.*, **2003**, *22*, 1273.

63 DONALDSON, K. M., YIN, H., GEKAKIS, N., SUPEK, F., and JOAZEIRO, C. A., Ubiquitin signals protein trafficking via interaction with a novel ubiquitin

binding domain in the membrane fusion regulator, Vps9p, *Curr. Biol.*, **2003**, *13*, 258.

64 KANG, R. S., DANIELS, C. M., FRANCIS, S. A., SHIH, S. C., SALERNO, W. J., HICKE, L., and RADHAKRISHNAN, I., Solution structure of a CUE-ubiquitin complex reveals a conserved mode of ubiquitin binding, *Cell*, **2003**, *113*, 621.

65 PRAG, G., MISRA, S., JONES, E. A., GHIRLANDO, R., DAVIES, B. A., HORAZDOVSKY, B. F., and HURLEY, J. H., Mechanism of ubiquitin recognition by the CUE domain of Vps9p, *Cell*, **2003**, *113*, 609.

66 BEATRIX, B., SAKAI, H., and WIEDMANN, M., The alpha and beta subunit of the nascent polypeptide-associated complex have distinct functions, *J. Biol. Biol.*, **2000**, *275*, 37838.

67 KAWASHIMA, T., BERTHET-COLOMINAS, C., WULFF, M., CUSACK, S., and LEBERMAN, R., The structure of the Escherichia coli EF-Tu.EF-Ts complex at 2.5 A resolution, *Nature*, **1996**, *379*, 511.

68 GRANT, R. P., NEUHAUS, D., and STEWART, M., Structural basis for the interaction between the Tap/NXF1 UBA domain and FG nucleoporins at 1A resolution, *J. Mol. Biol.*, **2003**, *326*, 849.

69 YOUNG, P., DEVERAUX, Q., BEAL, R. E., PICKART, C. M., and RECHSTEINER, M., Characterization of two polyubiquitin binding sites in the 26 S protease subunit 5a, *J. Biol. Biol.*, **1998**, *273*, 5461.

70 LAM, Y. A., LAWSON, T. G., VELAYUTHAM, M., ZWEIER, J. L., and PICKART, C. M., A proteasomal ATPase subunit recognizes the polyubiquitin degradation signal, *Nature*, **2002**, *416*, 763.

71 ELSASSER, S., GALI, R. R., SCHWICKART, M., LARSEN, C. N., LEGGETT, D. S., MULLER, B., FENG, M. T., TUBING, F., DITTMAR, G. A., and FINLEY, D., Proteasome subunit Rpn1 binds ubiquitin-like protein domains, *Nat. Cell Biol.*, **2002**, *4*, 725.

72 HOFMANN, K. and FALQUET, L., A ubiquitin-interacting motif conserved in components of the proteasomal and lysosomal protein degradation systems, *Trends Biochem. Sci.*, **2001**, *26*, 347.

73 KATZMANN, D. J., ODORIZZI, G., and EMR, S. D., Receptor downregulation and multivesicular-body sorting, *Nat. Rev. Mol. Cell Biol.*, **2002**, *3*, 893.

74 WENDLAND, B., Epsins: adaptors in endocytosis?, *Nat. Rev. Mol. Cell Biol.*, **2002**, *3*, 971.

75 DI FIORE, P. P., POLO, S., and HOFMANN, K., When ubiquitin meets ubiquitin receptors: a signalling connection, *Nat. Rev. Mol. Cell Biol.*, **2003**, *4*, 491.

76 POLO, S., CONFALONIERI, S., SALCINI, A. E., and DI FIORE, P. P., EH and UIM: endocytosis and more, *Sci. STKE*, **2003**, *213*, re17.

77 SCHEEL, H., TOMIUK, S., and HOFMANN, K., Elucidation of ataxin-3 and ataxin-7 function by integrative bioinformatics, *Hum. Mol. Genet.*, **2003**, *12*, 2845.

78 BURNETT, B., LI, F., and PITTMAN, R. N., The polyglutamine neuro-degenerative protein ataxin-3 binds polyubiquitylated proteins and has ubiquitin protease activity, *Hum. Mol. Genet.*, **2003**, *12*, 3195.

79 SWANSON, K. A., KANG, R. S., STAMENOVA, S. D., HICKE, L., and RADHAKRISHNAN, I., Solution structure of Vps27 UIM-ubiquitin complex important for endosomal sorting and receptor downregulation, *Embo J.*, **2003**, *22*, 4597.

80 FISHER, R. D., WANG, B., ALAM, S. L., HIGGINSON, D. S., ROBINSON, H., SUNDQUIST, W. I., and HILL, C. P., Structure and ubiquitin binding of the ubiquitin-interacting motif, *J. Biol. Biol.*, **2003**, *278*, 28976.

81 FUJIWARA, K., TENNO, T., SUGASAWA, K., JEE, J. G., OHKI, I., KOJIMA, C., TOCHIO, H., HIROAKI, H., HANAOKA, F., and SHIRAKAWA, M., Structure of the ubiquitin-interacting motif of S5a bound to the ubiquitin-like domain of HR23B, *J. Biol. Biol.*, **2003**, *29*, 29.

82 MUELLER, T. D. and FEIGON, J., Structural determinants for the binding of ubiquitin-like domains to the proteasome, *Embo J.*, **2003**, *22*, 4634.

83 KOONIN, E. V. and ABAGYAN, R. A., TSG101 may be the prototype of a class of dominant negative ubiquitin regulators, *Nat. Genet.*, **1997**, *16*, 330.

84 PONTING, C. P., CAI, Y. D., and BORK, P., The breast cancer gene product TSG101: a regulator of ubiquitination?, *J. Mol. Med.*, **1997**, *75*, 467.

85 HOFMANN, R. M. and PICKART, C. M., Noncanonical MMS2-encoded ubiquitin-conjugating enzyme functions in assembly of novel polyubiquitin chains for DNA repair, *Cell*, **1999**, *96*, 645.

86 MORAES, T. F., EDWARDS, R. A., MCKENNA, S., PASTUSHOK, L., XIAO, W., GLOVER, J. N., and ELLISON, M. J., Crystal structure of the human ubiquitin conjugating enzyme complex, hMms2-hUbc13, *Nat. Struct. Biol.*, **2001**, *8*, 669.

87 VANDEMARK, A. P., HOFMANN, R. M., TSUI, C., PICKART, C. M., and WOLBERGER, C., Molecular insights into polyubiquitin chain assembly: crystal structure of the Mms2/Ubc13 heterodimer, *Cell*, **2001**, *105*, 711.

88 KATZMANN, D. J., BABST, M., and EMR, S. D., Ubiquitin-dependent sorting into the multivesicular body pathway requires the function of a conserved endosomal protein sorting complex, ESCRT-I, *Cell*, **2001**, *106*, 145.

89 GARRUS, J. E., VON SCHWEDLER, U. K., PORNILLOS, O. W., MORHAM, S. G., ZAVITZ, K. H., WANG, H. E., WETTSTEIN, D. A., STRAY, K. M., COTE, M., RICH, R. L., MYSZKA, D. G., and SUNDQUIST, W. I., Tsg101 and the vacuolar protein sorting pathway are essential for HIV-1 budding, *Cell*, **2001**, *107*, 55.

90 PORNILLOS, O., ALAM, S. L., RICH, R. L., MYSZKA, D. G., DAVIS, D. R., and SUNDQUIST, W. I., Structure and functional interactions of the Tsg101 UEV domain, *Embo J.*, **2002**, *21*, 2397.

91 SHIBA, Y., KATOH, Y., SHIBA, T., YOSHINO, K., TAKATSU, H., KOBAYASHI, H., SHIN, H. W., WAKATSUKI, S., and NAKAYAMA, K., GAT (GGA and Tom1) domain responsible for ubiquitin binding and ubiquitination, *J. Biol. Biol.*, **2003**, *2*, 2.

92 PUERTOLLANO, R., RANDAZZO, P. A., PRESLEY, J. F., HARTNELL, L. M., and BONIFACINO, J. S., The GGAs promote ARF-dependent recruitment of clathrin to the TGN, *Cell*, **2001**, *105*, 93.

93 SHIBA, T., KAWASAKI, M., TAKATSU, H., NOGI, T., MATSUGAKI, N., IGARASHI, N., SUZUKI, M., KATO, R., NAKAYAMA, K., and WAKATSUKI, S., Molecular mechanism of membrane recruitment of GGA by ARF in lysosomal protein transport, *Nat. Struct. Biol.*, **2003**, *10*, 386.

94 SUER, S., MISRA, S., SAIDI, L. F., and HURLEY, J. H., Structure of the GAT domain of human GGA1: a syntaxin amino-terminal domain fold in an endosomal trafficking adaptor, *Proc. Natl. Acad. Sci. USA*, **2003**, *100*, 4451.

95 COLLINS, B. M., WATSON, P. J., and OWEN, D. J., The structure of the GGA1-GAT domain reveals the molecular basis for ARF binding and membrane association of GGAs, *Dev. Cell*, **2003**, *4*, 321.

96 WANG, B., ALAM, S. L., MEYER, H. H., PAYNE, M., STEMMLER, T. L., DAVIS, D. R., and SUNDQUIST, W. I., Structure and ubiquitin interactions of the conserved zinc finger domain of Npl4, *J. Biol. Biol.*, **2003**, *278*, 20225.

97 MEYER, H. H., WANG, Y., and WARREN, G., Direct binding of ubiquitin conjugates by the mammalian p97 adaptor complexes, p47 and Ufd1-Npl4, *Embo J.*, **2002**, *21*, 5645.

98 YASEEN, N. R. and BLOBEL, G., Two distinct classes of Ran-binding sites on the nucleoporin Nup-358, *Proc. Natl. Acad. Sci. USA*, **1999**, *96*, 5516.

99 HOOK, S. S., ORIAN, A., COWLEY, S. M., and EISENMAN, R. N., Histone deacetylase 6 binds polyubiquitin through its zinc finger (PAZ domain) and copurifies with deubiquitinating enzymes, *Proc. Natl. Acad. Sci. USA*, **2002**, *99*, 13425.

100 SEIGNEURIN-BERNY, D., VERDEL, A., CURTET, S., LEMERCIER, C., GARIN, J., ROUSSEAUX, S., and KHOCHBIN, S., Identification of components of the murine histone deacetylase 6 complex: link between acetylation and ubiquitination signaling pathways, *Mol. Cell Biol.*, **2001**, *21*, 8035.

101 HUIBREGTSE, J. M., SCHEFFNER, M., BEAUDENON, S., and HOWLEY, P. M., A family of proteins structurally and functionally related to the E6-AP ubiquitin-protein ligase, *Proc. Natl. Acad. Sci. USA*, **1995**, *92*, 2563.

102 DESHAIES, R. J., SCF and Cullin/Ring H2-based ubiquitin ligases, *Annu. Rev. Cell Dev. Biol.*, **1999**, *15*, 435.

103 HALDEMAN, M. T., XIA, G., KASPEREK, E. M., and PICKART, C. M., Structure and function of ubiquitin conjugating enzyme E2–25K: the tail is a core-dependent activity element, *Biochemistry*, **1997**, *36*, 10526.

104 WILKINSON, K. D., TASHAYEV, V. L., O'CONNOR, L. B., LARSEN, C. N., KASPEREK, E., and PICKART, C. M., Metabolism of the polyubiquitin degradation signal: structure, mechanism, and role of isopeptidase T, *Biochemistry*, **1995**, *34*, 14535.

105 AMERIK, A., SWAMINATHAN, S., KRANTZ, B. A., WILKINSON, K. D., and HOCHSTRASSER, M., In vivo disassembly of free polyubiquitin chains by yeast Ubp14 modulates rates of protein degradation by the proteasome, *Embo J.*, **1997**, *16*, 4826.

106 HARTMANN-PETERSEN, R., SEEGER, M., and GORDON, C., Transferring substrates to the 26S proteasome, *Trends Biochem. Sci.*, **2003**, *28*, 26.

107 HIYAMA, H., YOKOI, M., MASUTANI, C., SUGASAWA, K., MAEKAWA, T., TANAKA, K., HOEIJMAKERS, J. H., and HANAOKA, F., Interaction of hHR23 with S5a. The ubiquitin-like domain of hHR23 mediates interaction with S5a subunit of 26 S proteasome, *J. Biol. Biol.*, **1999**, *274*, 28019.

108 SCHAUBER, C., CHEN, L., TONGAONKAR, P., VEGA, I., LAMBERTSON, D., POTTS, W., and MADURA, K., Rad23 links DNA repair to the ubiquitin/proteasome pathway, *Nature*, **1998**, *391*, 715.

109 RAASI, S. and PICKART, C. M., Rad23 ubiquitin-associated domains (UBA) inhibit 26 S proteasome-catalyzed proteolysis by sequestering lysine 48-linked polyubiquitin chains, *J. Biol. Biol.*, **2003**, *278*, 8951.

13

The COP9 Signalosome: Its Possible Role in the Ubiquitin System

Dawadschargal Bech-Otschir, Barbara Kapelari, and Wolfgang Dubiel

13.1
Introduction

The COP9 signalosome (CSN) is a multimeric, highly-conserved protein complex [1]. Just like the ubiquitin system it occurs in all studied eukaryotic cells. Following its 1994 discovery in plant cells the complex was postulated to function in signal transduction [2]. Originally described as a regulator of light-dependent growth in plants [3, 4], identification and characterization of the CSN from mammalian cells led to the discovery of sequence homologies between CSN subunits and subunits of the 26S proteasome lid complex [5, 6] as well as subunits of the translation-initiation complex eIF3 [7]. Significant progress has been made towards understanding its structure and function by analyzing different eukaryotic organisms. The complex is involved in developmental processes of plants [8] and *Drosophila* [9] and is essential for embryogenesis in mice [10]. It seems to participate in processes such as DNA repair [11], cell-cycle regulation [12] and angiogenesis [13]. At the moment the pleiotropic effects of the CSN can be explained by its regulatory impact on the ubiquitin system. Here we provide a summary of current knowledge of CSN function in the ubiquitin system.

13.2
Discovery of the CSN

Deng and co-workers discovered the CSN in *Arabidopsis thaliana* when they characterized mutants of light-dependent development, and they called it the COP9 complex [2]. Morphogenesis of germinating seedlings is light-dependent. Light triggers a developmental process called photomorphogenesis. A number of mutations in the *Arabidopsis* COP/DET/FUS loci result in the loss of the COP9 complex accompanied with *cop* phenotypes in which germinating seedlings exhibit light-independent expression of light-induced genes [3]. Therefore the complex was originally hypothesized to be a repressor of photomorphogenesis [14]. The mam-

Protein Degradation. Edited by J. Mayer, A. Ciechanover, M. Rechsteiner
Copyright © 2005 WILEY-VCH Verlag GmbH & Co. KGaA, Weinheim
ISBN: 3-527-30837-7

malian CSN complex was independently isolated during preparations of the 26S proteasome and called the JAB1-containing signalosome [15]. The same studies identified proteins such as JAB1 [15] and TRIP15 [16] as components of the complex and revealed homologies between subunits of the CSN and components of the 26S proteasome lid complex. Purification and analysis of the complex from *Arabidopsis*, pork spleen [6, 17] and human red blood cells [5, 18] led to the conclusion that each subunit of the CSN has its paralog subunit in the 26S proteasome lid complex. These data suggested a common origin for the two complexes during evolution. Because they have similar architectures, the two complexes have been postulated to perform related functions (see below). Unfortunately there is only limited information on the structure or function of the eIF3 complex, and its relationship to the CSN and the lid is not well understood [7].

Studies have revealed that the CSN possesses both intrinsic and extrinsic (associated) activities, which will be reviewed in detail below. Historical gene names of the CSN have been summarized before [1]. In this article we use the unified nomenclature of the CSN [1].

13.3
Architecture of the CSN

13.3.1
CSN Subunit–Subunit Interactions

The CSN is composed of eight subunits called CSN1 to CSN8, which are highly conserved in eukaryotes, although only six of them occur in fission yeast. Two hybrid screens and biochemical methods such as far westerns, pull downs and co-precipitation defined a number of CSN subunit–subunit interactions. Figure 13.1 illustrates known subunit–subunit interactions. Initial insight into the architecture of CSN came from the first 2D electron microscopic analysis of purified CSN from human red blood cells [19] (see also Figure 13.2 below).

The CSN architecture shows similarity to that of the lid. Both complexes have an asymmetric arrangement of their subunits and exhibit a central groove structure [19]. Whether the structural similarity of the two complexes is connected with similar functions remains unclear. The exact arrangement of CSN and lid subunits within their complexes remains uncertain in the absence of high-resolution crystal structures for the two complexes.

Interestingly, the occurrence of smaller CSN sub-complexes apart from the large 500-kDa CSN complex has been described in different species such as *Arabidopsis*, *Drosophila*, *Schizosaccharomyces pombe* and mammalian cells (for a review see Ref. [20]). At the moment the physiological function of CSN sub-complexes is unclear. It can be speculated that a controlled equilibrium exists between the large and small CSN complexes. The small complexes may have a function in shuttling between nucleus and cytoplasma and/or between large multi-subunit complexes such as the 26S proteasome and cullin-based Ub ligases.

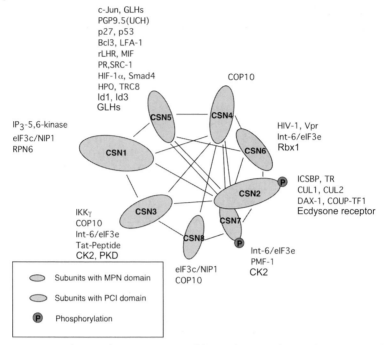

Fig. 13.1. Subunit–subunit interactions of the CSN and interactions of CSN subunits with other proteins. Subunits are numbered according to the unified nomenclature [1]. CSN subunit–subunit interactions have been published before [19]. Darker shading indicates subunits with MPN domains and lighter those with PCI domains. Known phosphorylated subunits are indicated. Details on CSN subunit interactions with other proteins can be found in the text.

13.3.2
CSN-subunit Interactions With Other Proteins

Apart from subunit–subunit interactions within the CSN, a considerable number of cellular proteins interact with CSN subunits (see Figure 13.1). Although the physiological relevance of many of the identified interactions is questionable, most of them might be attributed to a role of the CSN complex in signal transduction and ubiquitin-dependent proteolysis.

CSN1 formerly called Gps1 was first described as a signal transduction repressor in *Arabidopsis* [21]. Over-expression of CSN1 suppresses the activated JNK signaling pathway and also inhibits UV- and serum-induced c-fos expression as well as MEKK-activated AP1-activity in mammalian cells [21, 22]. It remains unclear whether overexpressed CSN1 plays a role as dominant negative regulator when in the CSN complex. Whereas the N-terminal region of CSN1 is sufficient for repression, the C-terminal region is necessary for its integration into the complex and for the stability of the CSN complex [22]. Curiously, the N-terminal region of CSN1 may be not required for the CSN-associated deneddylation of cullin 1 (CUL1) and cullin 3 (CUL3), components of cullin-based E3 Ub ligases in *Arabidopsis*, although it appears to be one of the binding sites of the CSN for cullin-based complexes [23]. Moreover, CSN1 is the receptor site for the interaction of the CSN with inositol 1,3,4-trisphosphate 5/6 kinase [24]. In addition, CSN1 represents the interactor for a subunit of the translation-initiation factor 3, eIF3c/NIP1 [25], and for the 26S proteasome non-ATPase subunit Rpn6 [26]. Possible functions of these interactions are discussed later.

CSN2 also known as alien [27] is perhaps an important regulatory subunit of the CSN. Firstly, CSN2 was identified as Trip15 (*t*hyroid hormone *r*eceptor-*i*nteracting *p*rotein) using a yeast-two-hybrid screen [16]. The binding site of CUL1 and CUL2 is located at the N-terminal region of CSN2. This interaction is important for the CSN-mediated deneddylation of cullin-based complexes that regulate their Ub-ligase activity [28]. Additionally, CSN2 binds to the transcription factor, ICSBP (*i*nterferon *c*onsensus *s*equence *b*inding *p*rotein), which modulates interferon-directed gene expression [29]. Moreover it interacts with the nuclear receptors DAX-1, COUP1-TF1 and ecdysone receptor [27, 30]. Interestingly, CSN2 is phosphorylated by the CSN-associated kinases CK2 and PKD [19, 31]. However, the phosphorylation sites and their physiological function remain unclear.

The CSN3 subunit interacts with IKKγ, a component of the IκB-kinase complex controlling NF-κB activity [32]. Additionally, it is the binding site for the CSN-associated kinases CK2 and PKD [31]. The subunit of the translation-initiation factor 3 complex, Int6/eIF3e, and the ubiquitin-conjugating enzyme variant, COP10, have been identified as other cellular interactors [33, 34]. Also the HIV-1 Tat protein interacts with CSN3 (our unpublished data).

CSN4 is a poorly studied subunit of the CSN. Only one interactor of CSN4 has been identified, the ubiquitin-conjugating enzyme COP10 [34].

CSN5 appears to be a most important subunit both in terms of interactions with other cellular proteins and because it is a component with intrinsic metalloprotease activity (see below). The binding of CSN5 to cellular proteins including the transcription factors p53 [35] and c-Jun [15], the cell-cycle regulator protein p27 [36], rLHR (lutropin/choriogonadotropin receptor precursor) [37], Smad4 (TGF-β signaling pathway common effector) [38] and HIF1α (*h*ypoxia-*i*nducible *f*actor 1) [39] appears to regulate their metabolic stability. In many cases the CSN5-interacting proteins are phosphorylated by the CSN-associated kinases, which determines the speed of their destruction [40]. In contrast, Id1 and Id3 binding to the CSN complex via CSN5 leads to their stabilization, not to their phosphorylation [41].

The interaction of CSN5 to the member of the IκB multigene family Bcl3, the progesterone receptor PR, and the steroid receptor co-activator SRC-1, leads to stabilization of Bcl3–p50 and PR–SRC-1 complexes and enhances transcriptional activity [42, 43]. Whereas AP-1 activity is stimulated by interaction of CSN5 with the integrin adhesion receptor LFA-1 [44], the opposite effect was reported in the case of the cytokine migration inhibitor factor, MIF [45]. Additionally, there are other published interactors of CSN5 including the membrane-associated RING-finger Ub ligase TRC8 [46], hepatopoietin (HPO) [47], germ-line RNA helicases (GLHs) [48], and the ubiquitin C-terminal hydrolase PGP9.5 [49]. However, the exact role of these interactions remains unclear.

Several groups reported the occurrence of a free CSN5 subunit [50] or CSN5 as a component of a smaller complex [51], although the exact physiological function of the different CSN5 forms is so far unclear. It is also unknown whether the occurrence of the different CSN5 forms is regulated. Moreover, little is known about CSN5 interactions *in vivo*, how they are regulated and under what circumstances they take place.

CSN6 like CSN8 exists in eukaryotes except in *S. pombe* [52]. There are only a few published interactions of CSN6 with other cellular proteins. It binds to the HIV-1 Vpr protein affecting cell-cycle-associated signaling [53] and to the RING-finger protein of the SCF-complex, Rbx1 [54, 55]. In addition, CSN6 is another binding site for Int-6/eIF3e [33].

Interestingly, in mammalian cells two homologs of CSN7, CSN7a and CSN7b, have been found [6]. *S. pombe* contains only one form of CSN7 whereas *Arabidopsis* contains two alternative splicing variants, CSN7i and CSN7ii [52, 56]. CSN7 interacts with the *p*olyamine-*m*odulated *f*actor PMF-1 [57]. Interestingly, CSN7 also binds the protein kinase CK2, one of the CSN-associated kinases, which phosphorylates CSN7 [31]. Whether the phosphorylated form of CSN7 is necessary for CSN complex assembly or for other regulatory events is unclear.

Little is known of CSN8 interactions. CSN8 like CSN3 and CSN4 binds to COP10 [34].

13.3.3
PCI and MPN

Six of the CSN subunits contain PCI (*p*roteasome, *C*OP9 signalosome, *i*nitiation factor 3) domains and two contain MPN (*M*pr-*P*ad1-*N*-terminal) domains [58]. These two characteristic domains have been found in three protein complexes: the CSN, the 26S proteasome lid complex (lid) and the eIF3 complex. The two domains are composed of about 150 to 200 amino acids at the N- or C-terminus of the CSN subunits. Apparently, the PCI domain has been shown to be important for interactions between CSN subunits. Thus, it might have a scaffolding function [22, 59].

The CSN subunit CSN5 has been shown to contain a metalloprotease motif localized on its MPN domain, which is essential for the cleavage of the ubiquitin-like modifier NEDD8 from cullins [60] (see below). Apart from the catalytic activity of the MPN domain of CSN5 it appears to be the receptor for different cellular proteins associated with the CSN complex (see above and Figure 13.1). Interestingly, an MPN domain similar to that of CSN5 is located in the N-terminal region of CSN6. However, this MPN domain has no deneddylation catalytic center like CSN5. The function of the CSN6 MPN domain remains obscure.

13.4
Biochemical Activities Associated With the CSN

13.4.1
Deneddylation Activity

Studies in fission yeast and *Arabidopsis* have revealed that the CSN has a role in the cleavage of NEDD8 from cullins [54, 55, 61]. The MPN domain of CSN5, like its paralog subunit Rpn11 of the 26S proteasome lid complex, possesses a highly conserved pattern of four charged amino acid residues: one glutamate, two histidines and one aspartate. This pattern represents a new type of metalloprotease motif called the JAMM (*Ja*b1/*M*PN domain *m*etalloenzyme) or MPN+ motif [62, 63]. In CSN5 the catalytic region is important for the cleavage of the ubiquitin-like modifier NEDD8 from its targets. Mutations in the conserved histidine and aspartate residues of CSN5 led to suppression of its deneddylation activity [60]. Crystal-structure analysis obtained with bacterial CSN5/MPN+ domain-containing AF2198 protein confirmed the metal-ion-dependent hydrolytic activity of CSN5, although it was inhibited by the alkylating agent NEM, an inhibitor of cysteine proteases [64].

NEDD8 is activated by a heterodimeric complex of APP-B1 and Uba3 and is conjugated to target proteins by the conjugating enzyme Ubc12. So far, the only known targets are cullin-family proteins (CUL1–5), which are components of the cullin-based E3 ligase complexes. The covalent linkage of NEDD8 to cullins *in vivo* is thought to activate Ub-ligase complex activity by facilitating ubiquitin-conjugating enzyme E2 recruitment [65]. Deneddylation of cullins inactivates ubiquitination *in vitro*, but seems to stimulate the Ub E3 ligase complex activity *in vivo* [66, 67]. In cell lysates only a small fraction of CUL1 is neddylated, but in *csn* deletion cells 100% of CUL1 is modified by NEDD8. The purified CSN complex is able to deneddylate, although recombinant CSN5 protein cannot. Obviously CSN5 deneddylation activity is dependent on its assembly into the CSN complex [28, 54]. The fact that null mutants in most CSN subunits lack the deneddylation activity in the presence of excess CSN5 supports the fact that CSN5 alone is inactive in deneddylation [61]. So far, the exact role of deneddylation is questionable (see below).

13.4.2
Protein Kinases

The CSN is associated with enzymes such as kinases, proteases and Ub ligases, which perhaps, besides the intrinsic deneddylation activity, determine the specific function of the CSN in the Ub system. Here we summarize the associated (extrinsic) activities of the CSN shown in Figure 13.2.

13.4.2.1 Associated Protein Kinases
Originally, a protein kinase was the first enzyme identified with the CSN purified from human erythrocytes. The CSN-associated kinase activity phosphorylated sev-

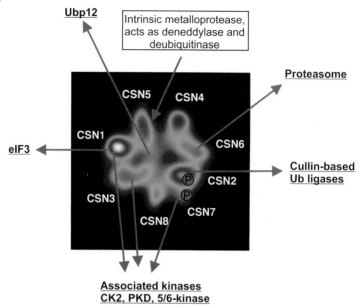

Fig. 13.2. Association of the CSN complex with enzymes. The Figure shows an electron-microscopy image of purified CSN complex from human erythrocytes. As indicated by arrows the CSN is associated with the Ub-specific protease Ubp12, the proteasome, presumably with most of the cullin-based Ub-ligase complexes, with a number of kinases, and with subunits of the translation initiation complex eIF3. In addition, subunit CSN5 has an intrinsic metalloprotease activity, which deneddylates cullins and also removes Ub conjugated to other proteins (for details see text).

eral serine and threonine residues in the N-terminal region of c-Jun [5] resulting in stabilization of c-Jun and increased AP-1 transcriptional activity. The pathway responsible for this c-Jun stabilization/activation was called CSN-directed c-Jun signaling [68]. It was subsequently shown that the CSN-directed c-Jun signaling pathway controls most of the VEGF (*v*ascular *e*ndothelial *g*rowth *f*actor) production in tumor cells [13]. VEGF is essential for tumor angiogenesis (see below).

In contrast to c-Jun, phosphorylation of the tumor suppressor p53 by CSN-associated kinases targets the protein for degradation by the Ub system [35]. For p53 stability, modification on Thr155 is most important as shown by mutational analysis [35] and by using different p53 peptides [31]. Mutation of Thr155 to Val led to stabilization of the transiently expressed p53 mutant in HeLa as well as in HL60 cells [35]. Inhibitors of CSN-associated kinases such as curcumin [18] caused stabilization of cellular p53 followed by massive cell death [35].

In addition to p53 and c-Jun, p27, ICSBP (*i*nterferon *c*onsensus *s*equence *b*inding *p*rotein) and IκBα were identified as substrates of the CSN-associated kinases (for a review see Ref. [40]). Similar to p53, the phosphorylation of p27 results in accelerated degradation of the cyclin-dependent kinase inhibitor p27 by the Ub system (our unpublished data). In the case of ICSBP and IκBα, it is still unclear whether CSN-mediated phosphorylation influences their stability. Interestingly, two of the CSN subunits, CSN2 and CSN7, are phosphorylated by the associated kinases [19, 69]. The physiological relevance of these modifications is currently obscure.

Identification of associated protein kinases Based on phosphopeptide analyses it became clear that associated kinases modify principally serine and threonine residues. Moreover, the analysis of putative phosphorylation-specific consensus sequences of p53, c-Jun, p27, ICSBP and IκBα revealed that the protein kinase CK2 and a member of the protein kinase C family might be associated with the CSN. It has been shown by immunoblotting that CK2 and the protein kinase Cμ (also called *p*rotein *k*inase *D*, PKD) co-purify with the CSN from human erythrocytes [31]. In addition, the two kinases co-immunoprecipitated together with the CSN from HeLa cells. Interaction of CK2 as well as PKD with the CSN is mediated by CSN3, as is the interaction between CK2 and the CSN7 subunit [31]. Interestingly, CSN7 itself is phosphorylated.

Majerus and co-workers have published work on the co-purification of inositol 1,3,4-trisphosphate 5/6-kinase (5/6-kinase) with the CSN from bovine brain [24, 70]. Although the 5/6-kinase was not detected in the final preparation of the CSN from human erythrocytes [31], it cannot be excluded that the enzyme is associated with another pool of CSN particles. The enzyme phosphorylates c-Jun, IκBα as well as p53 and is sensitive to curcumin. These characteristics are very similar to those described for CK2 and PKD. It has been shown that the 5/6-kinase interacts with CSN1 and that over-expression of CSN1 inhibits its activity [24]. It might be that it interacts with the N-terminal part of CSN1, which has been shown to suppress activation of an AP-1 promoter [22]. Future studies will show whether additional kinases besides 5/6-kinase, CK2 and PKD can interact with the CSN under certain circumstances. For example, an interaction of CSN3 with IKKγ, a component of the IKK kinase complex, has been published [32].

Functions of associated protein kinases Phosphorylation of a number of Ub-dependent substrates by CSN-associated kinases regulates the stability of the proteins towards the Ub system [40], presumably by promoting substrate ubiquitination. Most of the proteins bind to the CSN via CSN5, are phosphorylated and subsequently channeled to the associated Ub ligase for ubiquitination (see below). Modification of p53 induces a conformational change of the tumor suppressor, which leads to tighter binding to the Ub ligase [35]. In addition, there is evidence that phosphorylation might directly affect Ub-ligase activity. The transcriptional regulator Id3 interacts with the CSN, but is not phosphorylated. Nevertheless, inhibitors of CSN-associated kinases induce ubiquitination and degradation of the Id3 protein [41].

Because of associated kinases and their function in ubiquitination the CSN has been described as a complex "at the interface between signal transduction and ubiquitin-dependent proteolysis" [40]. This becomes even more significant if upstream regulation of the associated kinases is taken into account. Unfortunately at the moment little is known about the receptors or signal-transduction pathways leading to modification of the CSN and its associated kinase activities. It is also unclear whether there are interactions between the kinases and other associated activities of the CSN.

13.4.3
Deubiquitinating Enzymes

To date two deubiquitinating activities associated with the CSN have been identified. By mutational analysis one deubiquitinating activity has been mapped to the metalloprotease motif His–X–His–X10–Asp of the JAMM or MPN+ domain of CSN5 [11]. The conserved Asp residue of that motif was mutated and the mutant Flag-CSN5 was integrated into the CSN. The mutated CSN lost its ability to remove ubiquitin from the isopeptide bond of a mono-ubiquitinated conjugate [11]. Obviously the same intrinsic metalloprotease activity is responsible for deneddylation of mono-neddylated cullins [60], which is not surprising because of the homologies between Ub and NEDD8. It would be interesting to test whether the lid subunit Rpn11, which exhibits a deubiquitinating MPN+ domain [62], is able to deneddylate mono-neddylated proteins.

In addition, another deubiquitinating activity associated with the CSN disassembles poly-Ub chains [11, 71]. This activity is catalyzed in fission yeast by the Ub-specific protease Ubp12, which is a CSN-associated enzyme [71]. The interaction of Ubp12 with the CSN is required for Ubp12 transport to the nucleus. Presumably in the nucleus, S. pombe Pcu1- and Pcu3-based Ub-ligase activities are inhibited by Ubp12 enzyme, since the deubiquitinating enzyme protects a specific adapter protein, Pop1p, from autocatalytic destruction [71]. Thus it seems that the CSN has dual activity in suppressing cullin-based Ub-ligase reactions: one is the intrinsic deneddylation and the other is deubiquitination via associated Ubp12; both reactions serve to inhibit cullin-based Ub ligases *in vitro*.

Data have accumulated showing that CSN-associated deneddylation and deubiquitination are required for Ub-ligase activity *in vivo*. It has been hypothesized, therefore, that CSN-mediated inhibition of cullin-based ubiquitination might be necessary for the assembly of new cullin-based Ub-ligase complexes. After release from the CSN the new cullin-based complex would be active. It has to return to the CSN for re-assembly or is degraded after auto-ubiquitination [71, 72]. For example, p27 has to be degraded at the transition from G1 to S phase. In a first step p27 may bind to the CSN which signals, perhaps by phosphorylation, the assembly of the required SCF complex containing the specific F-box protein Skp2. After formation of the p27-specific SCF complex both p27 and the Ub ligase might be released from the CSN perhaps again by phosphorylation which then results in ubiquitination and complete degradation of p27. Finally the Skp2-containing SCF complex is auto-ubiquitinated and degraded unless additional substrate appears.

13.4.4
Ubiquitin Ligases

Data have been accumulated demonstrating interactions of the CSN with Ub ligases, in particular with the cullin-based Ub ligases. Cullins 1 to 7 (CUL1–CUL7) form a protein family detected in all eukaryotic cells, which is involved in protein ubiquitination. It is known that CUL1 to CUL5 interact with the RING-domain protein Rbx1, the Ub ligase of the cullin-based complexes. So far it has been shown that the CSN interacts with CUL1 to CUL4 [11, 12, 54, 55, 61, 73]. Binding studies with CUL1 and with CUL2 revealed that the two cullin proteins bind via CSN2 to the complex [28, 54, 55]. In addition, Rbx1 seems to interact with CSN6 [54, 55]. Moreover, CUL1 interacts with Skp1, which makes the connection to a substrate-specific F-box protein. Therefore, CUL1-based Ub ligases are called SCF complexes (*S*kp1–*C*DC53/CUL1–*F*-box protein) (for a review see Ref. [74]). CUL2 can be linked to the substrate-adapter protein the von Hippel–Lindau tumor suppressor via elongin C and elongin B forming the so called VCB (*v*on Hippel–Lindau–elongin *C*–elongin *B*) complex (for a review see Ref. [74]). BTB/POZ-domain proteins have been identified as possible substrate-specific adaptors of CUL3-based Ub ligases [73, 75, 76]. There are more than 200 putative BTB/POZ-domain proteins expressed in mammalian cells and together with the large number of possible F-box proteins one can estimate that several hundreds of different cullin-based Ub-ligase complexes with different substrate specificities can be formed. The CUL4–Rbx1 complex has been characterized, and seems to be important for checkpoint control [12], DNA repair [11] and ubiquitination of c-Jun [77]. Most likely all cullin-based complexes interact with the CSN. In other words, the CSN is associated with ubiquitinating activity (see Figure 13.2).

There are just a few data on interactions of the CSN with other Ub ligases besides the cullin-based complexes. For example, Mdm2, the RING domain Ub ligase of the tumor suppressor p53, binds to the CSN and is modified by CSN-associated kinases (our unpublished data). Whether Mdm2 is also modified by other CSN-associated activities has to be tested in the future. In addition, COP1, a putative RING-domain Ub ligase, which probably cooperates with the *C*OP1-*i*nteracting *p*rotein 8 (CIP8) also binds to the CSN (for a review see Ref. [4]). However, some data indicate that COP1 is associated with a CUL4A complex in which it acts together with DET1 as a heterodimeric substrate adapter [77]. In this complex the CSN interacts with both the CUL4A and the COP1.

**13.5
Association of the CSN With Other Protein Complexes**

13.5.1
The eIF3 Complex

MPN and PCI domains have been also found in subunits of the eIF3 complex. Because MPN and PCI domains are most likely involved in protein–protein interactions (see above), it is not surprising that there are also cross-interactions between subunits of the CSN, the eIF3 and the lid. It has been reported that eIF3e/INT6 possessing a PCI domain interacts with CSN7 [33, 78]. Another eIF3 subunit eIF3c/p105 co-immunoprecipitated with eIF3e/INT6, eIF3b, CSN1 and CSN8 [78]. eIF3e/INT6 was used as bait in a two-hybrid screen that revealed possible interactions with the 26S proteasome ATPase Rpt4, CSN3 and CSN6 but also with CSN7 [33]. Interestingly, the subunit of the CSN-like complex in *Saccharomyces cerevisiae* Pci8/CSN11 [79] seems also to be a subunit of the budding yeast eIF3 complex and perhaps plays a similar role to eIF3e/INT6 in eukaryotic cells [80]. It has been speculated that these interactions allow the CSN to control translation.

Interactions between eIF3e/INT6 or eIF3i with the 26S proteasome have also been described [33, 81]. It has been shown that eIF3e/INT6 interacts with Rpn5 of the lid complex. This has an impact on 26S proteasome activity/localization, presumably affecting cell division and mitotic fidelity [82]. Perhaps there exists a network of "PCI complexes" as suggested [83], which shares polypeptides and communicates via proteins such as eIF3e/INT6.

13.5.2
The Proteasome

In 1998 it was reported that the CSN co-fractionates with the 26S proteasome from human cells [5]. A yeast two-hybrid screen revealed that the C-terminal domain of the Arabidopsis atCSN1 subunit interacts with atRpn6 of the 26S proteasome lid [26]. Recently gel-filtration size-fractionation of material from *Arabidopsis* in the presence of ATP and phosphatase inhibitors indicated that the CSN1 and CSN6 subunits co-elute in the same fractions as subunits of the 26S regulatory complex [84]. Based on these data it has been speculated that the CSN might be an alternative lid of the 26S proteasome [85]. The "alternative lid hypothesis", however, makes little sense if the CSN interacts with the 26S proteasome via the lid component Rpn6 [26]. CSN pull-down experiments and subsequent mass-spectrometry analysis of co-precipitated proteins also revealed the presence of proteasome subunits in the precipitate [73]. However, since proteasome subunits are very abundant in cells, one has to be cautious with this type of data. So far there is no systematic binding study showing physical interaction of the CSN with sub-complexes of the 26S proteasome. Moreover, up to now there is no functional evidence for such a CSN/26S proteasome interaction.

13.6
Biological Functions of the CSN

13.6.1
Regulation of Ubiquitin Conjugate Formation

In general, and including all its activities, intrinsic as well as associated, the CSN seems to be a regulator of ubiquitination. Deneddylation, deubiquitination as well as CSN-mediated phosphorylation (at least with c-Jun and Id3 as substrates) cause inhibition of ubiquitination. It is likely that suppression of ligase activity is an essential step in the dynamic process of specific E3 complex assembly/reassembly. According to the model of Wolf et al. [72] cullin-based Ub-ligase complexes might assemble/reassemble in a protected environment produced by the CSN. In the CSN-associated-state, binding of any E2 to the Ub ligase is prevented, perhaps by deneddylation [65], self-ubiquitination is blocked by continuous deubiquitination [71] and substrate binding could be inhibited by phosphorylation [31]. Only under these conditions can the Ub ligase reassemble without itself being destroyed. For example, an SCF complex might associate with another F-box protein, or a CUL3-Ub ligase with another BTB/POZ-domain protein, as an adaptation to the next phase of cell cycle or signal transduction upon the appearance of a new substrate, which has to be degraded. Following this argument a major question arises. How does the substrate signal the assembly of the required Ub ligase performing its ubiquitination? Is it by binding to the CSN and subsequent signaling via specific kinases?

In the case of the SCF complexes, another protein called CAND1/Tip120A seems to be involved in the dynamic assembly/reassembly process of the E3 [86]. CAND1 binds to the deneddylated CUL1 and inhibits Ub-ligase activity by competing for the Skp1–F-box-protein unit of the SCF complex [87]. After the release of CAND1, a new Skp1–F-box-protein unit can dock to the CUL1–Rbx1 unit to form an SCF complex possessing the necessary substrate specificity. Now the freshly formed Ub ligase has to be released from the CSN to become active. At the moment it is unclear how the Ub ligase might be released from the CSN. The attractive model of CSN-assisted Ub-ligase-complex assembly has to be tested in the future. In this model the CSN would function as a platform for Ub-ligase assembly.

Interestingly, there are no reports of interactions between the 26S proteasome lid complex and Ub ligases. Known E3s directly interacting with the 26S proteasome seem to bind via base ATPases [88, 89]. This is an interesting functional difference between the CSN and the lid developed during evolution.

In an alternative model the CSN might be the platform for complete proteolysis. It forms supercomplexes consisting of both the ubiquitinating and the proteolytic machineries. According to this model, the substrate first binds to the CSN, is then ubiquitinated by the associated Ub ligase and finally directly channeled into the 26S proteasome. Deneddylation, deubiquitination and phosphorylation are necessary to maintain the supercomplex, to protect the intermediates and to stimulate proteolysis.

13.6.1.1 Cell-cycle and Checkpoint Control

Initial insight of the role of CSN in cell-cycle control came from the finding that *csn1* and *csn2* deletion *S. pombe* strains have an S-phase delay [52]. Interestingly, this effect did not occur in strains missing other CSN subunits. The S-phase delay was caused by the accumulation of the cell-cycle inhibitor Spd1 (*S-p*hase *d*elayed 1), which is involved in the misregulation of the *r*ibo*n*ucleotide *r*eductase (RNR). RNR catalyzes the production of deoxyribonucleotides for DNA synthesis and is composed of four subunits including Suc22. Activation of RNR is regulated by nuclear export of Suc22, which is suppressed by Spd1 [12]. Upon DNA damage or during S phase Spd1 is rapidly degraded, presumably leading to the RNR-dependent production of dNTPs. However, in *csn1* and *csn2* deletion mutants, Spd1 accumulates, causing Suc22-dependent suppression of RNR connected with the S-phase delay and DNA-damage sensitivity [12, 15].

In mammalian cells, binding of HIV-1 Vpr-protein to the CSN6 results in cell-cycle arrest at the G2/M phase [53]. Additionally, CSN is involved in the cell cycle via the nuclear export of cell-cycle kinase inhibitor $p27^{kip1}$ (p27). CSN5 binds to p27 and promotes its nuclear export followed by its proteasome-dependent degradation. The over-expression of CSN5 in mouse fibroblasts counteracts cell-cycle arrest induced by serum depletion [36, 51]. Microinjection of the purified CSN complex into synchronized G1 cells blocks the S-phase entry in a deneddylation-dependent manner [28]. Furthermore, the reduction of CSN subunit expression by RNAi in *Caenorhabditis elegans* causes the failure of Mei-1 degradation by regulation of its specific Ub-ligase CUL3-based complex, which leads to severe effects during mitotic cell division [76].

Moreover, the CSN is involved in checkpoint control. The double deletions of *csn1* and *csn2* mutants crossed with checkpoint pathway mutants such as *rad3*, *chk*, and *cds1* are synthetically lethal in *S. pombe* [52]. Cds1 kinase is constitutively activated in *csn1* mutants. Similarly, loss of *csn5* in *Drosophila* results in activation of Mei-41, one of the ATM/ATR family kinases involved in meiotic checkpoint upon DNA damage [90].

13.6.1.2 DNA Repair

Two papers have assigned the CSN a function in DNA repair. One study reports on the existence of two different complexes containing human CSN and either one of the two nucleotide-excision-repair proteins, DDB2 or CSA. DDB2 is involved in the global *g*enome-*r*epair *p*athway (GGR) and CSA functions in the *t*ranscription-*c*oupled *r*epair pathway (TCR). Additionally, these complexes possess Ub-ligase activity and contain cullin-based Ub-ligase components such as CUL4 and Rbx1/Roc1, and DDB1, a UV-damage DNA-binding protein [11]. However, so far their targets remain unclear. CSN differentially regulates the ubiquitin-ligase activity of the DDB2- and CSA-containing complexes in response to UV irradiation. In support of direct involvement of the CSN is the finding that knockdown of CSN5 with RNAi causes a failure in NER mechanisms [11]. Similarly, CSN in combination with the CUL4–Rbx1 complex is involved in Ub-dependent degradation of CDT1, a licensing factor of the pre-replication complex (preRC), after UV- or γ-irradiation. Knockdown of CSN completely suppresses CDT1 degradation, causing a defect G1 checkpoint in response to DNA damage [91].

13.6.1.3 Developmental Processes

Although CSN is not essential in yeast, the *csn1* and *csn2* *S. pombe* deletion mutants display slow growth and sensitivity to UV- and γ-irradiation [52]. Other csn mutants did not show significant phenotypes apart from the loss of cullin's deneddylation activity [61]. In mutants of CSN-like complexes in the budding yeast *S. cerevisiae* the sensitivity to the DNA-damage reagents is not affected [92]. In some *S. cerevisiae* mutants such as *csn5*, *csn9* and *csn12* deletions, increased mating efficiency and enhanced pheromone response has been observed [63].

In *Drosophila*, mutations of CSN causes lethality in early larval stages and defects during oogenesis or photoreceptor R cell differentiation [9, 90, 93, 94]. More specifically, lack of CSN5 leads to the activation of a DNA double-strand-break-dependent checkpoint mediated by Mei-41. This effect is caused by CSN5-dependent inhibition of gurken (Grk) protein translation [90]. In *C. elegans*, knockdown of CSN5 by RNAi resulted in a sterile phenotype, which could be explained by CSN interaction with germ-line RNA helicases [48].

The best studied physiological role of the CSN in developmental processes is derived from studies on *Arabidopsis*. *Csn* mutants can survive embryogenesis, but they die soon after germination. The *csn* mutants exhibit a defect in photomorphogenesis, a light-dependent developmental process of germinating seedlings. Even in total darkness the mutants display a light-dependent morphology and signal-independent expression of light-induced genes [3, 14, 95]. One key mechanism is the CSN-dependent regulation of the stability of the transcription factor HY5, a positive regulator of light-induced genes. In the dark it is degraded by the Ub system [8]. It has been suggested that in darkness the RING-finger protein COP1 ubiquitinates HY5 and triggers its degradation by the 26S proteasome (for a review see Ref. [4]). In the light, COP1 is relocated to the cytoplasm allowing expression of genes through HY5. Although the exact mechanism remains unclear, the CSN may be required for relocation of COP1 from cytoplasm to the nucleus in darkness. Identical phenotypes caused by different *csn* mutants in *Arabidopsis* could be explained by a role of the CSN as a whole complex (for a review see Ref. [20]).

There is accumulating evidence for cooperation of the CSN and cullin-based complexes in specific developmental processes [96]. First insight has been provided by studies on auxin response where the CSN interacts with SCF^{TIR1}, modulating its activity [55]. Similarly, binding of the CSN to other cullin-based complexes regulates their activity in mediating various developmental processes such as flower development, and plant defense responses [97, 98].

13.6.2
Tumor Angiogenesis

Tumor angiogenesis is the vascularization of solid tumors, an essential requirement for tumor growth and metastasis. After a solid tumor has reached a size of approximately 2 mm^3, it needs nutrient supply from blood vessels, otherwise it dies from necrosis. Many tumor cells are able to induce angiogenesis. In an initiation phase the tumor cells produce large amounts of pro-angiogenic factors such as *v*ascular *e*ndothelial *g*rowth *f*actor, VEGF. During proliferation and invasion VEGF stimulates migration of endothelial cells. Finally, after a maturation phase, vascularization of solid tumors is completed. Now the tumor can grow, and some tumor cells penetrate through vessel membranes and spread via the circulation. Therefore, inhibition of tumor angiogenesis has become an important strategy in tumor therapy.

There is functional cooperation between the CSN and the Ub system in tumor angiogenesis [13]. It has been known for some time that curcumin is an inhibitor of angiogenesis [99]. However, only in 2001 did it become clear that it acts via inhibition of CSN-associated kinases [13]. It has been demonstrated that overexpression of CSN2 subunit leads to elevated amounts of *de novo* assembled CSN complex connected with increased c-Jun levels and enhanced AP-1 transactivation activity [68]. This c-Jun activation/stabilization is independent of the JNK and the MAP kinase pathway and is called CSN-directed c-Jun signaling [68]. This process can be inhibited by curcumin or other inhibitors of CSN-associated kinases (Figure 13.3). The CSN-directed c-Jun signaling controls up to 75% of VEGF production in tumor cells [13]. In addition, Id1 and Id3 are also essential factors of tumor angiogenesis [100] and are degraded in the presence of CSN-associated kinase inhibitors in an Ub-dependent manner just like c-Jun [41]. Therefore, specific inhibition of CSN-associated kinases might become important for tumor therapy. The application of CSN-associated kinase inhibitors in tumor therapy could be beneficial owing to another effect of curcumin-like compounds, namely they stabilize cellular

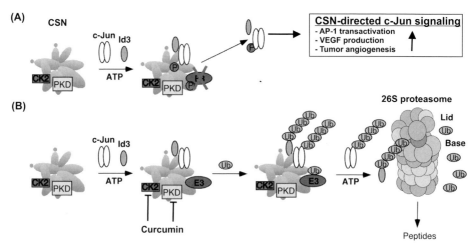

Fig. 13.3. The CSN-directed c-Jun signaling pathway. (A) The active CSN-directed c-Jun signaling pathway is shown. In case of active CSN-associated kinases c-Jun is phosphorylated, which stabilizes the transcription factor towards the Ub system. In addition, phosphorylation of the responsible E3 might inactivate the enzyme. In this situation Id1 and Id3 are also stabilized. Stable/active c-Jun causes enhanced AP-1 transactivation connected with an increase of VEGF production by tumor cells

(see text). VEGF is a major pro-angiogenic factor produced by many tumor cells. Id1 and Id3 are transcriptional regulators essential for tumor angiogenesis. (B) In the presence of curcumin or other kinase inhibitors the responsible Ub ligase is most likely active and ubiquitinates both c-Jun and Id3. In addition, unphosphorylated c-Jun might have higher affinity to its Ub ligase. This leads to quick degradation of the proteins by the Ub system.

p53 and, at least in tumors with wild-type p53 protein, massive cell death can be observed [35].

13.7
Concluding Remarks

The CSN is a regulatory complex of the Ub system. Physically it interacts with the proteasome and with Ub ligases. Although the exact mechanism remains obscure, the CSN regulates ubiquitination of important cell-cycle factors and transcriptional regulators. Its intrinsic deneddylating as well as the associated kinase and deubiquitinating activities seem to be required for determining protein stability towards the Ub system. As a major regulator of the Ub system the CSN is involved in processes such as DNA repair, cell-cycle progression and development. Its role in tumor angiogenesis makes the complex attractive for future tumor therapies.

Acknowledgment

We thank Rasmus Hartmann-Petersen for critical reading of the manuscript. The work was supported by research grants from the Deutsche Forschungsgemeinschaft and from the German–Israeli Foundation to W. D.

References

1 DENG, X. W., DUBIEL, W., WEI, N., HOFMANN, K., MUNDT, K., COLICELLI, J., KATO, J., NAUMANN, M., SEGAL, D., SEEGER, M., CARR, A., GLICKMAN, M., CHAMOVITZ, D. A. Unified nomenclature for the COP9 signalosome and its subunits: an essential regulator of development. *Trends Genet.* 2000, *16*, 202–203.

2 WEI, N., CHAMOVITZ, D. A., DENG, X. W. Arabidopsis COP9 is a component of a novel signaling complex mediating light control of development. *Cell* 1994, *78*, 117–124.

3 WEI, N., DENG, X. W. Making sense of the COP9 signalosome. A regulatory protein complex conserved from Arabidopsis to human. *Trends Genet.* 1999, *15*, 98–103.

4 SERINO, G., DENG, X. W. The COP9 signalosome: regulating plant development through the control of proteolysis. *Annu. Rev. Plant Biol.* 2003, *54*, 165–182.

5 SEEGER, M., KRAFT, R., FERRELL, K., BECH-OTSCHIR, D., DUMDEY, R., SCHADE, R., GORDON, C., NAUMANN, M., DUBIEL, W. A novel protein complex involved in signal transduction possessing similarities to 26S proteasome subunits. *FASEB J.* 1998, *12*, 469–478.

6 WEI, N., TSUGE, T., SERINO, G., DOHMAE, N., TAKIO, K., MATSUI, M., DENG, X. W. The COP9 complex is conserved between plants and mammals and is related to the 26S proteasome regulatory complex. *Curr. Biol.* 1998, *8*, 919–922.

7 GLICKMAN, M. H., RUBIN, D. M., COUX, O., WEFES, I., PFEIFER, G., CJEKA, Z., BAUMEISTER, W., FRIED, V. A., FINLEY, D. A subcomplex of the proteasome regulatory particle

required for ubiquitin-conjugate degradation and related to the COP9-signalosome and eIF3. *Cell* **1998**, *94*, 615–623.

8 SCHWECHHEIMER, C., DENG, X. W. COP9 signalosome revisited: a novel mediator of protein degradation. *Trends Cell Biol.* **2001**, *11*, 420–426.

9 FREILICH, S., ORON, E., KAPP, Y., NEVO-CASPI, Y., ORGAD, S., SEGAL, D., CHAMOVITZ, D. A. The COP9 signalosome is essential for development of Drosophila melanogaster. *Curr. Biol.* **1999**, *9*, 1187–1190.

10 LYKKE-ANDERSEN, K., SCHAEFER, L., MENON, S., DENG, X. W., MILLER, J. B., WEI, N. Disruption of the COP9 signalosome Csn2 subunit in mice causes deficient cell proliferation, accumulation of p53 and cyclin E, and early embryonic death. *Mol. Cell. Biol.* **2003**, *23*, 6790–6797.

11 GROISMAN, R., POLANOWSKA, J., KURAOKA, I., SAWADA, J., SAIJO, M., DRAPKIN, R., KISSELEV, A. F., TANAKA, K., NAKATANI, Y. The ubiquitin ligase activity in the DDB2 and CSA complexes is differentially regulated by the COP9 signalosome in response to DNA damage. *Cell* **2003**, *113*, 357–367.

12 LIU, C., POWELL, K. A., MUNDT, K., WU, L., CARR, A. M., CASPARI, T. Cop9/signalosome subunits and Pcu4 regulate ribonucleotide reductase by both checkpoint-dependent and -independent mechanisms. *Genes Dev.* **2003**, *17*, 1130–1140.

13 POLLMANN, C., HUANG, X., MALL, J., BECH-OTSCHIR, D., NAUMANN, M., DUBIEL, W. The constitutive photo-morphogenesis 9 signalosome directs vascular endothelial growth factor production in tumor cells. *Cancer Res.* **2001**, *61*, 8416–8421.

14 OSTERLUND, M. T., WEI, N., DENG, X. W. The roles of photoreceptor systems and the COP1-targeted destabilization of HY5 in light control of Arabidopsis seedling development. *Plant Physiol.* **2000**, *124*, 1520–1524.

15 CLARET, F. X., HIBI, M., DHUT, S., TODA, T., KARIN, M. A new group of conserved coactivators that increase the specificity of AP-1 transcription factors. *Nature* **1996**, *383*, 453–457.

16 LEE, J. W., CHOI, H. S., GYURIS, J., BRENT, R., MOORE, D. D. Two classes of proteins dependent on either the presence or absence of thyroid hormone for interaction with the thyroid hormone receptor. *Mol. Endocrinol.* **1995**, *9*, 243–254.

17 WEI, N., DENG, X. W. Characterization and purification of the mammalian COP9 complex, a conserved nuclear regulator initially identified as a repressor of photomorphogenesis in higher plants. *Photochem. Photobiol.* **1998**, *68*, 237–241.

18 HENKE, W., FERRELL, K., BECH-OTSCHIR, D., SEEGER, M., SCHADE, R., JUNGBLUT, P., NAUMANN, M., DUBIEL, W. Comparison of human COP9 signalsome and 26S proteasome lid'. *Mol. Biol. Rep.* **1999**, *26*, 29–34.

19 KAPELARI, B., BECH-OTSCHIR, D., HEGERL, R., SCHADE, R., DUMDEY, R., DUBIEL, W. Electron microscopy and subunit–subunit interaction studies reveal a first architecture of COP9 signalosome. *J. Mol. Biol.* **2000**, *300*, 1169–1178.

20 WEI, N., DENG, X. W. The COP9 signalosome. *Annu. Rev. Cell Dev. Biol.* **2003**, *19*, 261–286.

21 SPAIN, B. H., BOWDISH, K. S., PACAL, A. R., STAUB, S. F., KOO, D., CHANG, C. Y., XIE, W., COLICELLI, J. Two human cDNAs, including a homolog of Arabidopsis FUS6 (COP11), suppress G-protein- and mitogen-activated protein kinase-mediated signal transduction in yeast and mammalian cells. *Mol. Cell. Biol.* **1996**, *16*, 6698–6706.

22 TSUGE, T., MATSUI, M., WEI, N. The subunit 1 of the COP9 signalosome suppresses gene expression through its N-terminal domain and incorporates into the complex through the PCI domain. *J. Mol. Biol.* **2001**, *305*, 1–9.

23 WANG, X., KANG, D., FENG, S., SERINO, G., SCHWECHHEIMER, C., WEI, N. CSN1 N-terminal-dependent activity is required for Arabidopsis development but not for Rub1/Nedd8

deconjugation of cullins: a structure-function study of CSN1 subunit of COP9 signalosome. *Mol. Biol. Cell* **2002**, *13*, 646–655.

24 SUN, Y., WILSON, M. P., MAJERUS, P. W. Inositol 1,3,4-trisphosphate 5/6-kinase associates with the COP9 signalosome by binding to CSN1. *J. Biol. Chem.* **2002**, *277*, 45759–45764.

25 KARNIOL, B., YAHALOM, A., KWOK, S., TSUGE, T., MATSUI, M., DENG, X. W., CHAMOVITZ, D. A. The Arabidopsis homologue of an eIF3 complex subunit associates with the COP9 complex. *FEBS Lett.* **1998**, *439*, 173–179.

26 KWOK, S. F., STAUB, J. M., DENG, X. W. Characterization of two subunits of Arabidopsis 19S proteasome regulatory complex and its possible interaction with the COP9 complex. *J. Mol. Biol.* **1999**, *285*, 85–95.

27 DRESSEL, U., THORMEYER, D., ALTINCICEK, B., PAULULAT, A., EGGERT, M., SCHNEIDER, S., TENBAUM, S. P., RENKAWITZ, R., BANIAHMAD, A. Alien, a highly conserved protein with characteristics of a corepressor for members of the nuclear hormone receptor super-family. *Mol. Cell. Biol.* **1999**, *19*, 3383–3394.

28 YANG, X., MENON, S., LYKKE-ANDERSEN, K., TSUGE, T., DI, X., WANG, X., RODRIGUEZ-SUAREZ, R. J., ZHANG, H., WEI, N. The COP9 signalosome inhibits p27(kip1) degradation and impedes G1-S phase progression via deneddylation of SCF Cul1. *Curr. Biol.* **2002**, *12*, 667–672.

29 COHEN, H., AZRIEL, A., COHEN, T., MERARO, D., HASHMUELI, S., BECH-OTSCHIR, D., KRAFT, R., DUBIEL, W., LEVI, B. Z. Interaction between interferon consensus sequence-binding protein and COP9/signalosome subunit CSN2 (Trip15). A possible link between interferon regulatory factor signaling and the COP9/signalosome. *J. Biol. Chem.* **2000**, *275*, 39081–39089.

30 ALTINCICEK, B., TENBAUM, S. P., DRESSEL, U., THORMEYER, D., RENKAWITZ, R., BANIAHMAD, A. Interaction of the corepressor Alien

with DAX-1 is abrogated by mutations of DAX-1 involved in adrenal hypoplasia congenita. *J. Biol. Chem.* **2000**, *275*, 7662–7667.

31 UHLE, S., MEDALIA, O., WALDRON, R., DUMDEY, R., HENKLEIN, P., BECH-OTSCHIR, D., HUANG, X., BERSE, M., SPERLING, J., SCHADE, R., DUBIEL, W. Protein kinase CK2 and protein kinase D are associated with the COP9 signalosome. *EMBO J.* **2003**, *22*, 1302–1312.

32 HONG, X., XU, L., LI, X., ZHAI, Z., SHU, H. CSN3 interacts with IKKgamma and inhibits TNF- but not IL-1-induced NF-kappaB activation. *FEBS Lett.* **2001**, *499*, 133–136.

33 HOAREAU ALVES, K., BOCHARD, V., RETY, S., JALINOT, P. Association of the mammalian proto-oncoprotein Int-6 with the three protein complexes eIF3, COP9 signalosome and 26S proteasome. *FEBS Lett.* **2002**, *527*, 15–21.

34 SUZUKI, G., YANAGAWA, Y., KWOK, S. F., MATSUI, M., DENG, X. W. Arabidopsis COP10 is a ubiquitin-conjugating enzyme variant that acts together with COP1 and the COP9 signalosome in repressing photomor-phogenesis. *Genes Dev.* **2002**, *16*, 554–559.

35 BECH-OTSCHIR, D., KRAFT, R., HUANG, X., HENKLEIN, P., KAPELARI, B., POLLMANN, C., DUBIEL, W. COP9 signalosome-specific phosphorylation targets p53 to degradation by the ubiquitin system. *EMBO J.* **2001**, *20*, 1630–1639.

36 TOMODA, K., KUBOTA, Y., KATO, J. Degradation of the cyclin-dependent-kinase inhibitor p27Kip1 is instigated by Jab1. *Nature* **1999**, *398*, 160–165.

37 LI, S., LIU, X., ASCOLI, M. p38JAB1 binds to the intracellular precursor of the lutropin/choriogonadotropin receptor and promotes its degradation. *J. Biol. Chem.* **2000**, *275*, 13386–13393.

38 WAN, M., CAO, X., WU, Y., BAI, S., WU, L., SHI, X., WANG, N. Jab1 antagonizes TGF-beta signaling by inducing Smad4 degradation. *EMBO Rep.* **2002**, *3*, 171–176.

39 BAE, M. K., AHN, M. Y., JEONG, J. W.,

Bae, M. H., Lee, Y. M., Bae, S. K., Park, J. W., Kim, K. R., Kim, K. W. Jab1 interacts directly with HIF-1alpha and regulates its stability. *J. Biol. Chem.* **2002**, *277*, 9–12.

40 Bech-Otschir, D., Seeger, M., Dubiel, W. The COP9 signalosome: at the interface between signal transduction and ubiquitin-dependent proteolysis. *J. Cell Sci.* **2002**, *115*, 467–473.

41 Berse, M., Bounpheng, M. A., Huang, X., Christy, B. A., Pollmann, C., Dubiel, W. Ubiquitin-dependent degradation of Id1 and Id3 is mediated by the COP9 signalosome. *J. Mol. Biol.* **2004**, *343*, 361–370.

42 Dechend, R., Hirano, F., Lehmann, K., Heissmeyer, V., Ansieau, S., Wulczyn, F. G., Scheidereit, C., Leutz, A. The Bcl-3 oncoprotein acts as a bridging factor between NF-kappaB/Rel and nuclear co-regulators. *Oncogene* **1999**, *18*, 3316–3323.

43 Chauchereau, A., Georgiakaki, M., Perrin-Wolff, M., Milgrom, E., Loosfelt, H. JAB1 interacts with both the progesterone receptor and SRC-1. *J. Biol. Chem.* **2000**, *275*, 8540–8548.

44 Bianchi, E., Denti, S., Granata, A., Bossi, G., Geginat, J., Villa, A., Rogge, L., Pardi, R. Integrin LFA-1 interacts with the transcriptional co-activator JAB1 to modulate AP-1 activity. *Nature* **2000**, *404*, 617–621.

45 Kleemann, R., Hausser, A., Geiger, G., Mischke, R., Burger-Kentischer, A., Flieger, O., Johannes, F. J., Roger, T., Calandra, T., Kapurniotu, A., Grell, M., Finkelmeier, D., Brunner, H., Bernhagen, J. Intracellular action of the cytokine MIF to modulate AP-1 activity and the cell cycle through Jab1. *Nature* **2000**, *408*, 211–216.

46 Gemmill, R. M., Bemis, L. T., Lee, J. P., Sozen, M. A., Baron, A., Zeng, C., Erickson, P. F., Hooper, J. E., Drabkin, H. A. The TRC8 hereditary kidney cancer gene suppresses growth and functions with VHL in a common pathway. *Oncogene* **2002**, *21*, 3507–3516.

47 Lu, C., Li, Y., Zhao, Y., Xing, G.,

Tang, F., Wang, Q., Sun, Y., Wei, H., Yang, X., Wu, C., Chen, J., Guan, K. L., Zhang, C., Chen, H., He, F. Intracrine hepatopoietin potentiates AP-1 activity through JAB1 independent of MAPK pathway. *FASEB J.* **2002**, *16*, 90–92.

48 Smith, P., Leung-Chiu, W. M., Montgomery, R., Orsborn, A., Kuznicki, K., Gressman-Coberly, E., Mutapcic, L., Bennett, K. The GLH proteins, Caenorhabditis elegans P granule components, associate with CSN-5 and KGB-1, proteins necessary for fertility, and with ZYX-1, a predicted cytoskeletal protein. *Dev. Biol.* **2002**, *251*, 333–347.

49 Caballero, O. L., Resto, V., Patturajan, M., Meerzaman, D., Guo, M. Z., Engles, J., Yochem, R., Ratovitski, E., Sidransky, D., Jen, J. Interaction and colocalization of PGP9.5 with JAB1 and p27(Kip1). *Oncogene* **2002**, *21*, 3003–3010.

50 Chamovitz, D. A., Segal, D. JAB1/CSN5 and the COP9 signalosome. A complex situation. *EMBO Rep.* **2001**, *2*, 96–101.

51 Tomoda, K., Kubota, Y., Arata, Y., Mori, S., Maeda, M., Tanaka, T., Yoshida, M., Yoneda-Kato, N., Kato, J. Y. The cytoplasmic shuttling and subsequent degradation of p27Kip1 mediated by Jab1/CSN5 and the COP9 signalosome complex. *J. Biol. Chem.* **2002**, *277*, 2302–2310.

52 Mundt, K. E., Porte, J., Murray, J. M., Brikos, C., Christensen, P. U., Caspari, T., Hagan, I. M., Millar, J. B., Simanis, V., Hofmann, K., Carr, A. M. The COP9/signalosome complex is conserved in fission yeast and has a role in S phase. *Curr. Biol.* **1999**, *9*, 1427–1430.

53 Mahalingam, S., Ayyavoo, V., Patel, M., Kieber-Emmons, T., Kao, G. D., Muschel, R. J., Weiner, D. B. HIV-1 Vpr interacts with a human 34-kDa mov34 homologue, a cellular factor linked to the G2/M phase transition of the mammalian cell cycle. *Proc. Natl. Acad. Sci. USA* **1998**, *95*, 3419–3424.

54 Lyapina, S., Cope, G., Shevchenko, A., Serino, G., Tsuge, T., Zhou, C.,

WOLF, D. A., WEI, N., DESHAIES, R. J. Promotion of NEDD-CUL1 conjugate cleavage by COP9 signalosome. *Science* **2001**, *292*, 1382–1385.

55 SCHWECHHEIMER, C., SERINO, G., CALLIS, J., CROSBY, W. L., LYAPINA, S., DESHAIES, R. J., GRAY, W. M., ESTELLE, M., DENG, X. W. Interactions of the COP9 signalosome with the E3 ubiquitin ligase SCFTIRI in mediating auxin response. *Science* **2001**, *292*, 1379–1382.

56 FU, H., REIS, N., LEE, Y., GLICKMAN, M. H., VIERSTRA, R. D. Subunit interaction maps for the regulatory particle of the 26S proteasome and the COP9 signalosome. *EMBO J.* **2001**, *20*, 7096–7107.

57 WANG, Y., DEVEREUX, W., STEWART, T. M., CASERO, R. A., JR. Polyamine-modulated factor 1 binds to the human homologue of the 7a subunit of the Arabidopsis COP9 signalosome: implications in gene expression. *Biochem. J.* **2002**, *366*, 79–86.

58 HOFMANN, K., BUCHER, P. The PCI domain: a common theme in three multiprotein complexes. *Trends Biochem. Sci.* **1998**, *23*, 204–205.

59 KIM, T., HOFMANN, K., VON ARNIM, A. G., CHAMOVITZ, D. A. PCI complexes: pretty complex interactions in diverse signaling pathways. *Trends Plant Sci.* **2001**, *6*, 379–386.

60 COPE, G. A., SUH, G. S., ARAVIND, L., SCHWARZ, S. E., ZIPURSKY, S. L., KOONIN, E. V., DESHAIES, R. J. Role of predicted metalloprotease motif of Jab1/Csn5 in cleavage of Nedd8 from Cul1. *Science* **2002**, *298*, 608–611.

61 ZHOU, C., SEIBERT, V., GEYER, R., RHEE, E., LYAPINA, S., COPE, G., DESHAIES, R. J., WOLF, D. A. The fission yeast COP9/signalosome is involved in cullin modification by ubiquitin-related Ned8p. *BMC BioChem.* **2001**, *2*, 7.

62 VERMA, R., ARAVIND, L., OANIA, R., McDONALD, W. H., YATES, J. R., 3rd, KOONIN, E. V., DESHAIES, R. J. Role of Rpn11 metalloprotease in deubiquitination and degradation by the 26S proteasome. *Science* **2002**, *298*, 611–615.

63 MAYTAL-KIVITY, V., REIS, N., HOFMANN, K., GLICKMAN, M. H. MPN+, a putative catalytic motif found in a subset of MPN domain proteins from eukaryotes and prokaryotes, is critical for Rpn11 function. *BMC BioChem.* **2002**, *3*, 28.

64 TRAN, H. J., ALLEN, M. D., LOWE, J., BYCROFT, M. Structure of the Jab1/MPN domain and its implications for proteasome function. *Biochemistry* **2003**, *42*, 11460–11465.

65 KAWAKAMI, T., CHIBA, T., SUZUKI, T., IWAI, K., YAMANAKA, K., MINATO, N., SUZUKI, H., SHIMBARA, N., HIDAKA, Y., OSAKA, F., OMATA, M., TANAKA, K. NEDD8 recruits E2-ubiquitin to SCF E3 ligase. *EMBO J.* **2001**, *20*, 4003–4012.

66 PODUST, V. N., BROWNELL, J. E., GLADYSHEVA, T. B., LUO, R. S., WANG, C., COGGINS, M. B., PIERCE, J. W., LIGHTCAP, E. S., CHAU, V. A Nedd8 conjugation pathway is essential for proteolytic targeting of p27Kip1 by ubiquitination. *Proc. Natl. Acad. Sci. USA* **2000**, *97*, 4579–4584.

67 READ, M. A., BROWNELL, J. E., GLADYSHEVA, T. B., HOTTELET, M., PARENT, L. A., COGGINS, M. B., PIERCE, J. W., PODUST, V. N., LUO, R. S., CHAU, V., PALOMBELLA, V. J. Nedd8 modification of cul-1 activates SCF(beta(TrCP))-dependent ubiquitination of IkappaBalpha. *Mol. Cell. Biol.* **2000**, *20*, 2326–2333.

68 NAUMANN, M., BECH-OTSCHIR, D., HUANG, X., FERRELL, K., DUBIEL, W. COP9 signalosome-directed c-Jun activation/stabilization is independent of JNK. *J. Biol. Chem.* **1999**, *274*, 35297–35300.

69 KARNIOL, B., MALEC, P., CHAMOVITZ, D. A. Arabidopsis FUSCA5 encodes a novel phosphoprotein that is a component of the COP9 complex. *Plant Cell* **1999**, *11*, 839–848.

70 WILSON, M. P., SUN, Y., CAO, L., MAJERUS, P. W. Inositol 1,3,4-trisphosphate 5/6-kinase is a protein kinase that phosphorylates the transcription factors c-Jun and ATF-2. *J. Biol. Chem.* **2001**, *276*, 40998–41004.

71 ZHOU, C., WEE, S., RHEE, E.,

NAUMANN, M., DUBIEL, W., WOLF, D. A. Fission yeast COP9/signalosome suppresses cullin activity through recruitment of the deubiquitylating enzyme Ubp12p. *Mol. Cell* **2003**, *11*, 927–938.

72 WOLF, D. A., ZHOU, C., WEE, S. The COP9 signalosome: an assembly and maintenance platform for cullin ubiquitin ligases? *Nat. Cell Biol.* 2003, 5, 1029–1033.

73 GEYER, R., WEE, S., ANDERSON, S., YATES, J., WOLF, D. A. BTB/POZ domain proteins are putative substrate adaptors for cullin 3 ubiquitin ligases. *Mol. Cell* 2003, *12*, 783–790.

74 TYERS, M., JORGENSEN, P. Proteolysis and the cell cycle: with this RING I do thee destroy. *Curr. Opin. Genet. Dev.* 2000, *10*, 54–64.

75 XU, L., WEI, Y., REBOUL, J., VAGLIO, P., SHIN, T. H., VIDAL, M., ELLEDGE, S. J., HARPER, J. W. BTB proteins are substrate-specific adaptors in an SCF-like modular ubiquitin ligase containing CUL-3. *Nature* 2003, *425*, 316–321.

76 PINTARD, L., WILLIS, J. H., WILLEMS, A., JOHNSON, J. L., SRAYKO, M., KURZ, T., GLASER, S., MAINS, P. E., TYERS, M., BOWERMAN, B., PETER, M. The BTB protein MEL-26 is a substrate-specific adaptor of the CUL-3 ubiquitin-ligase. *Nature* 2003, *425*, 311–316.

77 WERTZ, I. E., O'ROURKE, K. M., ZHANG, Z., DORNAN, D., ARNOTT, D., DESHAIES, R. J., DIXIT, V. M. Human De-etiolated-1 regulates c-Jun by assembling a CUL4A ubiquitin ligase. *Science* 2004, *303*, 1371–1374.

78 YAHALOM, A., KIM, T. H., WINTER, E., KARNIOL, B., VON ARNIM, A. G., CHAMOVITZ, D. A. Arabidopsis eIF3e (INT-6) associates with both eIF3c and the COP9 signalosome subunit CSN7. *J. Biol. Chem.* 2001, *276*, 334–340.

79 MAYTAL-KIVITY, V., PICK, E., PIRAN, R., HOFMANN, K., GLICKMAN, M. H. The COP9 signalosome-like complex in S. cerevisiae and links to other PCI complexes. *Int. J. Biochem. Cell Biol.* 2003, *35*, 706–715.

80 SHALEV, A., VALASEK, L., PISE-

MASISON, C. A., RADONOVICH, M., PHAN, L., CLAYTON, J., HE, H., BRADY, J. N., HINNEBUSCH, A. G., ASANO, K. Saccharomyces cerevisiae protein Pci8p and human protein eIF3e/Int-6 interact with the eIF3 core complex by binding to cognate eIF3b subunits. *J. Biol. Chem.* **2001**, *276*, 34948–34957.

81 DUNAND-SAUTHIER, I., WALKER, C., WILKINSON, C., GORDON, C., CRANE, R., NORBURY, C., HUMPHREY, T. Sum1, a component of the fission yeast eIF3 translation initiation complex, is rapidly relocalized during environmental stress and interacts with components of the 26S proteasome. *Mol. Biol. Cell* **2002**, *13*, 1626–1640.

82 YEN, H. C., GORDON, C., CHANG, E. C. Schizosaccharomyces pombe Int6 and Ras homologs regulate cell division and mitotic fidelity via the proteasome. *Cell* **2003**, *112*, 207–217.

83 VON ARNIM, A. G., CHAMOVITZ, D. A. Protein homeostasis: a degrading role for Int6/eIF3e. *Curr. Biol.* **2003**, *13*, R323–325.

84 PENG, Z., SHEN, Y., FENG, S., WANG, X., CHITTETI, B. N., VIERSTRA, R. D., DENG, X. W. Evidence for a physical association of the COP9 signalosome, the proteasome, and specific SCF E3 ligases *in vivo*. *Curr. Biol.* **2003**, *13*, R504–505.

85 LI, L., DENG, X. W. The COP9 signalosome: an alternative lid for the 26S proteasome? *Trends Cell Biol.* **2003**, *13*, 507–509.

86 COPE, G. A., DESHAIES, R. J. COP9 signalosome: a multifunctional regulator of SCF and other cullin-based ubiquitin ligases. *Cell* **2003**, *114*, 663–671.

87 ZHENG, J., YANG, X., HARRELL, J. M., RYZHIKOV, S., SHIM, E. H., LYKKE-ANDERSEN, K., WEI, N., SUN, H., KOBAYASHI, R., ZHANG, H. CAND1 binds to unneddylated CUL1 and regulates the formation of SCF ubiquitin E3 ligase complex. *Mol. Cell* **2002**, *10*, 1519–1526.

88 XIE, Y., VARSHAVSKY, A. UFD4 lacking the proteasome-binding region catalyses ubiquitination but is

impaired in proteolysis. *Nat Cell Biol.* **2002**, *4*, 1003–1007.

89 CORN, P. G., McDONALD, E. R., 3rd, HERMAN, J. G., EL-DEIRY, W. S. Tat-binding protein-1, a component of the 26S proteasome, contributes to the E3 ubiquitin ligase function of the von Hippel-Lindau protein. *Nat. Genet.* **2003**, *35*, 229–237.

90 DORONKIN, S., DJAGAEVA, I., BECKENDORF, S. K. CSN5/Jab1 mutations affect axis formation in the Drosophila oocyte by activating a meiotic checkpoint. *Development* **2002**, *129*, 5053–5064.

91 HIGA, L. A., MIHAYLOV, I. S., BANKS, D. P., ZHENG, J., ZHANG, H. Radiation-mediated proteolysis of CDT1 by CUL4-ROC1 and CSN complexes constitutes a new checkpoint. *Nat Cell Biol.* **2003**, *5*, 1008–1015.

92 WEE, S., HETFELD, B., DUBIEL, W., WOLF, D. A. Conservation of the COP9/signalosome in budding yeast. *BMC Genet.* **2002**, *3*, 15.

93 ORON, E., MANNERVIK, M., RENCUS, S., HARARI-STEINBERG, O., NEUMAN-SILBERBERG, S., SEGAL, D., CHAMOVITZ, D. A. COP9 signalosome subunits 4 and 5 regulate multiple pleiotropic pathways in Drosophila melanogaster. *Development* **2002**, *129*, 4399–4409.

94 SUH, G. S., POECK, B., CHOUARD, T., ORON, E., SEGAL, D., CHAMOVITZ, D. A., ZIPURSKY, S. L. Drosophila JAB1/CSN5 acts in photoreceptor cells to induce glial cells. *Neuron* **2002**, *33*, 35–46.

95 MA, L., ZHAO, H., DENG, X. W. Analysis of the mutational effects of the COP/DET/FUS loci on genome expression profiles reveals their overlapping yet not identical roles in regulating Arabidopsis seedling development. *Development* **2003**, *130*, 969–981.

96 SCHWECHHEIMER, C., SERINO, G., DENG, X. W. Multiple ubiquitin ligase-mediated processes require COP9 signalosome and AXR1 function. *Plant Cell* **2002**, *14*, 2553–2563.

97 FENG, S., MA, L., WANG, X., XIE, D., DINESH-KUMAR, S. P., WEI, N., DENG, X. W. The COP9 signalosome interacts physically with SCF COI1 and modulates jasmonate responses. *Plant Cell* **2003**, *15*, 1083–1094.

98 WANG, X., FENG, S., NAKAYAMA, N., CROSBY, W. L., IRISH, V., DENG, X. W., WEI, N. The COP9 signalo-some interacts with SCF UFO and participates in Arabidopsis flower development. *Plant Cell* **2003**, *15*, 1071–1082.

99 ARBISER, J. L., KLAUBER, N., ROHAN, R., van LEEUWEN, R., HUANG, M. T., FISHER, C., FLYNN, E., BYERS, H. R. Curcumin is an *in vivo* inhibitor of angiogenesis. *Mol. Med.* **1998**, *4*, 376–383.

100 BENEZRA, R., RAFII, S., LYDEN, D. The Id proteins and angiogenesis. *Oncogene* **2001**, *20*, 8334–8341.

Subject Index

Protein Degradation. Edited by J. Mayer, A. Ciechanover, M. Rechsteiner
Copyright © 2005 WILEY-VCH Verlag GmbH & Co. KGaA, Weinheim
ISBN: 3-527-30837-7